Electronic Communications Systems

William D. Stanley
School of Engineering
Old Dominion University
Norfolk, Virginia

Reston Publishing Company, Inc.
A Prentice-Hall Company
Reston, Virginia

To my mother, Sallie Stanley

Library of Congress Cataloging in Publication Data

Stanley, William D.
  Electronic communications systems.

  1. Telecommunication. 2. Information theory.
I. Title.
TK5101.S66   621.38   81-15837
ISBN 0-8359-1666-9   AACR2

© 1982 by
Reston Publishing Company, Inc.
A Prentice-Hall Company
Reston, Virginia

10   9   8   7   6   5   4   3   2

Printed in the United States of America

Contents

Preface

The primary objective of this book is to present the general principles of electronic communications at a systems level. The emphasis is on the signal processing functions of various modulation and demodulation operations. The book is suited for individuals who need to understand and utilize some of the mathematical models for communications systems analysis and design, but in which the emphasis is more toward the applied design and operational approach. Before discussing further what the book actually does, it is appropriate to first discuss what the book does *not* do.

The book is not a highly theoretical and rigorous treatment of communications theory aimed at the systems-level analyst. There are a number of fine texts available at the advanced undergraduate and graduate engineering levels which serve that purpose. At the other extreme, the book is not a detailed collection of communications electronic circuits. Again, there are a number of good books available for this purpose, many of which are written at the two-year engineering technology level.

Since the vast majority of communication books are aimed at either the more mathematical and analytical part of the engineering spectrum or the more hardware part of the engineering technology spectrum, this book was developed as an attempt to fill the void between these two extremes. As such, some of the important mathematical concepts of communications theory are employed. However, such results are presented from as simplified a viewpoint as possible. Abstract developments are avoided, and an attempt is made to justify each analytical tool in a practical framework whenever possible.

Although not a communications electronics book, the book does definitely consider many of the problems associated with communications hardware. The dominating emphasis is concerned with the strategy for implementing a particular operation, rather than the details of a particular circuit. Even though the actual circuits are constantly changing in industry as new developments occur, many fundamental strategies from a signal-processing point of view have not changed that significantly.

It is expected that the book will be of primary interest for the following purposes: (1) as an upper-division (junior or senior level) course text or supplement in applied engineering programs, (2) as an upper-division course text for a baccalaureate engineering technology program, and (3) as a book for applied communications engineers and technologists who deal with a combination of system-level considerations and actual hardware implementation and operation. The large number of example problems and exercises should enhance the suitability of the book for the latter group.

It is anticipated that the book can be adapted to a number of different levels depending on the perspective of the instructor or the reader. Indeed, many of the analytical concepts could be expanded if desired, and a more rigorous treatment could be achieved by a reasonable expansion of the book. Conversely, many of the mathematical techniques could be "down-played," and the emphasis could be aimed at a more intuitive and qualitative approach, if desired. An attempt has been made to make as much of the final important results as possible to be reasonably independent of the more sophisticated mathematical tools.

A brief overview of the book will now be given. Chapter 1 provides a general overview of the concepts of communications systems, information, and modulation. With the exception of Section 1–6, all this material is qualitative in nature. This latter section provides some details on decibel signal and gain computations.

Chapter 2 is rather long and provides a general treatment of spectral analysis from both the Fourier series and Fourier transform points of view. Depending on the intended level of the book, this material could be covered in complete detail or portions could be either omitted or considered only qualitatively.

The fundamentals of signal transmission are covered in Chapter 3. The spectral analysis concepts introduced in the preceding chapter will be used to predict the results of finite bandwidth on signals.

In treating the various forms of modulation, the ideal noise-free situation is assumed in Chapters 4 through 7. In this manner, the modulation processes can be first understood without the additional complexity of the presence of noise.

Chapter 4 is devoted to amplitude modulation and its various forms (including single sideband and double sideband). A corresponding development of angle modulation (including both frequency and phase modulation) is provided in Chapter 5.

Chapter 6 deals with the sampling theorem and with various forms of pulse modulation. In addition, time-division and frequency-division multiplexing concepts are introduced. Digital-modulation methods are considered in Chapter 7.

Statistical methods are introduced in Chapter 8, and these methods are used to model noise. A more detailed treatment of noise effects in systems, particularly receivers, is given in Chapter 9.

The derivations of the relative performances of several important communications systems in the presence of noise are given in Chapter 10. Both analog and digital systems are considered. The various derivations are quite detailed, and much of this material may be omitted by readers desiring to focus on the results and their applications.

Chapter 11 provides an overall view of communications systems. The performance results of Chapter 10 are summarized for the benefit of readers who did not follow the details of the derivations, and emphasis is directed toward the applications of these results to complete systems.

The author wishes to express his deep appreciation to Mrs. Estelle B. Walker, who typed the final copy of the manuscript, and Mrs. Ann Reid, who typed much of the initial draft. Both deserve very special recognition for their outstanding work. I would also like to thank Professor L. A. Hobbs, who taught from various portions of the manuscript, for providing some valuable comments and suggestions.

William D. Stanley
Norfolk, Virginia

General
1 Considerations

1-0 INTRODUCTION

The primary objectives of this introductory chapter are to introduce the reader to some of the basic considerations in communications systems and to provide an overview of the remainder of the book. A few of the terms and definitions used in describing communications systems are introduced in the chapter, some of which are already familiar to many readers. In virtually all cases, the material presented here will be expanded and developed in greater depth later in the book. The intent of this chapter is simply to provide an initial broad overview before pursuing the many fine points throughout the book.

1-1 SIMPLIFIED COMMUNICATIONS SYSTEM

A block diagram illustrating a simplified model of a communications system is shown in Fig. 1-1. In spite of its simplicity, however, the elements of this model represent some of the major factors that appear in virtually all communications systems. The block on the left represents a certain source of *information,* which must be "delivered" or transmitted to a particular *destination* over a *channel.* The channel contains random noise, which adds to or otherwise disturbs the signal in some way.

The presence of noise in the channel adds both frustration and excitement to the design and operation of the communications system, depending on the point of view. Without the presence of noise, operation of the system would be much more straightforward, and the intended objective of information transmission could be accomplished with very little difficulty. Indeed, many of the more elaborate and sophisticated communications techniques were developed specifically to enhance the signal readability relative to the noise level.

At one extreme (e.g., high power and/or close range), the effect of noise may be insignificant, and the signal reaches the destination with no undesira-

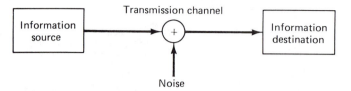

Figure 1-1. Simplified model of a communications system.

ble alteration. At another extreme condition, the signal may be so masked or overshadowed with noise that it is totally unintelligible, and nothing can be done to retrieve the signal. The former case represents complete success, and the latter represents complete failure. Most cases of practical importance fall somewhere between these extremes, and it is in this range that communications engineers and technologists "earn their keep" by designing, building, and operating systems that work, even though significant noise is present.

1-2 INFORMATION

The term "information" was used in describing the communications systems model in the last section. Most readers likely accepted the term on a purely qualitative basis, since it is a very common word whose definition is generally understood in a qualitative sense. However, in communications system theory, it is necessary to be more precise about what we mean by "information." After all, if the transmission of information is the major objective of a communications system, how can we express the relative performance of systems unless we can measure the volume of information delivered? For example, how can we compare the relative information content of a commercial television system transmitting a football game and a terminal transferring data on charge accounts to a central computer?

In the late 1940s, Claude E. Shannon of Bell Telephone Laboratories proposed some basic definitions for information and the rate that it could be transmitted without error. From this work and related developments, the science of *information theory* has evolved. Much of this field is highly mathematical and somewhat more abstract than the average "hardware engineer" or technologist is able to utilize in everyday problems. Indeed, many aspects of information theory have still to be implemented in actual hardware systems. Nevertheless, some significant aspects of information theory have been successfully utilized in practical systems, and some of these concepts will be encountered in the book.

One of the most fundamental questions addressed in information theory is the definition of information itself. On a qualitative basis, information is associated with uncertainty. Consider, for example, information associated with the knowledge that a given event has occurred. If it is very likely that the event will occur, very little information is attributed to our learning that the event has actually occurred. However, if the event has a very low likelihood of occurrence, the information content associated with the event occurring is very high. For example, if we learn that it snowed in Vermont in January, very little information would be received since that event is almost certain anyway. However, if we learned that it snowed in Miami, Florida, the information content would be considerably higher since the event is so unlikely.

From the preceding association of uncertainty with information, a mathematical definition of information has been formulated. This concept can be used in the construction of signaling schemes that attempt to maximize the average information per message. We do not expect the reader to appreciate fully such a concept at this point, but merely to accept the fact that some elegant results have been achieved from such philosophical considerations.

Many of the earlier communications systems evolved without the benefit of information theory, and very little mention will be made of this theory when we encounter such systems in the book. However, some of the more modern systems, such as digital communications techniques, have benefited from the utilization of information theory for optimum signal and system design.

1-3   BANDWIDTH AND SPECTRUM

All signals that are processed in a communications system can be expressed in terms of sinusoidal components. Some readers may have studied Fourier series and will undoubtedly recognize this concept from that experience. The relative weights of the sinusoidal components that comprise the signal can be thought of as the *frequency spectrum*. For example, it is known that the human voice can be represented by sinusoidal components whose frequencies range from well under 100 hertz (Hz) to several kilohertz (kHz). Conversely, the video signal in a television receiver has frequency components that extend above 4 megahertz (MHz).

Knowledge of the frequency content of the signal is very important in estimating the bandwidth required in the appropriate communications *channel*. Conversely, the bandwidth capability of a given channel can limit the bandwidth of any given signal that could be transmitted over the channel. For example, assume that a given channel is an unequalized telephone line having a frequency response from direct current (dc) to several kilohertz. The channel might be perfectly adequate for voice data, but would hardly suffice for video data. Thus, the frequency response of the channel must be adequate over the frequency range encompassing the spectrum of the signal, or severe distortion and degradation of the signal will result.

In some cases, it is possible to transmit information over a narrower channel by slowing down the information rate to a level compatible with the channel. For example, in some space communications systems, the bandwidth has been kept quite low to keep the total noise power to a minimum. In some cases, higher frequency data can be first recorded and then played back at a very low rate, which artifically reduces the signal bandwidth. In this case, a longer processing time is being substituted for a wider bandwidth.

1-4 TRANSMISSION CONSIDERATIONS

The frequency components of most message signals tend to be concentrated in the lower portion of the overall electromagnetic spectrum in a relative sense. The highest frequency contained in a message spectrum may be quite high in some cases (e.g., video data), but the typical frequency range starts near dc (or at dc in some cases), so in a relative sense such spectra quite often are concentrated in the lower portion of the spectrum. The term *baseband* refers to the general frequency range under such typical conditions.

The simplest type of electrical communications system is one in which the transmission of information takes place directly at baseband frequencies. The communications channel in such a case is often a direct wire or cable link between the source and the destination. For example, if two telephones are connected directly through wires and associated circuitry, the voice signal at the transmitter is converted to electrical impulses that are transmitted directly down the telephone line. There are obvious limitations on the volume and type of data that could be transmitted directly with a wire link.

A much more general transmission process makes use of electromagnetic waves and their propagation characteristics. One of the greatest scientific discoveries of all time was the deduction and verification of the existence of electromagnetic waves, which led to the development of radio transmission. The early mathematical work of James Clark Maxwell in England, followed by the experimental work of Heinrich Hertz in Germany, led Guglielmo Marconi in Italy to construct the first "wireless" system, and a whole new era of technology was initiated.

Electromagnetic fields are present in any circuit in which time-varying voltages and currents are present. However, in well-designed circuits in which radiation is not desired, the magnitudes of such fields can be controlled to acceptable levels by good component placement and shielding. When electromagnetic radiation is specifically desired, an interface between the circuit and the outside propagation medium is established in a manner to enhance the propagating field. The device that accomplishes this purpose is the *antenna*. Antennas are rather complex, but they all have the objective of providing a direct transition between the transmitter or receiver circuit, in which the radiation is hoped to be negligible, and the outside transmission medium, which is usually air or free space.

Contrary to popular belief, electromagnetic waves can theoretically be generated at any frequency, even in the audio range. The practical difficulty in implementation at low frequencies, however, is the cause of the widespread notion that radio propagation can be achieved only at much higher frequencies.

The frequency limitation is related to the requirements of practical antennas. While there are many different kinds of antennas, virtually all are characterized by the fact that the size must be an appreciable fraction of a wavelength at the operating frequency in order to "transfer" the electrical circuit power to the desired propagating waves. Some antennas, for example, require a length equal to a half-wavelength, while others require a length equal to a quarter-wavelength.

Let λ represent the wavelength in meters and let f represent the frequency in hertz. In free space, the velocity c of wave propagation is very nearly $c = 3 \times 10^8$ meters per second (m/s). While the velocity in air is slightly lower, very little error results from assuming the free space propagation velocity. The wavelength in meters can then be determined from the formula

$$\lambda = \frac{c}{f} = \frac{3 \times 10^8}{f} \qquad (1\text{-}1)$$

Consider then the problems that would be encountered if electromagnetic transmission were desired at, for example, 10 kHz. The wavelength is determined from Eq. (1-1) as $\lambda = 30$ kilometers (km). Any antenna having even an appreciable fraction of this wavelength would be enormous in size. On the other hand, the wavelength at 10 MHz, for example, is $\lambda = 30$ m, a much more reasonable size.

As it turns out, a few special-purpose communications systems have been constructed to operate in the range of tens of kilohertz, primarily for operating with submerged submarines. The depth of penetration of electromagnetic waves in water increases as the frequency decreases, so operation at very low frequencies is necessary in such cases. The antennas at the ground transmitting sites of such systems literally occupy many acres of real estate and are usually located in remote areas.

The vast majority of systems employing electromagnetic propagation operate at frequencies ranging from several hundred kilohertz to tens of gigahertz. In the higher frequency range, the dimensions of the antennas are so small that highly directional antennas can be constructed so as to focus or beam all the energy in one direction. In contrast, antennas operating in the lower portion of the electromagnetic spectrum tend to be less directional.

To summarize some points in the preceding discussion, direct baseband transmission of a data signal is usually practical only if there is a direct wire link between the source and destination. To utilize electromagnetic radiation, it is necessary in a practical sense that the frequency range be much higher than that of the typical message signal. This can be achieved through the process of modulation, as will be discussed in the next section.

1-5 MODULATION AND DEMODULATION

Modulation may be defined as the process of shifting the frequency spectrum (while possibly changing the form) of a message signal to a frequency range in which more efficient transmission can be achieved. At the receiver, *demodulation* (or *detection* as it is commonly called) is the process of shifting the spectrum back to the original baseband frequency range and reconstructing the original form if necessary. While the message frequency range is referred to as the baseband range, the term *radio frequency* (denoted RF) is commonly used to describe the frequency range of the modulated signal.

Numerous modulation methods have been developed over the years, and it is difficult to classify them in a simple nonambiguous manner. On the broadest scale, modulation may be roughly classified in one of three general forms as (1) analog (or continuous-time) modulation, (2) pulse (or discrete-time) modulation, and (3) digital modulation.

Analog modulation is the oldest category and includes all methods in which some characteristic of the high-frequency signal represents the modulating signal on a continuous basis (i.e., an "analog" of the modulating signal). This includes such traditional and widespread methods as *amplitude modulation* (along with variations such as *double sideband, single sideband,* etc.) and *frequency modulation.*

Pulse modulation involves sampling of the analog signal in discrete samples and in transmitting the intelligence by the samples. As will be seen later, it is absolutely necessary that a minimum sampling rate be employed if the signal is to be reconstructed. Among the methods of pulse modulation are *pulse amplitude modulation, pulse width modulation,* and *pulse position modulation.*

Digital modulation is the latest and most rapidly growing form of modulation. Digital modulation is closely akin to pulse modulation in that samples of the message signal are first taken at a certain minimum sampling rate. However, digital modulation takes the process one step farther in that the amplitude levels are encoded into binary words consisting of ones and zeros. The binary words are then transmitted as a combination of ones and zeros in some fashion. At one extreme, the binary words may be transmitted directly as a combination of positive and negative (or on and off) pulses at baseband in some cases. However, the more common case of interest for our discussion involves high-frequency encoding. This may take the form of shifting between two different frequencies to represent the two possible states or in shifting between two different phases. The first method is called *frequency-shift keying,* and the second is called *phase-shift keying.*

The preceding discussion has been sprinkled with a few "buzzwords" concerning different modulation methods, and many more will appear throughout the book. The intent here is not to confuse the reader, but to provide a brief overview. Much of the book is devoted to developments of these concepts in a more gradual and detailed fashion.

One problem in trying to categorize the different modulation methods is that there is often an overlap in the actual processes. For example, when a digital signal is converted to high-frequency form, it assumes some of the character of an analog modulated signal. As another example, some multiplexed systems involve combinations of several categories of modulation. Because of such ambiguities, some of the classifications will occasionally seem arbitrary and ambiguous.

1-6 DECIBEL RATIOS

A section devoted to a topic as specific as decibel ratios may seem out of place compared with the rather broad generalities covered elsewhere in this chapter. However, because of the widespread usage of decibel ratios and levels throughout the communications field, it is desirable to establish such usage early in the book, and this chapter is quite appropriate for that purpose. Historically, decibel terminology was first introduced in the telephone industry, so it is a concept having a well-established tradition in the communications field.

The original concept of decibel measurement was based on a comparison of two power levels. Using signal gain as the means for establishing the comparison, consider the block diagram shown in Fig. 1-2, which represents a linear amplifier. The input signal is assumed to deliver a power P_1 to the amplifier input, and the amplifier in turn delivers a power P_2 to the external load. The *absolute power gain G* is defined as

$$G = \frac{P_2}{P_1} \tag{1-2}$$

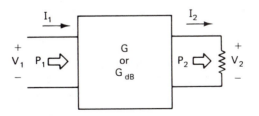

Figure 1-2. Block diagram of amplifier system used to define gain.

The *decibel power gain* G_{dB} in decibels (abbreviated as dB) is defined by

$$G_{dB} = 10 \log_{10} G = 10 \log_{10} \frac{P_2}{P_1} \qquad (1\text{-}3)$$

When $P_2 > P_1$ and $G > 1$ (i.e., there is a true gain), the decibel gain value is positive. However, if $P_2 < P_1$ and $G < 1$ (i.e., there is a loss), the decibel gain value turns out to be negative.

For systems in which the output power is usually less than the input power (e.g., attenuator networks and many filters), it is often more convenient to deal with *loss* rather than gain. The *absolute loss L* can be defined as

$$L = \frac{P_1}{P_2} = \frac{1}{G} \qquad (1\text{-}4)$$

The decibel loss L_{dB} can then be defined as

$$L_{dB} = 10 \log_{10} L = 10 \log_{10} \frac{P_1}{P_2} \qquad (1\text{-}5)$$

Before comparing the loss and gain expressions, let us pause briefly to summarize several properties of logarithms that will be used in several developments in this section. The pertinent properties are

$$\log_{10} \frac{1}{x} = -\log_{10} x \qquad (1\text{-}6)$$

$$\log_{10} x^2 = 2 \log_{10} x \qquad (1\text{-}7)$$

$$\log_{10} xy = \log_{10} x + \log_{10} y \qquad (1\text{-}8)$$

At this point, application of (1-6) to (1-5) and a comparison with (1-3) yield

$$L_{dB} = -G_{dB} \qquad (1\text{-}9)$$

Thus, a negative decibel gain is a positive decibel loss, and vice versa. A transmission cable having a signal loss of 20 dB, for example, could also be described as having a gain of -20 dB. The following comparison should help the reader clarify the preceding concepts:

Absolute Gain	dB Gain	dB Loss
> 1	$+$	$-$
< 1	$-$	$+$

If the value of decibel gain is given, it is necessary to invert the form of Eq. (1-3) to determine the corresponding absolute gain G. This is most readily done by first dividing both sides by 10, as follows:

$$\frac{G_{dB}}{10} = \log_{10} G \qquad (1\text{-}10)$$

Both sides of (1-10) are then equated as powers of 10:

$$10^{G_{dB}/10} = 10^{\log_{10} G} \qquad (1\text{-}11)$$

The right side of (1-11) is simply G, since by definition $\log_{10} G$ is a value such that when 10 is raised to that power, the result is G. Thus,

$$G = 10^{G_{dB}/10} \qquad (1\text{-}12)$$

A similar relationship can be deduced for L in terms of L_{dB}.

$$L = 10^{L_{dB}/10} \qquad (1\text{-}13)$$

Most modern scientific calculators have a y^x function, so operations of the forms of (1-12) and (1-13) can be easily achieved.

While the most basic decibel form is concerned with power ratios, it is also possible to convert voltage and current ratios to equivalent decibel forms if some care is exercised. Assume for the amplifier model previously discussed that the output power is delivered to a resistance R. Assume that the input resistance is also R. Let V_2 represent the output effective voltage, and let V_1 represent the input effective voltage. The decibel power gain is then

$$G_{dB} = 10 \log_{10} \frac{V_2^2/R}{V_1^2/R} = 10 \log_{10} \left(\frac{V_2}{V_1}\right)^2 \qquad (1\text{-}14)$$

in which R was canceled out in the expression. Application of (1-7) to (1-14) results in the form

$$G_{dB} = 20 \log_{10} \frac{V_2}{V_1} \qquad (1\text{-}15)$$

A similar development expressed in terms of the effective currents I_2 and I_1 for the case of equal resistances yields

$$G_{dB} = 20 \log_{10} \frac{I_2}{I_1} \qquad (1\text{-}16)$$

Comparing (1-3), (1-15), and (1-16), we see that **10** is the "magic" constant for computing decibels from power levels and **20** is the corresponding value for working with voltage or current. The corresponding inverse relationships for voltage and current are

$$\frac{V_2}{V_1} = 10^{G_{dB}/20} \qquad (1\text{-}17)$$

$$\frac{I_2}{I_1} = 10^{G_{dB}/20} \qquad (1\text{-}18)$$

Recall that, in developing the decibel forms for voltage and current, it was assumed that the input and load resistances were equal, and this is the proper assumption for the corresponding decibel value to represent a true power ratio. However, there has evolved a somewhat casual usage of decibel notation, and it is very common to find computations based on the forms of (1-15) or (1-16) when the resistances are not equal. Many voltmeters, for example, have decibel scales that simplify measurements of relative voltage levels through the use of decibel forms without regard to the resistance levels. The intent here is neither to praise nor to condemn the practice, but merely to alert the reader to its usage so that it will be clear that true power measurements are not obtained in such cases. This book will follow the traditional standard that voltage and current ratios will be used only when the resistance levels are the same.

The reader should carefully note that a decibel is not an absolute unit; rather it compares one quantity with another. Thus, to indicate that a certain signal level is 6 dB is meaningless unless a reference level has been established. However, to say that a signal level is 6 dB above a reference level of 1 milliwatt (mW) is proper. (The reader can verify that the level of such a signal is very nearly 4 mW.) In actual usage, different segments of the communications field have established certain reference levels, and workers often will refer to absolute levels in decibels when they really mean decibel levels referred to the standard. This is, of course, confusing to outsiders.

Certain decibel measurements have become so standardized about specific reference levels that the reference levels have been incorporated in the abbreviations. For example, many decibel measurements utilize 1 mW as a standard reference, and the abbreviation dBm is used to designate power levels relative to 1 mW. Thus,

$$\text{power level (dBm)} = 10 \log_{10} \frac{\text{power level (mW)}}{1 \text{ mW}} \qquad (1\text{-}19a)$$

$$= 10 \log_{10} \frac{\text{power level (W)}}{10^{-3} \text{ W}} \qquad (1\text{-}19b)$$

In higher-power applications such as transmitters, power levels relative to 1 watt (W) are more common, and the abbreviation dBW refers to dB values based on this level.

$$\text{power level (dBW)} = 10 \log_{10} \frac{\text{power level (W)}}{1 \text{ W}} \qquad (1\text{-}20a)$$

$$= 10 \log_{10} \text{power level (W)} \qquad (1\text{-}20b)$$

A more recent decibel standard that is coming into usage in specifying the very small power levels at receiver terminals is the power level relative to 1 femtowatt (fW). Note that 1 fW $= 10^{-15}$ W. The value in decibels above 1 fW is denoted as dBf.

$$\text{power level (dBf)} = 10 \log_{10} \frac{\text{power level (fW)}}{1 \text{ fW}} \qquad (1\text{-}21a)$$

$$= 10 \log_{10} \frac{\text{power level (W)}}{10^{-15} \text{ W}} \qquad (1\text{-}21\text{b})$$

One of the most useful aspects of working with decibel terminology is in analyzing a cascade of many successive stages of amplification and attenuation, as illustrated by the block diagram in Fig. 1-3. Assume that all the blocks are represented by equivalent "gains" (meaning that losses are represented as gains < 1). The overall gain G is

$$G = G_1 G_2 G_3 \ldots G_n \qquad (1\text{-}22)$$

The decibel gain can then be calculated and the property of Eq. (1-8) can be readily utilized. This leads to the simple result that

$$G_{\text{dB}} = G_{1\,\text{dB}} + G_{2\,\text{dB}} + \ldots + G_{n\text{dB}} \qquad (1\text{-}23)$$

where each of the terms on the right refers to the decibel gain of a corresponding stage. A given block exhibiting an actual loss would be represented by a negative gain in this context.

Example 1-1

A certain cable signal transmission system is shown in Fig. 1-4. The nominal level at the signal source output is 10 dBm. The signal is amplified before transmission by the line amplifier whose gain is 13 dB. The first section of cable (section A) has a loss of 26 dB, and a booster amplifier with a gain of 20 dB is placed at the output of that section. The second section of cable (section B) has a loss of 29 dB. The system is operated on a matched impedance basis with all effective sources and terminations having a resistive impedance of 600 ohms (Ω), and the characteristic impedance of the cable is the same. (a) Determine the signal levels both in dBm and in volts at various points in the system. (b) Determine the gain of the receiver amplifier such that its nominal output level is 6 volts (V) across a 600-Ω termination.

Solution

(a) The most convenient way to deal with the effects of various gains and losses in a system such as this is by decibel notation. The output of the signal source is given in dBm (decibels above 1 mW), so that system of decibel levels

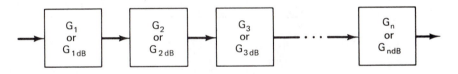

Figure 1-3. Block diagram of cascade system.

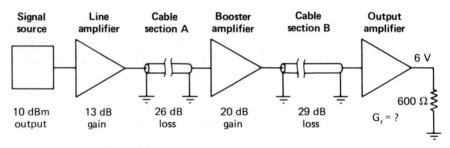

Figure 1-4. Cable system of Example 1-4.

will be employed. A convenient tabulation of the various results of the computations is provided in Table 1-1. Starting with the output level of the source, the gain in decibels of each component of the system is successively algebraically added to the input level to determine the output level in dBm. Note that *actual losses* are expressed as *negative decibel gains*. Note also that *signal levels* are expressed in *dBm,* but *component gains (or losses)* must be expressed simply as *decibels*. A dBm value refers to a signal level above a particular reference, while a gain in decibels represents a shift in the signal level. A decibel gain produces the same shift in the signal level whether the signal is expressed in dBm, dBW, or any other decibel level.

As can be seen from the numbers in the table, working with decibel values simplifies greatly the process of combining the effects of a number of different gain and loss factors. The problem also calls for the determination of the signal level in volts at various points. We can start with Eq. (1-19b) and solve for the actual power level. The result is

$$\text{power level (W)} = 10^{-3} \times 10^{\text{ power level (dBm)}/10} \tag{1-24}$$

The power produced by an effective (rms) value of voltage v across a resistance R is v^2/R. Equating this form to the power level, and setting $R = 600\ \Omega$, the voltage can be determined from

$$v = \sqrt{600 \times \text{power level (W)}} \tag{1-25}$$

The various values are tabulated in Table 1-1.

Table 1-1. Computations of Example 1-1.

	Gain (dB)	Output Level (dBm)	Output Level (V)
Signal source		10	2.449
Line amplifier	13	23	10.941
Cable section *A*	−26	−3	0.5484
Booster amplifier	20	17	5.484
Cable section *B*	−29	−12	0.1946

(b) The required output effective signal voltage level is 6 V. The power P in a 600-Ω load is $P = (6)^2/600 = 60$ mW. This corresponds to a level of about 17.78 dBm. The required gain G_r (dB) for the receiver amplifier is thus

$$G_r \text{ (dB)} = 17.78 \text{ dBm} - (-12 \text{ dBm}) = 29.78 \text{ dB} \qquad (1\text{-}26)$$

It may seem initially strange that the units of the two quantities in the middle of (1-26) are in dBm, but that the result of the subtraction is in dB, but this is due to the fact that this result represents a gain, rather than a signal level.

An alternative way to achieve this result is to first observe from Table 1-1 that the output of the line is 0.1946 V, and the required level is 6 V. The voltage gain A_v is then

$$A_v = \frac{6}{0.1946} = 30.832 \qquad (1\text{-}27)$$

The decibel gain G_r (dB) of the receiving amplifier is then

$$G_r \text{ (dB)} = 20 \log_{10} 30.832 = 29.78 \text{ dB} \qquad (1\text{-}28)$$

which agrees with the result obtained in (1-26).

PROBLEMS

1-1 Convert the following absolute power gains to decibel power gains: (a) $G = 1$, (b) $G = 2$, (c) $G = 4$, (d) $G = 2^n$, (e) $G = 10$, (f) $G = 100$, (g) $G = 10^n$, (h) $G = 0.1$, (i) $G = 750$.

1-2 Convert the following decibel power gains to absolute power gains: (a) $G_{dB} = 0$ dB, (b) $G_{dB} = 12$ dB, (c) $G_{dB} = 30$ dB, (d) $G_{dB} = 10n$ dB, (e) $G_{dB} = -40$ dB.

1-3 Convert the following absolute voltage gains (denoted by A_v) to decibel power gains, assuming that the resistance levels are the same at input and output: (a) $A_v = \sqrt{2}$, (b) $A_v = 2$, (c) $A_v = 2^n$, (d) $A_v = \sqrt{10}$, (e) $A_v = 10$, (f) $A_v = 10^n$.

1-4 Convert the decibel power gains in Prob. 1-2 to absolute voltage gains (denoted by A_v) if the resistance levels are the same at input and output.

1-5 Express the following absolute power levels as decibel levels referenced to 1 mW (i.e., dBm): (a) 1 mW, (b) 10 mW, (c) 1 W, (d) 1 kW, (e) 1 picowatt (pW), (f) 1 nanowatt (nW), (g) 1 fW.

1-6 Express the following absolute power levels as decibel levels referenced to 1 W (i.e., dBW): (a) 1 W, (b) 100 kW, (c) 1 mW, (d) 1 fW.

1-7 Express the following absolute power levels as decibel levels referenced to 1 fW (i.e., dBf): (a) 1 fW, (b) 1 pW, (c) 1 nW, (d) 1 microwatt (μW), (e) 1 mW, (f) 1 W, (g) 1 kW.

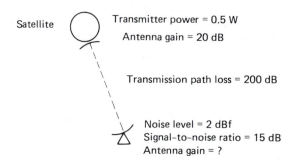

Satellite — Transmitter power = 0.5 W

Antenna gain = 20 dB

Transmission path loss = 200 dB

Noise level = 2 dBf
Signal-to-noise ratio = 15 dB
Antenna gain = ?

Figure P1-11

1-8 Show that power level (dBm) = power level (dBW) + 30

1-9 Show that power level (dBf) = power level (dBW) + 150

1-10 Perform all the computations of Ex. 1-1 if the system parameters are as follows: signal source output = -3 dBm, line amplifier gain = 18 dB, cable section A loss = 22 dB, booster amplifier gain = 26 dB, cable section B loss = 25 dB.

1-11 A certain satellite communications link is illustrated in Fig. P1-11. The output power of the satellite transmitter is 0.5 W (-3 dBW). This power level is effectively increased in the direction of transmission by the transmitter antenna, which has a gain of 20 dB. The RF signal is attenuated by a path loss of 200 dB. The total noise level at the receiver input is estimated to be 2 dBf. The desired signal-to-noise ratio at the receiver input is 15 dB. Determine the minimum receiving antenna gain G_r (dB) that will meet the specifications.

2 Spectral Analysis

2-0 INTRODUCTION

Spectral analysis is the process of determining the frequency content of a given signal in terms of the frequencies present in the signal, as well as their relative amplitudes and, in some cases, their phases. Spectral analysis is widely used in signal analysis and transmission because it provides a means of determining the bandwidth requirements for any processing system. For modulated signals that are to be transmitted via electromagnetic waves, information concerning the spectrum is very important in establishing channel compatibility with other systems.

The various mathematical techniques for spectral analysis will be discussed in this chapter. The development will begin with the Fourier series representation, which is most readily applied to periodic signals. This concept will then be extended to nonperiodic signals by means of the Fourier transform. Several sections will be devoted to the spectral properties of various types of pulse waveforms. Pulse functions occur in communications systems both as natural signals to be transmitted (e.g., digital data) and as switching waveforms for modulators.

To make this chapter reasonably complete for a range of possible objectives of the book, a considerable amount of detail on the properties of Fourier theory is included. For a complete and in-depth study of communications systems at a fundamental level, this material is necessary. The more analytical aspects of Fourier theory are needed for a deeper study of modulation systems, particularly when noise spectral properties are considered. On the other hand, many "hardware-oriented" aspects of communications systems operation can be understood without all the material of this chapter.

For readers desiring only a minimum treatment of spectral analysis at this point, the following sections are recommended:

2-1 Signal Terminology (concentrate on general overview)
2-2 Fourier Series (the exponential form can be deemphasized for minimal coverage)

A general treatment of these sections should allow the reader to understand
(at least qualitatively) most of the book. However, a full analytical treatment
of communications systems would necessitate a deeper comprehensive understanding of this chapter. Communications theory is mathematical in nature
and demands a reasonably rigorous approach before full insight is obtained.

2-1   SIGNAL TERMINOLOGY

The term "signal" will appear many times throughout the book. In a
collective sense, the term will be used to represent any of the many different
kinds of waveforms that are encountered in communications system analysis.
This will include both information or message waveforms and waveforms
generated directly in the communications hardware.

In the vast majority of cases, signals appear as voltages or currents in
different parts of the system. Rather than use the standard symbols for
voltage and current, which tend to be a bit too specific for the purpose, more
general notation will be employed except where specific circuits are being
studied. Thus, symbols such as x and y will be used to represent the forms of
signals that arise in a system. In this manner, a more general representation
of the results can be achieved.

The signals that arise in communications systems generally vary with time.
The standard mathematical notation to express a quantity x as a function of
time is the functional form $x(t)$, where t represents time in seconds (s). Thus,
a signal $x(t)$ represents the instantaneous value of the waveform as a function
of time. The signal could be described either by an equation or by a graph.

Signals that arise in communications systems can be classified in different
ways depending on the type of representation being used. We will study three
different types of classifications in this section, each of which will arise in
different parts of the book. Neither of the three breakdowns has a particularly
good, simple descriptive name for the process, so we will describe them in
terms of the categories used in the classifications.

The first classification is to categorize signals as either (1) *periodic* or (2)
nonperiodic. A *periodic* signal is one that repeats itself in a predictable
fashion, as illustrated by the example of Fig. 2-1a. The *period T* (in seconds)
of the periodic signal is the smallest value of time such that $x(t + T) = x(t)$
for all t. In simple terms, the period is simply the shortest length of time over

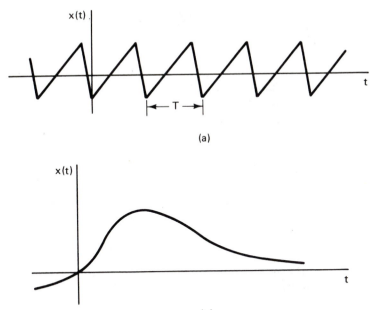

x(t)

t

|← T →|

(a)

x(t)

t

(b)

Figure 2-1. Examples of (a) a periodic signal and (b) a nonperiodic signal.

which the complete form of the signal is established. An interval of one period is also referred to as *one cycle* of the signal.

A nonperiodic signal can be thought of as a limiting case of a periodic signal in which the period approaches infinity. In more formal terms, a nonperiodic signal is one in which there is no finite T that satisfies the periodic constraint established in the preceding paragraph. An example of a nonperiodic signal is shown in Fig. 2-1b.

Since we can never look at a signal over an infinite interval of time, one could argue that an apparent nonperiodic signal might repeat itself after a sufficiently long time and thus actually be periodic. Conversely, one could also argue that an apparent periodic signal might stop repeating itself after a sufficiently long time. From a practical point of view, such idealistic arguments need not cause any difficulty, since it is over the usual time interval of interest in which these properties are important.

The second type of classification that will be used is to identify signals as being either (1) *deterministic* signals or (2) *random* signals. A *deterministic* signal is one whose instantaneous value as a function of time can be totally predicted. This includes all signals whose instantaneous values can be predicted with mathematical equations, but it also includes functions that do not have describing equations but are represented by graphs. The point is that the exact value of a deterministic signal can be predicted at a given time in

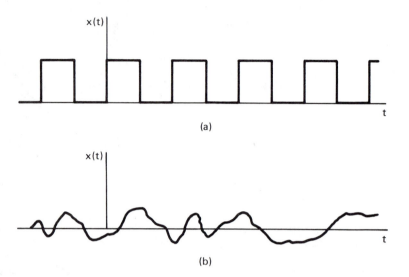

Figure 2-2. Examples of (a) a deterministic signal and (b) a random signal.

advance. An example of a deterministic signal (a square wave) is shown in Fig. 2-2a.

A *random* signal, in simplest terms, is one whose instantaneous value cannot be predicted at any given time. While the exact value at any given time is not known, many random signals encountered in communications systems have certain reasonably well behaved properties that can be described in statistical or probabilistic terms. Much of the treatment of random signals will be delayed until Chapter 8. An example of a random signal is shown in Fig. 2-2b.

Based on the relationship between information and uncertainly as discussed in Chapter 1, one can readily argue that only random signals provide true information, and deterministic signals provide no information since all possible values can be totally predicted in advance. This is true at the broadest level of analysis, and indeed virtually all data signals processed over a communications system are random in nature (at least as far as the destination is concerned). However, from the viewpoint of system design, analysis, testing, and operation, it is both necessary and desirable to use deterministic signals for predicting system performance. Such signals have well-known properties that are easy to deal with and will thus be employed throughout the developments in the book.

The third type of classification is according to whether the signal is (1) a *power signal* or (2) an *energy signal*. A *power* signal contains a *finite nonzero* limiting value of average power. An *energy* signal contains a *finite nonzero* amount of energy. In applying these definitions, it is necessary to visualize the signal as extending from $t = -\infty$ to $t = +\infty$.

Some background work is required to properly explain these last definitions. First, a common assumption made in communications system analysis is to use a 1-Ω reference for power and energy calculations. This may seem strange and would, of course, be completely incorrect if one were actually calculating the true power, for example, in a 50-Ω antenna. However, at a more abstract level of analysis, where no specific circuits are being considered, this assumption often simplifies comparisons, as we will see shortly.

Now we will determine an expression for the instantaneous power $p(t)$. If the power were dissipated in a resistance R, the power would be computed as either v^2/R or i^2R, depending on whether the signal were a voltage or a current. However, because the waveform is an arbitrary signal $x(t)$, and because the load resistance is 1 Ω, the instantaneous power is simply

$$p(t) = x^2(t) \tag{2-1}$$

The reader should pause long enough to ensure that this concept is solid before continuing since it will arise frequently throughout the book. By assuming a 1-Ω reference for power and energy comparisons, and since $x(t)$ may represent either voltage or current, the instantaneous power is simply the square of the signal. Because of the fact that this power is based on a reference load of 1 Ω, it will be referred to as *normalized power* whenever the 1-Ω reference is to be emphasized.

It will be recalled that power is the rate of performing work, that is, the rate of change of energy as a function of time. Conversely, the energy is the area under the power curve as a function of time. Letting $w(t)$ represent the net energy expended up to time t, we have

$$w(t) = \int_{-\infty}^{t} p(t)\, dt \tag{2-2}$$

For a normalized power function, the corresponding normalized energy function is simply

$$w(t) = \int_{-\infty}^{t} x^2(t)\, dt \tag{2-3}$$

Now let us return to the classification scheme of interest. The major criterion is to determine whether the average power or the total energy is finite and non-zero. The following intuitive process will work in a large number of cases: Mentally visualize the process of squaring the signal. If the resulting function has a finite amount of area under the curve (i.e., the energy is finite), the signal is an energy signal. Conversely, if the area (energy) approaches infinity, it is a power signal. As mentioned earlier, it is necessary to assume that the signal extends over an infinite interval in both directions, since, otherwise, the area would always be finite.

We will concentrate first on power signals. As noted previously, the area under the squared function curve becomes infinitely large over an infinite time interval. This means, of course, that the total energy contained in a power signal becomes infinite in the limit. However, the power or rate of energy change is finite, as we will see shortly.

One of the easiest power signals to recognize is the case of a periodic signal whose energy always increases without bound over an infinite time interval. Thus, *all periodic signals are necessarily power signals.* However, the converse is not true; there are many power signals that are not periodic. On the other hand, a definition of power similar to that for periodic signals can be applied to nonperiodic signals, as will be shown shortly.

Consider a periodic signal $x(t)$ and a 1-Ω reference. Let P represent the *average* power (or average normalized power to be precise), which can be evaluated by determining the area of $p(t)$ over a cycle and dividing by the period; that is,

$$P = \frac{1}{T} \int_{-(T/2)}^{T/2} p(t) \, dt = \frac{1}{T} \int_{-(T/2)}^{T/2} x^2(t) \, dt \qquad (2\text{-}4)$$

Because the area under one particular cycle is the same as that of any other cycle, the same result as (2-4) is obtained by integrating over n cycles (n is an integer) and dividing by nT as follows:

$$P = \frac{1}{nT} \int_{-(nT/2)}^{nT/2} x^2(t) \, dt \qquad (2\text{-}5)$$

Since n is arbitrary, it can assume any integer value (except 0) since the result is independent of the number of cycles. Consider then the process of letting n approach infinity, meaning that the interval of integration extends over an infinite limit. Although the area under the curve is increasing without limit, the average power as given by (2-5) remains finite as before. Thus, while the *energy* is approaching infinity, the *average power* remains finite at a nonzero value, which is the required character of a power signal. In a mathematical sense, nT increases at the same rate as the integral, and the ratio remains finite.

x(t)

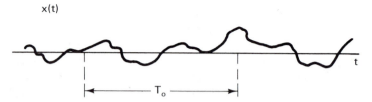

Figure 2-3. Process used in defining the average power in a nonperiodic power signal.

Let us now extend this argument to a nonperiodic power signal. Consider the signal $x(t)$ illustrated in Fig. 2-3. While the signal is nonperiodic, an arbitrary interval T_0 is first established for the signal as shown, and a power P_0 is computed as follows:

$$P_0 = \frac{1}{T_0} \int_{-(T_0/2)}^{T_0/2} x^2(t)\, dt \qquad (2\text{-}6)$$

The result is, of course, the average power over the given interval, and this result will vary with the particular interval selected. By the same argument used following (2-5), the integral of (2-6) should remain finite as T_0 increases, although its value would likely fluctuate over finite ranges of T_0.

Let P represent the limit of (2-6) as T_0 approaches infinity; that is,

$$P = \lim_{T_0 \to \infty} P_0 = \lim_{T_0 \to \infty} \frac{1}{T_0} \int_{-(T_0/2)}^{T_0/2} x^2(t)\, dt \qquad (2\text{-}7)$$

Many nonperiodic signals do, in fact, have a limiting constant value of P as $T_0 \to \infty$, even though such signals exhibit totally nonperiodic and even random behavior. *When we speak of the power in a nonperiodic random power signal throughout the text, it will be in the framework of a limiting value such as indicated by (2-7).*

To summarize this discussion, a power signal is one whose total energy would be infinite, but which possesses a non-zero limiting average power P resulting from dividing the energy by the time interval before allowing the time interval to increase without limit. A periodic signal is always a power signal in the classification of power and energy signals.

The energy signal is one in which the net total energy $w(\infty)$ contained in the signal is a nonzero constant value. If the average power definition just discussed were applied to an energy signal, the resulting average power would always approach zero as the period increases without limit. Two simple examples of energy signals are the single pulse and the decaying exponential function shown in Fig. 2-4. It is obvious that the first waveform has finite

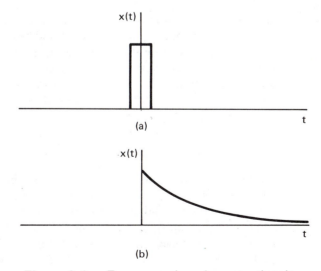

Figure 2-4. Two examples of energy signals.

energy, but the second case may not be so obvious. This latter function will be analyzed in Ex. 2-1.

Example 2-1

Consider the decaying exponential function given by

$$
\begin{aligned}
x(t) &= A\epsilon^{-\alpha t}, && \text{for } t > 0 \\
&= 0, && \text{for } t < 0
\end{aligned}
\tag{2-8}
$$

(The form of this function was illustrated by the second example in Fig. 2-4.) Classify this signal in the three different ways discussed in this section.

Solution

The first two forms of classifications are readily determined by inspection. The signal is *nonperiodic,* and it is obviously *deterministic.* On the other hand, since the curve theoretically extends over an infinite time range, it may not be obvious whether this function has finite or infinite energy. To check this point, we first compute the instantaneous normalized power $p(t)$ as

$$
p(t) = A^2\epsilon^{-2\alpha t}
\tag{2-9}
$$

We next determine if the area under this curve, that is, the net energy $w(\infty)$, is finite.

$$
\begin{aligned}
w(\infty) &= \int_0^\infty A^2\epsilon^{-2\alpha t}dt \\
&= \frac{A^2\epsilon^{-2\alpha t}}{-2\alpha}\bigg]_0^\infty = \frac{A^2}{2\alpha}
\end{aligned}
\tag{2-10}
$$

This result is finite for $\alpha > 0$, so the signal is an *energy signal.* If the area (or energy) had been infinite, the function would be a power signal.

This example illustrates the point that, just because a function extends over an infinite time range, it does not necessarily have infinite energy. The decaying exponential function approaches zero so quickly that the area in the range where t is reasonably large is completely negligible.

2-2 FOURIER SERIES

In this portion of the book, the emphasis will be directed primarily toward deterministic signals. As previously noted, most real-life communications signals are actually random in nature. However, the mathematic tools for dealing with random signals are somewhat more complex, so their consideration will be postponed until Chapter 8. Fortunately, many of the practical aspects of spectral analysis can be inferred from the analysis performed with deterministic signals. Thus, while the many deterministic examples that we consider in this chapter will obviously not represent true information signals, we stress again that the results obtained from such cases serve to establish many general guidelines for all types of signals.

In this section, the emphasis will be directed toward the concept of *Fourier series*. The Fourier series spectral representation is best explained by initially assuming only periodic signals. From the definitions in the preceding section, such signals are necessarily power signals. Thus, the initial emphasis here is directed toward deterministic periodic signals, which necessarily contain finite nonzero levels of power.

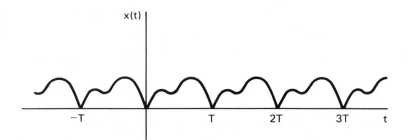

Figure 2-5.   Example of an arbitrary periodic signal for which a Fourier series representation could be obtained.

An example of an arbitrary signal of the type for which the Fourier series will be discussed is shown in Fig. 2-5. According to Fourier theory, this signal may be represented by a series of sinusoidal components, that is, sine and/or cosine terms, plus a dc term. The lowest frequency (other than dc) of the sinusoidal components is a frequency f_1 given by

$$f_1 = \frac{1}{T} \qquad (2\text{-}11)$$

This frequency is referred to as the *fundamental* component, and it is the same as that of the waveform itself. Thus, a periodic signal with a period of 1 millisecond (ms) will have a fundamental component with a frequency $f_1 = 1$ kHz.

All other frequencies in the signal will be integer multiples of the fundamental. These various components are referred to as *harmonics,* with the *order* of a given harmonic indicated by the ratio of its frequency to the fundamental frequency. Thus, for the waveform suggested in the preceding paragraph, frequencies appearing in the signal above 1 kHz would, in general, be a second harmonic at 2 kHz, a third harmonic at 3 kHz, a fourth harmonic at 4 kHz, and so on.

Any transmission system through which a given signal passes must have a bandwidth sufficiently large to pass all significant frequencies of the signal.

In a purely mathematical sense, many common waveforms theoretically contain an infinite number of harmonics. The preceding two sentences would then lead one to believe that an infinite bandwidth would be required to process such signals, which is impossible. The key to this apparent contradiction is the term "significant" in the first sentence.

It is easy to predict the theoretical frequencies present in a given periodic signal, as we have just seen. However, it is much more difficult to predict the relative magnitudes of the different components in order to determine which components are significant and which are not. From a signal-transmission point of view, if we can predict the frequency range over which the magnitudes are significant in size, we can then estimate the bandwidth requirements. In this sense, we simply ignore all harmonics of the signal that are outside the range and assume that their exclusion produces no noticeable degradation on the signal quality.

We now turn to the mathematical forms of the Fourier series. We will categorize the Fourier series in three separate forms, all of which produce identical information, but in which the results are exhibited in different ways. They will be denoted by the following three designations: (1) the sine-cosine form, (2) the amplitude-phase form, and (3) the complex exponential form. The first two forms are referred to as *one-sided* spectral representations since the resulting functions are real functions defined in terms of positive frequencies. (This point will be explained in depth later.) The third form is referred to as a *two-sided* spectral representation since it requires the use of both positive and negative frequencies (whatever that means) to make the results mathematically correct. Each of the three forms will now be discussed individually.

1. **Sine-Cosine Form.** The sine-cosine form is the one most commonly presented first in circuits and mathematics texts, so most readers having familiarity with Fourier series will likely have encountered it. This form represents a periodic signal $x(t)$ as a sum of sines and cosines in the form

$$x(t) = A_0 + \sum_{n=1}^{\infty} (A_n \cos n\omega_1 t + B_n \sin n\omega_1 t) \qquad (2\text{-}12)$$

where

$$\omega_1 = 2\pi f_1 = \frac{2\pi}{T} \qquad (2\text{-}13)$$

is the fundamental *angular* frequency in radians per second (rad/s). The nth harmonic *cyclic* frequency (in hertz) is nf_1, and the corresponding angular frequency is $n\omega_1$. As a result of the equalities in (2-13), the argument for either the sine or cosine function in (2-12) can be expressed in either of the following forms:

$$n\omega_1 t = 2\pi n f_1 t = \frac{2\pi n t}{T} \qquad (2\text{-}14)$$

The term A_0 represents the dc term and, as we will see shortly, it is simply the average value of the signal over one cycle. (*Note:* Some authors define the dc value as $A_o/2$ in order to make some of the later formulas apply to this case, but the A_0 form is easier to interpret.) Other than dc, there are two components appearing at a given harmonic frequency in the most general case: (1) a cosine term with an amplitude A_n and (b) a sine term with an amplitude B_n.

The major task involved in a Fourier analysis is in determining the A_n and B_n coefficients. The dc term A_0 is simply the average value and is given by

$$A_0 = \frac{1}{T} \int_0^T x(t)\ dt = \frac{\text{area under curve in one cycle}}{\text{period } T} \qquad (2\text{-}15)$$

Formulas for A_n and B_n are derived in applied mathematics books and are summarized here as follows:

$$A_n = \frac{2}{T} \int_0^T x(t) \cos n\omega_1 t\ dt \qquad (2\text{-}16)$$

for $n \geq 1$ but *not* for $n = 0$.

$$B_n = \frac{2}{T} \int_0^T x(t) \sin n\omega_1 t\ dt \qquad (2\text{-}17)$$

for $n \geq 1$. In all three of the preceding integrals, the limits of integration may be changed for convenience provided that the interval of integration is over one complete cycle in a positive sense. A common alternative range is from $-T/2$ to $T/2$.

The formulas for A_n and B_n indicate that the quantities are determined by first multiplying the signal by a cosine or sine term of the corresponding frequency at which the coefficient is desired and then determining the area of the resulting product function over one cycle. In many cases, a general expression for A_n or B_n can be determined from one integration with n appearing as a parameter. In other cases, particularly for experimental data, a separate integration may be required at each frequency. Computational means have been developed for performing such operations efficiently on a computer, and many scientific computer systems have programs available for determining the Fourier series of experimental data signals. If a signal $x(t)$ is defined in different forms over different parts of a cycle, the evaluation of either (2-16) or (2-17) may require expansion into several integrals over shorter portions of the cycle.

2. **Amplitude-Phase Form.** The sine-cosine form of the Fourier series given in (2-12) is usually the easiest form from which to evaluate the coefficients, and it is the form most commonly tabulated in reference books. However, it suffers from the fact that there are, in a sense, two separate components at a given frequency, each of which has a separate amplitude. When we measure the magnitude of a given spectral component with a

frequency-selective instrument, which component do we obtain? The fact is that the actual magnitude that would be measured with most instruments would be neither A_n nor B_n, but rather a special combination of the two, as will be seen shortly.

The amplitude-phase form of the Fourier series can be expressed as

$$x(t) = C_0 + \sum_{n-1}^{\infty} C_n \cos{(n\omega_1 t + \theta_n)} \tag{2-18}$$

The first term C_0 is the dc value and is the same as given by (2-15); that is, $C_0 = A_0$. It has been redefined here as C_0 in order to maintain a consistent form of notation.

In the form of (2-18), a given C_n represents the *net* amplitude of a given component at the frequency nf_1, and θ_n represents the corresponding phase. It should be mentioned that the assumption of the cosine is arbitrary, and one could just as easily assume a sine function. The resulting amplitudes would be identical, but the phase values resulting from a sine function representation would be different.

The easiest way to determine C_n and θ_n is to first determine A_n and B_n by use of (2-16) and (2-17) and then to employ the following relationships:

$$C_n = \sqrt{A_n^2 + B_n^2} \tag{2-19}$$

$$\theta_n = \tan^{-1}\left(\frac{-B_n}{A_n}\right) \tag{2-20}$$

The signs of $-B_n$ and A_n in (2-20) must be separately noted before the inverse tangent is computed since the resulting angle can be in either of the four quadrants.

If C_n and θ_n are known, the values of A_n and B_n may be determined from the following relationships:

$$A_n = C_n \cos{\theta_n} \tag{2-21}$$

$$B_n = -C_n \sin{\theta_n} \tag{2-22}$$

A straightforward means for dealing with the conversion from the sine-cosine form to the amplitude-phase form and vice versa will now be given. This method is particularly easy to apply with modern scientific calculators having rectangular-to-polar and polar-to-rectangular conversion functions. Basically, it involves the representations of sine and cosine functions as complex phasors, although one need not view it in that way. The relative phase sequence of sinusoidal components is shown in Fig. 2–6. In this figure, the + cosine function is represented by the positive real ($+x$) axis, the − sine function by the positive imaginary ($+y$) axis, and so on. The values of A_n and B_n with their respective signs considered are interpreted as phasor components along the appropriate axes. The complex result is then converted to polar form. The magnitude of the polar form is C_n and the angle (referred to

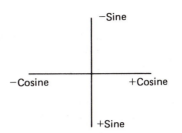

Figure 2-6. Relative phase sequence of sinusoids.

the positive real axis, of course) is θ_n. If C_n and θ_n are given, the reverse process is performed; that is, the given polar form is converted to rectangular form. The real (x) component is A_n and the imaginary (y) component is B_n. The signs of various terms must be carefully noted in this process. This process is illustrated in Fig. 2-7 for the case where A_n and B_n are both positive. The resulting value of θ_n in this case is negative.

We return now to the question of which component is measured with a frequency-selective instrument. With the majority of common instruments of this type, the reading is proportional to C_n, the net amplitude. (The instrument may actually be calibrated to read the rms value of C_n, which is $C_n / \sqrt{2}$.) There are, however, certain special phase-sensitive instruments that can be used to obtain A_n and B_n separately, so the pecularities of such instruments should be understood before measurements are made.

There are a number of cases of well-known wave forms in which either A_n or B_n (but obviously not both) is identically zero at all possible frequencies for the signal. Indeed, as we will see later in the text, it is possible in some cases to choose the time origin in a way that will force this result. With such signals, the distinctions between the sine-cosine form and the amplitude-phase form are no longer very significant. In fact, for the case where $B_n = 0$ at all frequencies (corresponding to cosine terms only in the series), the two forms

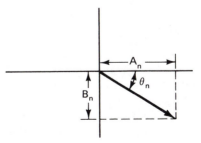

Figure 2-7. Relationship between sine-cosine and amplitude-phase components at a given frequency. (A_n and B_n are both positive and θ_n is negative in this example.)

are exactly the same. If instead $A_n = 0$ at all frequencies (corresponding to sine terms only in the series), the two forms vary only in the phase terms. In either event, the amplitude values C_n are the same as either A_n or B_n in such cases. In short, the lines of distinction between the forms are no longer of much importance, so when such signals arise later in the book, very little will be said about categorizing the signals in either of the forms discussed.

3. **Complex Exponential Form.** The complex exponential form of the Fourier series is the most difficult form for many persons to perceive, primarily because it represents a step away from the domain of real signals into a domain of complex mathematical representations. Because this book is aimed at a practical level of application, we will tend to avoid this form wherever appropriate and remain with one of the two earlier forms. On the other hand, there are some developments that can be done much easier with the exponential form than with either of the earlier forms. In particular, the exponential form is a direct link with the concept of the Fourier transform, which is very important in dealing with nonperiodic signals, as we will see later. At any rate, anyone dealing with communications theory at a systems level of analysis cannot avoid dealing with the complex exponential series at some point or another, so a complete treatment of Fourier series must necessarily include this form.

The exponential form of the Fourier series is related to the fact that both the sine and cosine functions can be expressed in terms of exponential functions with purely imaginary arguments. The basis for this is Euler's formula, which is written in two separate forms as

$$\epsilon^{jn\omega_1 t} = \cos n\omega_1 t + j \sin n\omega_1 t \qquad (2\text{-}23a)$$

$$\epsilon^{-jn\omega_1 t} = \cos n\omega_1 t - j \sin n\omega_1 t \qquad (2\text{-}23b)$$

Alternate addition and subtraction of (2-23a) and (2-23b) result in the following two expressions for cosine and sine functions:

$$\cos n\omega_1 t = (\epsilon^{jn\omega_1 t} + \epsilon^{-jn\omega_1 t})/2 \qquad (2\text{-}24)$$

$$\sin n\omega_1 t = (\epsilon^{jn\omega_1 t} - \epsilon^{-jn\omega_1 t})/2j \qquad (2\text{-}25)$$

Note that both forms contain an exponential function with a $(jn\omega_1 t)$ argument and an exponential function with a $(-jn\omega_1 t)$ argument. The first term may be thought of as a "positive frequency" term corresponding to a frequency nf_1 (assuming n is positive), and the second term may be considered as a "negative frequency" term corresponding to a frequency $-nf_1$. Both terms are required to completely describe the sine or cosine function.

The exponential form can be developed by expanding the sine and cosine functions according to these exponential definitions and regrouping. This general process is somewhat detailed and will not be given here. The interested reader is referred to one of the many intermediate or advanced engineering circuit analysis or applied mathematics texts for complete details. We will concentrate here on the results and the corresponding interpretations.

The general form of the complex exponential form of the Fourier series can be expressed as

$$x(t) = \sum_{n=-\infty}^{\infty} \overline{X}_n \, \epsilon^{jn\omega_1 t} \qquad (2\text{-}26)$$

where the bar above $\overline{X}_n$ indicates that it is, in general, a complex value. The Fourier coefficient at a given frequency nf_1 is the complex quantity $\overline{X}_n$. An expression for determining $\overline{X}_n$ is

$$\overline{X}_n = \frac{1}{T} \int_0^T x(t) \, \epsilon^{-jn\omega_1 t} \, dt \qquad (2\text{-}27)$$

Some interpretation of the preceding results is in order. Note in (2-26) that the exponential series is summed over both negative and positive frequencies (or negative and positive values of n), as previously discussed. At a given real frequency kf_1, $(k > 0)$, the spectral representation consists of

$$\overline{X}_k \, \epsilon^{jk\omega_1 t} + \overline{X}_{-k} \, \epsilon^{-jk\omega_1 t}$$

The first term is thought of as the "positive frequency" contribution, while the second is the corresponding "negative frequency" contribution. Although either one of the two terms is a complex quantity, they add together in such a manner as to create a real function, and this is why both terms are required to make the mathematical form complete. On the other hand, all the spectral information can be deduced from either the $\overline{X}_k$ term or the $\overline{X}_{-k}$ term, since there is a direct relationship between them. Let $\overline{X}^*$ represent the complex conjugate of $\overline{X}$. Then it can be shown that

$$\overline{X}_{-n} = \overline{X}_n^* \qquad (2\text{-}28)$$

Thus, the negative frequency coefficient is the complex conjugate of the corresponding positive frequency coefficient.

While the coefficient $\overline{X}_n$ can be calculated from (2-27) directly, it turns out that $\overline{X}_n$ can also be calculated directly from A_n and B_n of the sine-cosine form. The relationship reads

$$\overline{X}_n = \frac{A_n - jB_n}{2}, \qquad \text{for } n \neq 0 \qquad (2\text{-}29)$$

Even though A_n and B_n are interpreted only for positive n in the sine-cosine form, their functional forms may be extended for both positive and negative n in applying (2-29). Alternatively, (2-29) may be applied for positive n, and (2-28) may be used for determining the corresponding coefficients for negative n. The dc component X_0 is simply

$$\overline{X}_0 = \frac{1}{T} \int_0^T x(t) \, \epsilon^{-jn\omega_1 t} \, dt = A_0 = C_0 \qquad (2\text{-}30)$$

which is the same in all the Fourier forms.

In many situations involving the complex form of the Fourier series, it is desirable to express $\overline{X}_n$ as a magnitude and an angle. Since $\overline{X}_n$ is a complex quantity having a real and an imaginary part, this can be readily achieved. Thus $\overline{X}_n$ can be expressed in polar form as

$$\overline{X}_n = X_n \, \epsilon^{j\phi_n} = X_n \, \underline{/\phi_n} \qquad (2\text{-}31)$$

The quantity $X_n = |\overline{X}_n|$ represents the magnitude of the complex Fourier coefficient at a frequency nf_1, and ϕ_n is the corresponding angle or phase expressed in the polar form. For example, if $\overline{X}_n = 3 + j4$ at a given frequency, it can also be expressed as $\overline{X}_n = 5\underline{/53.13°}$ with $X_n = 5$ and $\phi_n = 53.13°$.

To end this section, we refer the reader to Appendix A for later reference. In this appendix are tabulated a number of common waveforms and their Fourier series forms. It turns out that all the waveforms shown are either even or odd so that the sine-cosine and amplitude-phase forms need not be tabulated separately. As previously noted, they are essentially the same except for the difference in phase when the function is odd. The exponential

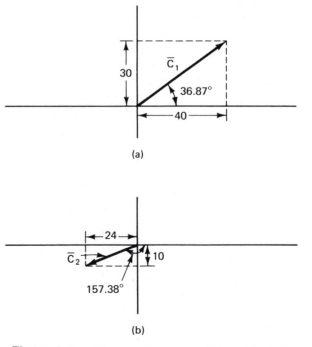

(a)

(b)

Figure 2-8. Phasor diagrams of Example 2-2.

representations are not shown, but they can readily be determined from the results of this section.

Example 2-2

This example will have as a primary objective the conversion between different forms of the Fourier series. A certain periodic band-limited signal has only three frequencies in its Fourier series representation: dc, 1 kHz, and 2 kHz. The signal can be expressed in sine-cosine form as

$$x(t) = 18 + 40 \cos 2000\pi t - 30 \sin 2000\pi t$$
$$- 24 \cos 4000\pi t + 10 \sin 4000\pi t \quad (2\text{-}32)$$

Express the signal in (a) amplitude-phase form and (b) complex exponential form.

Solution

(a) The amplitude-phase form desired for the signal will read

$$x(t) = 18 + C_1 \cos (2000\pi t + \theta_1) + C_2 \cos (4000\pi t + \theta_2) \quad (2\text{-}33)$$

where $C_0 = 18$ has been noted by obvious inspection. We will determine C_1 and C_2 from the phasor concept suggested by Figs. 2-6 and 2-7. The resulting phasor diagrams for this example are shown in Fig. 2-8. Let $\overline{C}_1$ represent the phasor associated with 1000 Hz as shown in Fig. 2-8a. We have

$$\overline{C}_1 = 40 + j30 = 50\underline{/36.87°} \quad (2\text{-}34)$$

Let $\overline{C}_2$ represent the corresponding phasor associated with 2000 Hz as shown in Fig. 2-8b. We have

$$\overline{C}_2 = -24 - j10 = 26\underline{/-157.38°} \quad (2\text{-}35)$$

The final form of the amplitude-phase representation is then expressed as

$$x(t) = 18 + 50 \cos (2000\pi t + 36.87°)$$
$$+ 26 \cos (4000\pi t - 157.38°) \quad (2\text{-}36)$$

(b) The complex exponential form will be determined from the formula (2-29) in conjunction with the conjugate relationship of (2-28). Thus

$$\overline{X}_1 = \frac{40 - j(-30)}{2} = 20 + j15 = 25\underline{/36.87°} \quad (2\text{-}37)$$

$$\overline{X}_{-1} = \overline{X}_1^* = 20 - j15 = 25\underline{/-36.87°}$$

$$\overline{X}_2 = \frac{-24 - j10}{2} = -12 - j5 = 13\underline{/-157.38°}$$

$$\overline{X}_{-2} = \overline{X}_2^* = -12 + j5 = 13\underline{/157.38°}$$

The series may then be expressed as

$$x(t) = 18 + (20 + j15) \, \epsilon^{j2000\pi t} + (20 - j15) \, \epsilon^{-j2000\pi t} \qquad (2\text{-}38)$$
$$+ (-12 - j5) \, \epsilon^{j4000\pi t} + (-12 + j5) \, \epsilon^{-j4000\pi t}$$

in which the coefficients have been expressed in rectangular forms. Alternatively, the polar forms of the coefficients may be used in the expansion in which the expression becomes

$$x(t) = 18 + 25 \, \epsilon^{j(2000\pi t + 36.87°)} + 25 \, \epsilon^{-j(2000\pi t + 36.87°)} \qquad (2\text{-}39)$$
$$+ 13 \, \epsilon^{j(4000\pi t - 157.38°)} + 13 \, \epsilon^{-j(4000\pi t - 157.38°)}$$

Example 2-3

Determine the Fourier series representation for the wave form shown in Fig. 2-9. Express in each of the following forms: (a) sine-cosine, (b) amplitude-phase, and (c) complex exponential.

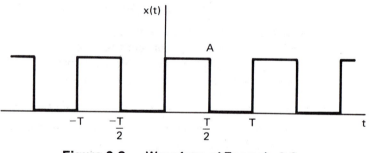

Figure 2-9. Waveform of Example 2-3.

Solution

(a) The sine-cosine form is usually the easiest form for determining the coefficients directly. The signal $x(t)$ over one complete cycle can be expressed as

$$x(t) = A, \qquad \text{for } 0 < t < T/2 \qquad (2\text{-}40)$$
$$= 0, \qquad \text{for } T/2 < t < T$$

The dc component A_0 can be determined by inspection since it is the average value. Thus,

$$A_0 = \frac{\text{area under curve in one cycle}}{T} = \frac{(AT/2)}{T} = \frac{A}{2} \qquad (2\text{-}41)$$

The coefficients A_n can be determined from (2-16). Note that, while integration over a complete cycle is required, the function is zero over half of the cycle, so the integral reduces to

$$A_n = \frac{2}{T} \int_0^{T/2} A \cos n\omega_1 t \, dt \qquad (2\text{-}42)$$

$$= \frac{2A}{n\omega_1 T} \sin n\omega_1 t \Big]_0^{T/2} = \frac{2A}{n\omega_1 T} \left(\sin \frac{n\omega_1 T}{2} - 0 \right)$$

In performing Fourier coefficient evaluations, the product $n\omega_1 T$ appears in virtually every case. When this occurs, the substitution $n\omega_1 T = 2\pi n$ is recommended to the reader as a suggested simplification. This equality is readily verified by noting that $n\omega_1 T = n2\pi f_1 T = n2\pi \, (1/T)T = 2\pi n$. This substitution will be made in various examples that follow without further justification.

The preceding evaluation now reduces to

$$A_n = \frac{2A}{2\pi n} \sin n\pi = 0, \qquad \text{for } n \neq 0 \qquad (2\text{-}43)$$

since $\sin n\pi = 0$ for n an integer. Thus, in this particular example, the coefficients of all the cosine terms in the sine-cosine form are zero (except dc).

The B_n coefficients are now determined from the following development:

$$B_n = \frac{2}{T} \int_0^{T/2} A \sin n\omega_1 t \, dt \qquad (2\text{-}44)$$

$$= \frac{-2A}{n\omega_1 T} \cos n\omega_1 t \Big]_0^{T/2}$$

$$= \frac{-2A}{2\pi n} \left(\cos \frac{n\omega_1 T}{2} - 1 \right) = \frac{A}{n\pi} (1 - \cos n\pi)$$

The quantity $\cos n\pi$ satisfies

$$\cos n\pi = -1 \quad \text{for } n \text{ odd}$$
$$+1 \quad \text{for } n \text{ even}$$

When n is even, there is a cancellation inside the parentheses of the last term in (2-44), while for n odd, the term $-(-1) = +1$ adds to the other term. Thus, the expression for B_n reduces to

$$B_n = \frac{2A}{n\pi}, \qquad \text{for } n \text{ odd} \qquad (2\text{-}45)$$

$$= 0, \qquad \text{for } n \text{ even}$$

The resulting sine-cosine form of the Fourier series representation of $x(t)$ can be expressed as

$$x(t) = \frac{A}{2} + \frac{2A}{\pi} \sin \omega_1 t + \frac{2A}{3\pi} \sin 3\omega_1 t \qquad (2\text{-}46\text{a})$$

$$+ \frac{2A}{5\pi} \sin 5\omega_1 t + \frac{2A}{7\pi} \sin 7\omega_1 t + \cdots$$

$$= \frac{A}{2} + \sum_{\substack{n-1 \\ n \text{ odd}}}^{\infty} \frac{2A}{n\pi} \sin n\omega_1 t \qquad (2\text{-}46\text{b})$$

(b) Since the A_n coefficients in the sine-cosine form are zero, the amplitude-phase form of the Fourier series is not much different from the sine-cosine form in this particular example. In fact, if the sine function instead of the cosine function had been used to establish the basic form in (2-18), the amplitude-phase form would be perfectly identical with the sine-cosine form in this case. However, since the cosine was used in establishing the basic form, we will convert the results of part (a) to that form to be consistent with our approach.

The magnitude of the components in the amplitude-phase form are readily expressed as

$$C_0 = A_0 = \frac{A}{2} \qquad (2\text{-}47)$$

$$C_n = \sqrt{A_n^2 + B_n^2} = \sqrt{0 + B_n^2} = B_n \qquad (2\text{-}48)$$

$$\theta_n = -90° \qquad \text{(from relative phase sequence)} \qquad (2\text{-}49)$$

The amplitude-phase form can then be expressed as

$$x(t) = \frac{A}{2} + \frac{2A}{\pi} \cos (\omega_1 t - 90°) + \frac{2A}{3\pi} \cos (3\omega_1 t - 90°) \qquad (2\text{-}50)$$

$$+ \frac{2A}{5\pi} \cos (5\omega_1 t - 90°) + \cdots$$

(c) The complex exponential form will now be developed. While we could determine the coefficients $\overline{X}_n$ directly from the sine-cosine terms making use of (2-29), it will be more instructive for the reader's sake to start over again using the defining relationship for $\overline{X}_n$ as given by (2-27). First, we note that the dc component is again

$$\overline{X}_0 = A_0 = \frac{A}{2} \qquad (2\text{-}51)$$

The general coefficient $\overline{X}_n$ is given by

$$\overline{X}_n = \frac{1}{T} \int_0^{T/2} A\epsilon^{-jn\omega_1 t}\, dt = \frac{-A}{jn\omega_1 T} \epsilon^{-jn\omega_1 t}\Big]_0^{T/2}$$

$$= \frac{-A}{j2n\pi} (\epsilon^{-jn\omega_1 T/2} - 1) = \frac{A}{j2n\pi}(1 - \epsilon^{-jn\pi}) \qquad (2\text{-}52)$$

$$= \frac{A}{j2n\pi}(1 - \cos n\pi + j \sin n\pi)$$

where Euler's formula was used in the last step. This result can be readily simplified by first noting that $\sin n\pi = 0$ for n an integer. Furthermore, $1 - \cos n\pi = 2$ for n odd and $1 - \cos n\pi = 0$ for n even. Thus, $\overline{X}_n$ reduces to

$$\overline{X}_n = \frac{A}{jn\pi} = \frac{-jA}{n\pi}, \qquad \text{for } n \text{ odd} \qquad (2\text{-}53)$$

The reader can readily verify that this same result is obtained quite quickly by applying (2-29) to the previous results of the sine-cosine form.

The complex Fourier series for $x(t)$ can now be expressed as

$$x(t) = \frac{A}{2} - j\frac{A}{\pi} \epsilon^{j\omega_1 t} - j\frac{A}{3\pi} \epsilon^{j3\omega_1 t} - \cdots \qquad (2\text{-}54)$$

$$= + j\frac{A}{\pi} \epsilon^{-j\omega_1 t} + j\frac{A}{3\pi} \epsilon^{-j3\omega_1 t} + \cdots$$

2-3 FREQUENCY-SPECTRUM PLOTS

One of the most useful forms for displaying the Fourier series of a signal is by means of a graphical plot showing the relative strengths of the components as a function of frequency. Such a plot is loosely referred to as the *frequency spectrum* of the given signal. A frequency-spectrum plot permits a quick visual determination of the frequencies present in the signal and their relative magnitudes. This is the basis for the spectrum analyzer, which is a widely used instrument in communications systems work providing a cathode-ray tube (CRT) display of the spectrum in much the same form as the graphical technique that will be discussed in this section.

In principle, either of the three forms could be shown graphically. In practice, however, the sine-cosine form is less desirable for this purpose since both A_n and B_n would have to be plotted separately (unless, of course, one of the two is zero). For the same reason, spectral displays involving the complex exponential form generally focus on the polar form (magnitude and angle) rather than the real and imaginary representation. Consequently, we will focus on the amplitude-phase form and the magnitude and angle representation of the complex exponential form in our developments.

Both the amplitude-phase form and the complex exponential form have two quantities to be specified at each frequency (i.e., the amplitude and phase of the spectral components). While both quantities are necessary for mathematical reconstruction of the signal, in practical spectral displays, the amplitude is almost always the quantity that is emphasized. The reason for this is the simple fact that the relative amplitudes of spectral components are most significant in determining the bandwidth, while the phase only indicates the relative time shift of a given component relative to others. In fact, a simple time shift of the signal will readily change the phase terms without affecting the amplitude terms. Consequently, throughout the book when frequency spectra are being discussed and graphical displays are shown, only amplitude plots will be shown in most cases. This form will be often referred to as an *amplitude frequency-spectrum plot.*

One point about the terms "amplitude" and "magnitude" should be noted. By convention, these terms normally indicate a positive, real value. There are instances, however, when it is convenient to allow a given amplitude or magnitude spectrum to assume negative real values, and this concept will appear later in the chapter. This will be done only as a convenience in simplifying the mathematical form of the function. In the sense of complex numbers, this is permissible providied the phase is adjusted accordingly. For example, $5/-150°$ can be expressed as $-5/30°$. The first form involves a "positive amplitude," and the second involves a "negative amplitude," but the two results are identical in the sense of complex numbers. Thus, the reader should not be disturbed by negative values appearing in some of the amplitude spectra in the book.

As a result of the discussion this far, it would seem then that a frequency-spectrum plot would most likely consist of a graph of either C_n from the amplitude-phase form or X_n from the complex exponential form. It will be recalled that the C_n terms are defined only for $n \geq 0$, while the X_n terms are defined for both positive and negative n (as well as $n = 0$). For this reason, the plot of C_n as a function of frequency is called a *one-sided spectrum,* while the plot of X_n as a function of frequency is called a *two-sided spectrum.* The terms "one-sided" and "two-sided" will be used extensively in this book since they are easy to remember and they quickly alert the reader to the required mathematical form.

Referring back to Eq. (2-29) momentarily and evaluating the magnitude of $\overline{X}_n$, the result obtained is

$$X_n = |\overline{X}_n| = \sqrt{\frac{A_n^2 + B_n^2}{4}} = \frac{\sqrt{A_n^2 + B_n^2}}{2}, \qquad \text{for } n \neq 0 \qquad (2\text{-}55)$$

Comparing this result with (2-19), we see that

$$X_n = \frac{C_n}{2}, \qquad \text{for } n \neq 0 \tag{2-56}$$

and, of course, $X_0 = C_0$ as previously noted.

The result of (2-56) may clear away at least a portion of the mystery surrounding the complex exponential form of the Fourier series. On an amplitude-spectrum basis, the magnitudes of the components in the two-sided spectral form (except for dc) are exactly one-half the values in the one-sided form. An artificial, but rather easy way to remember this is that, when the terms are displayed on both sides, the one-sided terms are "cut in half" in order to provide the components for the other side. The dc term appears only in one place, so its value does not change.

When a signal is described as a function of time, that is, as $x(t)$, the result is said to be a *time-domain* representation. Conversely, when the spectral information is provided, the result is said to be a *frequency-domain* representation. Either form completely describes the signal provided that both amplitude and phase are given in the frequency domain.

A typical example of a one-sided frequency spectrum is shown in Fig. 2-10. The corresponding two-sided spectrum for the same signal is shown in Fig. 2-11. Note that the lengths of all the components except dc in the second case are half the values in the first case, as expected. The dc component, however, is the same in both cases.

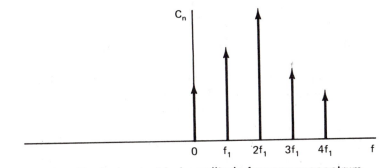

Figure 2-10.   Typical one-sided amplitude frequency spectrum.

An important summary point concerning the frequency spectrum will now be made. The frequency spectrum of a periodic signal is a *discrete* or *line* spectrum; that is, it contains components only at integer multiples of the repetition frequency of the signal (including dc).

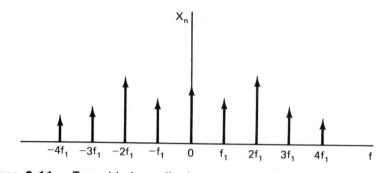

Figure 2-11. Two-sided amplitude spectrum corresponding to Figure 2-10.

Example 2-4

Consider the band-limited signal of Ex. 2-2, whose Fourier series was defined in Eq. (2-32). Plot (a) one-sided and (b) two-sided frequency spectra for this signal.

Solution

The one-sided spectrum is obtained from the amplitude-phase form of the series, which was developed in Eq. (2-36). The amplitudes of the components on a one-sided basis and their frequencies are summarized as follows:

Frequency (Hz)	0	1000	2000
Amplitude	18	50	26

The one-sided plot is shown in Fig. 2-12a.

While the two-sided plot could be readily determined directly from the one-sided plot, for convenience, the results will be tabulated again. These results could be deduced from (2-56) or from the expansion of (2-39)..The results are summarized as follows:

Frequency (Hz)	0	± 1000	± 2000
Amplitude	18	25	13

The two-sided plot is shown in Fig. 2-12b.

Example 2-5

Plot (a) the one-sided amplitude frequency spectrum and (b) the two-sided amplitude frequency spectrum for the square-wave signal considered in Ex. 2-3.

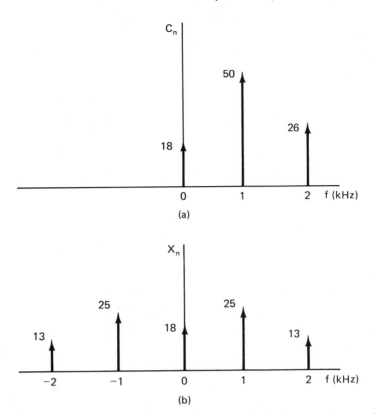

Figure 2-12.   One-sided and two-sided amplitude frequency spectra of Examples 2-2 and 2-4.

Solution

The one-sided form is determined by noting the C_n coefficients in the sine-cosine expansion as given by either (2-47) and (2-48) or by (2-50). Incidentally, since $A_n = 0$ in this case (except $n = 0$), the coefficients could have been determined just as easily directly from either (2-41) and (2-45) or from (2-46). The one-sided spectrum out to the ninth harmonic is shown in Fig. 2-13a.

The two-sided form can be determined either directly from the one-sided form or from the results of (2-54). The form of the two-sided spectrum is shown in Fig. 2-13b.

2-4 FOURIER SERIES SYMMETRY CONDITIONS

The various equations developed in the preceding several sections may, in theory, be applied to any signal to determine its spectrum. On the other hand, there are certain properties that may be used to simplify the computation of

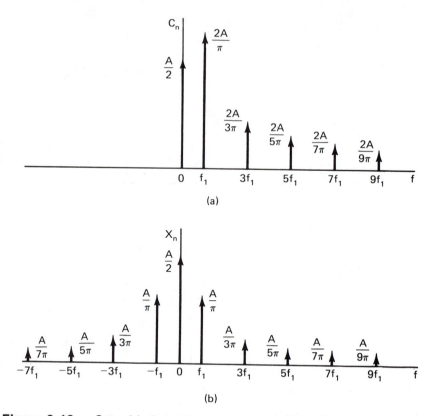

Figure 2-13. One-sided and two-sided amplitude frequency spectra of square wave. (See Example 2-5.)

the spectrum in many cases. Furthermore, some of these conditions permit many important properties of certain waveforms to be obtained by simple inspection procedures. For some applications, this information might be sufficient without having to compute the spectrum at all. In most cases, the use of these conditions will at least provide some information about the spectrum from a direct inspection.

The type of criteria to be studied in this section are the *symmetry* conditions. All the various symmetry conditions to be considered in this book are summarized in Table 2-1. The proofs of these conditions are given in various texts on applied mathematics. The emphasis in the book will be directed toward the practical interpretation and application of these various properties.

The equations at the top of the table are the various forms of the Fourier series that were developed in the preceding sections along with the relationships for converting from one form to another. The first row of the table provides the general relationships developed in the preceding sections, which

Table 2-1. Fourier Series Symmetry Conditions

Sine-cosine form: $x(t) = A_0 + \sum_{n=1}^{\infty}(A_n \cos n\omega_1 t + B_n \sin n\omega_1 t)$, $\qquad \omega_1 = 2\pi f_1 = \dfrac{2\pi}{T}$

Amplitude-phase form: $x(t) = C_0 + \sum_{n=1}^{\infty} C_n \cos(n\omega_1 t + \theta_n)$, $\qquad C_n = \sqrt{A_n^2 + B_n^2}$, $\qquad \theta_n = \left(\tan^{-1}\dfrac{-B_n}{A_n}\right)$

Complex exponential form: $x(t) = \sum_{n=-\infty}^{\infty} \bar{X}_n \epsilon^{jn\omega_1 t}$, $\qquad \bar{X}_n = \dfrac{A_n - jB_n}{2}$, $\qquad$ for $n \neq 0$ $\qquad \bar{X}_0 = A_0$

Condition	A_n (except $n=0$)	B_n	$\bar{X}_n$	Comments
General	$\dfrac{2}{T}\displaystyle\int_0^T x(t)\cos n\omega_1 t\, dt$	$\dfrac{2}{T}\displaystyle\int_0^T x(t)\sin n\omega_1 t\, dt$	$\dfrac{1}{T}\displaystyle\int_0^T x(t)\,\epsilon^{-jn\omega_1 t}\, dt$	
Even function $x(-t) = x(t)$	$\dfrac{4}{T}\displaystyle\int_0^{T/2} x(t)\cos n\omega_1 t\, dt$	0	$\dfrac{2}{T}\displaystyle\int_0^{T/2} x(t)\cos n\omega_1 t\, dt$	One-sided forms have only cosine terms. $\bar{X}_n$ terms are real.
Odd function $x(-t) = -x(t)$	0	$\dfrac{4}{T}\displaystyle\int_0^{T/2} x(t)\sin n\omega_1 t\, dt$	$\dfrac{-2j}{T}\displaystyle\int_0^{T/2} x(t)\sin n\omega_1 t\, dt$	One-sided forms have only sine terms. $\bar{X}_n$ terms are imaginary.
Half-wave symmetry $x\left(t + \dfrac{T}{2}\right) = -x(t)$	$\dfrac{4}{T}\displaystyle\int_0^{T/2} x(t)\cos n\omega_1 t\, dt$	$\dfrac{4}{T}\displaystyle\int_0^{T/2} x(t)\sin n\omega_1 t\, dt$	$\dfrac{2}{T}\displaystyle\int_0^{T/2} x(t)\,\epsilon^{-jn\omega_1 t}\, dt$	Odd-numbered harmonics only.
Full-wave symmetry $x\left(t + \dfrac{T}{2}\right) = x(t)$	$\dfrac{4}{T}\displaystyle\int_0^{T/2} x(t)\cos n\omega_1 t\, dt$	$\dfrac{4}{T}\displaystyle\int_0^{T/2} x(t)\sin n\omega_1 t\, dt$	$\dfrac{2}{T}\displaystyle\int_0^{T/2} x(t)\,\epsilon^{-jn\omega_1 t}\, dt$	Even-numbered harmonics only.

can be used in all cases. Subsequent cases apply whenever the waveform possesses one or more symmetry conditions. Let us consider each case individually.

Even Function

A function $x(t)$ is said to be *even* if

$$x(-t) = x(t) \tag{2-57}$$

An example of an even function is shown in Fig. 2-14. It can be shown that in this case only cosine terms appear in the spectrum. Furthermore, the integral used in determining A_n has the property that the area under the curve in one half of a cycle is the same as that in the other half. Hence, we need only integrate over half a cycle and double the result as indicated in the table. Finally, the integral for $\overline{X}_n$ reduces to an integral involving the product of the time signal and the cosine function rather than the complex exponential. The coefficients $\overline{X}_n$ are all *real* in this case.

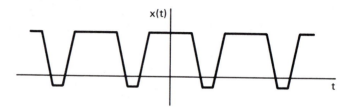

Figure 2-14.   Example of an even function.

Odd Function

A function $x(t)$ is said to be *odd* if

$$x(-t) = -x(t) \tag{2-58}$$

An example of an odd function is shown in Fig. 2-15. It can be shown that in this case, only sine terms appear in the spectrum. The integral for B_n need be evaluated only over half a cycle and doubled, and the integral used in determining $\overline{X}_n$ reduces to the product of the time signal and a sine function. Note that all of the $\overline{X}_n$ terms are purely imaginary in this case.

This type of symmetry is one of two types that can be "disguised" by the presence of a dc component. Consider the waveform shown in Fig. 2-16, which is identical to that of Fig. 2-15 except that a dc component has been added. Certainly, all logic tells us that the dc component should not affect any other portion of the spectrum except at zero frequency, so there should be only sine terms in the remaining part of the spectrum. However, the basic odd

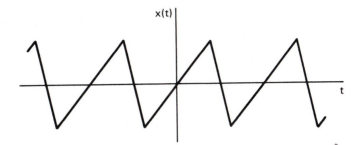

Figure 2-15. Example of an odd function.

function condition of (2-58) is not satisfied. The dc component can be thought of as a limiting case of a cosine function of zero frequency, and this one "cosine" function obscures the symmetry.

The way around this problem is to inspect each waveform by mentally shifting it up or down to see if the symmetry condition can be achieved by this process. If so, a new function $x_1(t)$ can be formed as follows:

$$x_1(t) = x(t) - A_0 \qquad (2-59)$$

where A_0 is the dc component of the signal. The function $x_1(t)$ now satisfies the pertinent symmetry condition and can be integrated according to the form given in the table. Note that if the symmetry condition is employed, the new function $x_1(t)$ should be integrated rather than the original function.

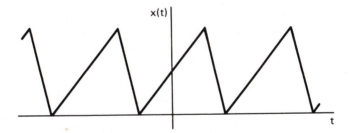

Figure 2-16. Function of Figure 2-15 with dc component added.

Half-Wave Symmetry

A function is said to possess half-wave symmetry if

$$x\left(t + \frac{T}{2}\right) = -x(t) \qquad (2-60)$$

A typical function having half-wave symmetry is shown in Fig. 2-17. It can be shown that a function satisfying this condition will have only *odd-numbered*

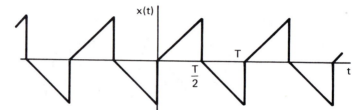

Figure 2-17. Example of function with half-wave symmetry.

harmonics (i.e., $n = 1, 3, 5, 7, \ldots$). However, unless one of the previous two conditions is also satisfied, there will be both sine and cosine terms in the expansion. As in the case of even and odd functions, integration need only be performed over half a cycle and the result is doubled.

One point of confusion regarding terms should be considered. The words *even* and *odd* were used in a different sense entirely for the previous two symmetry conditions as compared with the present condition and the next one. In the former case, even and odd referred to definitions regarding the image of a function projected around the vertical axis. In this case and in the next one, even and odd refer to the numbers of the harmonics. The two meanings are entirely different. For example, we can have an *even* function that has only *odd*-numbered harmonics.

As in the case of an odd function, the presence of a dc component can disguise half-wave symmetry. In this case the dc component is an even-numbered harmonic ($n = 0$), and its presence obscures the symmetry. A function satisfying this property is shown in Fig. 2-18. The procedure for handling this case is the same as for the previous condition. The dc component is subtracted, and the symmetry condition is applied to the new function.

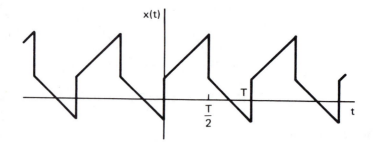

Figure 2-18. Function of Figure 2-17 with dc component added.

Full-Wave Symmetry

A function is said to possess full-wave symmetry if

$$x\left(t + \frac{T}{2}\right) = x(t) \qquad (2\text{-}61)$$

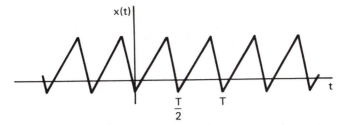

Figure 2-19. Example of function with full-wave symmetry.

A typical function having full-wave symmetry is shown in Fig. 2-19. It can be shown that a function satisfying this condition will have only *even-numbered harmonics* (i.e., $n = 0, 2, 4, 6, \ldots$).

Actually, the reader may see a flaw in the preceding discussion. From (2-61) and Fig. 2-19, the question arises as to whether the period is really T as assumed, or whether the period is, in fact, $T/2$. To get to the point, the period is really $T/2$, and we could avoid discussing the concept of this symmetry condition altogether by redefining the period as $T/2$. However, this situation frequently arises in conjunction with nonlinear operations on signals where the period is effectively halved. In such cases, it is often desirable to maintain the original base period as T. If we redefined the period as $T/2$, the fundamental frequency would be $2/T$, which is the second harmonic of the original reference fundamental frequency. However, with respect to the original fundamental, there are only even-numbered harmonics.

A given waveform may possess no more than two of the preceding symmetry conditions. The function may be either even or odd (but not both), and the function may possess either half-wave or full-wave symmetry (but not both).

Other properties pertaining to Fourier series will be discussed after the Fourier transform is introduced. The reason we postpone such considerations now is that most of them apply equally well to the Fourier transform.

Example 2-6

Consider again the square wave that was analyzed in Exs. 2-3 and 2-5. Analyze this waveform to determine any symmetry conditions present, and use such symmetry conditions to simplify the computation of the spectrum.

Solution

The waveform is repeated in Fig. 2-20a for convenience. As it appears, the reader should verify that neither of the four symmetry conditions is satisfied. However, suppose the dc value $A/2$ is subtracted, and a new signal $x_1(t)$ is

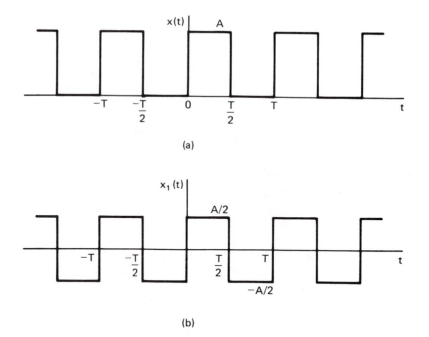

Figure 2-20. Removal of dc component from square wave to allow use of symmetry condition.

formed as shown in Fig. 2-20b; that is,

$$x_1(t) = x(t) - \frac{A}{2} \qquad (2\text{-}62)$$

In performing this shift, we understand, of course, that the new function $x_1(t)$ will not have a dc component, so the dc component of $x(t)$ has been recognized already and should be tabulated. All other terms in $x_1(t)$ should be the same as those in $x(t)$ since only the dc level has been affected.

Let us now check the symmetry conditions. We quickly establish that $x_1(t)$ is an odd function since it satisfies (2-58). This indicates that only sine terms will appear in the spectrum for $x_1(t)$, so $A_n = 0$.

It is also noted that the half-wave symmetry condition of (2-60) is satisfied. This means that only odd-numbered harmonics will appear in the spectrum for $x_1(t)$.

Relating the observation for $x_1(t)$ back to $x(t)$, it can be concluded that the Fourier series for $x(t)$ consists of a dc component and sine components at odd-numbered harmonic frequencies (including the fundamental, i.e., $n = 1$). This conclusion is in obvious agreement with the result of Ex. 2-3, as the reader may quickly verify by referring back to that example. However, this inspection process provides valuable information in advance of a detailed calculation and can save some of the steps involved.

As a final step in this example, let us verify that application of the symmetry condition in the computation of the B_n coefficients for the modified function $x_1(t)$ produces the same results as were obtained in Ex. 2-3 for $x(t)$. According to Table 2-1, we should integrate only over half a cycle and double the result. However, the amplitude of $x_1(t)$ is half that of $x(t)$, so the integral is

$$B_n = \frac{4}{T} \int_0^{T/2} \frac{A}{2} \sin n\omega_1 t \, dt \qquad (2\text{-}63a)$$

$$= \frac{2A}{T} \int_0^{T/2} \sin n\omega_1 t \, dt \qquad (2\text{-}63b)$$

Comparison of (2-63b) with the first expression of (2-44) reveals that the two expressions are exactly the same, so we need not proceed further. The "doubling" in front of the integral cancelled the "halving" of the function amplitude after shifting, so that the B_n coefficients are the same as before.

2-5   FOURIER TRANSFORM

The emphasis on spectral analysis thus far has been directed toward periodic power signals whose spectra consist of discrete spectral components at integer multiples of the fundamental repetition frequency. We will now turn to the case of nonperiodic signals. In particular, we will assume at this time that the nonperiodic signals of interest are energy signals. Thus, our focus will be shifted shortly from periodic power signals, as considered in the past several sections with Fourier series analysis, to nonperiodic energy signals as appropriate for Fourier transform analysis.

To illustrate qualitatively the development of the Fourier transform, the reader is asked to look ahead momentarily to Fig. 2-33 in Sec. 2-8. A periodic pulse signal $x(t)$ and its amplitude spectrum X_n are shown in Fig. 2-33a. (These functions will be developed in Sec. 2-8). In successive parts of the figure, the period is allowed to increase, but the pulse shape is preserved. In each case, $\overline{X}_n$ is determined by integrating $x(t) \cos 2\pi n f_1 t$ over one cycle. (The time function is even.) Since $x(t)$ remains the same within a cycle, the only quantity within the integrand that changes is $f_1 = 1/T$, which becomes increasingly smaller as the period is allowed to increase.

The effect of this trend is that the relative shape of the envelope of the $\overline{X}_n$ coefficients remains the same, but the number of components in a given frequency interval increases as the period increases. (The level of the envelope changes but the *relative* shape remains constant.) Simultaneously, the increment between successive frequency components f_1 decreases. In the limit as $T \to \infty$, as shown in Fig. 2-33d, the frequency difference approaches zero.

In this limiting form, the spectral lines all merge together, so it no longer is desirable to display the spectrum as a group of lines. Instead, the points representing the amplitudes of the lines effectively all merge together and form a continuous curve. Thus, in the limit as the period approaches infinity, the spectrum becomes a continuous spectrum. Thus, while the spectrum of a *periodic* signal is *discrete,* and components appear only at integer multiples of the repetition rate, the spectrum of a *nonperiodic* signal is *continuous* and could, in theory, appear at any frequency. Of course, a band-limited signal would necessarily have a spectrum only over a finite frequency range, but the point is that there are no restrictions of the type that exist for the periodic signal. Note also that as the period approaches infinity the signal changes from a power signal (infinite energy but finite power) to an energy signal, since, in the limit, only one pulse remains, and the resulting energy would be finite.

The *Fourier transform* is the commonly used name for the mathematical function that provides the frequency spectrum of a nonperiodic signal. Assume that a nonperiodic time signal $x(t)$ is given. The Fourier transform is designated as $\overline{X}(f)$, and f is the cyclic frequency in hertz. [The overbar on $\overline{X}(f)$ emphasizes that it is a complex quantity.] Thus, $\overline{X}(f)$ is the mathematical function expressed as a function of frequency that indicates the relative spectrum, which could presumably exist at any arbitrary frequency f. $\overline{X}(f)$ for the nonperiodic signal corresponds to the two-sided spectrum $\overline{X}_n$ for the periodic case. While it would be possible to obtain Fourier transform forms that would relate more directly to the one-sided discrete spectral forms, almost all the available results are more directly related to the two-sided exponential form, and that approach will be followed here.

The process of Fourier transformation of a time function is designated symbolically as

$$\overline{X}(f) = \mathcal{F}[x(t)] \qquad (2\text{-}64)$$

The inverse operation is designated symbolically as

$$x(t) = \mathcal{F}^{-1}[\overline{X}(f)] \qquad (2\text{-}65)$$

The actual mathematical processes involved in these operations are as follows:

$$\overline{X}(f) = \int_{-\infty}^{\infty} x(t)\epsilon^{-j\omega t}\, dt \qquad (2\text{-}66)$$

$$x(t) = \int_{-\infty}^{\infty} \overline{X}(f)\epsilon^{j\omega t}\, df \qquad (2\text{-}67)$$

Note that the argument of $\overline{X}(f)$ and the differential of (2-67) are both expressed in terms of the cyclic frequency f (in hertz), but the arguments of the exponentials in both (2-66) and (2-67) are expressed in terms of the radian frequency ω, where $\omega = 2\pi f$. These are the most convenient forms for manipulating and expressing the given functions. Most spectral displays are made in terms of f, while the analytical expressions are often easier to deal with in terms of ω. This should present no serious problem as long as the 2π scale factor is understood: $\omega = 2\pi f$.

The Fourier transform $\overline{X}(f)$ is, in general, a complex function and has both a magnitude and an angle. Thus, $\overline{X}(f)$ can be expressed as

$$\overline{X}(f) = X(f)\epsilon^{j\phi(f)} = X(f)\underline{/\phi(f)} \qquad (2\text{-}68)$$

where $X(f)$ represents the *amplitude spectrum* and $\phi(f)$ is the *phase spectrum*. A typical amplitude spectrum is shown in Fig. 2-21.

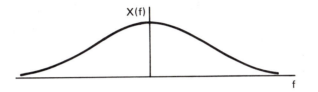

Figure 2-21. Typical amplitude spectrum of nonperiodic signal.

Again, a point that should be stressed for the *nonperiodic* signal is that its spectrum is *continuous,* and, in general, it consists of components at *all* frequencies in the range over which the spectrum is present.

As in the case of periodic signals, certain symmetry conditions can be applied to aid in the computation of the Fourier transform. However, not all the periodic signal properties are applicable to a nonperiodic signal. Furthermore, because of established convention, the results will be interpreted in slightly different forms. For example, the coefficients A_n and B_n are widely used in dealing with Fourier series, but the corresponding forms in the Fourier transform (which would be proportional to the real and imaginary parts of the transform) are not used nearly as often in signal analysis and will not be discussed here.

The symmetry conditions to be considered are summarized in Table 2-2. These results indicate that for either an even or an odd function, one need integrate only over half the total interval and double the result. Furthermore, the form of the integrand is different in each case. Finally, if $x(t)$ is even, $\overline{X}(f)$ is real and an even function of f. Notice the similarity between the forms of the Fourier transform and Fourier series integrals when the function is either even or odd.

The Fourier transforms of several common waveforms have been tabulated in Appendix B. Some of these pairs will be developed later in this chapter, and the general results will appear throughout the text.

Table 2-2. Fourier Transform Symmetry Conditions

Condition	$\overline{X}(f)$	Comments
General	$\int_{-\infty}^{\infty} x(t)\,\epsilon^{-j\omega t}\,dt$	
Even function $x(-t) = x(t)$	$2\int_{0}^{\infty} x(t)\cos\omega t\,dt$	$X(f)$ is an even real function of f.
Odd function $x(-t) = -x(t)$	$-2j\int_{0}^{\infty} x(t)\sin\omega t\,dt$	$X(f)$ is an odd imaginary function of f.

Example 2-7

Derive the Fourier transform of the exponential function given by

$$x(t) = A\epsilon^{-\alpha t} \qquad \text{for } t > 0$$
$$= 0, \qquad \text{for } t < 0 \tag{2-69}$$

where $\alpha > 0$.

Solution

The function is shown in Fig. 2-22a. Application of the definition of the Fourier transform as given by (2-66) yields

$$\overline{X}(f) = \int_{0}^{\infty} A\epsilon^{-\alpha t}\epsilon^{-j\omega t}\,dt$$
$$= \frac{A\epsilon^{-(\alpha+j\omega)t}}{-(\alpha + j\omega)}\Bigg]_{0}^{\infty} = 0 + \frac{A}{\alpha + j\omega} \tag{2-70}$$

The result is a complex function as expected. The amplitude and phase functions are found by determining the magnitude and angle associated with the final result of (2-70). These functions are

$$X(f) = \frac{A}{\sqrt{\alpha^2 + \omega^2}} = \frac{A}{\sqrt{\alpha^2 + (2\pi f)^2}} \tag{2-71}$$

$$\phi(f) = -\tan^{-1}\frac{\omega}{\alpha} = -\tan^{-1}\frac{2\pi f}{\alpha} \tag{2-72}$$

These functions are shown in Fig. 2-22b and c for some arbitrary value of α. As in the case of discrete spectra, the phase spectrum is not as useful as the amplitude spectrum and will usually not be shown in most developments. However, it was included here to make the development complete.

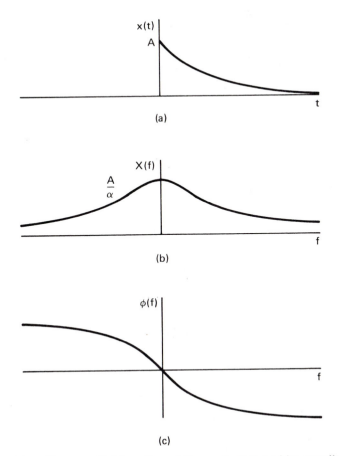

$$x(t)$$

(a)

$$X(f)$$

(b)

$$\phi(f)$$

(c)

Figure 2-22.   Exponential function of Example 2-7 and its amplitude and phase spectra.

Example 2-8

The impulse function $\delta(t)$, occurring at $t = 0$, is defined as follows:

$$\delta(t) = 0, \qquad \text{for } t \neq 0 \qquad (2\text{-}73a)$$

$$\int_{-\infty}^{\infty} g(t)\, \delta(t)\, dt = g(0) \qquad (2\text{-}73b)$$

where $g(t)$ is any continuous function. Derive the Fourier transform of the impulse function. This hypothetical function is used to model spikelike phenomena, such as a sudden noise burst.

Solution

An impulse function is illustrated by a line at the point where the impulse occurs, as shown in Fig. 2-23a. The number beside the line indicates the weight of the impulse, which is unity in this case.

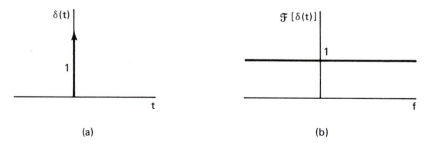

Figure 2-23. Impulse function and its spectrum as developed in Example 2-8.

Application of the definition of the Fourier transform yields

$$\mathcal{F}[\delta(t)] = \int_{-\infty}^{\infty} \delta(t)\,\epsilon^{-j\omega t}\,dt \qquad (2\text{-}74)$$

This integral is readily evaluated by the second part of the definition of the impulse function as given by (2-73b).

$$\mathcal{F}[\delta(t)] = 1 \qquad (2\text{-}75)$$

This result is shown in Fig. 2-23b.

This result is rather interesting and is one that should catch the reader's attention. Although the impulse function is more of an ideal mathematical model than a real physical waveform, it is often used to represent noise phenomena (e.g., a function that appears as a sharp pulse with near-zero duration). The spectrum of such a signal is extremely wide and, in the theoretical limit, would contain components at all frequencies over an infinite frequency interval.

To illustrate one more point regarding this type of phenomenon, the reader has probably observed interference produced on a radio receiver due to "impulse" sources in the immediate area (e.g., electrical appliances, automobile ignition systems). These "impulses" create broad spectra that can be heard on radio receivers over a wide frequency range, as noted from the nature of the Fourier transform of the function.

2-6 FOURIER TRANSFORM OPERATIONS AND SPECTRAL ROLLOFF

Communications signals are altered in form as they pass through various stages of a signal-processing system or a signal-transmission channel. These operations may be used to deliberately change the form of the signal in some cases, or certain of the operations may arise as a result of natural limitations of a system and could distort the signal. In either event, it is worthwhile to study the effects on the resulting signal from a spectral point of view.

The primary Fourier transform operation pairs of interest for our purposes are summarized in Table 2-3. Certain of these pairs will be derived in example problems at the end of this section; others will be left as exercises for analytically inclined readers.

The practical significance of all these operation pairs will now be discussed. The following notational form will be used here and in certain subsequent sections:

$$x(t) \Longleftrightarrow \overline{X}(f) \qquad (2\text{-}76)$$

This notation indicates that $x(t)$ and $\overline{X}(f)$ are a corresponding transform pair; that is, $\overline{X}(f) = \mathcal{F}[x(t)]$. The equation numbers of the operation pairs will correspond to those of Table 2-3.

Superposition Principle (0-1). The first transform pair in the table is

$$ax_1(t) + bx_2(t) \Longleftrightarrow a\overline{X}_1(f) + b\overline{X}_2(f) \qquad (0\text{-}1)$$

This result specifies the basic property that the Fourier transform integral is a linear operation and thus obeys the principle of superposition as far as the level of a signal and the combination of several signals are concerned.

Table 2-3. Fourier Transform Operation Pairs

$x(t)$	$\overline{X}(f) = \mathcal{F}[x(t)]$	
$ax_1(t) + bx_2(t)$	$aX_1(f) + bX_2(f)$	(0-1)
$\dfrac{dx(t)}{dt}$	$j2\pi f \overline{X}(f)$	(0-2)
$\displaystyle\int_{-\infty}^{t} x(t)\,dt$	$\dfrac{\overline{X}(f)}{j2\pi f}$	(0-3)
$x(t - \tau)$	$\epsilon^{-j2\pi f\tau}\,\overline{X}(f)$	(0-4)
$\epsilon^{j2\pi f_0 t}\,x(t)$	$\overline{X}(f - f_0)$	(0-5)
$x(at)$	$\dfrac{1}{a}\overline{X}\left(\dfrac{f}{a}\right)$	(0-6)

Differentiation (0-2). The differentiation Fourier transform pair is

$$\frac{dx(t)}{dt} \longleftrightarrow j2\pi f \overline{X}(f) \qquad (0\text{-}2)$$

This theorem, which will be derived in Ex. 2-9, indicates that each time a signal is differentiated the spectrum is multiplied by $j2\pi f$. Multiplication by $j2\pi f$ has the effect of decreasing the relative level of the spectrum at low frequencies and increasing the relative level at higher frequencies. Note that a pure dc component is eliminated. A sketch illustrating the general effect on the amplitude spectrum resulting from differentiating a time signal is shown in Fig. 2-24.

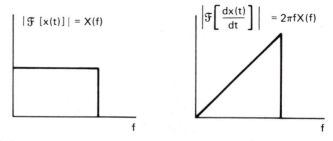

Figure 2-24. Effect on the spectrum of differentiating a time signal.

Integration (0-3). The integration Fourier transform pair is

$$\int_{-\infty}^{t} x(t)\, dt \longleftrightarrow \frac{\overline{X}(f)}{j2\pi f} \qquad (0\text{-}3)$$

This theorem, which is the reverse of (0-2), indicates that when a signal is integrated the amplitude spectrum is divided by $2\pi f$. Division by $2\pi f$ has the effect of increasing the relative level of the spectrum at low frequencies and decreasing the relative level at higher frequencies. A sketch illustrating the general effect on the amplitude spectrum resulting from integrating a time signal is shown in Fig. 2-25. In a sense, integration is a form of low-pass filtering in that high-frequency components of the spectrum are attenuated.

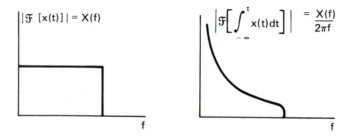

Figure 2-25. Effect on the spectrum of integrating a time signal.

However, a pure integrator is not often used exactly as a low-pass filter due to the pronounced accentuation effect at very low frequencies. Nevertheless, there are situations in which the integrator is considered as a form of a low-pass filter.

Time Delay (0-4). The time delay transform operation, which is derived in Ex. 2-10, is

$$x(t - \tau) \Longleftrightarrow \epsilon^{-j2\pi f \tau} \overline{X}(f) \qquad (0\text{-}4)$$

The function $x(t - \tau)$ represents the delayed version of a signal $x(t)$ as illustrated in Fig. 2-26. This operation could occur as a result of passing a signal through an ideal delay line with delay τ, for example. It can be readily shown that the amplitude spectrum is not changed by the shifting operation, but the phase spectrum is shifted by $-2\pi f \tau$ radians. Certainly, one would expect the amplitude spectrum of a given fixed signal to be independent of the time at which the signal occurs, but the phase shifts of all the components are increased to reflect the result of the time delay.

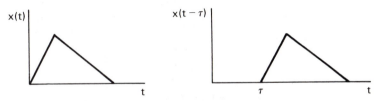

Figure 2-26. A time signal and its delayed form.

Modulation (0-5). This operation bears the same relationship to the frequency domain as the time-delay theorem does to the time domain. This theorem, whose derivation will be left as an exercise (Prob. 2-36), is

$$\epsilon^{j2\pi f_0 t} x(t) \Longleftrightarrow \overline{X}(f - f_0) \qquad (0\text{-}5)$$

Actually, this result has profound implications in the study of amplitude modulation, so it will appear throughout the text in a variety of forms. However, for the moment we will concentrate on the basic form as given. If a time signal is multiplied by a complex exponential, the spectrum is translated to the right by the frequency of the exponential, as shown in Fig. 2-27. In practical cases, complex exponentials occur in pairs with a term of the form of (0-5) along with its conjugate, as will be seen later.

Time Scaling (0-6). The time scaling operation transfer pair is

$$x(at) \Longleftrightarrow \frac{1}{a} \overline{X}\left(\frac{f}{a}\right) \qquad (0\text{-}6)$$

The derivation of this transform pair will be left as an exercise (Prob. 2-37). If $a > 1$, $x(at)$ represents a "faster" version of the original signal, while if $a < 1$,

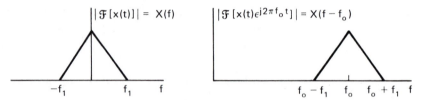

Figure 2-27.   Effect on the spectrum of the modulation operation.

$x(at)$ represents a "slower" version. In the former case, the spectrum is broadened, while in the latter case, it is narrowed. These concepts are illustrated in Fig. 2-28.

We now turn to the problem of determining the *spectral rolloff rate* or, in more formal mathematical terms, the *convergence* of the Fourier spectrum. We have seen that the Fourier transforms and series for many common waveforms appear to be infinitely wide. Obviously, in the "real world" this result cannot be true. In reality, spectral components above a certain frequency range are simply ignored since their contributions are negligible, and the practical bandwidth is established at the minimum level for reasonable reproduction of the signal.

An important factor that can be used qualitatively in estimating the relative bandwidths of different signals is the spectral rolloff rate. The rolloff rate is an upper-bound (worst case) measure of the rate at which the spectral components diminish with increasing frequency. The basic way to specify the rolloff rate is a $1/f^k$ variation for a Fourier transform or a $1/n^k$ variation for a Fourier series, where k is an integer. As k increases, the spectrum diminishes more rapidly. Thus, a signal with a $1/f^3$ rolloff rate would normally have a narrower bandwidth than a signal with a $1/f^2$ rate.

A common practical way to specify the rolloff rate is in decibels/octave in somewhat the same manner as for Bode plots. A rolloff rate of $1/f^k$ can be readily shown to correspond to a slope of $-6k$ dB/octave, where an octave corresponds to a doubling of the frequency. Thus, a $1/f^3$ rolloff rate corresponds to -18 dB/octave, and a $1/f^2$ rate corresponds to -12 dB/octave.

The rolloff rate refers to the worst case or upper bound of the magnitude spectrum of the signal. It is an estimate of the worst-case effect and should not be interpreted as an exact formula for predicting spectral components. For example, the spectrum of a signal may contain components that have a -6 dB/octave rolloff rate, plus components that have a -12 dB/octave rolloff rate. Eventually, the latter components will be so small that they may be ignored, and the -6 dB/octave components will dominate. The tests we will present would predict a -6 dB/octave rolloff rate, and the result has to be interpreted as a worst-case bound.

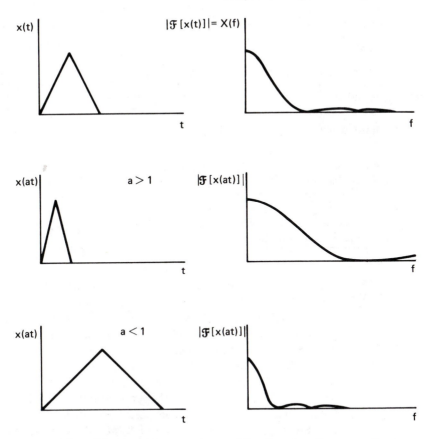

Figure 2-28. Effect on the spectrum of the time scaling operation.

Before discussing the particular tests, the following qualitative points are very worthwhile for dealing with signals in a general sense and should be carefully noted:

1. Time functions that are relatively "smooth" (i.e., no discontinuities or jumps in the signal or its lower-order derivatives) tend to have higher rolloff rates and corresponding narrower bandwidths.
2. Time functions with discontinuities in the signal tend to have lower rolloff rates and corresponding wider bandwidths.

An example of a very smooth signal is the sinusoid whose bandwidth is so narrow that it has only one component. Conversely, a square wave has finite discontinuities in each cycle, and its spectrum is very wide.

The rolloff rates for a wide variety of common signals may be estimated from the information provided in Table 2-4. The left column provides a test to

Table 2-4. Spectral Rolloff Rates of Fourier Transforms and Fourier Series

Smoothness of Function	Rolloff Rate	
	Fourier Transform	Fourier Series
$x(t)$ has impulses	No spectral rolloff	No spectral rolloff
$x(t)$ has finite discontinuties	$\frac{1}{f}$ or -6 dB/octave	$\frac{1}{n}$ or -6 dB/octave
$x(t)$ is continuous $x'(t)$ has finite discontinuities	$\frac{1}{f^2}$ or -12 dB/octave	$\frac{1}{n^2}$ or -12 dB/octave
$x(t)$ is continuous $x'(t)$ is continuous $x''(t)$ has finite discontinuities	$\frac{1}{f^3}$ or -18 dB/octave	$\frac{1}{n^3}$ or -18 dB/octave

apply to the signal, and the right columns provide the appropriate rolloff rates. Observe that one column on the right is applicable to nonperiodic signals analyzed with the Fourier transform, while the other is applicable to periodic signals analyzed with the Fourier series. Some of the information in the table will now be discussed.

The first condition is hypothetical and applies when the signal is assumed to contain one or more ideal impulse functions. This situation could never actually exist in practice, but there are situations in which the assumption of ideal impulse functions is convenient. In such a case, the spectrum will contain a portion having no rolloff at all; that is, the spectrum would theoretically be infinite in bandwidth.

If the signal has finite discontinuities or jumps, the spectrum has a -6 dB/octave rolloff rate. The ideal square wave fits this case, as will be illustrated in Ex. 2-11.

If the signal is continuous (i.e., no jumps), but the first derivative or slope changes abruptly at one or more points, the spectrum has a -12 dB/octave rolloff rate. The triangular wave is a good example of this type. (See Probs. 2-12b and 2-13b.)

If the signal and its first derivative are both continuous, but the second derivative has a finite discontinuity, the spectrum has a -18 dB/octave rolloff rate. It is nearly impossible to detect this condition visually, but a knowledge of the fact may occur as a result of other information. (See Prob. 2-14.)

The table could be continued, but the range shown covers most of the normal requirements for common waveforms. In general, if a function and its first $k-1$ derivatives are continuous, but its kth derivative has a finite discontinuity, the spectral rolloff rate will be $-6(k+1)$ dB/octave.

Example 2-9

Derive the Fourier transform of the first derivative of a time signal (i.e., pair 0-2 of Table 2-3).

Solution

The theorem is best derived by considering the inverse transform of $\overline{X}(f)$.

$$x(t) = \int_{-\infty}^{\infty} \overline{X}(f)\, \epsilon^{j\omega t}\, df \qquad (2\text{-}77)$$

Both sides of this equation are now differentiated with respect to time. This yields

$$\frac{dx(t)}{dt} = \int_{-\infty}^{\infty} j\omega \overline{X}(f)\, \epsilon^{j\omega t} df \qquad (2\text{-}78)$$

By comparison with (2-67), the quantity $j\omega \overline{X}(f)$ is seen to represent the Fourier transform of the derivative on the left. By induction, this result is readily extended to the case of the nth derivative.

Example 2-10

Derive operation pair (0-4) of Table 2-3.

Solution

Application of the definition of the Fourier transform to $x(t - \tau)$ yields

$$\mathscr{F}[x(t - \tau)] = \int_{-\infty}^{\infty} x(t - \tau)\, \epsilon^{-j\omega t}\, dt \qquad (2\text{-}79)$$

A change in variables will now be made. Let $u = t - \tau$, which results in $du = dt$. Substitution of these values yields

$$\mathscr{F}[x(t - \tau)] = \int_{-\infty}^{\infty} x(u)\epsilon^{-j\omega u}\epsilon^{-j\omega \tau}\, du \qquad (2\text{-}80)$$

$$= \epsilon^{-j\omega \tau} \int_{-\infty}^{\infty} x(u)\epsilon^{-j\omega u}\, du = \epsilon^{-j\omega \tau} \overline{X}(f)$$

$$= \epsilon^{-j2\pi f \tau}\, \overline{X}(f)$$

Example 2-11

Using the results of Table 2-4, predict the rolloff rates of the spectra for the the functions of (a) Ex. 2-7, (b) Ex. 2-8, and (c) Ex. 2-3. Verify by comparing with the results of those examples.

Solution

(a) The function of Ex. 2-7 has a finite discontinuity at $t = 0$. Hence, the spectrum should display a rolloff rate of $1/f$ or -6 dB/octave. From Eq. (2-71), the high-frequency asymptotic behavior of $X(f)$ approaches

$$X(f) \approx \frac{A}{2\pi f}, \qquad \text{for } f \gg \frac{\alpha}{2\pi} \tag{2-81}$$

which is a $1/f$ form, as predicted.

(b) The function of Ex. 2-8 is an impulse, which should display no rolloff at all. Indeed, this prediction is readily verified by Eq. (2-75).

(c) The function of Ex. 2-3 has two finite discontinuities in each cycle, a positive jump and a negative jump. The function is periodic, so it should display a rolloff rate of $1/n$. Observe from (2-45) that this property is true, so the spectrum has a -6 dB/octave rolloff. The absence of the even harmonics in the spectrum should not confuse the issue, since the rolloff rate is correct for those components present.

Example 2-12

Derive the Fourier transform of the two-sided exponential function shown in Fig. 2-29a having a damping factor α. Sketch the form of the spectrum.

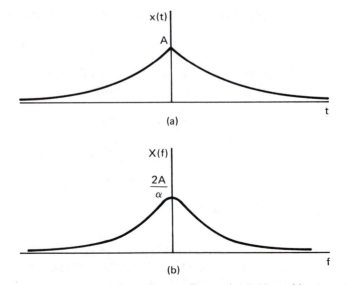

Figure 2-29. Exponential function of Example 2-12 and its transform.

Solution

This function may be expressed as

$$x(t) = A\epsilon^{-\alpha|t|} \tag{2-82}$$

or

$$x(t) = A\epsilon^{\alpha t}, \qquad \text{for } t < 0 \tag{2-83}$$
$$= A\epsilon^{-\alpha t}, \qquad \text{for } t > 0$$

where $\alpha > 0$. Since the function is even, the transform may be evaluated by integrating only over positive time, and this operation is

$$\overline{X}(f) = 2 \int_0^\infty A\epsilon^{-\alpha t} \cos \omega t \, dt \tag{2-84}$$

This integral may be evaluated by parts or the result may be found in a standard integral table. The result is

$$\overline{X}(f) = \frac{2A\epsilon^{-\alpha t}(-\alpha \cos \omega t + \omega \sin \omega t)}{\alpha^2 + \omega^2} \Bigg]_0^\infty \tag{2-85}$$

$$= \frac{2\alpha A}{\alpha^2 + \omega^2} = \frac{2\alpha A}{\alpha^2 + 4\pi^2 f^2}$$

It is interesting to point out that the spectrum is real and is an even function of frequency as expected. Furthermore, the function itself is continuous, but its first derivative has a finite discontinuity at $t = 0$. This property indicates that the spectrum should have a rolloff rate of $1/f^2$, which is readily observed in (2-85). A sketch of the spectrum is shown in Fig. 2-29b. This two-sided exponential function and its spectrum can be compared with the one-sided exponential function and its spectrum as considered in Ex. 2-7. In the present example, the spectrum has a rolloff rate of $1/f^2$ (or -12 dB/octave); in the earlier example, the discontinuity at $t = 0$ resulted in a $1/f$ (or -6 dB/octave) rolloff rate.

2-7 POWER AND ENERGY RELATIONSHIPS

In this section, we will investigate certain important relationships concerning the power and energy contained in signals as a function of their spectral forms. We will see that the power or energy associated with a signal can be determined in either the time domain or the frequency domain. In the frequency-domain form, we will see that the power or energy level at different frequencies is a function of the spectrum.

As explained in Sec. 2-1, a 1-Ω reference will be chosen for most of the power and energy computations. Thus, unless some other specific value of resistance is indicated, the developments given here will be based on *normalized power* and *normalized energy*.

All the spectral forms considered earlier in the chapter will be collectively referred to as *linear* spectral forms. This definition is associated with the fact that such terms as A_n, B_n, C_n, $\overline{X}_n$, and $\overline{X}(f)$ were obtained directly from a time-varying signal $x(t)$, which is usually a voltage or a current wave form (rather than power or energy). In contrast, power and energy are nonlinear quantities, and the process of spectral representation is somewhat different.

As discussed in Sec. 2-1, signals may be classified as either *power signals* or *energy signals* according to whether they exhibit finite nonzero power or finite nonzero energy, respectively. As we will see shortly, it is possible to define either a *power spectrum* or an *energy spectrum*. As one might logically suspect from the terminology, the *power spectrum* applies to the *power signal,* and the *energy spectrum* applies to the *energy signal.* The power or energy spectrum shows a relative distribution of the power or energy as a function of frequency.

We will concentrate first on *power signals,* which were defined in Sec. 2-1 as signals that exhibit a finite nonzero level of power. Within the category of power signals, the initial emphasis will be on periodic signals. Thus, consider a periodic signal $x(t)$. The instantaneous normalized power $p(t)$ associated with $x(t)$ is

$$p(t) = x^2(t) \tag{2-86}$$

The average normalized power P is then obtained by a time average of $p(t)$ over a period T.

$$P = \frac{1}{T} \int_0^T p(t)\, dt = \frac{1}{T} \int_0^T x^2(t)\, dt \tag{2-87}$$

The result of (2-87) can be used to determine the average power directly in the time domain.

The preceding formula also allows us to introduce the standard definition of the *effective* or *root-mean-square (rms)* value X_{rms}. The rms value is defined as a constant value that produces the same average power as the given time-varying signal. Thus, in the 1-Ω reference,

$$P = X_{rms}^2 = \frac{1}{T} \int_0^T x^2(t)\, dt \tag{2-88}$$

This leads to

$$X_{rms} = \sqrt{\frac{1}{T} \int_0^T x^2(t)\, dt} \tag{2-89}$$

which is appropriate for the definition in that the result is the square *root* of the *mean* of the *square* of the signal.

Since we are considering periodic power signals at this time, we will observe the result of (2-87) when a series form is substituted for $p(t)$. While either form could be used, the amplitude-phase form will be used for convenience, which is given as

$$x(t) = C_0 + \sum_1^\infty C_n \cos(n\omega_1 t + \theta_n) \qquad (2\text{-}90)$$

This signal must be squared and substituted in the integral on the right side of (2-87). At first glance, this may seem like a horrendous chore, since it involves squaring an infinitely long Fourier series. Fortunately, the results may be reduced much more easily because all the products of cosine terms at different frequencies integrate to zero over the period of the fundamental. The only terms that contribute to the result are those that correspond to the squares of the respective cosine components. Without putting in the details, the result turns out to be simply

$$P = \frac{1}{T} \int_0^T x^2(t)\, dt = C_0^2 + \frac{1}{2} \sum_{n-1}^\infty C_n^2 \qquad (2\text{-}91)$$

This result is a form of *Parseval's theorem* and should be discussed before proceeding further. This theorem indicates that the average power can be computed in either the time domain or the frequency domain. In the time domain, the average power is obtained by determining the average of the instantaneous power over a cycle. On the other hand, in the frequency domain, the total average power is obtained by summing the power levels associated with each frequency. Thus, C_0^2 is the normalized dc power, $C_1^2/2$ is the power in the fundamental, $C_2^2/2$ is the power in the second harmonic, and so on. The ½ factor associated with each of the sinusoidal components is a result of the fact that, when the power in a sine or cosine function is calculated in terms of the peak value, this factor is required. An alternative approach is to define rms values for the different components as $C_{rms,n} = C_n/\sqrt{2}$ for $n \neq 0$ and $C_{rms,0} = C_0$, in which case (2-91) reduces to

$$P = \frac{1}{T} \int_0^T x^2(t)\, dt = \sum_{n-0}^\infty C_{rms,n}^2 \qquad (2\text{-}92)$$

In simple terms, the squared values of the rms components at different frequencies add together to determine the net normalized power.

Several variations of this form may be developed by use of the sine-cosine form and the complex exponential form. Rather than develop them all individually, the results will be summarized here for reference. Thus, the following equations all represent different forms of Parseval's theorem.

$$P = \frac{1}{T} \int_0^T x^2(t) \, dt = A_0^2 + \frac{1}{2} \sum_{n-1}^{\infty} (A_n^2 + B_n^2) \qquad \text{(2-93a)}$$

$$= C_0^2 + \frac{1}{2} \sum_{n-1}^{\infty} C_n^2 \qquad \text{(2-93b)}$$

$$= A_0^2 + \sum_{n-1}^{\infty} (A_{rms,n}^2 + B_{rms,n}^2) \qquad \text{(2-93c)}$$

$$= C_0^2 + \sum_{n-1}^{\infty} C_{rms,n}^2 \qquad \text{(2-93d)}$$

$$= \sum_{n=-\infty}^{\infty} \overline{X}_n \overline{X}_n^* = \sum_{n=-\infty}^{\infty} X_n^2 \qquad \text{(2-93e)}$$

The expression of (2-93a) permits the power to be expressed in terms of the peak values of the separate sine and cosine terms, while in (2-93b) the peak values are those obtained after combining separate sine and cosine components. The expressions of (2-93c) and (2-93d) involve the rms or effective values of the various components. Note that $C_{rms,n}^2 = A_{rms,n}^2 + B_{rms,n}^2$, which is a well-known result for sinusoidal components of the same frequency but with a phase difference of 90°.

All the preceding forms are *one-sided* in that the various series are summed from zero to infinity. The expression of (2-93e) is a *two-sided* form in that the power is assumed to be associated with both positive and negative spectral components, with one-half assigned to each. The various one-sided forms are more often used in dealing with practical measurements, since most instruments usually measure such properties as the rms or peak values. On the other hand, the two-sided exponential form is very useful in analytical developments, so its importance should be emphasized.

Consider now the case of an *energy signal*. The normalized energy was defined in (2-3). The energy in the signal can be related to the spectrum by the relationship

$$W = \int_{-\infty}^{\infty} x^2(t) \, dt = \int_{-\infty}^{\infty} \overline{X}(f) \overline{X}^*(f) \, df = \int_{-\infty}^{\infty} X^2(f) \, df \quad \text{(2-94)}$$

This result is called *Rayleigh's energy theorem* and is analogous to Parseval's theorem for periodic signals.

In developing the concept of a power or energy spectrum, it is convenient to have some special terminology that will apply with slight modification to all cases. This is desirable since there are certain results that apply to all cases, irrespective of whether the signals are periodic or nonperiodic or whether they are power or energy signals. We will define the symbol $S(f)$ as a general designation for either power or energy spectra. It will be understood that $S(f)$ represents a power spectrum when we are dealing with a power signal, and it represents an energy spectrum when we are dealing with an energy signal.

If two or more signals such as $x(t)$ and $y(t)$ appear in the analysis, subscripts will be used to indicate which power or energy function is being represented. Thus, $S_x(f)$ and $S_y(f)$ represent the power or energy spectra of $x(t)$ and $y(t)$, respectively. Both one- and two-sided spectra will be included in the notation. When it is clear which is being considered, no special emphasis will be made. However, the adjectives "one-sided" and "two-sided" will appear as modifiers to the terms "power spectrum" and "energy spectrum" when it is necessary to emphasize the particular form.

A special comment about the power spectrum of a periodic power spectrum should be made. Since the linear spectrum of a periodic power spectrum is a series of lines at discrete frequencies, the power spectrum is likewise concentrated at the same frequencies. The most elaborate way to handle this mathematically is to define a set of impulse functions in the frequency domain occurring at the discrete frequencies, with the weight of a given impulse representing the power associated with that particular frequency. Formal integration rules utilizing the impulses can also be formulated. We will, however, sidestep this issue by not specifically expressing $S(f)$ as an equation when the signal has discrete frequency terms. However, the power spectrum can be readily shown visually when the signal has discrete components, and this will be illustrated in Ex. 2-14.

A summary of the different forms of the power and energy spectra will now be made.

Periodic Power Signal

One-sided form. The one-sided power spectrum $S(f)$ is a series of lines at dc and positive integer multiples of f_1. The weight of each power component is $C_{rms,n}^2 = C_n^2/2$.

Two-sided form. The two-sided power spectrum $S(f)$ is a series of lines at dc, positive, and negative integer multiples of f_1. The weight of each power component is X_n^2.

For periodic power signals, the "integral" used to obtain the total power is the summation as established by Parseval's theorem.

Nonperiodic Power Signal

Rigorously, $S(f)$ is obtained as a limiting form by defining a "period" T_0 and allowing T_0 to increase without limit. For our purposes, we need not be concerned about actually carrying out this operation. In all places where it occurs, the resulting one- or two-sided power spectrum $S(f)$ will be given. An example of such a signal is random thermal noise, which will be first encountered in Chapter 8 and used in the remainder of the book.

Energy Signal

One-sided form. $S(f) = 2X^2(f), 0 < f < \infty$.

Two-sided form. $S(f) = X^2(f), -\infty < f < \infty$.

Recall that $X(f)$ is the magnitude of the complex Fourier transform $\overline{X}(f)$.

Example 2-13

For the square wave considered in Exs. 2-3, 2-5, 2-6, and 2-11c, determine the fraction of the total power that is contained in the combination of the dc component, the fundamental, and the third harmonic.

Solution

Let P_k represent the normalized power (1-Ω reference) associated with the kth harmonic. The amplitudes of the harmonics can be observed directly from the Fourier series of (2-46a). The dc power P_0 is

$$P_0 = A_0^2 = \left(\frac{A}{2}\right)^2 = \frac{A^2}{4} \tag{2-95}$$

The power P_1 in the fundamental component is

$$P_1 = \frac{1}{2} B_1^2 = \frac{1}{2}\left(\frac{2A}{\pi}\right)^2 = \frac{2A^2}{\pi^2} \tag{2-96}$$

There is, of course, no second harmonic. The power P_3 in the third harmonic is

$$P_3 = \frac{1}{2} B_3^2 = \frac{1}{2}\left(\frac{2A}{3\pi}\right)^2 = \frac{2A^2}{9\pi^2} \tag{2-97}$$

Observe that the normalized dc power is simply the square of the dc value, but all the sinusoidal component power values require the ½ factor since the peak values of the sinusoids are used. The net power in the three components will be denoted as P_{013} and it is

$$P_{013} = \frac{A^2}{4} + \frac{2A^2}{\pi^2} + \frac{2A^2}{9\pi^2} = 0.47516A^2 \tag{2-98}$$

We now must determine the total power P. One could attempt to sum a large number of power values from the Fourier series, as we have already done for three components, and this would eventually converge to a fixed value by Parseval's theorem. However, the easiest way to obtain the total power is to return to the time domain and evaluate the power by the basic relationship

$$P = \frac{1}{T} \int_0^T x^2(t)dt = \frac{1}{T} \int_0^{T/2} A^2 dt = \frac{A^2}{2} \tag{2-99}$$

The fraction of the total power is then

$$\frac{P_{013}}{P} = \frac{0.47516A^2}{0.5A^2} = 0.9503 \tag{2-100}$$

Thus, about 95% of the total power is contained in the combination of the dc component, the fundamental, and the third harmonic.

Example 2-14

For the square wave considered in Exs. 2-3, 2-5, 2-6, 2-11c, and 2-13, plot (a) the one-sided power spectrum and (b) the two-sided power spectrum.

Solution

(a) For convenience, $S_1(f)$ will be used to denote the one-sided form and $S_2(f)$ will denote the two-sided form. The one-sided form is determined directly from the coefficients appearing in the conventional Fourier series of (2-46a). The dc power is $A_0^2 = (A/2)^2 = A^2/4$, the fundamental power is $A_1^2/2 = 2A^2/\pi^2$, the third harmonic power is $A_3^2/2 = 2A^2/9\pi^2$, and so on. The spectrum $S_1(f)$ is shown in Fig. 2-30a. Observe how rapidly the components approach zero when plotted on a power basis.

(b) The two-sided power spectrum is determined directly from the complex exponential Fourier series of (2-54). Note that in this case no division by 2 is necessary for the component power values. Furthermore, we must associate power both with positive and negative frequencies. Thus, the power at f_1 is $X_1^2 = (A/\pi)^2 = A^2/\pi^2$, and the power at $-f_1$ has the same value. The power spectrum is shown in Fig. 2-30b.

2-8 BASEBAND PULSE FUNCTIONS

A signal that finds extensive application in communications systems is the pulse function. Due to its importance, both this section and the next one are devoted to discussing many of the significant features of various types of pulse waveforms and their spectra. In this section, we will concentrate on baseband pulse functions. Baseband signals are those whose spectra tend to be concentrated at lower frequencies and that have not been shifted in frequency by modulation.

Consider first the single nonperiodic pulse $x(t)$ shown in Fig. 2-31a. The

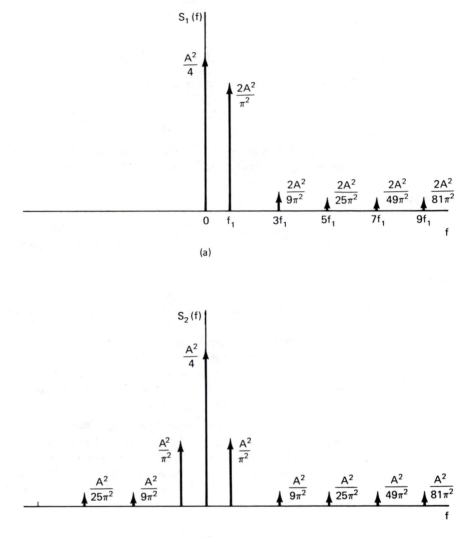

Figure 2-30. One-sided and two-sided spectra of square wave. (See Example 2-14.)

Fourier transform of $x(t)$ is $\overline{X}(f)$, and this function can be readily determined as

$$\overline{X}(f) = \int_{-\tau/2}^{\tau/2} x(t)\epsilon^{-j\omega t}dt = 2 \int_{0}^{\tau/2} A \cos \omega t\, dt \qquad (2\text{-}101)$$

$$= \frac{2A}{\omega} \sin \frac{\omega\tau}{2}$$

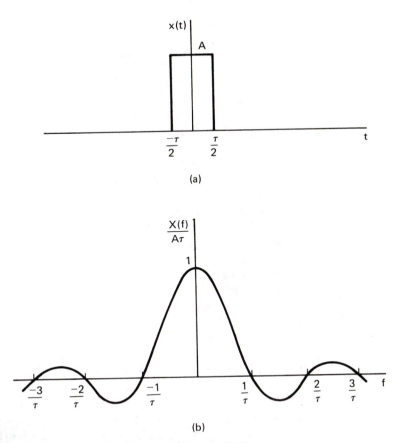

Figure 2-31. Single pulse and its spectrum.

By letting $\omega = 2\pi f$ and making a slight rearrangement of terms, the following form is obtained:

$$\overline{X}(f) = A\tau \frac{\sin \pi f\tau}{\pi f\tau} \qquad (2\text{-}102)$$

This functional form appears frequently in communications theory, and a common reference to it will be denoted as the sin x/x form.

The function $\overline{X}(f)$ is real, but it can be either positive or negative. While the strict definition of magnitude is a positive value, it is convenient to display the form of the result in its natural form, which is a function that can assume both positive and negative values. Thus, we will interpret the result (2-102) as

$$\overline{X}(f) = X(f)\underline{/0^\circ} \qquad (2\text{-}103)$$

where

$$X(f) = A\tau \frac{\sin \pi f\tau}{\pi f\tau} \tag{2-104}$$

is a "magnitude" spectrum that is allowed to assume both positive and negative values.

A sketch of the spectrum (normalized to have a maximum value of unity) is shown in Fig. 2-31b. Several important properties can be deduced:

1. The spectrum is continuous as is to be expected of a nonperiodic signal. The spectral range theoretically extends over an infinite frequency range.
2. The spectrum is bounded by a $1/f$ or a -6 dB/octave rolloff rate. This could have been predicted in advance as a result of the finite discontinuity in the time function.
3. The major lobe of the spectrum is located in the lower frequency range. Since no system has infinite bandwidth, it is necessary in a practical system to band-limit a baseband pulse to some finite lower-frequency range. Appropriate approximations will be developed in Chapter 3.
4. Zero crossings of the spectrum occur at integer multiples of $f_0 = 1/\tau$. This is readily verified by setting $\pi f_0\tau = k\pi$ in (2-104).

The last property will be expanded further as the reader must fully comprehend intuitively the concept involved. A wide pulse (i.e., large τ) is characterized by a lower value of the first zero-crossing frequency, and the major part of the spectrum will be in a lower frequency range. Conversely, a very narrow pulse is characterized by a higher value for the first zero-crossing frequency, and the spectrum will be broader in frequency. Thus, a shorter pulse will require a relatively larger bandwidth.

Having discussed the nonperiodic single baseband pulse, we will now consider the periodic baseband pulse train. Consider the pulse train $x(t)$ shown in Fig. 2-32a. Since the function is periodic, it can be described by a Fourier series. All the properties in which we are interested can be deduced from the exponential form of the series, so we will consider only that form. The coefficients $\overline{X}_n$ can be obtained as follows:

$$\overline{X}_n = \frac{1}{T} \int_{-\tau/2}^{\tau/2} x(t) \epsilon^{-jn\omega_1 t}\, dt = \frac{2}{T} \int_0^{\tau/2} A \cos \frac{2\pi nt}{T}\, dt \tag{2-105}$$

$$= \frac{A}{n\pi} \sin \frac{\pi n\tau}{T}$$

The concept of the duty cycle will be introduced. The duty cycle d is defined as

$$d = \frac{\tau}{T} \tag{2-106}$$

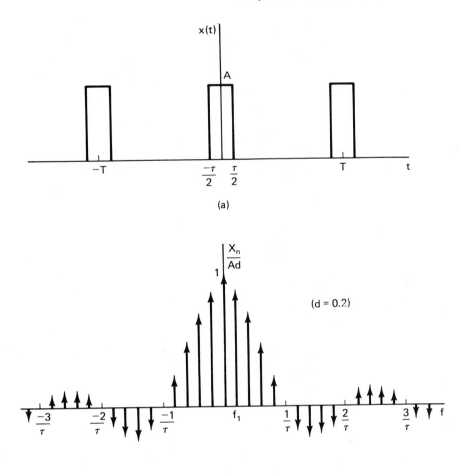

Figure 2-32. Periodic pulse train and its spectrum ($d = 0.2$).

The duty cycle is the ratio of the time interval the pulse is on to the total period. Making use of this definition and rearranging terms in (2-105), we obtain

$$\overline{X}_n = Ad\, \frac{\sin n\pi d}{n\pi d} = X_n\,\underline{/0^\circ} \tag{2-107}$$

where

$$X_n = Ad\, \frac{\sin n\pi d}{n\pi d} \tag{2-108}$$

The function of (2-108) is applicable only for integer values of n, so it is a discrete spectrum, as expected. However, the mathematical form is the same as for the nonperiodic pulse, that is, a sin x/x function. Thus, the envelope of the spectrum follows the properties discussed earlier in relationship to the spectrum of the nonperiodic pulse. The major difference now is that the spectral components will be zero except at dc, f_1, and integer multiples of f_1. The general form of the spectrum is illustrated in Fig. 2-32b. This spectrum corresponds to $d = 0.2$. In general, the number of spectral components from dc to the first zero crossing (including dc) is the smallest integer greater than or equal to $1/d$. In this example, $1/d = 0.2 = 5$. From the figure, it is immediately noted that there are five spectral lines from dc to the first zero crossing (including dc).

Example 2-15

A certain periodic pulse train has a fixed pulse width τ and a pulse amplitude A, but the period is variable. Sketch the spectrum for each of the following values of duty cycle: (a) $d = 0.5$, (b) $d = 0.25$, (c) $d = 0.1$. (d) Sketch the spectrum also for the case of a single nonperiodic pulse.

Solution

Although sketches of the nature indicated can be made by inspection, we will tabulate the mathematical forms of the spectrum for each case in this example. Assuming a pulse centered at the origin, the first three cases utilize (2-107) and (2-108) with appropriate values of d.

(a) $\quad X_n = 0.5A \dfrac{\sin 0.5n\pi}{0.5n\pi}$ \hfill (2-109)

(b) $\quad X_n = 0.25A \dfrac{\sin 0.25n\pi}{0.25n\pi}$ \hfill (2-110)

(c) $\quad X_n = 0.1A \dfrac{\sin 0.1n\pi}{0.1n\pi}$ \hfill (2-111)

(d) The spectrum of the nonperiodic pulse is obtained directly from (2-103) and (2-104). We have

$$X(f) = A\tau \frac{\sin \pi f \tau}{\pi f \tau} \tag{2-112}$$

The four pulse functions and their corresponding spectra are shown in Fig. 2-33. This figure was referred to in Sec. 2-5 during the development of the Fourier transform. It was used to illustrate how the spectrum changes from a discrete form to a continuous form as the period increases without limit.

The actual amplitude levels for each of the four spectra are different. However, the scales for the results were chosen so that the four results could

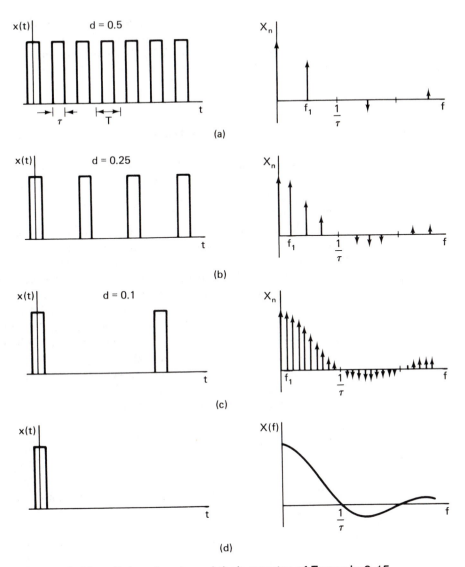

Figure 2-33.   Pulse signals and their spectra of Example 2-15.

be compared more closely. In many spectral analysis problems, it is the *relative* amplitudes of components in a given spectrum that are most significant rather than their exact amplitudes. Indeed, if X_n/Ad or $X(f)/A\tau$ is plotted, the normalized levels are then the same.

Including the dc component, observe that for $d = 0.5$, there are $1/0.5 = 2$ components below the first zero crossing; for $d = 0.25$, there are $1/0.25 = 4$ components below the first zero crossing; and for $d = 0.1$, there are $1/0.1 = 10$ components below the first zero crossing.

It should be stressed again that the general form of the spectra can be determined with only simple arithmetic. The zero-crossing frequency is determined as $1/\tau$. In the case of a periodic function, the spacing between frequencies is $f_1 = 1/T$, and the form of the spectrum can be readily sketched.

2-9 RADIO-FREQUENCY PULSE FUNCTIONS

In this section, the spectra of modulated pulse wave forms will be developed and discussed. Modulated pulse signals are also referred to as RF (radio frequency) pulse functions. Due to popular usage, the latter terminology will be employed in most discussions that follow.

Consider first the single nonperiodic modulated pulse function $x_{rf}(t)$ shown in Fig. 2-34c. This function can be considered as the product of the baseband pulse function $x_{bb}(t)$ shown in Fig. 2-34a and the infinitely long sinusoid shown in Fig. 2-34b. (It does not matter how the wave form was actually generated, but it is *equivalent* to having been generated as the product of the two functions shown, and this interpretation is very useful.)

For the purpose of this analysis, we will assume that the sinusoidal function is of the form $\cos 2\pi f_c t$, where f_c is the frequency of the oscillation. The result can be extended to different phase angles as required. We have also shown an example in the figure in which the pulse width is an integer number of cycles, but this restriction is not necessary.

The RF pulse can then be expressed as

$$x_{rf}(t) = x_{bb}(t) \cos 2\pi f_c t$$

$$= \frac{x_{bb}(t)}{2} \epsilon^{j2\pi f_c t} + \frac{x_{bb}(t)}{2} \epsilon^{-j2\pi f_c t} \qquad (2\text{-}113)$$

The Fourier transform is determined by an application of operation (0–5) of Table 2-3. The result is

$$\overline{X}_{rf}(f) = \frac{1}{2}\overline{X}_{bb}(f - f_c) + \frac{1}{2}\overline{X}_{bb}(f + f_c) \qquad (2\text{-}114)$$

This result is best explained by parts (d), (e), and (f) of Fig. 2-34. The amplitude spectrum of the baseband signal is shown in Fig. 2-34d, and two discrete lines at $\pm f_c$, representing the spectrum of the sinusoid, are shown in Fig. 2-34e. The final RF amptitude spectrum shown in Fig. 2-34f is generated by shifting the baseband spectrum both to the right and to the left by f_c.

For clarity in visualizing this result, a portion of the amplitude spectrum of the RF pulse function is expanded in Fig. 2-35. The portion of the spectrum shown is the center of the amplitude corresponding to the term $\overline{X}_{bb}(f - f_c)/2$ as given in (2-114). The term $\overline{X}_{bb}(f + f_c)/2$ is the symmetrical image shifted

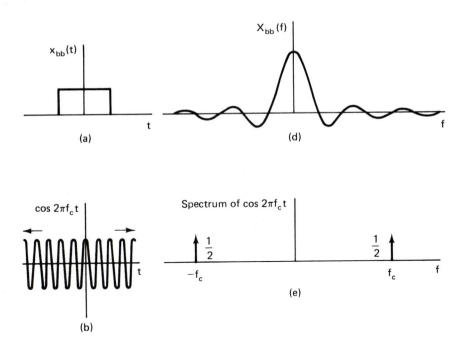

Figure 2-34. Development of the spectrum of an RF pulse.

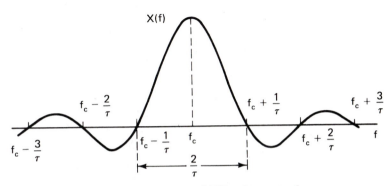

Figure 2-35. Expanded spectrum of RF pulse near f_c.

to the left, and its center would be in the negative frequency region, so that portion is not shown on this figure.

From the function shown in Fig. 2-35, several important properties can be deduced:

1. The spectrum of the RF pulse has the same form for each of its two components as the baseband RF pulse, that is, the sin x/x function. However, while the center of the spectrum for the baseband pulse is dc, the center of the spectrum for the RF pulse is f_c for one component (and $-f_c$ for the other). Thus, the spectrum is centered at the frequency of the sinusoidal oscillation.
2. The zero-crossing frequencies are located at integer multiples of $1/\tau$ on either side of the center frequency f_c.
3. The width (in hertz) of the main lobe of the spectrum as measured between corresponding zero crossings is $2/\tau$.

During the time the pulse is on, the signal is actually a sinusoid. This might lead one naively to expect only a single line component in the spectrum. However, the fallacy in this logic is that a sinusoid must be assumed to exist for *all* time in order to have a single line component. The signal under discussion is gated on for a specific interval of time and then turned off. In this case, the spectrum is broadened, although the center of the spectrum is indeed at the frequency of the sinusoid. As the width of the sinusoidal burst increases, the zero crossings "close in" on the center frequency and approach, in some sense, a single line in the limit.

Consider now the case of a periodic RF pulse function as shown for a typical case in Fig. 2-36a. The signal will be defined over a cycle as

$$x(t) = A \cos 2\pi f_c t, \qquad \text{for } \frac{-\tau}{2} < t < \frac{\tau}{2} \qquad (2\text{-}115)$$

$$= 0, \qquad\qquad \text{for remainder of period}$$

Making use of even function symmetry, the complex coefficients $\overline{X}_n$ can be determined from the integral

$$\overline{X}_n = \frac{2}{T} \int_0^{\tau/2} A \cos 2\pi f_c t \cos 2\pi n f_1 t \, dt \qquad (2\text{-}116)$$

where $f_1 = 1/T$. The details of this development will be left as an exercise (Prob. 2-33). The result can be expressed as

$$\overline{X}_n = X_n \underline{/0^\circ} \qquad (2\text{-}117)$$

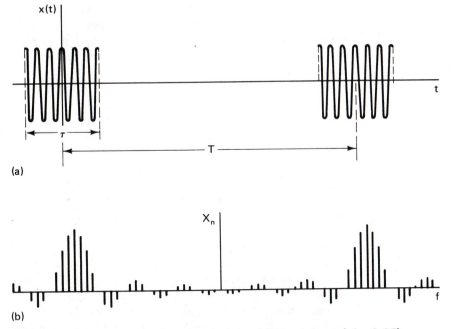

Figure 2-36. Periodic RF pulse train and its spectrum ($d = 0.25$).

where

$$X_n = \frac{Ad}{2}\left[\frac{\sin\,(nd - f_c\tau)\pi}{(nd - f_c\tau)\pi} + \frac{\sin\,(nd + f_c\tau)\pi}{(nd + f_c\tau)\pi}\right] \qquad (2\text{-}118)$$

and d is the duty cycle as defined in (2-106).

The form of a typical spectrum is shown in Fig. 2-36b, which in this case corresponds to the RF pulse in Fig. 2-36a. The duty cycle in this case is $d = 0.25$. The spectrum shown follows a pattern similar to that of the nonperiodic RF pulse in that it is centered at f_c. The envelope exhibits the same form as for the nonperiodic case. However, since the present function is periodic, the spectrum is discrete and appears only at integer multiples of the fundamental repetition frequency $f_1 = 1/T$.

For the sake of clarity, we have assumed that the RF frequency f_c is an integer multiple of the fundamental frequency f_1. Successive components of the spectrum are displaced at integer multiples of f_1 on either side of the center.

For both the nonperiodic and periodic cases, we have assumed a cosine function, and we have assumed that the pulse width represents an integer number of cycles. Furthermore, for the periodic case, we have assumed that the period T is an integer multiple of the pulse width τ. These assumptions represent the "cleanest" situation as far as visualizing the spectrum is concerned, and the results may often be used to closely approximate the spectrum in cases when these assumptions are not quite met. Specifically, however, what are the significant differences in the spectrum when these conditions are not met?

In the case where the sinusoidal function is not a pure cosine, the major difference is that each of the components in (2-114) will contain complex multiplicative factors. It turns out that the magnitudes of each of the two components remain the same as before, but the manner in which they add is now affected by the complex factors.

Actually, as long as $f_c \gg 1/\tau$, the effect produced from adding corresponding spectral terms from the two components is usually insignificant since the main lobes are far apart. Hence, the practical effect on the amplitude spectrum caused by varying the phase of the sinusoidal component is significant only when there are relatively few cycles of the RF sinusoid in the pulse interval.

The assumption of an integer number of cycles in the pulse width τ for the nonperiodic case results in the center frequency f_c being an integer multiple of the zero-crossing frequency $1/\tau$. This means that the tail of the negatively shifted component has its zero crossings exactly at the same points as the positively shifted component, and definite zero crossings are preserved. However, if there is not an integer number of cycles in a pulse interval, the spectrum will no longer have precise zero crossings due to the presence of small contributions from the negatively shifted component. As in the previous case, this effect is quite insignificant when $f_c \gg 1/\tau$, and it does not really change the primary nature of the spectrum.

For the periodic pulse train, an additional slight complication results when the period is not an integer multiple of the pulse width. However, the effects can be predicted by first drawing the spectrum of a nonperiodic pulse having the given pulse width and sinusoidal frequency. Then, starting at dc, components will appear at integer multiples of f_1. If f_c is not an integer multiple of f_1, the largest line component in the spectrum will not be exactly at f_c, but it will be displaced slightly. Various patterns are possible depending on the duty cycle and the number of sinusoidal cycles in a pulse width.

One more important point should be made before leaving this section. All RF pulses that were considered assume instantaneous switching as indicated by the rectangular pulse shown in Fig. 2-34. This finite discontinuity in the time function leads to the -6 db/octave spectral rolloff on either side of the center frequency. The result is a rather slowly converging spectrum that is spread over a wide (theoretically infinite) frequency range. From the standpoint of electromagnetic transmission, this means that the required transmission bandwidth is quite large. Thus, the instantaneously gated RF pulse is not desirable from a practical radio-transmission point of view.

This problem can be alleviated by turning on and off the RF carrier by a different type of pulse waveform than the instantaneous rectangular pulse. A typical realistic pulse waveform would have nonzero rise and fall times and some rounding on the edges. This would ensure that the pulse function and one or more lower-order derivatives would be continuous, which would lead to a narrower spectrum.

Example 2-16

A certain high-frequency sinusoid is periodically gated on for an interval of 0.1 μs and then turned off for 0.9 μs. The carrier frequency is 500 MHz. Sketch the spectrum over the range of three zero crossings on either side of the peak and label significant frequencies. Assume instantaneous rectangular switching.

Solution

The period of the high-frequency sinusoid is $T_c = 1/(500 \times 10^6) = 2$ ns. This means that there are $(0.1 \times 10^{-6})/(2 \times 10^{-9}) = 50$ cycles of the sinusoid during the interval that the sinusoid is gated on. For all practical purposes, then, we can ignore the effect of the tail of the negatively shifted component as far as the main lobe around 500 MHz is concerned.

The function is periodic with overall period $T = 1$ μs, or $f_1 = 1$ MHz. The spectrum is discrete with components spaced apart by 1 MHz. The zero crossings on either side of the main lobe are spaced apart by $1/\tau = 1/(0.1 \times 10^{-6}) = 10$ MHz. The main part of the amplitude spectrum is then sketched using the pertinent parameters calculated and is shown in Fig. 2-37. All the coefficients have been plotted as positive quantities in this case (as they would be viewed on a spectrum analyzer).

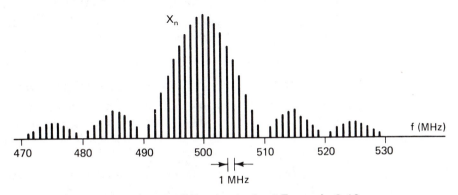

Figure 2-37. Spectrum of RF pulse train of Example 2-16.

A complete mathematical expression for the spectrum has been deliberately avoided here in order to illustrate to the reader how the essential properties of a pulse spectrum can often be determined by relatively simple calculations, provided that the form of the spectrum is understood.

PROBLEMS

2-1 Classify the signals in (a) through (h) in each of the following three ways: (1) periodic or nonperiodic, (2) deterministic or random, and (3) power signal or energy signal: (a) square wave of Fig. 2-9 continuing as shown; (b) the square wave of Fig. 2-9 turned on for the first three cycles and then turned off; (c) the pulse of Fig. 2-4a plus a second pulse of the same form occurring t_1 seconds later; (d) the exponential function shown in Fig. 2-29a; (e) the output of a broadcast studio microphone operated on a continuous basis; (f) the output of a microphone operated for a short period and turned off; (g) continuous thermal noise produced by a natural noise source; (h) the function $x(t) = t\epsilon^{-\alpha t}$ for $0 < t < \infty$.

2-2 A certain periodic band-limited signal has only three components in its Fourier representation: dc, 100 Hz, and 200 Hz. The signal can be expressed in sine-cosine form as

$$x(t) = 12 + 15 \cos 200\pi t + 20 \sin 200\pi t - 5 \cos 400\pi t - 12 \sin 400\pi t$$

Express the signal in (a) amplitude-phase form and (b) complex exponential form.

2-3 A certain periodic band-limited signal has the following Fourier series amplitude-phase representation:

$$x(t) = 10 + 20 \cos (100\pi t + 30°) + 30 \cos (200\pi t - 120°)$$

Express the signal in (a) sine-cosine form and (b) complex exponential form.

2-4 A certain periodic band-limited signal has the following Fourier series complex exponential representation:

$$x(t) = (6 - j8)\epsilon^{-j100\pi t} + 12 + (6 + j8)\epsilon^{j100\pi t}$$

Express the signal in (a) sine-cosine form and (b) amplitude-phase form.

2-5 Derive the sine-cosine form of the Fourier series representation for the periodic square wave of Appendix A using any applicable symmetry conditions. Observe that the amplitude-phase form is virtually the same in this case.

2-6 Determine the complex exponential form of the series of Prob. 2-5.

2-7 Consider the square wave shown in Fig. P2-7, which could be considered as a special case of a pulse train with $d = 0.5$. However, a different

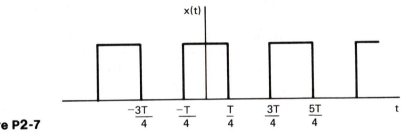

Figure P2-7

approach will be utilized here to emphasize the half-wave symmetry condition.

(a) Making use of the obvious even nature of the function, which allows integration over one half-cycle, determine the form of the Fourier series representation.

(b) Define a new function $x_1(t) = x(t) -$ (dc value). Show that the half-wave symmetry condition applies to $x_1(t)$. Does the implication of this symmetry condition agree with the results of part (a)?

2-8 Derive the sine-cosine form of the Fourier series for the triangular wave of Appendix A. Observe that the amplitude-phase form is the same in this case.

2-9 In deriving the Fourier transform of a single pulse in Eq. (2-101), use of the even-function symmetry condition was employed. Repeat the derivation without employing this symmetry condition.

2-10 Derive the Fourier transform of the triangular pulse of Appendix B.

2-11 Derive the Fourier transform of the pulse of Fig. P2-11 two ways: (a) by direct application of the transform to the function, and (b) by using the result of Eq. (2-102) for a pulse centered at the origin and employing an appropriate transform operation from Table 2-3.

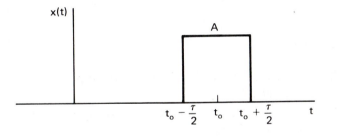

Figure P2-11

2-12 By inspection, determine the spectral rolloff rates for the periodic signals tabulated in Appendix A: (a) square wave, (b) triangular wave, (c) sawtooth wave, (d) half-wave rectified wave, (e) full-wave rectified wave, and (f) pulse train. Express the results both as $1/n^k$ and in decibels per octave. Check conclusions with the tabulated results.

2-13 By inspection, determine the spectral rolloff rates for the nonperiodic signals tabulated in Appendix B: (a) rectangular pulse, (b) triangular pulse, (c) sawtooth pulse, and (d) half-cycle cosine pulse. Express the results both as $1/f^k$ and in decibels per octave. Check your conclusions with the tabulated results.

2-14 Visual determination of a discontinuity of the second derivative of a function is nearly impossible. However, consider the integrator circuit shown in Fig. P2-14 and assume that the input $x(t)$ is the triangular wave given.

    (a)   Starting at $t = 0$ with $y(0) = 0$, sketch the form of the output $y(t)$ paying close attention to the curvature of the signal at various points.

    (b)   Express $x(t)$ in the form of a Fourier series using the results of Appendix A and integrate this result. (Don't worry about the dc term as the wave form has been "fixed" to yield a zero resulting dc value.)

    (c)   From the result of part (b), what is the rolloff rate of the output signal $y(t)$? What does this condition predict as far as the continuity of $y(t)$ and its derivatives is concerned?

    (d)   Comment on the possible use of this circuit to generate an approximate sinusoid from a triangular wave. If the per unit distortion is approximated as the ratio of the third harmonic to the fundamental, determine the approximate harmonic distortion for $y(t)$.

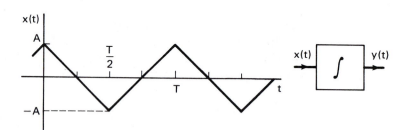

Figure P2-14

2-15 Plot (a) one-sided and (b) two-sided power spectra for the square wave of Appendix A.

2-16 Plot (a) one-sided and (b) two-sided power spectra for the triangular wave of Appendix A.

2-17 Determine expressions for the (a) one-sided and (b) two-sided energy spectra for the baseband pulse waveform of Appendix B.

2-18 Determine expressions for the (a) one-sided and (b) two-sided energy spectra for the triangular pulse waveform of Appendix B.

2-19 Extend the results of Ex. 2-13 to determine the percentage of the total power contained in all components up through the ninth harmonic.

2-20 Determine the average *normalized* power dissipated by the waveform of Prob. 2-2 in a 1-Ω resistor.

2-21 Repeat Prob. 2-20 if $x(t)$ is a voltage waveform and the resistance is 50 Ω.

2-22 Determine the average *normalized* power dissipated by the waveform of Prob. 2-3 in a 1-Ω resistor.

2-23 Determine the average *normalized* power dissipated by the waveform of Prob. 2-4 in a 1-Ω resistor.

2-24 Write an equation for and sketch the amplitude spectrum of a single nonperiodic baseband pulse having a width of 0.4 s and an arbitrary height A. Label significant frequencies.

2-25 Repeat Prob. 2-24 for a pulse having a width of 50 μs.

2-26 Repeat Prob. 2-24 for a pulse having a width of 20 ns.

2-27 A periodic baseband pulse train has a fixed pulse repetition rate of 1 kHz, but the pulse width τ can be varied. Write an equation for and sketch the amplitude spectrum for each of the following values of τ, and label the pertinent frequencies in each case: (a) $\tau = 0.5$ ms, (b) $\tau = 0.2$ ms, (c) $\tau = 0.1$ ms. Assume an arbitrary height A for the pulses.

2-28 A periodic baseband pulse train has a fixed pulse width of $\tau = 1$ ms, but the period T can be varied. Write an equation and sketch the amplitude spectrum for each of the following values of T, and label the pertinent frequencies in each case: (a) $T = 2$ ms, (b) $T = 10$ ms, (c) $T = 20$ ms. Assume an arbitrary height A for the pulses.

2-29 Consider a periodic baseband pulse train with duty cycle d. Let $r = 1/d$. Show that if r is an integer the rth harmonic and all integer multiples of the rth harmonic are zero.

2-30 A certain baseband pulse train with a fixed period of 200 ns is operating in the vicinity of a narrowband receiver tuned to 100 MHz. What is the lowest duty cycle at which the pulse train can be set in order to ensure that no interference will be heard on the receiver?

2-31 An RF oscillator operating at 10 MHz is gated on for an interval of 5 μs by a single baseband pulse. Sketch the form of the amplitude spectrum in the vicinity of the major positive frequency lobe. Label significant frequencies.

2-32 Repeat Prob. 2-31 if the oscillator is gated on for only 1 μs.

2-33 Perform the integration of Eq. (2-116), thus deriving the form of (2-117) and (2-118).

2-34 An RF transmitter operating at 500 MHz is gated on and off by a periodic baseband pulse train having a pulse repetition frequency of 1 MHz. Sketch the form of the amplitude spectrum in the vicinity of the major positive frequency lobe for each of the following values of τ, the duration of the RF pulse: (a) $\tau = 0.5$ μs, (b) $\tau = 0.2$ μs, (c) $\tau = 0.1$ μs. Label significant frequencies.

2-35 An RF transmitter operating at 2 gigahertz (GHz) is gated on by a pulse of width $\tau = 100$ ns. The pulse repetition rate is variable. Sketch the form of the amplitude spectrum for each of the following pulse repetition rates: (a) 5 MHz, (b) 1 MHz, (c) 10 kHz. Label significant frequencies.

2-36 Derive Fourier transform operation pair (0–5) of Table 2-3.

2-37 Derive Fourier transform operation pair (0–6) of Table 2-3.

2-38 Consider a periodic pulse train $p(t)$ of the form considered in Sec. 2-8 with amplitude A and duty cycle d. Compute the percentage of the total power contained in all spectral components up to the first zero crossing for the following values of d: (a) 0.5, (b) 0.25, (c) 0.2, (d) 0.1.

2-39 One of the most common sources of degradation in practical amplifiers is the phenomenon of *harmonic distortion,* which can be explained nicely with Fourier terminology. The level of harmonic distortion can be determined by applying an ideal sinusoid and measuring the levels of Fourier components at frequencies which are integer multiples of the fundamental. If the amplifier were perfectly linear, no harmonics would be created and the amplifier would be free of harmonic distortion. However, non-linearities in the gain characteristic create harmonics of the input frequency. The percent harmonic distortion (P.H.D.) can be defined as

$$\text{P.H.D.} = \frac{\sqrt{\sum_{2}^{\infty} C_{n,rms}^2}}{C_{1,rms}} \times 100\% = \frac{\sqrt{\sum_{2}^{\infty} C_n^2}}{C_1} \times 100\%$$

In actual practice, the terms usually converge rapidly so that the sum may be truncated after a few terms. The values of the various terms may be measured with a frequency selective voltmeter or a spectrum analyzer. For a particular amplifier, assume that with a 1 kHz single tone input, the following data are measured at the output:

Frequency	1 kHz	2 kHz	3 kHz	4 kHz
RMS Value	4 V	120 mV	40 mV	30 mV

All other harmonics are assumed to be negligible. Determine the percent harmonic distortion of the amplifier.

2-40 This problem will illustrate how the phenomenon of harmonic distortion defined in Prob. 2-39 arises from non-linear amplifier characteristics. Assume an amplifier having an input v_1 and an output v_2. If the amplifier were perfectly linear, the output would simply be $v_2 = A v_1$, where A is the gain, and no distortion would occur. Assume, however, that some second-degree non-linearity is present, and that the output v_2 is given by

$$v_2 = A_1 v_1 + A_2 v_1^2$$

For an input $v_1 = B \cos \omega t$, show that the harmonic distortion is all at the second harmonic, and that the percentage harmonic distortion is given by

$$\text{P.H.D.} = \frac{A_2 B}{2 A_1} \times 100\%$$

The dc term in the output may be disregarded for this purpose. (See Prob. 2-39 for definition of P.H.D.) In practical amplifiers, higher order harmonics are generated by higher degree terms in the gain characteristics.

2-41 The circuit shown in (a) of Fig. P2-41 is a simplified circuit diagram of one form of a *push-pull complementary symmetry common collector class B power amplifier*. The two transistors are both biased to near cutoff. (Assume cutoff for this analysis since the small forward biases of the two transistors cancel in the load.) The upper transistor conducts when $v_1 > 0$, while the lower transistor conducts when $v_1 < 0$. For large signal power amplifiers, it is virtually impossible to completely eliminate distortion, but this circuit offers certain advantages in distortion reduction. Assume that the input v_1 is a sinusoidal waveform. For the sake of illustration, the distortion produced in the first half-cycle has been *highly exaggerated* as shown in (b)

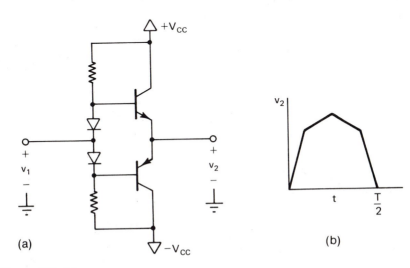

(a)   (b)

Figure P2-41

of Fig. P2-41. (a) Assuming that the two transistors have *exact complementary* characteristics, sketch the form of v_2 for the second half-cycle. (b) Using one of the Fourier symmetry conditions of Table 2-1, what can be concluded about the nature of the resulting harmonic distortion?

2-42 The circuit shown in Fig. P2-42 is a simplified circuit of one form of a *push-push frequency doubler*. The diodes may be assumed to be ideal. (a) For an input sinusoid $v_1 = A \sin \omega t$ having a frequency $f = \omega/2\pi$, sketch the form of the output v_2. (b) Using one of the Fourier symmetry conditions of Table 2-1, what can be concluded about the lowest frequency of the output? Is the name of the circuit appropriate? (In practice, the output signal would be filtered to eliminate all but the desired frequency component.)

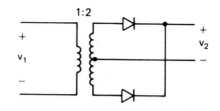

Figure P2-42

3 Signal Transmission

3-0 INTRODUCTION

The emphasis in this chapter will be on the consideration of signal transmission through linear networks. The transfer function concept will be introduced as an analytical means for predicting the output of a network or system from a knowledge of the input and the transmission characteristics of the system. The Fourier or steady-state transfer function will be used, since it can be applied directly to the spectrum of the signal as developed in Chapter 2. Thus, the present chapter complements the work of Chapter 2 in that the concept of the frequency spectrum is extended to include transmission systems and their effects on the spectrum of a signal. This interpretation is most important in predicting bandwidth requirements for signal transmission and in predicting signal distortion.

The concept of the ideal filter will be discussed, and several commonly used actual filter characteristics will be introduced. Simple approximations for predicting approximate bandwidth requirements for pulse transmission will be given. Finally, the effect of a transmission system on the power or energy spectral density will be discussed.

3-1 TRANSFER FUNCTION

The concept of the transfer function arises in many different areas of engineering system analysis and design. This powerful concept permits systems having complex derivative and/or integral relationships to be represented in terms of algebraic forms. The algebraic relationships of the transfer functions arise through the use of mathematical transforms.

Two transforms having particular significance in electrical circuits and systems are the (a) Laplace transform and the (b) Fourier transform. The Laplace transform is most useful when transient phenomena are to be studied. One area in which Laplace analysis is widely used is in feedback control system analysis and design.

While Laplace transforms may be applied in a number of different areas of communications, the Fourier transform is usually more appropriate for problems associated with communications systems. The reason is that the Fourier transform provides a direct link between the overall signal-processing characteristics of a system and the frequency spectrum associated with signals appearing in the system. We have already been exposed to the concept of the frequency spectrum as provided by the Fourier series and transforms in Chapter 2, so it is only necessary to extend the concept to input-output relationships.

We will restrict the consideration at this point to those systems that can be characterized as *continuous-time, linear, lumped,* and *time-invariant.* Formal and rather elegant definitions of these terms are given in various advanced texts on linear system and circuit theory, but simplified definitions are adequate for our purposes.

As a guide to these definitions, consider the block diagram shown in Fig. 3-1. The quantity $x(t)$ represents the *input* or *excitation,* and $y(t)$ represents the corresponding *output* or *response.* The block could represent any applicable signal-processing operation in a communications system. As a single example, the input could represent all the signals present at the antenna connected to a receiver, the block could represent the frequency-selection filter (tuning network) in the receiver, and $y(t)$ could represent the particular signal that the filter selects. Thus, the system block "operates" on an input $x(t)$ to produce an output $y(t)$. In most communications applications, $x(t)$ and $y(t)$ represent voltages or currents, but the general concepts apply to a large number of physical systems.

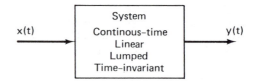

Figure 3-1.   Block diagram of system input-output relationship.

A brief discussion of the terms introduced will now be given. The term *continuous-time* is, for all practical purposes, a more formal description of what is casually called an *analog* system. The term indicates that the response could theoretically exist at any continuous range of time values. This is in contrast to a *discrete-time* system, in which the response can only exist at discrete values of time.

A *linear* continuous-time system is one whose input-output relationships can be described by a linear differential equation. In practical terms, this means that the parameters of the system do not vary with the signal levels. Nonlinear systems would be characterized by parameter values that are dependent on the levels of the signals present (e.g., nonlinear resistors, voltage variable capacitors, saturable core reactors).

A *lumped* system is one composed of a finite number of finite elements, such as resistors, inductors, and capacitors. This is in contrast to a *distributed parameter* system in which distributed effects have to be considered (e.g., a transmission line). The input-output characteristic of a lumped system can be described by an ordinary differential equation. In contrast, a distributed parameter system usually requires a partial differential equation for a full representation.

A *time-invariant* system is one in which the parameters of the system are fixed and do not vary with time.

In subsequent work, the transfer function concepts to be developed in this section will be applied to many different situations. We will not normally review the conditions just discussed due to the time and space involved, but it will be assumed that they apply in all such cases. Many signal processing operations encountered in communications systems do, in fact, meet all these requirements, and there is often a tendency to be casual about applying the concepts without qualification. The reader should remember, however, that the systems for which the transfer function methods apply are restricted, and the conditions stated should be reviewed when necessary.

Consider now a system with a single input $x(t)$ and a single output $y(t)$. For systems meeting all the various conditions just stated, the input-output relationship can always be expressed through a differential equation relationship of the form

$$b_m \frac{d^m y}{dt^m} + b_{m-1} \frac{d^{m-1} y}{dt^{m-1}} + \cdots + b_0 y \qquad (3\text{-}1)$$

$$= a_n \frac{d^n x}{dt^n} + a_{n-1} \frac{d^{n-1} x}{dt^{n-1}} + \cdots + a_0 x$$

In most physical systems, $m \geq n$. In this case, the integer m is said to be the *order* of the system.

Operation pair (0-2) of Table 2-3 provides a relationship for the Fourier transform of the derivative of a time function. By mathematical induction, this theorem can be extended to higher-order derivatives, and the result is

$$\mathscr{F}\left[\frac{d^k x}{dt^k}\right] = (j\omega)^k \overline{X}(f) \qquad (3\text{-}2)$$

This operation will be applied to both sides of (3-1). The function $\overline{Y}(f)$ can then be factored out on the left, and $\overline{X}(f)$ can be factored out on the right. The result is

$$\overline{Y}(f)\left[b_m(j\omega)^m + b_{m-1}(j\omega)^{m-1} + \cdots + b_0\right] \qquad (3\text{-}3)$$

$$= \overline{X}(f)\left[a_n(j\omega)^n + a_{n-1}(j\omega)^{n-1} + \cdots + a_0\right]$$

The differential equation of (3-1) has now been converted into an equation that is algebraic in form and can be manipulated by standard algebraic techniques.

We may now define a quantity called the *Fourier transfer function,* which will be denoted by $H(f)$. By manipulation of (3-3), we have

$$H(f) = \frac{\overline{Y}(f)}{\overline{X}(f)} = \frac{a_n(j\omega)^n + a_{n-1}(j\omega)^{n-1} + \cdots + a_0}{b_m(j\omega)^m + b_{m-1}(j\omega)^{m-1} + \cdots + b_0} \qquad (3\text{-}4)$$

For reasons that will be clearer later, the Fourier transfer function is also called the *steady-state transfer function.*

The Fourier or steady-state transfer function is seen to be a complex function (due to both real and imaginary terms) providing a ratio of the output Fourier transform $\overline{Y}(f)$ to the input Fourier transform $\overline{X}(f)$. Since ω appears as a variable in the transfer function, the value of the transfer function is obviously a function of frequency.

A few comments about notation are in order at this point. First, the arguments of the transfer function and input and output transforms are all expressed as f, while ω appears on the right of (3-4). The f argument is delineated due to the importance of frequency in hertz for communications calculations. However, since $\omega = 2\pi f$, it is more compact in long expressions such as the right side of (3-4) to leave the independent variable simply as ω with the 2π factor understood, and this practice will be used extensively throughout the book.

A second comment concerns the apparent inconsistency in notation in placing a bar above $\overline{X}(f)$ and $\overline{Y}(f)$ to label them as complex but in not doing so for $H(f)$. The reason is that the notation established in Chapter 2 defined $X(f)$ and $Y(f)$ as representing the magnitudes of $\overline{X}(f)$ and $\overline{Y}(f)$, respectively. However, a different form of notation will be established in the next section for the magnitude of the transfer function, so it is not necessary to place the bar above $H(f)$. In short, $H(f)$ will always be considered as complex, so no further labeling is required. These "fine points" of notation may not bother all readers, but there is a meaningful pattern for those who follow the notation closely.

Returning to the primary train of thought, if the transfer function of a system is known, the output transform can be determined by multiplying the input transform by the transfer function; that is,

$$\overline{Y}(f) = H(f)\overline{X}(f) \qquad (3\text{-}5)$$

A block diagram illustrating the input-output transfer function concept is shown in Fig. 3-2. In practical terms, the output frequency spectrum is the input frequency spectrum multiplied by the transfer function. For this reason, the Fourier transform function is sometimes referred to as the *spectral weighting function.* The Fourier transfer function description is a *frequency-domain* representation of the system.

Figure 3-2. Block diagram of transfer function input-output relationship.

The preceding development was aimed at presenting the basic concept through the differential-equation viewpoint, since this provides a broad and fundamental view of the approach. In actual practice, it is seldom necessary to deal directly with the differential-equation approach in applied communications systems problems. A technique for determining the transfer function more easily when the system is an electrical circuit will be given.

As we will see shortly, the Fourier transfer function concept is closely aligned with the familiar steady-state ac phasor approach, and the standard ac circuit phasor form may be used in determining a transfer function. There is one important difference, however. In the context of steady-state ac, the frequency ω is usually fixed. However, since the Fourier spectra of most signals may contain a large number of frequencies, the frequency ω must be considered as a variable and retained in that form until a computation at a specific frequency is required.

From the preceding discussion, the following procedure may be used for determining the Fourier transform $H(f)$ from a circuit diagram:

1. Convert each inductor L to a purely imaginary impedance $j\omega L$.
2. Convert each capacitor C to a purely imaginary impedance $1/j\omega C$ (or $-j/\omega C$).
3. Each resistor R remains as a real value R.
4. The input is represented by an arbitrary Fourier transform source $\overline{X}(f)$.
5. The desired output variable is represented by an arbitrary Fourier transform $\overline{Y}(f)$.
6. By standard circuit analysis and algebraic manipulations, the output function $\overline{Y}(f)$ is determined in terms of the input function $\overline{X}(f)$.
7. The transfer function is then determined as $H(f) = \overline{Y}(f)/\overline{X}(f)$.

The preceding procedure has involved a direct analysis in the frequency domain, which is usually easier to deal with than the derivative and/or integral relationships in the time domain.

In the case where a single frequency phasor analysis is desired, $\overline{X}(f)$ is replaced by a single frequency phasor $\overline{X}$, and $\overline{Y}(f)$ is replaced by a single frequency phasor $\overline{Y}$. This concept will be considered in the next section.

The determination of some representative transfer functions will now be illustrated with examples.

Example 3-1

The circuit shown in Fig. 3-3 is one of the simplest forms of a low-pass filter. The input is $v_1(t)$ and the output is $v_2(t)$. Working directly in the frequency domain, determine the Fourier transfer function $H(f) = \overline{V}_2(f)/\overline{V}_1(f)$.

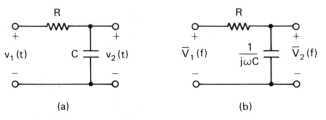

(a)   **(b)**

Figure 3-3.   Circuit of Example 3-1.

Solution

The circuit is converted to the frequency domain by the procedure of the preceding section. This form is shown in Fig. 3-3b. The output $\overline{V}_2(f)$ may be readily determined in terms of the input $\overline{V}_1(f)$ by application of the voltage-divider rule. We have

$$\overline{V}_2(f) = \frac{1/j\omega C}{R + 1/j\omega C} \times \overline{V}_1(f) = \frac{1}{1 + j\omega RC} \times \overline{V}_1(f) \qquad (3\text{-}6)$$

The transfer function is readily determined by forming the ratio of the output to the input:

$$H(f) = \frac{\overline{V}_2(f)}{\overline{V}_1(f)} = \frac{1}{1 + j\omega RC} = \frac{1}{1 + j2\pi f RC} \qquad (3\text{-}7)$$

A full interpretation of this result will be given in Ex. 3-3.

Example 3-2

Consider the *RLC* circuit shown in Fig. 3-4a. The input is $v_1(t)$ and the output is $v_2(t)$. (a) Working directly in the frequency domain, determine the Fourier transfer function $H(f) = \overline{V}_2(f)/\overline{V}_1(f)$. (b) Show that a differential equation of the form of (3-1) can be obtained directly in the time domain. (c) Using the result of part (b), determine the transfer function.

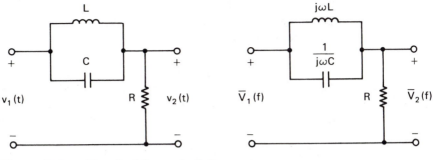

Figure 3-4. Circuit of Example 3-2.

Solution

(a) The circuit is first converted to the frequency domain by the procedure of this section. This form is shown in Fig. 3-4b. This circuit could be analyzed by several different standard analysis methods. To make the circuit analysis approach of this part different from the approach that will be used in the next part, the voltage-divider rule will be used. It is first necessary to determine an equivalent parallel combination for the inductive and capacitive impedances. Letting $\overline{Z}_1$ represent this quantity, we have

$$\overline{Z}_1 = j\omega L \parallel \frac{1}{j\omega C} = \frac{j\omega L \times \frac{1}{j\omega C}}{j\omega L + \frac{1}{j\omega C}} = \frac{j\omega L}{1 - \omega^2 LC} \tag{3-8}$$

The output voltage $\overline{V}_2(f)$ can now be expressed as

$$\overline{V}_2(f) = \frac{R\overline{V}_1(f)}{R + \overline{Z}_1} = \frac{R\overline{V}_1(f)}{R + \frac{j\omega L}{1 - \omega^2 LC}} = \frac{R(1 - \omega^2 LC)\overline{V}_1(f)}{R - R\omega^2 LC + j\omega L} \tag{3-9}$$

The transfer function is determined by forming the ratio $\overline{V}_2(f)/\overline{V}_1(f)$. In addition, the numerator and denominator of (3-9) will be divided by R. Thus,

$$H(f) = \frac{\overline{V}_2(f)}{\overline{V}_1(f)} = \frac{1 - \omega^2 LC}{1 - \omega^2 LC + \frac{j\omega L}{R}} \tag{3-10}$$

(b) A differential equation of the form desired is most readily determined by writing a node voltage equation at the output upper node. Currents leaving that node will be considered positive, and we have

$$\frac{v_2}{R} + C\frac{d(v_2 - v_1)}{dt} + \frac{1}{L}\int_{-\infty}^{t} (v_2 - v_1)\, dt = 0 \tag{3-11}$$

The first step is to remove the integral in (3-11). This is achieved by differentiating both sides of (3-11) with respect to t. In addition, the quantities in parentheses will be expanded. The result is

$$\frac{1}{R}\frac{dv_2}{dt} + C\frac{d^2v_2}{dt^2} - C\frac{d^2v_1}{dt^2} + \frac{v_2}{L} - \frac{v_1}{L} = 0 \tag{3-12}$$

The terms involving v_1 can now be moved to the right side of the equation. In addition, all terms will be multiplied by L for convenience. This results in

$$LC\frac{d^2v_2}{dt^2} + \frac{L}{R}\frac{dv_2}{dt} + v_2 = LC\frac{d^2v_1}{dt^2} + v_1 \tag{3-13}$$

This differential equation is readily seen to be in the form of (3-1). Incidentally, it was not possible to apply the voltage-divider rule in the time domain due to the derivative and integral relationships.

(c) The transfer function is determined by first applying (3-2) to all terms on both sides of (3-13). After factoring out $\overline{V}_2(f)$ and $\overline{V}_1(f)$, the form resulting is

$$\overline{V}_2(f)\left[LC(j\omega)^2 + \frac{L}{R}j\omega + 1\right] = \overline{V}_1(f)[LC(j\omega)^2 + 1] \tag{3-14}$$

The transfer function is then determined as $H(f) = \overline{V}_2(f)/\overline{V}_1(f)$. It is readily seen that the result will be the same as obtained in (3-10) after some simple algebraic manipulations are performed.

The reader will likely find the earlier approach of working directly in the frequency domain easier to apply. This approach will work with electrical circuits. However, if the transfer function of some other type of physical system is desired, it may not be possible to apply the standard circuit analysis methods. In such cases, it may be necessary to revert back to the basic differential equation approach.

3-2 AMPLITUDE AND PHASE FUNCTIONS

While the input-output spectral relationship of (3-5) applies for any complex signal input, the simplest and most widely employed interpretation of the Fourier transfer function is based on the assumption of a single-frequency, steady-state sinusoid. In this case, the Fourier transfer function reduces to the standard steady-state phasor form widely used in circuit analysis. This form will be considered next.

Assume an input sinusoid of the form

$$x(t) = X\sin(\omega t + \theta_x) \tag{3-15}$$

The output steady-state response will also be a sinusoid and will be of the form

$$y(t) = Y \sin(\omega t + \theta_y) \tag{3-16}$$

The input will be represented in complex phasor form as

$$\overline{X} = X\underline{/\theta_x} \tag{3-17}$$

The output will be represented in complex phasor form as

$$\overline{Y} = Y\underline{/\theta_y} \tag{3-18}$$

The *steady-state transfer function* can be expressed as

$$H(f) = \frac{\overline{Y}}{\overline{X}} = \frac{Y\underline{/\theta_y}}{X\underline{/\theta_x}} \tag{3-19}$$

The Fourier transfer function and the steady-state transfer function are essentially the same, so the symbol $H(f)$ is used for both. The major difference is in the interpretation. When a complex signal having many frequency components appears at the input, $H(f)$ is a weighting function applying to a range of frequencies. However, when the input is a steady-state sinusoid, the function $H(f)$ is evaluated at the specific frequency $f = \omega/2\pi$ corresponding to the frequency of the sinusoid.

The actual measurement of a Fourier transfer function may be most easily achieved through the use of a variable frequency sinusoid. At a given frequency, the function $H(f)$ is measured using techniques to be discussed shortly, and the frequency is incremented as many times as necessary to obtain the complete function. At a given frequency, it is necessary that steady-state conditions be reached before final measurements are made.

The most useful and intuitive form of the Fourier or steady-state transfer function is the amplitude-phase form. Since $H(f)$ is a complex function, it can be expressed in terms of a magnitude or amplitude and an angle. Both the magnitude and the angle will be functions of the frequency f. We thus express $H(f)$ as

$$H(f) = A(f)\epsilon^{j\beta(f)} = A(f)\underline{/\beta(f)} \tag{3-20}$$

The function $A(f) = |H(f)|$ is called the *amplitude* or *magnitude response* of the system, and the function $\beta(f)$, which is the angle, is called the *phase response*. Both $A(f)$ and $\beta(f)$ are real functions of frequency. It can be shown that $A(f)$ is an even function of frequency, and $\beta(f)$ is an odd function of frequency.

The amplitude-phase form will now be related to the steady-state phasor input-output relationships. From the definition of the transfer function we have

$$\overline{Y} = H(f)\overline{X} \tag{3-21}$$

Substitution of (3-17), (3-18), and (3-20) in (3-21) results in

$$Y\underline{/\theta_y} = [A(f)\underline{/\beta(f)}][X\underline{/\theta_x}] = A(f)X\underline{/\beta(f) + \theta_x} \tag{3-22}$$

We readily see that

$$Y = A(f)X \tag{3-23}$$

and

$$\theta_y = \beta(f) + \theta_x \tag{3-24}$$

The results of (3-23) and (3-24) are particularly important and are summarized as follows: (1) The amplitude of the output sinusoid (or phasor) is the amplitude of the input sinusoid (or phasor) multiplied by the amplitude response of the network evaluated at the particular sinusoidal frequency. (2) The phase of the output sinusoid is the phase of the input sinusoid plus the phase response of the system evaluated at the same frequency.

These concepts serve as the practical basis for measuring the amplitude and phase response functions for linear systems, as illustrated in Fig. 3-5. The frequency of a sinusoidal generator of variable frequency is incremented in steps. At each frequency, the input and output amplitudes are measured, and the phase shift produced by the system is measured. The ratio of the output amplitude to the input amplitude plotted as a function of frequency is $A(f)$, and the phase shift plotted as a function of frequency is $\beta(f)$.

It has become standard practice in many applications to express the amplitude response in terms of a relative decibel gain or loss. While there are several variations on these definitions, the forms given here should suffice for most purposes.

The *relative decibel amplitude response* $A_{dB}(f)$ will be defined as

$$A_{dB}(f) = 20 \log_{10} \frac{A(f)}{A_0} \tag{3-25}$$

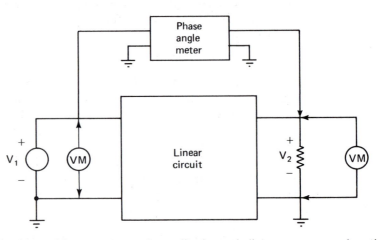

Figure 3-5. Measurement of amplitude and phase response functions for a circuit. (The input is V_1 and the output is V_2.)

where A_0 is some reference comparison level. In many cases, the convenient choice $A_0 = 1$ is made.

An alternative form is the *relative decibel loss function* $\alpha_{dB}(f)$, which will be defined as

$$\alpha_{dB}(f) = 20 \log_{10} \frac{A_0}{A(f)} \tag{3-26}$$

The two functions are readily seen to be related by

$$\alpha_{dB}(f) = -A_{dB}(f) \tag{3-27}$$

Note that if $A(f) > A_0$ then $A_{dB}(f)$ is positive and $\alpha_{dB}(f)$ is negative; while if $A(f) < A_0$, then A_{dB} is negative and $\alpha_{dB}(f)$ is positive.

When a complex input signal having a spectrum $\overline{X}(f)$ excites a system, the spectrum is weighted by the transfer function $H(f)$. The amplitude of the output spectrum is the amplitude of the input spectrum times the amplitude response of the system. The phase of the output spectrum is the phase of the input spectrum plus the phase response of the system. In a frequency range where the amplitude response of the system is relatively large, the output spectrum will be accentuated. Conversely, in a frequency range where the amplitude response of the system is relatively small or zero, the output spectrum will be attenuated or even eliminated. These concepts are extremely important in studying the transmission of signals through linear systems, so they should be thoroughly studied by the reader.

Example 3-3

Determine the amplitude and phase response functions for the simple low-pass filter of Ex. 3-1. Verify that the filter is a low-pass type.

Solution

The transfer function was derived in Ex. 3-1, and the result was given in Eq. (3-7). The amplitude and phase functions are determined by representing $H(f)$ in magnitude and phase forms. The computations follow:

$$A(f) = |H(f)| = \frac{1}{\sqrt{1 + (\omega RC)^2}} = \frac{1}{\sqrt{1 + (2\pi RCf)^2}} \tag{3-28}$$

$$\beta(f) = -\tan^{-1} \omega RC \tag{3-29}$$

The amplitude response $A(f)$ is observed to have a value $A(0) = 1$ at dc, and it decreases with increasing frequency. As f becomes increasingly large, it is readily seen from (3-28) that $A(f)$ becomes increasingly small. For notational convenience, we will simply define $A(\infty) = 0$ to indicate this limiting condition. Thus, the amplitude response of the filter varies from a value of unity at dc to a limiting value of zero at an infinite frequency, so the response is low pass in nature.

This simple low-pass response occurs so often that it is convenient to rearrange the basic transfer function in a more convenient form. A reference frequency f_c will be defined as

$$f_c = \frac{1}{2\pi RC} \qquad (3\text{-}30)$$

This definition applied to the transfer function of Eq. (3-7) results in

$$H(f) = \frac{1}{1 + j\dfrac{f}{f_c}} \qquad (3\text{-}31)$$

which is a common form for representing a simple low-pass response. (The term "simple" keeps appearing to indicate to the reader that this circuit has a very limited usefulness as a low-pass filter since its rolloff rate is not very large.)

The amplitude and phase functions expressed in terms of f_c are

$$A(f) = \frac{1}{\sqrt{1 + \left(\dfrac{f}{f_c}\right)^2}} \qquad (3\text{-}32)$$

$$\beta(f) = -\tan^{-1}\frac{f}{f_c} \qquad (3\text{-}33)$$

The forms of $A(f)$ and $\beta(f)$ are illustrated in Fig. 3-6a and b. The shapes of these functions have been made in accordance with the log and semilog techniques used with Bode plots for the benefit of readers familiar with that procedure.

Observe that $A(f_1) = 1/\sqrt{2}$, corresponding to the amplitude response being down by about 3 dB at $f = f_c$. Observe also that the phase shift changes from zero at dc to $-90°$ at an infinite frequency and is $-45°$ at $f = f_c$. (Zero frequency does not appear on the figures due to the assumed logarithmic scale.)

Example 3-4

The Fourier transfer function of a certain circuit is given by

$$H(f) = \frac{20j\omega(9 + 2j\omega)}{(1 + j\omega)(10 + j\omega)} \qquad (3\text{-}34)$$

Determine an expression for the amplitude response $A(f)$ and the phase response $\beta(f)$.

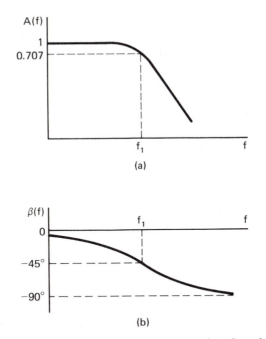

Figure 3-6.   Amplitude and phase responses for the simple low-pass filter of Examples 3-1 and 3-3 (logarithmic frequency scales).

Solution

It is convenient to first organize the factors of $H(j\omega)$ in polar form. We have

$$H(f) = \frac{(20\omega\underline{/90^\circ})[\sqrt{81 + 4\omega^2}\underline{/\tan^{-1}(2\omega/9)}]}{(\sqrt{1 + \omega^2}\underline{/\tan^{-1}\omega})[\sqrt{100 + \omega^2}\underline{/\tan^{-1}(\omega/10)}]} \qquad (3\text{--}35)$$

The magnitude and angle of this function can then be determined by the standard manipulative rules of complex algebra; that is, the net magnitude is the product of the numerator magnitudes divided by the product of the denominator magnitudes, and the net angle is the sum of the numerator angles minus the sum of the denominator angles. Thus,

$$A(f) = \frac{20\,\omega\,(\sqrt{81 + 4\omega^2})}{(\sqrt{1 + \omega^2})(\sqrt{100 + \omega^2})} \qquad (3\text{-}36)$$

$$\beta(f) = 90^\circ + \tan^{-1}\frac{2\omega}{9} - \tan^{-1}\omega - \tan^{-1}\frac{\omega}{10} \qquad (3\text{-}37)$$

Plots of these functions could be made with the aid of Bode-plot techniques. Routines for a programmable calculator could also be written to evaluate the functions at different frequencies. We will not pursue these points further since the major objective of this problem was to illustrate the formulation of the amplitude and phase functions from the transfer function, rather than the detailed evaluation.

3-3 IDEAL FREQUENCY-DOMAIN FILTER

In this section, we will investigate the behavior of the ideal frequency-domain filter. While such a filter is not realizable, the concept serves as a very useful model for comparing the behavior of actual filters.

All filters considered here are assumed to meet the general conditions stated in Sec. 3-1. Consider the block diagram shown in Fig. 3-7. Assume that the input can be expressed as $x(t) + z(t)$, where $x(t)$ represents a desired signal at the input and $z(t)$ represents an undesired signal or composite of signals. The purpose of the filter is to eliminate $z(t)$ while preserving $x(t)$ as closely as possible to its original form.

Figure 3-7. Block diagram of filter having desired plus undesired input.

The process of filtering results in a certain inherent delay and a possible change in the signal level, so the best we can hope for is that the output signal will be a delayed version of the desired signal with a possible change in the amplitude level. If the shape of the desired signal is preserved, the output can be expressed as

$$y(t) = Kx(t - \tau) \tag{3-38}$$

where K represents a level change and τ is the delay. This concept is illustrated in Fig. 3-8 and is a property of a distortionless ideal filter.

The frequency-domain interpretation of the ideal filter will now be investigated. Applying the Fourier transform to both sides of (3-38) and employing operation (0–4) of Table 2-3, we obtain

$$\overline{Y}(f) = K\epsilon^{-j\omega\tau}\overline{X}(f) \tag{3-39}$$

The steady-state transfer function is obtained as

$$H(f) = \frac{\overline{Y}(f)}{\overline{X}(f)} = K\epsilon^{-j\omega\tau} = K\underline{/-\omega\tau} \tag{3-40}$$

The amplitude and phase functions are determined as

$$A(f) = K \tag{3-41}$$

$$\beta(f) = -\omega\tau \tag{3-42}$$

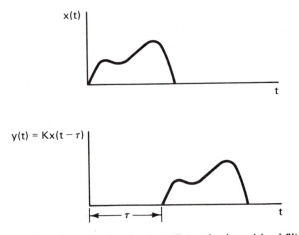

Figure 3-8. Input and output of distortionless ideal filter.

From these results, it can be seen that the amplitude response of the ideal filter should be constant, and the phase response should be a linear function of frequency. However, these conditions apply only to the frequency range of the desired signal $x(t)$. If the amplitude response were constant everywhere, the undesired signal would not be eliminated! In most communications applications involving frequency-domain filtering, the spectrum of the desired signal occupies a different frequency range than that of the undesired signal. The amplitude response of the filter is chosen in such a manner as to pass the desired signal and reject the other signals present at the input.

It can be seen then that an ideal frequency-domain filter is characterized by a constant-amplitude response and a linear phase response over the frequency band representing the desired signal. Outside this range, the amplitude response should drop toward zero very rapidly so as to reject or suppress frequency components not in the desired signal band. The frequency range in which a signal is transmitted through the filter is called the *passband,* and the frequency range in which a signal is rejected is called the *stopband.*

From certain theoretical developments of network theory, it can be shown that the attainment of both ideal constant amplitude and linear phase responses is physically impossible in a practical filter. It also turns out that as the amplitude approximation is improved, the phase response usually tends to be poorer, and vice versa. However, it is possible to provide approximations that approach the ideal conditions sufficiently close to satisfy most applications, and some of these approximations will be surveyed in Sec. 3-5.

Practical filters are characterized by a *transition band* between the passband and the stopband. The exact locations of the boundaries of these different bands are somewhat arbitrary. The forms of the amplitude and phase characteristics of a hypothetical filter having nearly ideal characteristics in the passband, but with a nonzero transition band, are illustrated in Fig.

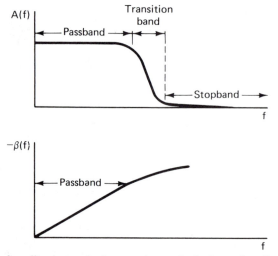

Figure 3-9. Amplitude and phase characteristics of a filter with ideal passband characteristics.

3-9. Since the phase shift of a low-pass filter is often negative, the function $-\beta(f)$ is shown for convenience.

In certain applications, the time delay of a signal passing through a filter has more significance than the phase shift. Two definitions of certain time-delay parameters will now be given. These are the *phase delay* T_p and the *group (or envelope) delay* T_g. These functions are defined as follows:

$$T_p(f) = \frac{-\beta(f)}{\omega} \tag{3-43}$$

$$T_g(f) = \frac{-d\beta(f)}{d\omega} \tag{3-44}$$

The graphical significance of these functions is illustrated in Fig. 3-10. The phase delay at a given frequency is proportional to the slope of the secant line from dc to the particular frequency and is a type of overall delay parameter. The group delay is proportional to the slope of the tangent at a particular frequency and represents a narrow-range delay parameter.

Consider now the case of a filter with a constant-amplitude response and a linear phase response as described by (3-41) and (3-42). It is readily seen that

$$T_p(f) = T_g(f) = \tau \tag{3-45}$$

For the ideal filter, the phase and group delays are identical and represent the exact delay of the signal, which has not been distorted in this case. In the more general case, where the amplitude response is not constant in the

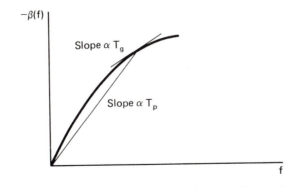

Figure 3-10. Graphical significance of phase delay and group delay.

passband and the phase response is not linear, it is more difficult to precisely define the exact delay, since a signal will undergo some distortion in passing through the filter. In fact, any attempt to define the exact delay will result in some variation of delay as different types of signals are applied to the filter. Nevertheless, the preceding definitions are very useful in determining the approximate delay characteristics of a filter.

The phase-delay parameter can be used to estimate the delay of a low-pass type signal, such as a basic pulse waveform, when it is passed through a low-pass filter. The phase delay is computed over the frequency range representing the major portion of the input signal spectrum. If the phase response does not deviate too far from linearity over the range involved, this value may represent a reasonable approximation to the actual delay of the waveform involved.

A case of significance involving both phase delay and group delay is that of a narrow-band modulated signal. It can be shown that when a narrow-band modulated signal is passed through a filter the carrier is delayed by a time equal to the phase delay, while the envelope (or intelligence) is delayed by a time approximately equal to the group delay. Since the intelligence represents the desired information contained in such signals, strong emphasis on good group delay characteristics is often made in filters designed for communications applications.

Returning to the ideal frequency-domain filter concept, it is convenient to consider several models representing the amplitude responses for different classes of filters, as illustrated in Fig. 3-11. The four models shown are the *low-pass, high-pass, band-pass,* and *band-rejection* ideal frequency-domain amplitude functions. The corresponding ideal phase functions should be linear over the passband in each case. It should be emphasized again that these exact ideal functions are not physically realizable, but they may be approximated sufficiently close to meet engineering requirements.

So far, we have studied the ideal filter concept only from the frequency-domain point of view. An alternative, and sometimes equally important, point

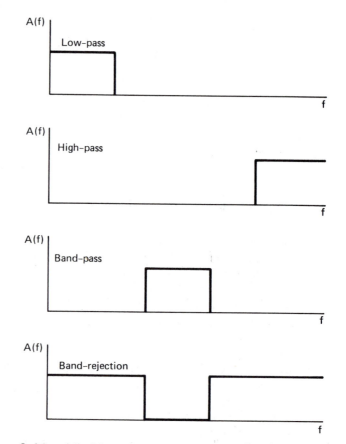

Figure 3-11.   Ideal frequency-domain amplitude response models.

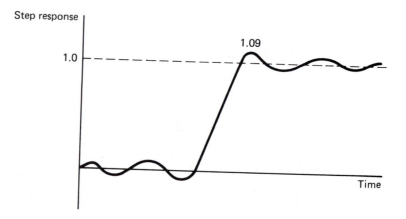

Figure 3-12.   Step response of ideal frequency-domain low-pass filter.

of view is the time domain or transient behavior of the filter. From that standpoint, the ideal filter considered here exhibits significant ringing and overshoot. This property is caused by the finite discontinuity in the assumed ideal block characteristic. The form of the step response of the ideal low-pass filter (with some assumed delay) is illustrated in Fig. 3-12. The actual transient response of a real filter results from a combination of both the amplitude and phase characteristics. As long as the phase response does not deviate too far from linearity, the transient response of a real filter will be superior to that of the ideal filter. Thus, while the ideal frequency-domain filter represents a goal that we constantly seek, it is not necessarily the best result we could achieve, particularly from the transient point of view.

3-4 IDEAL FILTER RESPONSES WITH COMMON WAVEFORMS

The concept of the ideal filter was introduced in the preceding section. This concept may be coupled with the Fourier series approach of Chapter 2 to enable one to determine the effect of filtering on a given signal. This would be achieved by summing those Fourier components at the output that fall in the filter passband and rejecting all others. Any phase shift produced by the filter would have to be included in the angles of the spectral components.

Consider the situation depicted in Fig. 3-13a in which a periodic input signal $x(t)$ excites a low-pass filter. For the purpose of this development, an *ideal low-pass filter* with constant-amplitude response and linear phase response will be assumed as shown in Fig. 3-13b and c, respectively. The assumed forms for $A(f)$ and $\beta(f)$ are then

$$A(f) = 1, \qquad \text{for } 0 \leq f < Nf_1$$
$$= 0, \qquad \text{elsewhere} \tag{3-46}$$

and

$$\beta(f) = -n\omega_1\tau \tag{3-47}$$

The input signal $x(t)$ can be expressed as an amplitude-phase Fourier series of the form

$$x(t) = \sum_{n=0}^{\infty} C_n \cos(n\omega_1 t + \theta_n) \tag{3-48}$$

The steady-state form of the output $y(t)$ is determined by summing those components of the input falling in the passband and applying the phase shifts to the arguments of the cosine terms. Since $A(f) = 1$ in the passband, the output spectral components will have the same magnitudes as the input components. Thus, $y(t)$ can be expressed as

$$y(t) = \sum_{n=0}^{N} C_n \cos(n\omega_1 t + \theta_n - n\omega_1\tau) \tag{3-49}$$

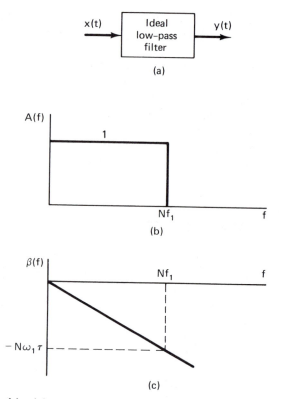

Figure 3-13. Ideal low-pass filter with amplitude and phase functions used in development of Section 3-4. (The frequency f_1 is the fundamental of the square-wave input.)

In practice, the actual computation of an expression like (3-49) is extremely tedious to perform manually. As previously explained, the insight and interpretation are the aspects most important in many common problems. The digital computer, however, can readily perform summations such as (3-49), and many scientific computer systems have routines for performing such operations.

To illustrate to the reader the relative effects of band limiting with an ideal low-pass filter, the results of (3-49) for a number of common waveforms with different bandwidths have been compiled and are presented graphically in Appendix C. These results were obtained with a special instructional computer program developed by Smolleck and Bergrab.* The reader is referred to Appendix C to support the following discussion. Incidentally, in all

*H. A. Smolleck and S. T. Bergrab, "An Interactive Computational/Graphical Capability for Fourier-Series Investigations," Presented at SOUTHEASCON 1979, Roanoke, Va., 1979.

the cases to be discussed, the phase shift of the ideal filter was assumed to be zero; that is, $\beta(f) = 0$. The effect of a nonzero linear phase would be to shift the waveforms to the right by the time-delay parameter τ, but the shapes would not be affected.

The first waveform to be considered is the ideal square wave arranged as an odd function. Various results of filtering the square wave are given in Figs. C-1 through C-5. At the first extreme, Fig. C-1 corresponds to a cutoff frequency just above $3f_1$ (allowing fundamental and third harmonic only), and at the other extreme, Fig. C-5 corresponds to a cutoff frequency just above $100f_1$, which allows all components up to the ninety-ninth harmonic to pass. (There are neither even harmonics nor a dc component in the spectrum of this waveform.) Observe that the square wave is seriously degraded at the lower bandwidth values but begins to resemble the ideal square wave as the bandwidth increases.

These results indicate that the time it takes the output "pulses" to rise or fall is very large at lower bandwidths, but the rise time becomes smaller as the bandwidth increases. However, in theory, it would take an infinite bandwidth to reproduce accurately a square wave or square pulse. We will establish a good working estimate later for relating bandwidth to rise time later, but, for the moment, concentrate on the qualitative result that rise time decreases as the bandwidth of the filter increases.

The difficulty in reproducing the square wave through the finite bandwidth of the filter is a direct result of the finite discontinuity of the signal. It was observed in Chapter 2 that the spectrum of a signal having a finite discontinuity converges only as $1/n$ (-6 dB/octave rolloff), and the resulting bandwidth for faithful reproduction is enormous. Even with a bandwidth $100f_1$, there is a considerable overshoot and some rise time, the latter of which is difficult to see due to the resolution of the computer plot.

The results of filtering a half-wave rectified sine wave are given in Figs. C-6 through C-10. These results show a much faster convergence. Even with five harmonics, the reproduction is reasonably good. The major factor here is the fact that the function is continuous. The first derivative has finite continuities, and the spectrum, therefore, has a -12 dB/octave rolloff rate.

The results of filtering a sawtooth waveform are given in Figs. C-11 through C-15. The poor convergence in this case is a direct result of the discontinuity, as was the case of the square wave. Observe that the convergence is poorer in the vicinity of the discontinuity.

3-5   SEVERAL FILTER APPROXIMATIONS

The ideal block filter characteristics considered in the last two sections cannot be attained with actual "real-life" filters. However, there are a number of filter functions that may be used to approximate the ideal cases sufficiently close to satisfy most practical engineering requirements. In this section, we will investigate a few of these functions.

Filter synthesis and design is a complete field of study in itself, and much work has been done in this area through the years. Any attempt to deal comprehensively with such a vast subject in a single section of this book would be futile. The intent here is merely to expose the reader to a few of the common forms of filter responses encountered in practical systems. The treatment should be sufficient to allow the serious reader to understand the types of assumptions made and the validity of such assumptions encountered in using the ideal filter models for communications system analysis.

All the functions considered in this section will be given in *low-pass* form. Such functions may be extended to *high-pass, band-pass,* and *band-rejection* forms by means of certain transformations.

Butterworth Response

The Butterworth or maximally flat amplitude response of order m is given by

$$A(f) = \frac{1}{\sqrt{1 + (f/f_c)^{2m}}} \tag{3-50}$$

The frequency f_c is called the *cutoff* frequency, and it is the frequency at which $A^2(f) = 1/2$ or $A(f) = 1/\sqrt{2}$ times the dc value. This level corresponds to an attenuation of 3.01 dB with respect to the dc response, but as a matter of common usage this value is rounded off to 3 dB.

The Butterworth amplitude response can be shown to be optimum at dc in the maximally flat sense. This means that the difference between the ideal

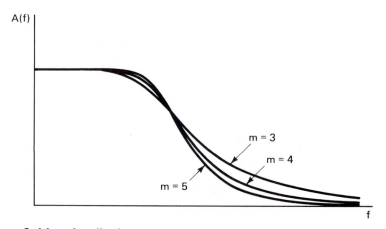

Figure 3-14. Amplitude response functions for several Butterworth filters.

amplitude response and the approximation and as many lower-order derivatives as possible are equated to zero at $f = 0$. The Butterworth response is a monotonically decreasing function of frequency in the positive frequency range. As the order m increases, the response becomes "flatter" in the passband, and the attenuation is greater in the stopband. Above cutoff, the Butterworth amplitude response of order m approaches a high-frequency asymptote having a slope of $-6m$ dB/octave. Several Butterworth functions are shown in Fig. 3-14.

Chebyshev Response

The Chebyshev or equiripple amplitude response is derived from the Chebyshev polynomials $C_m(x)$, which are a set of orthogonal functions possessing certain interesting and useful properties. Some of these properties are as follows: (1) The polynomials have equiripple amplitude characteristics over the range $-1 < x < 1$ with ripple oscillating between -1 and $+1$. (2) $C_m(x)$ increases more rapidly for $x > 1$ than any other polynomial of order m bounded by the limits stated in property 1.

The basic Chebyshev amplitude response is defined by

$$A(f) = \frac{\gamma}{\sqrt{1 + \epsilon^2 C_m^2(f/f_c)}} \tag{3-51}$$

where m represents both the order of the Chebyshev polynomials and the order of the corresponding transfer function. The quantity ϵ^2 is a parameter chosen to provide the proper passband ripple, and γ is a constant chosen to determine the proper dc gain level. The frequency f_c represents the *cutoff* frequency, which is the highest frequency at which the response crosses the passband ripple bound and moves into the transition band. This definition is consistent with that of the Butterworth response only for the case of 3 dB ripple.

Due to the fact that both the order and the passband ripple are parameters, a variety of Chebyshev responses can be created. Several representative cases are shown in Fig. 3-15. From these figures, it can be deduced that the number of maxima and minima in the passband is equal to the order of the filter.

As the order of the filter is increased, the attenuation in the stopband increases for a given passband ripple. For a given order, the stopband attenuation increases as the passband ripple is allowed to increase. Thus, there is a direct trade-off between the allowable passband ripple and the stopband attenuation for a given order. As in the case of the Butterworth response, the Chebyshev response of order m approaches a high-frequency asymptote having a slope of $-6m$ dB/octave.

Both the Butterworth and Chebyshev responses were obtained from approximations involving only the amplitude response, and no attention was

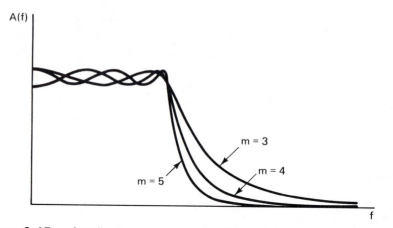

Figure 3-15. Amplitude response functions for several Chebyshev filters.

paid to the phase response in either case. In some applications, the phase (or time delay) response is more important than the amplitude response.

One approximation in which linear phase (or constant time delay) is optimized is the *maximally flat time delay* (MFTD) response. The amplitude response that results from the MFTD approximation has a low-pass shape with a monotonically decreasing behavior as the frequency is increased. However, the passband response is not as flat as for the Butterworth function, and the stopband attenuation is not as great at a given frequency as for either the Butterworth or Chebyshev response of the same order.

The MFTD filter is used where excellent phase shift (or time delay) characteristics are required, but where the amplitude response need not display a rapid attenuation increase just above cutoff. As in the case of both the Butterworth and Chebyshev functions, the high-frequency attenuation rate of the MFTD filter will eventually approach $6m$ dB/octave, but the total

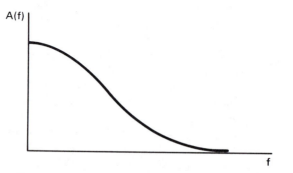

Figure 3-16. Form of amplitude response for maximally flat time-delay filter.

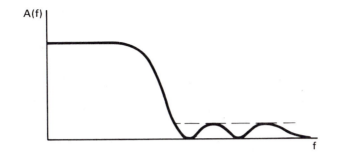

Figure 3-17. Form of amplitude response for inverted Chebyshev filter.

attenuation will not be as great. The general form of the amplitude response is illustrated in Fig. 3-16.

The preceding filter functions represent some very common types, but they are by no means the only ones available. A few others will be very briefly mentioned. The inverted *Chebyshev filter* is characterized by a maximally flat passband response and an equiripple stopband response, as illustrated in Fig. 3-17. The *Cauer*, or *elliptic function filter* has both an equiripple passband and an equiripple stopband, as illustrated in Fig. 3-18.

An earlier form of filter technology is that of *image parameter* filter design, which makes use of the *constant k* and *m-derived* filter sections. This technique is based somewhat on principles similar to transmission line theory and involves matching the impedances between successive sections. Most communication system filters designed before the mid-1950s or so were probably based on this concept. There are still situations where the image approach yields the most satisfactory solution, particularly in systems involving sections of transmission lines.

Finally, no modern discussion of filter classifications would be complete without some mention, however brief, of the concept of the *digital filter*.

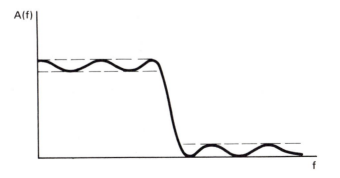

Figure 3-18. Form of amplitude response for Cauer or elliptic function filter.

There are few (if any) areas of the electrical field that have not been affected by the rapid evolution of modern computer technology, and filter technology is no exception. A digital filter is a numerical algorithm that transforms an input data signal into an output data signal, with a desired filtering objective accomplished through the data processing. Most digital filters employ linear constant coefficient difference equations that relate the output data stream to the input data stream. If the difference equation is designed properly, the same effect as can be accomplished with an analog filter can be implemented with arithmetic operations on the computer.

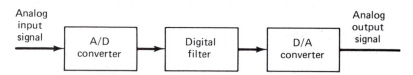

Figure 3-19. Illustration of a digital filter with analog input and output.

A possible system employing a digital filter is illustrated in Fig. 3-19. The input analog signal is applied to an A/D converter, which converts the analog signal into a sequence of digital numbers (or words). The signal is then processed by the digital filter, which might, for example, be a dedicated microcomputer. The output is then converted back to an analog signal by means of a D/A converter. With some systems, the input might be in digital form initially, and/or the output might be desired in a digital form, in which case one or more of the conversions could be eliminated. Using the accuracy attainable with modern digital circuitry, it is possible to realize responses not physically possible with conventional analog circuitry. It is also possible to take advantage of multiplexing techniques in which a large number of signals are processed by one digital signal-processing unit.

3-6 PULSE TRANSMISSION APPROXIMATIONS

From the material in the preceding several sections, we have seen that the effect of a transmission system with finite bandwidth is to reject some of the Fourier spectral components of wideband signals. The resulting output signals will then display some distortion, which can be significant in many applications. The limited bandwidth may be deliberately introduced in some applications (e.g., a channel separation filter), or it may occur as a natural undesirable property of many transmission components, such as amplifiers and cables. In either event, the communications engineer or technologist must continually be aware of the bandwidth requirements of signals as they are processed in communications systems.

Extensive and detailed system analysis and design at a fundamental level necessarily requires the use of complex Fourier theory to predict accurately the results of finite band limiting of complex signals. This level of analysis is usually not practical in applied communications hardware implementation due to the detail of the mathematics involved and the necessity for special computer programs to aid in the evaluation. Fortunately, there are some simple "rules of thumb" or estimates that permit the applied communications technologist or engineer to predict approximate bounds for certain bandwidth requirements, and our focus will be directed toward these estimates.

It should be stated at the outset that the formulas to be given are *not* intended to be interpreted as *exact* formulas. They are simply estimates that predict the approximate range of bandwidth requirements, and their use must be tempered with good judgment and allowances for worst-case conditions. Indeed, variations on these formulas appear in the literature, and the nature of the band limiting functions may affect the accuracy of the results. Nevertheless, the estimates are simple and easy to apply, and if interpreted properly, they may save a significant amount of time and effort.

The approximations to be developed will be applied to pulse transmission. While this may seem restrictive, it turns out that, since a pulse is a worst-case type of condition as far as bandwidth is concerned, the results can be adapted to other types of signals. Furthermore, the increasing use of digital signals certainly places pulse transmission as a most important condition to be considered in some detail.

We will consider two distinctly separate types of criteria: (1) approximate pulse reproduction, which will be denoted as *coarse* reproduction, and (2) exact reproduction, which will be denoted as *fine* reproduction. Consider for purposes of discussion the pulse shown in Fig. 3-20a.

The first condition is common in any type of transmission in which the presence of the pulse must be recognized and in which the amplitude level must be approximately preserved, but in which some smearing or rounding of the beginning and ending of the pulse is acceptable. This condition would apply to some of the pulse amplitude modulation techniques, as well as to digital data transmission and certain types of radar. In digital data transmission, for example, the transmitted "ones" and "zeros" may be restored perfectly at the receiver as long as the receiver can recognize which was transmitted. The coarse criterion of pulse reproduction is illustrated in Fig. 3-20b.

The second condition (fine reproduction) applies to systems in which a more exact pulse shape must be preserved. This condition is required in modulation systems in which the beginning and ending of pulses are important and in which no smearing or spreading of the pulse is acceptable. This second condition would also be the requirement for analog signals in which the pulse criterion is used for a worst-case condition. The fine criterion of reproduction is illustrated in Fig. 3-20c.

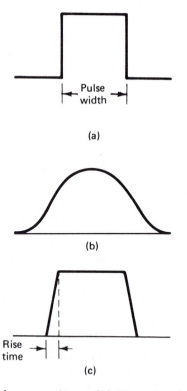

Figure 3-20. (a) Pulse waveform, (b) filtered pulse corresponding to "coarse" reproduction, and (c) filtered pulse corresponding to "fine" reproduction.

Extensive studies have been made of the bandwidth requirements for these conditions. Based on reasonable criteria, the following conclusions have been determined: First, the bandwidth B for coarse reproduction is inversely proportional to the width of a given pulse; that is,

$$B = \frac{K_1}{\text{pulse width}} \tag{3-52}$$

where K_1 is a constant that will be discussed shortly. Second, the bandwidth B for fine reproduction is inversely proportional to the allowable rise time (or fall time) of the pulse; that is,

$$B = \frac{K_2}{\text{rise time}} \tag{3-53}$$

where K_2 will also be discussed shortly.

Thus, the bandwidth required for coarse reproduction increases as the pulse width decreases, and the bandwidth required for fine reproduction increases as the rise time decreases.

We will next consider the constant K_1 and K_2. The first consideration will be for baseband pulses, in which B will be interpreted as a baseband bandwidth ranging from dc to a maximum frequency equal to B. The exact value of K_1 depends on the criterion for coarse reproduction, as well as on the sharpness of the band-limiting characteristic.

Certain theoretical developments in sampled-data theory show that, under special conditions with an ideal low-pass filter, the approximate form of a pulse may be reproduced in a bandwidth given by 0.5/pulse width. If the bandwidth is lowered below this value, the amplitude drops rapidly. This corresponds to the minimum bandwidth for the dc and fundamental component in a symmetrical square wave. For actual pulse-type signals and realistic filter rolloff characteristics, the actual bandwidth required for coarse reproduction may be greater. Nevertheless, there are some worthwhile reasons for using this idealized assumption, and thus $K_1 = 0.5$ will be used in most subsequent developments for baseband pulses.

As in the case of K_1, the constant K_2 will depend on the exact rolloff characteristics of the effective filter function and other factors. The reader may be familiar with a formula from electronics texts in which the rise time of an amplifier is expressed as 0.35/(3 dB bandwidth). If this formula were rearranged in the form in which we are interested, it would suggest that $K_2 = 0.35$. However, the formula given is based on a specific -6 dB/octave rolloff rate for the sytem, and it assumes the standard IEEE definition of rise time as the time between 10% and 90% points on the output pulse. For filters exhibiting sharper cutoff, the constant is usually greater than 0.35.

As a simplified approach to this problem, we will assume a value of $K_2 = 0.5$, which is usually adequate for baseband pulse transmission. This choice has the advantage that the constants K_1 and K_2 are both equal and can be easily remembered. We note, however, that the choice of K_1 was somewhat on the optimistic side, while K_2 was more of a worst-case choice. However, this is partially justified since coarse reproduction is open to more leeway in definition than fine reproduction, which should indicate a nearly perfect replica of the input.

In the case of radio-frequency (RF) pulses, the corresponding bandwidth values must be doubled for the same quality of reproduction. This is true because the spectrum appears on both sides of the carrier frequency, and both sidebands must be reproduced for the same degree of accuracy. Thus, the constants K_1 and K_2 must be doubled for RF pulses.

The preceding results can be summarized as follows:

Baseband Pulses

$$\text{Coarse reproduction:} \quad B \approx \frac{0.5}{\text{pulse width}} \tag{3-54}$$

$$\text{Fine reproduction:} \quad B \approx \frac{0.5}{\text{rise time}} \tag{3-55}$$

RF Pulses

$$\text{Coarse reproduction:} \quad B \approx \frac{1}{\text{pulse width}} \tag{3-56}$$

$$\text{Fine Reproduction:} \quad B \approx \frac{1}{\text{rise time}} \tag{3-57}$$

It should be understood that B for baseband pulses is a low-pass bandwidth, while B for RF pulses is a bandpass bandwidth centered at the center frequency of the RF spectrum.

In addition to determining bandwidth requirements for communications signal transmission, the preceding formulas may also be used in other electronic applications in which rise time and bandwidth trade-offs are required. For example, the formulas may be used to predict the approximate rise time introduced by an amplifier or filter with bandwidth B when a nearly perfect pulse is applied to the system. The formulas may also be used to estimate the bandwidth required in an oscilloscope to measure a given rise time of a pulselike signal. In this case, it is necessary to use a bandwidth much greater than the bandwidth corresponding to the given rise time. For example, suppose it is desired to measure a rise time in the neighborhood of 1 μs. The bandwidth associated with this rise time is determined from (3-55) for a baseband pulse as about 500 kHz. An oscilloscope with a much greater bandwidth should be selected, since the rise time introduced by the instrument should be small compared with the value being measured.

In concluding this section, it should be stressed once more that these formulas are estimates and are not intended as precise formulas. They are very useful in practice and provide some quick "ball-park" figures for simplifying an analysis that would otherwise be extremely cumbersome and difficult to perform.

Example 3-5

A certain baseband pulse generator produces pulses having widths of 2 μs. Determine the approximate bandwidths for (a) coarse reproduction and (b) fine reproduction based on rise and fall times of 10 ns.

Solution

(a) The approximate bandwidth for coarse reproduction is

$$B = \frac{0.5}{\text{pulse width}} = \frac{0.5}{2 \times 10^{-6}} = 250 \text{ kHz} \tag{3-58}$$

(b) The approximate bandwidth for fine reproduction is

$$B = \frac{0.5}{\text{rise time}} = \frac{0.5}{10 \times 10^{-9}} = 50 \text{ MHz} \tag{3-59}$$

3-7   ENERGY AND POWER SPECTRA TRANSMISSION

We have seen in previous sections that the output signal of an appropriate system may be determined from a knowledge of the input signal and the system transfer function. In this section, we will extend this concept to permit the determination of the output energy or power spectral density functions from similar functions defined at the input.

The concepts apply equally well to both energy and power spectral density functions and to both periodic and random power signals. However, the development is made easier by momentarily considering an energy signal $x(t)$.

Assume that the desired output is $y(t)$ and that the transfer function of the system is $H(f)$. From previous work, we know that

$$\overline{Y}(f) = H(f)\overline{X}(f) \tag{3-60}$$

We will now form the square of the magnitude of both sides of (3-60). Since the magnitude of the product of two complex numbers is the product of the magnitudes, we have

$$|\overline{Y}(f)|^2 = |H(f)\overline{X}(f)|^2 = |H(f)|^2|\overline{X}(f)|^2 \tag{3-61}$$

The input and output energy spectral density functions $S_x(f)$ and $S_y(f)$ are

$$S_x(f) = |\overline{X}(f)|^2 = X^2(f) \tag{3-62}$$

$$S_y(f) = |\overline{Y}(f)|^2 = Y^2(f) \tag{3-63}$$

The input-output relationship of (3-61) may now be expressed as

$$S_y(f) = |H(f)|^2 S_x(f) = A^2(f)S_x(f) \tag{3-64}$$

In the discussion that follows, the concept will be generalized to include both energy and power spectral functions.

The result of (3-64) indicates that the power or energy spectral density function at the output of an appropriate system is the input power or energy spectral density function multiplied by the amplitude response squared for the sytem. The amplitude response squared $A^2(f) = |H(f)|^2$ can be considered as the *power or energy spectral density weighting function*. A comparison between the block diagram forms representing the input and output linear spectra and the input and output power or energy spectra is shown in Fig. 3-21.

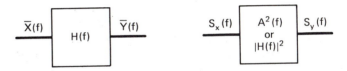

Figure 3-21.   Comparison between linear spectrum transmission and power or energy spectrum transmission.

No mention was made as to whether $S_x(f)$ and $S_y(f)$ were one- or two-sided functions in the development. It turns out that either form can be used for $S_x(f)$, and the same form will automatically be obtained for $S_y(f)$.

Example 3-6

Consider again the circuit of Exs. 3-1 and 3-3. Determine the power or energy spectral density weighting function.

Solution

The amplitude response can be obtained from the transfer function $H(f)$ as given in (3-7) or (3-31) by forming the magnitude squared, or it may be determined from the amplitude response function $A(f)$ in (3-28) or (3-32). Using the form of (3-32), the result is readily determined as

$$A^2(f) = \left[\frac{1}{\sqrt{1 + (f/f_c)^2}}\right]^2 = \frac{1}{1 + (f/f_c)^2} \qquad (3\text{-}65)$$

PROBLEMS

3-1 The circuit shown in Fig. P3-1 is the simplest form of a high-pass filter. The input is $v_1(t)$ and the output is $v_2(t)$.

(a) Working directly in the frequency domain, determine the Fourier transfer function $H(f) = \overline{V}_2(f)/\overline{V}_1(f)$.

(b) Determine the amplitude response $A(f)$ and the phase response $\beta(f)$.

(c) Compute $A(0)$ and $A(\infty)$ and comment on these values with respect to the filter's high-pass nature.

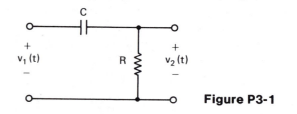

Figure P3-1

3-2 The circuit shown in Fig. P3-2 is a *series* resonant circuit, which represents one of the simplest forms of a band-pass filter. The input is $v_1(t)$ and the output is $v_2(t)$.

(a) Working directly in the frequency domain, determine the transfer function $H(f) = \overline{V}_2(f)/\overline{V}_1(f)$.

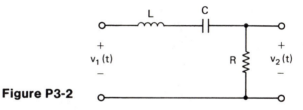

Figure P3-2

(b) Determine the amplitude response $A(f)$ and the phase response $\beta(f)$.

(c) The resonant frequency is defined as $f_0 = 1/(2\pi\sqrt{LC})$. Compute $A(f_0)$, $A(0)$, and $A(\infty)$ and comment on these values with respect to the filter's band-pass nature.

3-3   The circuit shown in Fig. P3-3 is a *parallel* resonant circuit, which represents one of the simplest forms of a band-pass filter. The input is $i(t)$ and the output is $v(t)$.

(a) Working directly in the frequency domain, determine the transfer function $H(f) = \overline{V}(f)/\overline{I}(f)$.

(b) Determine the amplitude response $A(f)$ and the phase response $\beta(f)$.

(c) The resonant frequency is defined as $f_0 = 1/(2\pi\sqrt{LC})$. Compute $A(f_0)$, $A(0)$, and $A(\infty)$ and comment on these values with respect to the filter's band-pass nature.

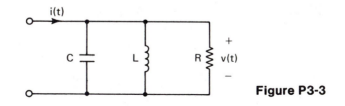

Figure P3-3

3-4 (a) Show that a differential equation of the form of Eq. (3-1) can be obtained for the circuit of Prob. 3-1 by writing a time-domain node voltage equation at the upper output node.

(b) Using the result of part (a), determine the transfer function.

3-5 (a) For the circuit of Prob. 3-2, write a time-domain mesh current equation in terms of a current variable $i(t)$. Show that by relating this current to the voltage $v_2(t)$ and differentiating all terms, a differential equation of the form of Eq. (3-1) can be obtained.

(b) Using the result of part (a), determine the transfer function.

3-6   For the circuit of Ex. 3-1, compute (a) the phase delay $T_p(f)$ and (b) the group delay $T_g(f)$.

3-7 For the circuit of Prob. 3-1, compute (a) the phase delay $T_p(f)$ and (b) the group delay $T_g(f)$. (*Note:* Any phase shift β_0 must be expressed in radians in order for β_0/ω to represent delay in seconds.)

3-8 A certain nearly ideal low-pass filter has a cutoff frequency just above 1 kHz. A square wave whose frequency can be adjusted is applied as the input to the filter. Using the results of Appendix C, predict which figure there would represent the closest approximation to the shape of the steady-state output of the low-pass filter for each of the following square wave frequencies: (a) 200 Hz, (b) 300 Hz, (c) 33 Hz, (d) 10 Hz, (e) 100 Hz.

3-9 A certain nearly ideal low-pass filter has an adjustable cutoff frequency. A square wave having a frequency of 1 kHz is applied as the input to the filter. Using the results of Appendix C, predict which figure there would represent the closest approximation to the shape of the steady-state output of the low-pass filter for each of the following filter cutoff frequencies (you may assume that the abrupt cutoff occurs just above the listed frequency): (a) 3 kHz, (b) 5kHz, (c) 10 kHz, (d) 30 kHz, (e) 100 kHz.

3-10 For the situation described in Prob. 3-9, sketch the form of the steady-state filter output for the following two filter cutoff frequencies: (a) 1.5 kHz, (b) 500 Hz. (*Note:* You will not find the results in Appendix C. Think about transmission through the filter from the Fourier point of view.)

3-11 The circuit of Fig. P3-11 represents a normalized passive low-pass three-pole Butterworth filter designed to be driven with an ideal voltage-source input. In this normalized form, the 3-dB frequency is 1 rad/s and a $1 - \Omega$ termination is used. (By procedures called frequency and impedance scaling, the circuit may be converted to any desired frequency and impedance levels.)

(a) For the normalized circuit given, determine the transfer function $H(f) = \overline{V}_2(f)/\overline{V}_1(f)$.

(b) Determine the amplitude response $A(f)$ and the phase response $\beta(f)$. Show that $A(f)$ is of the form of Eq. (3-50) with $m = 3$ and $f_c = 1/2\pi$ Hz.

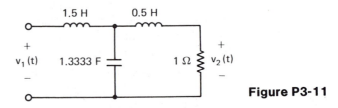

Figure P3-11

3-12 The circuit of Fig. P3-12 represents a normalized active low-pass two-pole Butterworth filter designed with a unity gain active voltage follower. (The active element eliminates the need for an inductor.) In this normalized form, the 3 dB frequency is 1 rad/s. (By procedures called frequency and

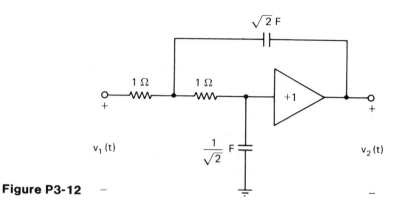

Figure P3-12

impedance scaling, the circuit may be converted to any desired frequency range and a more practical impedance level.)

(a) For the normalized circuit given, determine the transfer function $H(f) = \overline{V}_2(f)/\overline{V}_1(f)$.

(b) Determine the amplitude response $A(f)$ and the phase response $\beta(f)$. Show that $A(f)$ is of the form of Eq. (3-50) with $m = 2$ and $f_c = 1/2\pi$ Hz.

3-13 A certain baseband pulse generator produces pulses having widths of 0.5 μs. Determine the approximate baseband bandwidths for (a) coarse reproduction and (b) fine reproduction based on rise and fall times of 40 ns.

3-14 Assume that the pulses of Prob. 3-13 modulate an RF carrier, and the resulting signal is to be processed by high-frequency circuitry. Assuming the same values of pulse width and rise and fall times, determine the approximate band-pass bandwidths for (a) coarse reproduction and (b) fine reproduction.

3-15 A certain cable link is to be used to process digital data, which consists of a combination of pulses and spaces, representing binary ones and zeros. The amplitude response of the link is nearly flat from dc to 1 MHz, but it drops off rapidly above this frequency. Determine the approximate maximum data rate in bits per second that can be transmitted over this link. The bits are assumed to be transmitted in succession with no extra spaces inserted. (*Note:* For digital data, it is only necessary that the two levels be recognized at the receiver.)

3-16 For the cable link of Prob. 3-15, assume that some pulse position modulated signals are to be transmitted over the system. Assume that the rise time (or fall time) of each pulse cannot exceed 0.1% of the total time allocated to the pulse. Determine the approximate maximum data rate in pulses per second that would be permissible in this case. (*Note:* For pulse position modulated signals, it is necessary that the exact beginning and ending of each pulse be determined.)

3-17 An oscilloscope is to be selected to accurately measure some logic signals in a high-speed data system. The rise and fall times of the pulses are in the neighborhood of 10 ns. Suppose the measurement criteria are specified so that the rise time introduced by the oscilloscope should not exceed 10% of the rise time to be measured. Determine the approximate bandwidth required for the oscilloscope.

3-18 Determine the power or energy spectral density weighting function for the circuit of Prob. 3-1.

3-19 Determine the power or energy spectral density weighting function for the circuit of Prob. 3-2.

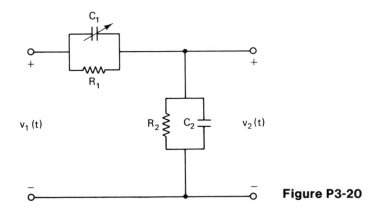

Figure P3-20

3-20 A circuit that can be used to partially compensate for the undesirable effects in input shunt capacitance, such as lead and vertical amplifier input capacitance in an oscilloscope, is shown in Fig. P3-20. The capacitance C_2 is the fixed circuit input capacitance, and C_1 is the adjustable probe capacitance. The resistances R_1 and R_2 are selected to form a fixed attenuation ratio.

    (a)   Determine the transfer function $H(f) = \overline{V}_2(f)/\overline{V}_1(f)$ in terms of arbitrary parameters.

    (b)   Show that if $R_1C_1 = R_2C_2$ the amplitude response $A(f)$ is a constant value independent of frequency.

    (c)   For the condition of part (b), assume that the resistors are selected such that a 10-to-1 attenuation results; that is, $A(f) = \frac{1}{10}$. Determine the relationship between C_1 and C_2 for this case.

Amplitude- Modulation Methods

4

4-0 INTRODUCTION

A short discussion concerning the purpose of modulation was given in Chapter 1. A brief listing of some of the modulation classifications was also presented. It was pointed out that most modulation processes involve a baseband message signal varying a higher frequency carrier in some manner that will allow more efficient transmission of the information.

The emphasis in this chapter will be directed toward a class of modulation techniques that will be collectively referred to as *amplitude-modulation* methods. In general, amplitude modulation will be defined as any process in which the instantaneous amplitude of a higher-frequency carrier is varied in accordance with the message signal. As we will see, there are several ways in which this process can be achieved with noticeable differences in generation, detection, efficiency, and performance.

The various modulation methods that can be properly classified under the heading of amplitude modulation include the following:

1. Conventional amplitude modulation (usually called simply AM).
2. Double sideband.
3. Single sideband.
4. Vestigial sideband.

Note that the term "amplitude modulation" is being used here to refer to both the general collection of methods under consideration as well as one particular method within the collection. It will usually be clear in which context the term is being used. However, we will normally use the term "conventional amplitude modulation" for the particular method usually referred to as AM.

While the various forms of amplitude modulation are important in themselves, the reader should bear in mind that some of the operations involved also frequently occur in conjunction with other modulation methods. Thus, it is difficult to understand fully any of the various modulation procedures in the remainder of the book without first having a reasonably good understanding of amplitude modulation in both the time domain and the frequency domain.

4-1 APPROACH AND TERMINOLOGY

The general approach and assumptions that will be made in this chapter as well as in other sections of the book for explaining various modulation and demodulation operations will now be discussed. The analysis of a given modulation or demodulation operation begins with the assumption of an input signal of the proper form. The operation of the modulator or demodulator is then represented by a certain mathematical process, and this function is applied to the input signal. The resulting signal is then analyzed in terms of its closeness to the desired output.

Both the time-domain and frequency-domain forms of the input and output signals of a modulator or demodulator are important. The time-domain forms are important because they display the instantaneous levels of signals that must be processed by the circuitry. Such waveforms can be displayed on an oscilloscope during hardware development, alignment, and testing, so the communications engineer or technologist should learn to recognize the various time-domain waveforms. The frequency-domain forms are important because they predict the exact frequency ranges involved, the relative magnitudes of the various spectral terms, and the bandwidth required. Spectrum analyzers are frequently used in testing communications systems, so it is important to be able to predict the proper forms in the frequency-domain.

In analyzing or explaining the various modulation or demodulation operations, it is desirable to make some simplified assumptions concerning the various signal and impedance levels in the circuits. Such assumptions allow the major operation of the particular function to be emphasized without the added "clutter" of various constants and odd signal and/or impedance levels, and the like. For that reason, many of the developments in the text will assume convenient signal levels such as unity, as well as normalized impedance levels of $1 \, \Omega$. (This latter concept has already been used in earlier chapters.) The reader should bear in mind the reason for doing this and should understand that in actual equipment the signal levels could be quite different.

An important consideration is the assumed form of the modulating signal that should be used in analyzing the operation of a modulator. In a sense, one can assume either a nonperiodic signal or a periodic signal. If a nonperiodic signal is assumed, the Fourier transform must be used to predict the spectrum. Conversely, if a periodic signal is assumed, the Fourier series can be used to predict the spectrum.

It is this writer's opinion that anyone dealing seriously with communication system level considerations should learn to work with both forms. It turns out that some operations are more easily explained or interpreted from the Fourier transform point of view, while others are more easily handled by the Fourier series point of view. Maximum flexibility is possible when one possesses the capability of dealing with both forms whenever possible.

To provide maximum insight, several of the developments in this chapter will be done with both the Fourier transform and Fourier series forms. Readers having limited background with Fourier transforms will be able to bypass portions of those developments, although it is strongly recommended that the graphical forms of the time-domain and frequency-domain plots be carefully surveyed and noted, as a minimum.

To establish a simplified form of reference, whenever a nonperiodic time signal requiring the use of a Fourier transform in the frequency domain is assumed, we will refer to the signal as a *continuous-spectrum signal*. Whenever a periodic signal requiring the use of a Fourier series representation is used, we will refer to the signal as a *discrete-spectrum signal*.

An arbitrary example of a continuous spectrum signal $x(t)$ and an assumed amplitude spectrum $X(f)$ are shown in Fig. 4-1. An example of a discrete-spectrum signal $x(t)$ and its assumed amplitude spectrum X_n (two-sided in this case) are shown in Fig. 4-2. The signals are both baseband in nature and are band-limited from dc to W hertz. (The symbol W has long been used in many communications theory references to represent the highest baseband cyclic frequency, in hertz, for a modulating signal, so we will use that convention throughout the book.) Observe that for the discrete-spectrum signal the frequency W occurs at a discrete frequency; that is, $W = Nf_1$, where N is an integer representing the order of the highest harmonic.

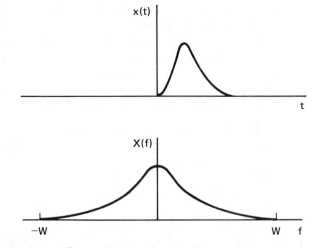

Figure 4-1.   Example of a signal with a continuous spectrum.

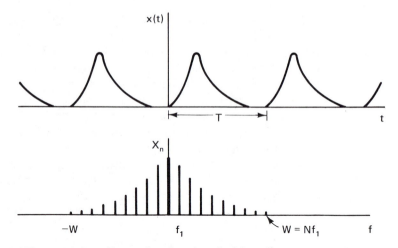

Figure 4-2. Example of a signal with a discrete spectrum.

The signals and their assumed spectra shown in Figs. 4-1 and 4-2 are not intended to represent any actual transform pairs, but the forms given were chosen simply to illustrate the concepts involved. Throughout the book, in general developments simplified forms for time-domain and spectral forms will be used to illustrate the nature of the phenomena involved, and in many cases the functions themselves will not necessarily represent any actual transform pairs. Of course, when specific functions are being considered, it will be clear that the results in those cases provide actual transform pairs.

As a final point, which may seem initially confusing, but which turns out to make the spectral descriptions easier to follow in some cases, we will sometimes actually show a continuous-spectrum display even when the assumed form of the signal is a Fourier series. The reason for this seemingly contradictory choice is very simple: It is easier to display a continuous curve than a number of separate spectral lines, and the result is often easier to interpret, particularly when there is a large number of components in the spectrum. When this is done, the reader should carefully recognize that it is done only as a visual convenience.

With the preceding concepts in mind, the general form of the notation that will be used in subsequent sections is defined as follows:

$x(t)$ = modulating or message signal at transmitter (usually at baseband)

$y(t)$ = modulated signal at transmitter (spectrum usually shifted to higher frequencies)

$y_r(t)$ = modulated signal at receiver

$x_d(t)$ = detected signal at receiver, which should be proportional to $x(t)$ at the transmitter

f_c = carrier frequency in hertz

ω_c = carrier frequency expressed in radians per second = $2\pi f_c$

$\cos \omega_c t$ = carrier reference in most developments

W = highest frequency in $x(t)$ or baseband bandwidth in hertz

B_T = transmission bandwidth of the modulated signal in hertz

Where it is necessary to have several steps in the development of a signal, various subscripts may be attached to these terms. However, a key point in the notation is that a function defined with the symbol x will usually be at baseband frequencies, while a function defined with the symbol y will usually have a spectrum that is shifted to a higher frequency range.

Several trigonometric identities will frequently arise in conjunction with modulation operations, particularly for the case of discrete spectra, and they will be summarized here:

$$\cos A \cos B = \tfrac{1}{2} \cos (A - B) + \tfrac{1}{2} \cos (A + B) \qquad (4\text{-}1)$$

$$\sin A \sin B = \tfrac{1}{2} \cos (A - B) - \tfrac{1}{2} \cos (A + B) \qquad (4\text{-}2)$$

$$\sin A \cos B = \tfrac{1}{2} \sin (A - B) + \tfrac{1}{2} \sin (A + B) \qquad (4\text{-}3)$$

In these expressions, A and B may each involve combinations of a carrier frequency angle such as $\omega_c t$, as well as arbitrary phase shifts. The operations occur when products of sinusoidal functions are formed and in which it is desired to expand such products as a series of single frequency terms. This concept arises repeatedly in analyzing modulator circuits. It is recommended that the reader either commit these results to memory or have them readily available for use in the remainder of the book.

4-2 DOUBLE-SIDEBAND MODULATION

We will begin the study of amplitude modulation methods by the analysis of *double-sideband (DSB)* modulation. Although DSB has certain disadvantages in usage, it is the easiest to analyze and will serve as the starting point for discussion.

A DSB signal can be generated by multiplying the baseband modulating signal $x(t)$ by a high-frequency carrier $\cos \omega_c t$. Multiplier circuits designed for modulation applications are readily available as off-the-shelf components. In addition, a number of other circuits may be adapted for this purpose, as will be discussed later in the chapter.

A multiplying device used to generate a DSB signal is called a *balanced modulator*. A block diagram of the functional relationship of the balanced modulator is shown in Fig. 4-3. In practice, some filtering is required at the output to reject spurious components resulting from imperfections in the output multiplication process, but for the moment an ideal multiplication will be assumed.

The instantaneous output signal $y(t)$ of the ideal balanced modulator is the product of the message signal $x(t)$ and the carrier reference $\cos \omega_c t$ and is

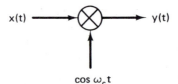

Figure 4-3. Block diagram form for a balanced modulator.

given by

$$y(t) = x(t) \cos \omega_c t \qquad (4\text{-}4)$$

where the carrier is normalized to unity magnitude for convenience. The function $y(t)$ is a DSB signal. The significance will be observed by first assuming that $x(t)$ is a continuous-spectrum signal and taking the Fourier transforms of both sides. Expressing the cosine function as a combination of exponential terms and applying the modulation theorem of Chapter 2, we have

$$y(t) = \tfrac{1}{2} x(t) \, \epsilon^{j2\pi f_c t} + \tfrac{1}{2} x(t) \, \epsilon^{-j2\pi f_c t} \qquad (4\text{-}5)$$

and

$$\overline{Y}(f) = \tfrac{1}{2} \overline{X}(f - f_c) + \tfrac{1}{2} \overline{X}(f + f_c) \qquad (4\text{-}6)$$

The process involved is illustrated in Fig. 4-4. An arbitrary modulating signal $x(t)$ is shown in Fig. 4-4a and its amplitude spectrum $X(f)$ is assumed to be of the form shown in Fig. 4-4b. Note that the baseband spectrum is assumed to be band-limited from near dc to W. (One would certainly be proper to say from $-W$ to $+W$, but the negative frequency range is automatically understood in the mathematics, so it is more customary to express the actual values in terms of the "real-world" frequency range of dc to W.)

The DSB signal corresponding to (4-4) is shown in Fig. 4-4c, and its spectrum corresponding to (4-6) is shown in Fig. 4-4d. The first term in (4-6) represents a translation or shift of the original spectrum to the right by f_c hertz and is the part of the spectrum shown in Fig. 4-4d whose center appears at f_c. Conversely, the second term in (4-6) represents a translation of the original spectrum to the left by f_c hertz and is the portion centered at $-f_c$. Concentrating momentarily on the positive frequency range, it is observed that the message spectrum appears on *both* sides of the carrier frequency f_c. Hence, the term *double-sideband* is quite appropriate. The portion above f_c is called the *upper sideband (USB)*, and the portion below f_c is called the *lower sideband (LSB)*. Note, however, that the carrier at f_c does not appear in the spectrum at all. This may seem a little strange considering that the composite signal is oscillating at a rate of f_c hertz. However, the successive phase reversals at the carrier rate caused by zero crossings of the envelope, as may

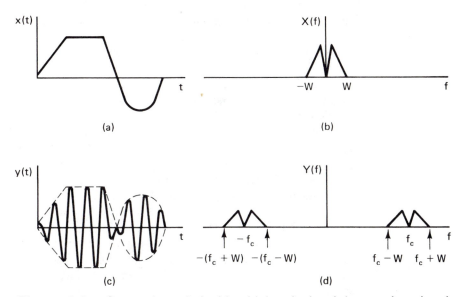

Figure 4-4. Generation of double-sideband signal from a baseband signal and the corresponding spectra.

be observed in Fig. 4-4c, result in a complete cancellation of the carrier in the ideal case. In practical balanced modulators, a small carrier component can usually be measured, but it is typically 40 dB or more below the sideband levels in well-designed modulators. In view of the carrier cancellation, this method is often called *suppressed-carrier* amplitude modulation. However, the shorter term, double-sideband, will be used throughout this book.

One erroneous interpretation of Eq. (4-6) describing the spectrum will now be noted. There is a tendency by some persons to label one of the two terms in (4-6) as the lower sideband and to label the other term as the upper sideband. This interpretation is *not correct*. The first term in (4-6) represents the portion of the spectrum shifted to the right, which in this case corresponds to positive frequencies in $Y(f)$, and the second term represents the portion of the spectrum shifted to the left, which in this case corresponds to negative frequencies. Each term contains a portion of the lower sideband and a portion of the upper sideband. An equation expressing only the lower sideband or the upper sideband is more difficult to formulate for the continuous-spectrum case and will not be considered at this time.

The result of the modulation operation is that the low-frequency baseband signal has been shifted upward in frequency and is now centered at f_c rather than at dc. Since f_c is quite arbitrary, it could very well be in the radio-frequency (RF) range. (Unfortunately, the scale of the figure does not permit us to illustrate typical frequency shifts, which could be, for example, several orders of magnitude times the bandwidth of the baseband signal.) By shifting the signal to the RF range, it is possible to take advantage of electromagnetic radiation to propagate the signal.

One point about the nature of the spectral shifting observation should be made. We discuss the concepts of two-sided and one-sided spectra a number of times throughout the book. In a sense, the definitions are arbitrary in that negative frequencies are used more for mathematical completeness than physical significance. On the other hand, with the continuous-spectrum modulation operation just developed, it can be seen that the upper sideband of the modulated signal for $f > 0$ results from shifting the positive frequency portion of the modulating signal to the right, while the lower sideband for $f > 0$ can be thought of as arising from the negative frequency portion of the baseband modulating signal shifted to the right. It is possible to develop a mathematical model to predict the same result from a one-sided continuous spectrum, but the result is not nearly as convenient.

The point of the preceding discussion is that when modulation operations are being performed on an arbitrary continuous spectrum $\overline{X}(f)$ and use is being made of the modulation theorem as given by (4-6), the two-sided spectrum should be employed before the shifting is performed. Some of the positive frequency components that appear in the final spectrum will seem to arise from negative frequencies in the original spectrum, but this is a result of the mathematical symmetry of the result and is to be expected.

Observe that while the amplitude level has changed, the *shape* of the spectrum is preserved perfectly by the translation process. As a result, this technique is sometimes called *linear modulation.* Depending on the application, various other terms have been employed to describe the modulation operation just discussed. Some of these terms are *frequency translation, frequency shifting, frequency conversion,* and *heterodyning,* depending on the specific application. With respect to the *frequency conversion* designation, the term *up-conversion* is used to describe the process of shifting to a higher frequency range (as was the case in Fig. 4-4), and *down-conversion* refers to the opposite type of shift (to be considered later). Finally, the term *mixing* is often used in place of the term *modulation,* and some modulator circuits are denoted as *mixers.* Some of these various terms will arise in specific applications later in the book.

Observe from Fig. 4-4c that the envelope of the DSB signal bears a direct relationship to the form of the modulating signal, but the relationship is clear to us only because we can see both signals at the same type. Because of the crossovers that occur, it is not possible to recover the intelligence from either the positive or the negative envelope of the DSB signal. This is an important distinction between DSB and conventional AM and will be discussed in more detail later.

Having considered the continuous-spectrum case, we now will investigate the nature of the modulation operation with an assumed discrete-spectrum signal. Assume a message signal $x(t)$ of the form

$$x(t) = \sum_{n-1}^{N} C_n \cos (n\omega_1 t + \theta_n) \qquad (4\text{-}7)$$

which represents a conventional Fourier series representation with the coefficients C_n representing the amplitudes of the spectral terms on a one-sided basis. For reasons that will be clearer later, a dc term has been omitted in (4-7) and the highest frequency is assumed to be $W = Nf_1$.

When the signal of (4-7) is applied to the balanced modulator, the output $y(t)$ is

$$y(t) = \cos \omega_c t \sum_{n-1}^{N} C_n \cos (n\omega_1 t + \theta_n) \qquad (4\text{-}8a)$$

$$= \sum_{n-1}^{N} C_n \cos \omega_c t \cos (n\omega_1 t + \theta_n) \qquad (4\text{-}8b)$$

The identity of (4-1) can be applied to each of the terms in the series of (4-8b), and the total number of terms is doubled. The result is

$$y(t) = \frac{1}{2} \sum_{n-1}^{N} C_n \cos [(\omega_c - n\omega_1)t - \theta_n] \qquad (4\text{-}9)$$

$$+ \frac{1}{2} \sum_{n-1}^{N} C_n \cos [(\omega_c + n\omega_1)t + \theta_n]$$

which describes a DSB discrete-spectrum signal.

The effect of the $\frac{1}{2}$ factor is to multiply all the terms by a common multiplier, so it does not change the relative sizes of the components. Thus, we will momentarily ignore its effect. It is observed that the frequencies of the components in the first series of (4-9) start at $f_c - f_1$ and appear at successive line frequencies down to $f_c - Nf_1 = f_c - W$. Ignoring the $\frac{1}{2}$ common factor, the relative magnitudes of the components over this range vary from C_1 at $f_c - f_1$ to C_N at $f_c - W$. This part of the modulated function represents the lower sideband and is "inverted" in frequency from the original spectrum. The frequencies of the components in the second series of (4-9) start at $f_c + f_1$ and appear at successive line frequencies up to $f_c + Nf_1 = f_c + W$. The relative magnitudes over this range vary from C_1 at $f_c + f_1$ to C_N at $f_c + W$. This part of the modulated function represents the upper sideband. Thus, unlike the case of the continuous spectrum, we can readily separate the upper and lower sidebands from the expression of (4-9). If we let $y_l(t)$ represent the lower sideband and $y_u(t)$ represent the upper sideband, we have

$$y_l(t) = \frac{1}{2} \sum_{n-1}^{N} C_n \cos [(\omega_c - n\omega_1)t - \theta_n] \qquad (4\text{-}10)$$

and

$$y_u(t) = \frac{1}{2} \sum_{n-1}^{N} C_n \cos [(\omega_c + n\omega_1)t + \theta_n] \qquad (4\text{-}11)$$

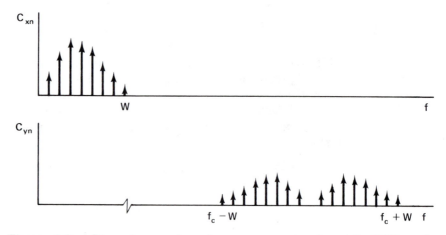

Figure 4-5. Discrete spectrum for message signal and its DSB modulated form.

While the figures already used for an assumed continuous spectrum can be applied directly to this case, we will choose a different form here for illustrative purposes. In this case, we will emphasize the line structure for both the baseband spectrum and the DSB spectrum, as shown in Fig. 4-5. The input and output one-sided spectra of the modulator are designated as C_{xn} and C_{yn}, respectively.

For both the continuous- and the discrete-spectrum cases, it can be readily observed that the minimum transmission bandwidth required for a DSB signal is

$$B_T = 2W \qquad (4\text{-}12)$$

Thus, the minimum transmission bandwidth for a DSB signal is twice the highest modulating frequency.

Example 4-1

A carrier with a frequency of 100 kHz is modulated in an ideal balanced modulator by the multitone signal

$$x(t) = 10 \cos 2\pi \times 10^3\, t + 8 \cos 4\pi \times 10^3\, t + 6 \cos 8\pi \times 10^3\, t \qquad (4\text{-}13)$$

(a) Using only simple arithmetic, list the frequencies (in hertz) appearing in the output of the modulator. (b) Develop an expression for the DSB output $y(t)$. Assume a cosine function with a normalized amplitude of unity for the carrier. (c) Plot the one-sided spectra for both the input and output.

Solution

(a) Expressed in hertz, the modulating signal contains three components: 1 kHz, 2 kHz, and 4 kHz. The output frequencies of the balanced modulator will be all sum and difference frequencies obtained from the combination of these three frequencies with the 100 kHz carrier component. The resulting values are summarized as follows:

USB: 101, 102, 104 kHz
LSB: 99, 98, 96 kHz

This part illustrates how the output frequencies can be determined very quickly by simply forming the sums and differences between the carrier frequency and the modulating frequencies. In some applications, these results may be all that is required. However, this simplified process does not tell us the relative magnitudes and/or phases associated with the spectrum. A more lengthy analysis is required for that purpose, as will be demonstrated shortly.

(b) The form of the carrier is $\cos 2\pi \times 10^5 t$. The output $y(t)$ is obtained by forming the product of $x(t)$ and the carrier. The result is expanded as follows:

$$y(t) = (10 \cos 2\pi \times 10^3 t + 8 \cos 4\pi \times 10^3 t \qquad (4\text{-}14)$$
$$+ 6 \cos 8\pi \times 10^3 t) \cos 2\pi \times 10^5 t$$

Each term is thus multiplied by the carrier. Application of (4-1) to all these products produces a sum and difference cosine function in each case. The result will be expressed in a symmetrical pattern about the carrier frequency. We have

$$y(t) = 5 \cos 2\pi \times 99 \times 10^3 t + 4 \cos 2\pi \times 98 \times 10^3 t \qquad (4\text{-}15)$$
$$+ 3 \cos 2\pi \times 96 \times 10^3 t + 5 \cos 2\pi \times 101 \times 10^3 t$$
$$+ 4 \cos 2\pi \times 102 \times 10^3 t + 3 \cos 2\pi \times 104 \times 10^3 t$$

(c) A plot of the one-sided spectrum C_{xn} for the input is shown in Fig. 4-6a, and the output one-sided spectrum C_{yn} is shown in Fig. 4-6b. For the assumed value of unity for the carrier magnitude, the magnitudes of the output spectral components are half the values of the corresponding input components.

4-3   SINGLE-SIDEBAND MODULATION

We observed in the preceding section that when the product of a baseband message signal and a higher-frequency carrier is generated in a balanced modulator, a double-sideband signal is obtained at the output. This process is

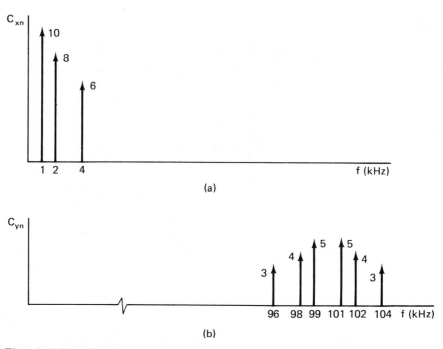

Figure 4-6. One-sided spectra for message signal and DSB modulated signal in Example 4-1.

reviewed in Fig. 4-7a and b. (For convenience, the translated components are shown with the same level as the baseband signal spectrum.) Due to the symmetry of the spectrum about f_c, it seems apparent that either of the two sidebands contains all the spectral information associated with the message. Of course, the lower sideband is inverted in direction, but, as we shall see later, the signal can be inverted again at the receiver.

The basic strategy in *single sideband (SSB)* is to eliminate one of the sidebands and transmit only one sideband. This concept is illustrated in Fig. 4-7d for the case in which the upper sideband (USB) is transmitted and in Fig. 4-7f for lower sideband (LSB). (The functions in Fig. 4-7c and e represent ideal filter functions that will be discussed shortly.) Observe that the transmission bandwidth is simply

$$B_T = W \qquad (4\text{-}16)$$

Thus, the bandwidth for SSB transmission is exactly the same as the baseband bandwidth. This is the smallest possible transmission bandwidth that can be achieved for any modulation method without resorting to certain complex bandwidth compression schemes, which are rather specialized and out of the context of the present discussion.

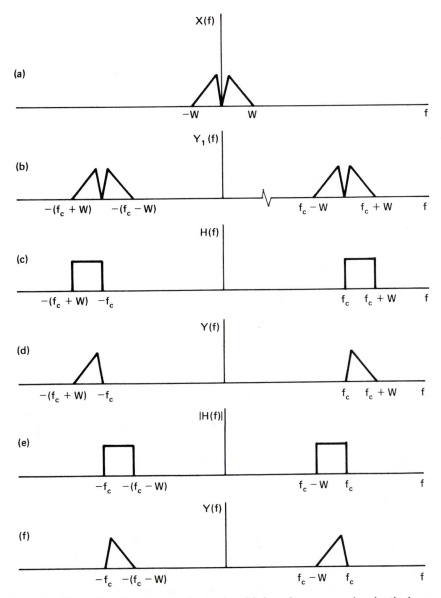

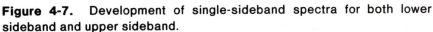

Figure 4-7. Development of single-sideband spectra for both lower sideband and upper sideband.

While there are several ways of producing an SSB signal, the most common approach, and by far the simplest to understand, is the *filter method,* which is illustrated in Fig. 4-8. The notation on this figure correlates with that of Fig. 4-7. A DSB signal $y_1(t)$ is first generated in a balanced modulator. A bandpass filter having a bandwidth W and a passband aligned exactly with

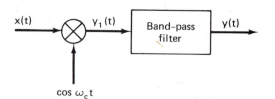

Figure 4-8. Block diagram of SSB generator using the filter method.

that of the desired sideband is used to pass the appropriate sideband and to reject the other sideband. Ideal block filters are illustrated in Fig. 4-7c and e. Thus, $y(t)$ at the output of the filter is an SSB signal.

The perceptive reader might wonder if a high-pass filter would be sufficient in the case of USB, and if a low-pass filter would be sufficient in the case of LSB. In theory, such filters would produce the desired results. However, in practice a bandpass filter is usually employed in order to eliminate additional spurious components that might be produced by imperfections in the balanced modulator.

The sideband filter is one of the most complex components in an SSB generator from a design point of view. The primary reason is the fact that for a perfect SSB signal it is necessary to pass one sideband completely and reject the other. Observe from Fig. 4-7 that if the message has components down near dc there is very little guard band for the filter to provide an appropriate transition region between the passband and the stopband. For example, suppose it were desired to preserve frequencies in the message signal down to 50 Hz. The lowest frequency in the upper sideband would then be 100 Hz higher than the highest frequency in the lower sideband. A band-pass filter with a transition region of 100 Hz operating in the RF range is no simple chore to design and adjust!

The actual implementation of sideband filters is eased by two approaches: (1) the SSB signal is usually generated at a relatively low frequency below the RF range, where it is easier to attain the required transition region. The modulated signal is then translated by one or more additional frequency shifts to the proper required frequency range. (2) Wherever practical, the lowest message signal frequency is set at as high a frequency as intelligibility and accuracy of reproduction will permit, thus easing the filtering burden. For example, if voice transmission is employed and perfect fidelity is not a major criterion, it may be possible to simply eliminate some of the very lowest frequencies in the modulator and still have intelligible conversations.

The equation of an SSB signal modulated by a continuous-spectrum message signal can only be expressed through the use of an operation called the Hilbert transform, which is not within the intended coverage of this book. However, the assumption of a discrete spectrum leads to an appropriate expression. In fact, we can readily make use of the results of the preceding section for the DSB signal. Assume that the message signal is

$$x(t) = \sum_{n=1}^{N} C_n \cos (n\omega_1 t + \theta_n) \qquad (4\text{-}17)$$

Adapting the results of (4-10) and (4-11), a USB signal is given by

$$\text{USB:} \quad y(t) = \frac{1}{2} \sum_{n=1}^{N} C_n \cos [(\omega_c + n\omega_1)t + \theta_n] \qquad (4\text{-}18)$$

An LSB signal is expressed by

$$\text{LSB:} \quad y(t) = \frac{1}{2} \sum_{n=1}^{N} C_n \cos [(\omega_c - n\omega_1)t - \theta_n] \qquad (4\text{-}19)$$

In both (4-18) and (4-19) the phase shift and any amplitude variation produced by the filter on the spectrum have been ignored at this time since they do not contribute to the present discussion. However, such effects may have to be considered in a complete system analysis.

Example 4-2

Consider the system described in Ex. 4-1, and assume that an SSB signal is to be generated. (a) For USB, list the frequencies present, and write an equation for $y(t)$. Ignore the phase shift produced by the sideband filter. (b) Repeat part (a) for LSB.

Solution

(a) For USB, the frequencies are simply the sum of the carrier frequency and the three frequencies present in $x(t)$. The resulting frequencies are

101, 102, 104 kHz

Assuming a perfect bandpass filter and no phase shift, we have

$$y(t) = 5 \cos 2\pi \times 101 \times 10^3 \, t + 4 \cos 2\pi \times 102 \times 10^3 \, t \qquad (4\text{-}20)$$
$$+ \, 3 \cos 2\pi \times 104 \times 10^3 \, t$$

(b) For LSB, the frequencies are

99, 98, 96 kHz

With the same assumptions as in part (a), the result is expressed as

$$y(t) = 5 \cos 2\pi \times 99 \times 10^3 \, t + 4 \cos 2\pi \times 98 \times 10^3 \, t \qquad (4\text{-}21)$$
$$+ \, 3 \cos 2\pi \times 96 \times 10^3 \, t$$

The one-sided input spectrum C_{xn} and the corresponding upper and lower sideband spectra C_{yn} are shown in Fig. 4–9.

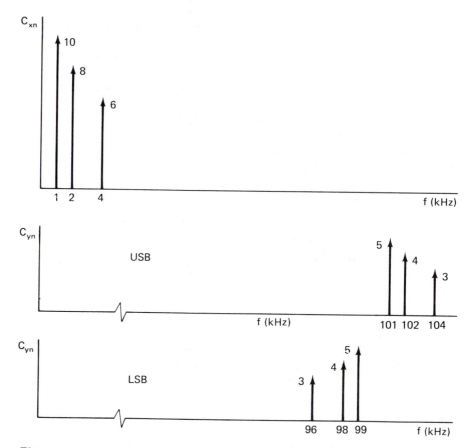

Figure 4-9. One-sided spectra for message signal and USB and LSB modulated signals in Example 4-2.

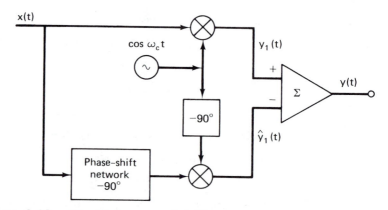

Figure 4-10. Block diagram of SSB generator using the phase-shift method as discussed in Example 4-3.

Example 4-3

The system shown in Fig. 4-10 represents the *phase-shift method* for generating an SSB signal. The audio phase shift-network is assumed to shift all components in the signal $x(t)$ by a constant $-90°$, while one component of the carrier is also shifted in phase by $-90°$. The summing circuit is a *differential* summer, which forms $y_1 - \hat{y}_1$. Beginning with a discrete spectrum of the form of Eq. (4-7), prove that the output $y(t)$ is a USB SSB signal.

Solution

The message signal is assumed to be of the form

$$x(t) = \sum_{n=1}^{N} C_n \cos (n\omega_1 t + \theta_n) \tag{4-22}$$

This signal is multiplied by $A \cos \omega_c t$ in the upper balanced modulator to yield $y_1(t)$. This function is a DSB signal and is identical in form with the result of Eq. (4-9). Hence,

$$y_1(t) = \frac{1}{2} \sum_{n=1}^{N} C_n \cos [(\omega_c - n\omega_1)t - \theta_n] \tag{4-23}$$

$$+ \frac{1}{2} \sum_{n=1}^{N} C_n \cos [(\omega_c + n\omega_1)t + \theta_n]$$

The audio phase-shift network is a special circuit that shifts the phase of all components of $x(t)$ by $-90°$. Calling this output $\hat{x}(t)$, we have

$$\hat{x}(t) = \sum_{n=1}^{N} C_n \cos (n\omega_1 t + \theta_n - 90°) \tag{4-24}$$

$$= \sum_{n=1}^{N} C_n \sin (n\omega_1 t + \theta_n)$$

where the identity $\cos (\phi - 90°) = \sin \phi$ has been employed. Since the lower carrier component is also shifted by $-90°$, the other input to the lower balanced modulator is $\sin \omega_c t$. The output $\hat{y}_1(t)$ of this balanced modulator is

$$\hat{y}_1(t) = \sin \omega_c t \sum_{n=1}^{N} C_n \sin (n\omega_1 t + \theta_n) \tag{4-25}$$

All the terms in the series can be multiplied by the carrier factor in front. By use of Eq. (4-2), each product term can be expanded into two separate terms. The result is readily expanded as

$$\hat{y}_1(t) = \frac{1}{2} \sum_{n=1}^{N} C_n \cos [(\omega_c - n\omega_1)t - \theta_n] \tag{4-26}$$

$$- \frac{1}{2} \sum_{n=1}^{N} C_n \cos [(\omega_c + n\omega_1)t + \theta_n]$$

The output differential amplifier generates $y(t) = y_1(t) - \hat{y}_1(t)$. Substitution of (4–23) and (4–26) into this function results in

$$y(t) = \sum_{n=1}^{N} C_n \cos \left[(\omega_c + n\omega_1)t + \theta_n \right] \qquad (4\text{-}27)$$

This result is readily seen to be an USB SSB signal. It will be left as an exercise for the reader to show that an LSB SSB signal is generated for either of the following changes in the system: (1) Either (but not both) of the two phase-shift networks provides $+90°$ rather than $-90°$. (2) The two functions $y_1(t)$ and $\hat{y}_1(t)$ are added at the output. (See Prob. 4-7.)

A few comments on this method are in order. While the phase-shift method has been used in some applications, its major drawback is the difficulty in designing a network that will provide a constant $-90°$ (or $+90°$) phase shift over a wide range of frequencies. The wider the bandwidth of the modulating signal, the more difficulty there is in designing such a network. No linear network is capable of producing such a phase shift exactly over a band of frequencies, but a number of rather complex networks have been developed that approximate the desired result. The required phase shift at the carrier frequency, however, is easy to achieve since this represents only one frequency.

Along with the filter method and the phase-shift method, there is still a third method of SSB generation called the Weaver method, whose analysis is left as an exercise for the reader (Prob. 4-27).

4-4 PRODUCT DETECTION OF DSB AND SSB

In the previous two sections, we have seen how DSB and SSB signals could be generated with a balanced modulator and associated circuitry. At the receiving end, it is necessary to perform an inverse process on the composite signal in order to recover the desired baseband message signal. The signal processing required to extract the original baseband signal from the composite modulated signal is called *detection* or *demodulation*.

The detection process normally used for DSB and SSB signals is referred to as *product detection*. Product detection is achieved by mixing a carrier generated at the receiver with the incoming modulated signal in a multiplier circuit followed by low-pass filtering. If the carrier used at the receiver is locked exactly in phase and frequency with the carrier used for generating the signal, the product-detection process is also referred to as *synchronous* or *coherent* detection. As we will see shortly, the circuitry required for detection is essentially of the same form as required for generating the signal.

As a first step, we will consider ideal synchronous or coherent detection. Consider a DSB signal appearing at the receiver having the form

$$y_r(t) = Ay(t) = Ax(t) \cos \omega_c t \qquad (4\text{-}28)$$

where A is a constant indicating that the signal level is different than at the transmitter. To keep the mathematical development as simple as possible, the time scale has been adjusted in (4-28) so that it has the same form as the signal generated at the transmitter as given by (4-4). Where it is necessary to maintain an exact reference in time with the transmitted signal, the time scale in (4-28) would have to be shifted. We will sidestep this point in subsequent developments and use the simplified time scale form as in (4-28).

Assume that the received signal is applied as one input to an ideal multiplier circuit, as shown in Fig. 4-11, and assume that a local carrier of the form $\cos \omega_c t$ is applied as the other input. Note that both *frequency* and *phase* of the local carrier are assumed to be *exactly* the same, a requirement for synchronous or coherent detection. The practical significance of this assumption will be discussed later. The oscillator used at the receiver for injecting the mixing carrier is often referred to as the *beat frequency oscillator (BFO)* in many communications receivers.

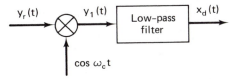

Figure 4-11.   Block diagram of product detector employing coherent reference.

The output $y_1(t)$ of the multiplier is

$$y_1(t) = y_r(t) \cos \omega_c t = Ax(t) \cos^2 \omega_c t \qquad (4\text{-}29)$$

Since $\cos^2 \theta = (1 + \cos 2\theta)/2$, we can expand 4-29) as

$$y_1(t) = \frac{A}{2} x(t) + \frac{A}{2} x(t) \cos 2\omega_c t \qquad (4\text{-}30)$$

Observe that the first term of (4-30) is the baseband message signal multiplied by a constant. For the present discussion, the constant is of little importance, since the signal probably has undergone a significant change in level since it left the transmitter. The point is that the term is proportional to the message signal $x(t)$.

The amplitude spectra $Y_r(f)$ and $Y_1(f)$ are shown in Fig. 4-12a and b. Along with the desired message, $Y_1(f)$ also contains a component representing the DSB product of the intelligence modulating the second harmonic of the carrier. Provided that $f_c \gg W$, which is usually the case, this latter component is easily removed with a low-pass filter. The detected output $x_d(t)$ of the low-pass filter is then

$$x_d(t) = \frac{A}{2} x(t) \qquad (4\text{-}31)$$

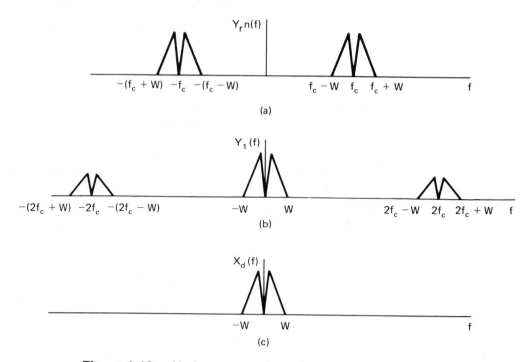

Figure 4-12. Various spectral terms arising in the detection process of a DSB signal (relative levels based on $A = 1$).

where the phase shift of the filter has not been considered. The form of the output amplitude spectrum $X_d(f)$ is shown in Fig. 4-12c. The magnitudes of the various components were selected to illustrate the *relative* sizes. The exact levels will depend on the constant A.

We will now consider the detection of an SSB signal using a multipler. Due to the difficulty in expressing the equation of an SSB signal for a continuous-spectrum message, we will deal exclusively with the discrete-spectrum form for analysis purposes in this case. We will assume a USB signal as an arbitrary choice, although the result applies equally well to either case. Consider then the USB signal described by the form of Eq. (4-18) with a constant multiplier A. We have

$$y_r(t) = Ay(t) = \frac{A}{2} \sum_{n=1}^{N} C_n \cos\left[(\omega_c + n\omega_1)t + \theta_n\right] \qquad (4\text{-}32)$$

The local carrier is assumed to be of the form $\cos \omega_c t$, as shown in Fig. 4-11. The output $y_1(t)$ of the multiplier is

$$y_1(t) = \frac{A}{2} \cos \omega_c t \sum_{n=1}^{N} C_n \cos\left[(\omega_c + n\omega_1)t + \theta_n\right] \qquad (4\text{-}33)$$

$$= \frac{A}{2} \sum_{n=1}^{N} C_n \cos \omega_c t \cos\left[(\omega_c + n\omega_1)t + \theta_n\right]$$

Each term in (4-33) may be expanded by (4-1), and the result may be divided into two series as follows:

$$y_1(t) = \frac{A}{4} \sum_{n=1}^{N} C_n \cos (n\omega_1 t + \theta_n) \tag{4-34}$$

$$+ \frac{A}{4} \sum_{n=1}^{N} C_n \cos [(2\omega_c + n\omega_1)t + \theta_n]$$

The first series of $y_1(t)$ is a constant times the message signal, as can be noted by comparing this series with Eq. (4-17). The second series represents an SSB signal located above $2f_c$. This last series is easily removed by filtering provided that $f_c \gg W$. The detected signal $x_d(t)$ is then of the form

$$x_d(t) = \frac{A}{4} x(t) \tag{4-35}$$

where the phase shift in the filter has been ignored. The message signal is thus recovered by the synchronous detector. The various steps involved in the detection process are illustrated in Fig. 4-13 with a *continuous*-spectrum signal shown for illustration.

Comparing Figs. 4-12 and 4-13, it is noted that the *relative* magnitude of the recovered spectrum as compared with the input spectrum appears to be twice as large for DSB as for SSB if all other factors are equal. The reason for this is that with DSB two translated sidebands overlap and add to each other coherently, while for SSB, there is only one component in each segment. The result is that, for a given signal level, DSB produces a larger detected signal

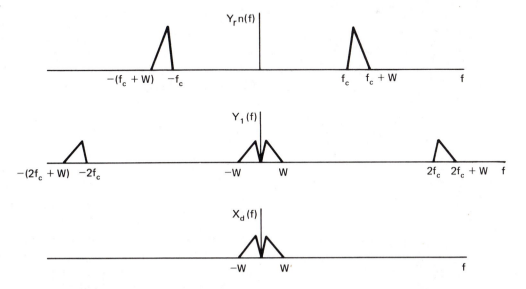

Figure 4-13.   Various spectral terms arising in the detection process of an SSB signal (relative levels based on $A = 1$).

than SSB. However, as will be seen in Chapters 10 and 11, the input noise power for the DSB signal is twice as great as for SSB, so the final signal-to-noise ratios for the two modulation methods are the same. The exact details of the comparison between the various levels will be deferred to Chapters 10 and 11; the intent here was simply to introduce the reader to the notion that two sidebands could actually overlap and combine in the detection process.

We have seen so far that synchronous or coherent detection can be theoretically used to recover the message from either a DSB or an SSB signal. Ignoring for the moment any differences in the levels of the detected signals, both modulation methods appear to be similar as far as the actual recovery of the signal is concerned. Indeed, this conclusion is essentially true if perfect coherent or synchronous detection is used. Remember, we assumed that a carrier was available at the receiver having exact frequency and phase coherency with that of the received signal.

Let us investigate now the detection process when the local carrier is *not exactly synchronized* in frequency and/or phase with the received signal. This would be the case when the local carrier is generated without the benefit of any synchronization, which is often the case in high-frequency communications since synchronization adds to additional design complexity. Try as we may, it is quite difficult to manually adjust a local oscillator to remain exactly locked with an incoming signal unless there is a synchronizing signal, since normal random drift produces a continual shift of the frequency and/or phase.

Assume then that the local carrier is of the form

$$c(t) = \cos\left[(\omega_c + \Delta\omega)t + \Delta\phi\right] \tag{4-36}$$

where $\Delta\omega$ represents a frequency difference and $\Delta\phi$ represents a phase difference. The DSB detection process is easier to analyze with the continuous-spectrum signal, and SSB detection is more readily analyzed with the discrete-spectrum signal. The steps in the analysis follow a pattern similar to the previous developments in the text, but we will leave the exact details to the interested reader (Probs. 4-15 and 4-16). The results after low-pass filtering are summarized as follows:

$$DSB: \quad x_d(t) = \frac{A}{2}x(t)\cos\left(\Delta\omega t + \Delta\phi\right) \tag{4-37}$$

$$SSB: \quad x_d(t) = \frac{A}{4}\sum_{n=1}^{N} C_n \cos\left[(n\omega_1 - \Delta\omega)t + \theta_n - \Delta\phi\right] \tag{4-38}$$

Inspection of (4-37) and (4-38) reveals some significant differences between the detected outputs in this case. The DSB detected output is no longer simply the intelligence signal $x(t)$, but rather it is the intelligence modulated by a low-frequency sinusoid. Since both $\Delta\omega$ and $\Delta\phi$ will drift with time, it is impossible to recover the desired signal as required.

On the other hand, the effect of the frequency difference on the SSB output appears as a shift in all the frequency components representing the signal. While this certainly represents a distortion of the original signal, it is not nearly as serious as the case of DSB, provided the frequency shift is small. Anyone who has tuned SSB voice signals on a communications receiver will undoubtedly be aware of this phenomenon. Depending on the side of the carrier one is tuning, it can result in a bass voice sounding like a soprano, or vice versa. Nevertheless, the signal may be perfectly intelligible provided that the local oscillator is at least moderately stable over a period of time. Of course, the possible shifts in the frequencies could limit the types of data that one would transmit with SSB, but for noncritical applications (e.g., simple point-to-point voice transmissions), SSB finds wide usage.

To review our findings, DSB and SSB provide comparable results under ideal synchronous detection. However, when local carrier coherency cannot be established, SSB is less susceptible to phase and frequency variations and may be quite usable for certain types of communications, while DSB is most troublesome and perhaps impossible to use for some applications. Thus, while SSB is more difficult to generate at the transmitter due to the complexity of the circuits involved, it is much easier to detect at the receiver. Of course, we have also seen that the transmission bandwidth for SSB is W, while for DSB it is $2W$, thus providing another advantage for SSB in terms of less transmission bandwidth. There is also a question of comparing the performance of DSB and SSB in the presence of noise, which will be considered in Chapter 10. From that viewpoint, the two techniques will be shown to have identical performances.

The preceding discussion might lead us to conclude that DSB is not very useful, but this is not the case at all. There are several situations in which DSB can be employed.

1. In some systems, a small pilot carrier is added and transmitted along with the DSB signal. By means of a phase locked loop (to be introduced in Chapter 5), the small carrier can be extracted and used to synchronize an oscillator at the receiver for coherent detection. This process should not be confused with conventional AM (to be studied in the next section), in which the carrier is large compared with the sidebands and in which the purpose is quite different from a coherent reference.

2. By means of some rather special signal processing, it may be possible to generate a carrier component directly from the sidebands. (This concept will be discussed in Chapter 7.)

3. Turning to the field of control systems, there are various control components of the "ac-carrier" type that employ DSB signals for controlling system response. These devices utilize DSB signals based on relatively low frequency carrier frequencies (e.g., 60 or 400 Hz). The use of modulated techniques allows ac-coupled amplifiers and components to be employed

where direct-coupled amplifiers would otherwise be needed for very low frequency error signals. Synchronization is no problem in this situation since the local carrier is readily available.

Both methods 1 and 2 place most of the burden on the receiver to establish the coherency required to properly detect a DSB signal, while in SSB, the major portion of the burden appears at the transmitter.

4-5 CONVENTIONAL AMPLITUDE MODULATION

Within the general class of amplitude-modulation methods that are being considered, the specific method that we have chosen to designate as "conventional amplitude modulation" will now be considered. (In most usage, it is simply referred to as "amplitude modulation.") Conventional amplitude modulation is the oldest form of modulation, and it served as the primary basis for commercial broadcasting for many years. Later, of course, frequency modulation and television broadcasting systems were established, but conventional amplitude modulation still remains as a stable and entrenched phase of the commercial broadcasting industry.

A significant boon to the utilization of amplitude modulation occurred with the rapid increase in citizen's band (CB) operators in the 1970s. The technology for implementing AM systems has been so well developed that it has been possible to mass produce CB transceivers for all the "good buddies" at amazingly low costs.

As we shall see in the next few sections, conventional AM is not nearly as efficient from a power-utilization point of view as either DSB or SSB, and it does not have the noise-reducing capabilities of frequency modulation or some of the other methods that will be encountered later. However, there are two factors that support its probable continuing usage: (1) As already noted, it was the first commercial method and it is now firmly established in usage. Millions of dollars have been invested in AM systems over the years, and any attempt to drastically change such systems would be met with strong public resistance. (2) A conventional AM signal can be detected much easier than most other forms of modulation. This fact has led to the mass production of small receivers that can be purchased for a few dollars.

We will illustrate the formulation of a conventional AM signal by a process that correlates with the methods already considered for DSB and SSB. Consider the system shown in Fig. 4-14. For reasons that will be clear shortly, we will designate the modulating signal momentarily as $x_1(t)$. For clarity in this development, assume that the maximum and minimum values of the message signal are equal in magnitude and that this magnitude is X_m. Thus, the maximum value of $x_1(t)$ is X_m, and the minimum value is $-X_m$. A dc bias equal to X_m is added to the modulating signal in a summing circuit. An attenuator in the signal path allows an adjustment to be made on the

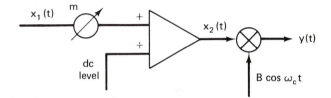

Figure 4-14. Block diagram of AM generator in which dc level is added to message signal ahead of balanced modulator.

modulating signal. The attenuator multiplies the level of the modulating signal by m, where $0 \le m \le 1$. The composite signal $x_2(t)$ can be expressed as

$$x_2(t) = X_m + mx_1(t) = X_m \left[1 + \frac{mx_1(t)}{X_m}\right] \tag{4-39}$$

The maximum magnitude of the ratio $x_1(t)/X_m$ is unity, and since $m \le 1$, the quantity in the brackets of (4-39) can never be negative. At this point the absolute values of the quantities in (4-39) are not nearly as important as the relative levels. It is very convenient from an analysis point of view to think of the ratio $x_1(t)/X_m$ as a normalized modulating signal. We will thus define

$$x(t) = \frac{x_1(t)}{X_m} \tag{4-40}$$

with $x(t)$ representing the modulating signal normalized to a maximum magnitude of unity.

The output of the summer is then multiplied by $B \cos \omega_c t$ to obtain the composite signal. Defining $A = BX_m$, the form of the composite signal can be expressed as

$$y(t) = A[1 + mx(t)] \cos \omega_c t \tag{4-41}$$

As previously noted and in subsequent developments concerning conventional AM, the modulating signal $x(t)$ is normalized to have a maximum magnitude of unity.

The function of (4-41) is a conventional amplitude-modulated (AM) signal. The various time- and frequency-domain waveforms involved are illustrated in Fig. 4-15. An arbitrary modulating signal is shown in Fig. 4-15a, and its assumed spectrum is shown in Fig. 4-15b. The signal plus the dc level is shown in Fig. 4-15c, and this adds a spectral line at dc to the spectrum, as shown in Fig. 4-15d. The output of the modulator and its spectrum are shown in Fig. 4-15e and f. Since A has not been specified at this point, the amplitude level of the modulated spectrum is arbitrary, but for convenience it is shown at the same level as the baseband signal and carrier in this figure.

148

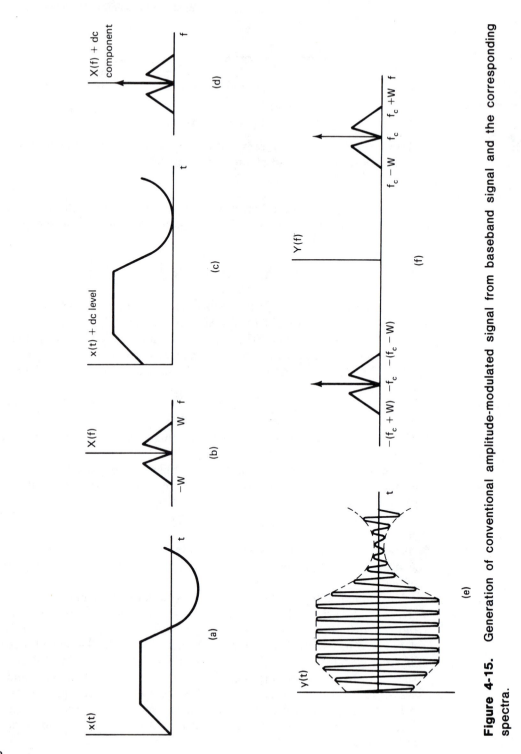

Figure 4-15. Generation of conventional amplitude-modulated signal from baseband signal and the corresponding spectra.

A most significant property has resulted from adding the dc component equal to or greater than the maximum magnitude of the modulating signal. This property is that the envelope of the modulated signal now carries the message directly. This can be seen by comparing Fig. 4-15a and e. As will be seen in the next section, envelope detection is one of the simplest forms of detection.

For the envelope to carry the message signal without distortion, it is necessary that $0 < m \leq 1$ in (4-41) for the normalized form of $x(t)$. The quantity m in this equation will be referred to as the *modulation factor*. More commonly, in commercial applications the term *percentage modulation* is employed. It is defined as

$$\text{percentage modulation} = m \times 100\% \qquad (4\text{-}42)$$

The percentage of modulation may vary from 0% to 100% but should never exceed 100%. The Federal Communications Commission has stringent rules prohibiting commercial radio stations from *overmodulating* (i.e., exceeding 100%).

An alternative way to express (4-41) is

$$y(t) = A \cos \omega_c t + mAx(t) \cos \omega_c t \qquad (4\text{-}43)$$

In this form, the signal is seen to be a DSB signal plus a carrier of amplitude A, with the carrier term having the same phase as the sinusoid in the DSB signal. This result suggests an alternative approach for generating an AM signal, as can be seen in Fig. 4-16. In this latter approach, a DSB signal is first generated in a balanced modulator, and the carrier component is then added to the DSB signal. Verification of this circuit is left as an exercise for the reader (Prob. 4-17). This particular implementation also has significance in conjunction with the generation of FM and will be encountered again in Chapter 5.

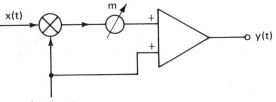

Figure 4-16. Block diagram of AM generator in which carrier is directly added to DSB signal.

Since the spectrum of a conventional AM signal contains two sidebands, the transmission bandwidth B_T is

$$B_T = 2W \qquad (4\text{-}44)$$

Thus, conventional AM and DSB both require the same bandwidth.

Example 4-4

Consider the two-tone signal

$$x_1(t) = 4 \cos 2\pi \times 10^3 t + \cos 6\pi \times 10^3 t \qquad (4\text{-}45)$$

The signal is to be used to modulate a 100 kHz carrier using conventional amplitude modulation. (a) Define the normalized form of the modulating signal in accordance with the text. (b) Assuming a final carrier amplitude of 20, determine an expression for the modulated signal in terms of an arbitrary modulation factor m. (c) Sketch the spectra for the two cases $m = 1$ and $m = 0.5$.

Solution

(a) The two frequencies contained in the given signal are 1 kHz and 3 kHz. It can be verified either by differential calculus or by sketching the two sinusoids that there are common points of both maxima and minima for the two functions. The maximum value is thus $4 + 1 = 5$ and the minimum value is $-4 - 1 = -5$. In accordance with the text, the normalized modulating signal is formed as follows:

$$x(t) = \frac{x_1(t)}{5} = 0.8 \cos 2\pi \times 10^3 t + 0.2 \cos 6\pi \times 10^3 t \qquad (4\text{-}46)$$

The resulting function thus varies between a maximum of $+1$ and a minimum of -1.

(b) For a carrier amplitude of 20, the modulated signal is

$$y(t) = 20 \, [1 + m \, (0.8 \cos 2\pi \times 10^3 t \qquad (4\text{-}47)$$
$$+ \, 0.2 \cos 6\pi \times 10^3 t)] \cos 2\pi \times 10^5 t$$

(c) The spectrum is best observed by expanding (4-47) and substituting the pertinent values of m. The two functions for $m = 1$ and $m = 0.5$ are as follows:

$m = 1$:

$$y(t) = 20 \cos 2\pi \times 10^5 t$$
$$+ \, 8 \cos 2\pi \times 99 \times 10^3 t + 2 \cos 2\pi \times 97 \times 10^3 t \qquad (4\text{-}48)$$
$$+ \, 8 \cos 2\pi \times 101 \times 10^3 t + 2 \cos 2\pi \times 103 \times 10^3 t$$

$m = 0.5$:

$$y(t) = 20 \cos 2\pi \times 10^5 t$$
$$+ \, 4 \cos 2\pi \times 99 \times 10^3 t + \cos 2\pi \times 97 \times 10^3 t \qquad (4\text{-}49)$$
$$+ \, 4 \cos 2\pi \times 101 \times 10^3 t + \cos 2\pi \times 103 \times 10^3 t$$

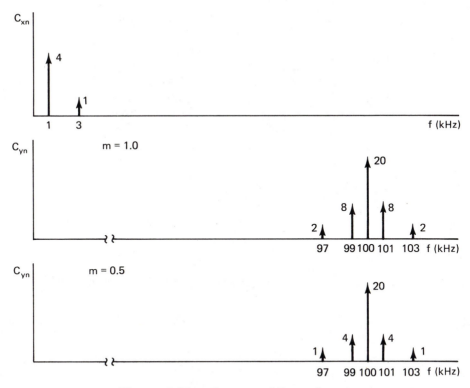

Figure 4-17. Spectra of Example 4-4.

The amplitudes of the various frequency components can be read directly from these equations, and the spectra are displayed in Figure 4-17.

4-6 ENVELOPE DETECTION OF CONVENTIONAL AMPLITUDE MODULATION

It was shown in the last section that a conventional AM signal displays the form of the message signal directly on the envelope, while it was established earlier that the message signal cannot be recovered from the envelope of either a DSB or an SSB signal. In this section, we will discuss the means by which envelope detection can be achieved, including some consideration of the actual circuitry involved. It should be pointed out here that envelope detection also arises in conjunction with other forms of modulation to be considered later, so the process is rather common.

For envelope detection to be practical, it is necessary that the carrier frequency be much higher in frequency (perhaps several orders of magnitude) than the highest modulating frequency. Let $T_c = 1/f_c$ represent the period of the carrier, and let $T_m = 1/W$ represent the shortest possible period of a component of the modulating signal. (The shortest period occurs at the highest modulating frequency W.)

Consider the circuit shown in Fig. 4-18 and the waveforms shown in Fig. 4-19. For the purpose of this discussion, it will be assumed that $R_0 \gg R$, in which case there is negligible loading effect from the part of the circuit containing R_0 and C_0 on the remainder of the circuit. With this assumption and with the diode reverse biased, the discharge path of the capacitor C is primarily through the resistor R.

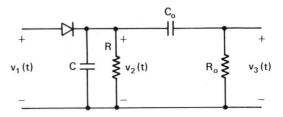

Figure 4-18.   Envelope detector circuit.

Let $\tau = RC$ represent the time constant of the parallel RC combination. Since the carrier frequency is several orders of magnitude higher than the highest message frequency, it is possible to select a time constant that will satisfy *both* of the following inequalities:

$$\tau \gg T_c \qquad (4\text{-}50)$$

$$\tau \ll T_m \qquad (4\text{-}51)$$

If both of these inequalities are satisfied, the following conditions exist: (1) The voltage $v_2(t)$ cannot follow the rapid variation of the carrier due to the restriction of (4-50). (2) The voltage $v_2(t)$ will, however, be able to follow variations of the message signal due to the restriction of (4-51).

The result of these conditions is that the voltage $v_2(t)$ represents very closely the positive envelope of the modulated signal. Recall that the envelope consisted of the message signal plus a dc bias component chosen to "lift" the baseband signal so that it could not be negative. The dc level can be readily removed by the simple high-pass filter consisting of C_0 and R_0 in Fig. 4-18, and the voltage $v_3(t)$ in Fig. 4-19 represents the message signal after transients have settled and C_0 has been charged to the dc value. Because of the need to add an extra dc level to the signal at the transmitter, conventional AM is usually not employed when dc data are to be transmitted. Most applications of conventional AM involve signals having no spectral components of interest at dc or at a few hertz, so the dc component can be freely added and extracted without significantly affecting the signal.

One final point is that the negative envelope could have just as easily been used, and the only change required in the circuit to extract it would be to reverse the direction of the diode. The resulting detected signal will be

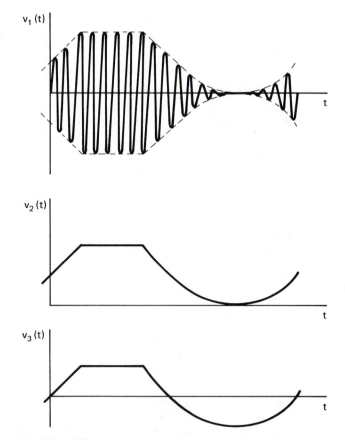

Figure 4-19. Waveforms occurring in envelope detector circuit.

inverted with respect to the positive envelope, and this fact might have to be considered in some applications.

4-7 SINGLE-TONE MODULATION

As a matter both of special interest and of practical importance, we will investigate each of the forms of modulation considered so far in this chapter for the case when the modulating signal is a single frequency sinusoid. Actually, a single steady-state tone does not really convey any information since it is totally predictable, but an analysis of tone modulation is important for several reasons:

1. Various tests on actual communications equipment can be readily performed with a single frequency modulating signal, and the results can be easily interpreted and analyzed.

2. The power relationships are the simplest of any type of modulating signal, and they provide a convenient base of reference for more complex wave forms.

3. The spectra for the different modulation methods have the simplest possible forms.

4. The time waveforms are readily visualized and additional insight is provided from a study of their properties.

Assume then that the modulating signal $x(t)$ is a unit amplitude sinusoid with frequency f_m of the form

$$x(t) = \cos \omega_m t \tag{4-52}$$

The modulating signal is shown in Fig. 4-20a, and the form of the one-sided spectrum C_{xn} is shown in Fig. 4-20b. The results for each of the previous modulation methods are summarized in the next few paragraphs.

DSB

The DSB signal is

$$y(t) = A \cos \omega_m t \cos \omega_c t \tag{4-53}$$
$$= \frac{A}{2} \cos (\omega_c - \omega_m)t + \frac{A}{2} \cos (\omega_c + \omega_m)t$$

where an arbitrary multiplicative factor A is used for relative comparison with other methods. This function is shown in Fig. 4-20c, and its one-sided spectrum C_{yn} is shown in Fig. 4-20d. Observe that the positive and negative envelopes cross every half-cycle of the modulating frequency. As mentioned earlier, this phenomenon results in periodic abrupt phase reversals of the oscillations at the carrier rate, and the carrier component is effectively eliminated from the spectrum.

SSB

An SSB signal can be obtained by selecting either of the sidebands in (4-53). However, in order that the peak value of the SSB time signal be the same as for the DSB signal, we will multiply the amplitude of the sideband by two. Arbitrarily selecting the USB form, we have

$$y(t) = A \cos (\omega_c + \omega_m)t \tag{4-54}$$

This signal is shown in Fig. 4-20e, and the form of its one-sided spectrum is shown in Fig. 4-20f. Recall that it was not possible to determine in a simple manner the shape of an SSB signal for an arbitrary modulating signal. However, for a sinusoidal modulating signal, the SSB signal is readily visualized and is also a sinusoid, but at a higher frequency.

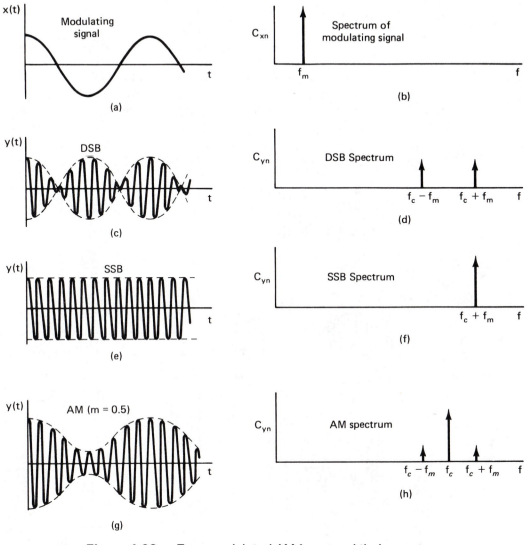

Figure 4-20. Tone-modulated AM forms and their spectra.

AM

The AM signal with a single frequency modulating signal is

$$
\begin{aligned}
y(t) &= A(1 + m \cos \omega_m t) \cos \omega_c t \\
&= A \cos \omega_c t + \frac{mA}{2} \cos (\omega_c - \omega_m)t + \frac{mA}{2} \cos (\omega_c + \omega_m)t
\end{aligned}
$$

(4-55)

A family of different AM signals and their corresponding spectra could be generated by assuming different values of m. A sketch of the time function for $m = 0.5$ is shown in Fig. 4-20g, and the corresponding form of the one-sided spectrum is shown in Fig. 4-20h.

As the modulation factor m increases, the maximum value of the envelope increases, and the minimum value decreases. This phenomenon is illustrated in Fig. 4-21 for several values of m, and the corresponding one-sided spectra

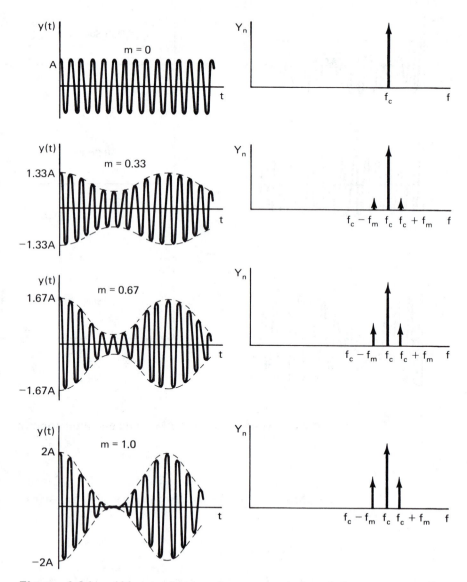

Figure 4-21. AM waveforms as m as varied and the corresponding spectra.

are also shown. The functions at the top correspond to the trivial situation in which $m = 0$; that is, there is no modulation.

It is of interest to note that there is a transmitted carrier present even without any modulation applied for conventional AM unless provisions are made for turning off the carrier during lull periods of the message signal. However, note that for both DSB and SSB, nothing is transmitted when the modulating signal is zero.

Two final observations will be made from Fig. 4-21 concerning the case of $m = 1$ (100% modulation). (1) The peak amplitude of the modulated signal is twice the value of the unmodulated carrier amplitude. (2) Each of the sidebands has one-half the amplitude of the carrier component.

4-8   POWER RELATIONSHIPS FOR TONE MODULATION

In this section, we will continue the analysis of tone modulation by investigating the relative magnitude and distribution of power for each of the three forms of AM previously considered. While the results based on tone modulation are not necessarily the same as for complex modulating signals, they serve to set representative bounds on power levels that are encountered in the general case. Furthermore, many power measurements are made and equipment power specifications are stated on the basis of tone modulation, so the results are quite useful.

In accordance with the approach of Chapter 2, the assumption of a 1-Ω reference will be made in developing the power relationships. The reader is reminded that this approach is universally used for convenience when working with arbitrary waveform functions, but when an actual voltage or current is given along with a resistance R, the proper power form ($i^2 R$ or v^2/R) must be used in practical problems.

Assume then a modulated signal $y(t)$ and a normalized 1-Ω reference base. The following power parameters are of interest in communications analysis:

1.   Instantaneous power $p(t)$

$$p(t) = y^2(t) \tag{4-56}$$

2.   Average power P

$$P = \frac{1}{T} \int_0^T y^2(t)\,dt \tag{4-57}$$

3.   Peak envelope power P_p

The peak envelope power P_p can be defined as the average power produced by a sinusoid whose *peak amplitude* is the same as the *peak* of $y(t)$. Thus, if $y(t)$ has a peak value A_p, the peak normalized envelope power is

$$P_p = \frac{A_p^2}{2} \tag{4-58}$$

The first two definitions are the same as used in Chapter 2, so the reader should already be familiar with them. On the other hand, the definition of peak envelope power (also called simply *peak power*) may seem a little strange, and it needs some further clarification. With amplitude modulated signals, the envelope of the complex signal may alternately be quite large and quite small. (Refer to Figs. 4-20 and 4-21.) Assuming that the carrier frequency is large compared with frequencies contained in the message signal, a relatively long interval compared to the carrier period may exist in which the signal appears to be very nearly a sinusoid with amplitude equal to the peak amplitude of the complex signal. During this interval, power is being produced at a rate approximately equal to a sine wave having this peak value. Hence, the peak envelope power provides a measure of the level of power required during peaks of the modulated signal.

One important aspect of the peak envelope power is that it helps to specify the power rating of the components required in processing the transmitted signal. For example, suppose a certain system produces an average power of 100 W and a peak power of 200 W. The transistors and other components in the power-amplifier stage must be designed to deal with the 200-W level. This means that the ratio of peak envelope power to average power could be used as some sort of "overrating factor" in specifying component power limits for a transmitter. As this ratio increases, the power ratings of the components must be increased relative to a given average power level. For equipment that must be very light (e.g., hand-carried walkie-talkies), this concept imposes certain restraints on the components.

We will now consider the power relationships for each of the three AM processes considered so far, assuming tone modulation. Parseval's theorem may be easily applied in the analysis, and the equations expressed in the last section will be freely used.

DSB

From (4-53), the average normalized power P is obtained by adding the power produced by each of the two components.

$$P = \frac{1}{2}\left(\frac{A}{2}\right)^2 + \frac{1}{2}\left(\frac{A}{2}\right)^2 = \frac{A^2}{4} \tag{4-59}$$

Since the peak of the envelope is A, the peak envelope power P_p is the average power that would be produced by a sinusoid of amplitude A and is

$$P_p = \frac{A^2}{2} \tag{4-60}$$

The ratio of peak envelope power to average power for a single tone modulating signal is

$$\frac{P_p}{P} = \frac{A^2/2}{A^2/4} = 2 \tag{4-61}$$

SSB

Using the form of (4-54) in which the peak value of the SSB signal was adjusted to be the same as that of the DSB signal, the average power is simply

$$P = \frac{A^2}{2} \tag{4-62}$$

However, since the function is a single sinusoid, the peak envelope power according to the definition is

$$P_p = \frac{A^2}{2} \tag{4-63}$$

Thus, the average and peak envelope powers are the same for SSB *when the modulating signal is a single tone.*

This result could be misleading, since it might lead one to incorrectly assume that the average and peak power levels are always the same for SSB. This equality is true only for a single-tone modulating signal. With more complex modulating signals, the peak power in SSB will be greater than the average power. For waveforms having sharp discontinuities (e.g., a square wave), the peak power with SSB may be significantly larger than the average power. The two-tone test is more meaningful in evaluating the power relationships for an SSB signal and will be considered next.

SSB (Two-Tone)

The two-tone test is a widely used means for evaluating the performance of an SSB transmitter and in specifying convenient power levels. Assume that an SSB generator is modulated by the two-tone baseband signal

$$x(t) = A \cos \omega_{m1} t + A \cos \omega_{m2} t \tag{4-64}$$

If the carrier magnitude is conveniently chosen as unity, the output of the SSB modulator for USB is of the form

$$y(t) = \frac{A}{2} \cos (\omega_c + \omega_{m1})t + \frac{A}{2} \cos (\omega_c + \omega_{m2})t \tag{4-65}$$

The average normalized power P is

$$P = \frac{1}{2}\left(\frac{A}{2}\right)^2 + \frac{1}{2}\left(\frac{A}{2}\right)^2 = \frac{A^2}{4} \tag{4-66}$$

To determine the peak power, it is necessary to investigate more closely the time-domain form of (4-65). By comparing (4-65) with (4-53), the following form can be obtained:

$$y(t) = \frac{A}{2} \cos (\omega_c' - \omega_m')t + \frac{A}{2} \cos (\omega_c' + \omega_m')t \tag{4-67}$$

where $\omega'_m = (\omega_{m2} - \omega_{m1})/2$ and $\omega'_c = \omega_c + \omega_{m1} + \omega'_m$. While this expression may seem initially awkward, it shows that the function is of the same mathematical form as a DSB signal arising from a single-tone modulating signal. Because of this similarity, $y(t)$ will look the same as a DSB signal in which the modulating signal is a single tone of frequency ω'_m and in which the reference carrier is ω'_c. Using again the form of (4-53) for comparison, (4-67) may be expressed as

$$y(t) = A \cos \omega'_m t \cos \omega'_c t \qquad (4\text{-}68)$$

The peak envelope power is readily determined from (4-68) as

$$P_p = \frac{A^2}{2} \qquad (4\text{-}69)$$

The ratio of peak envelope power to average power is

$$\frac{P_p}{P} = \frac{A^2/2}{A^2/4} = 2 \qquad (4\text{-}70)$$

Thus, the functional and power relationships for two-tone modulation with SSB are of the same form as for single-tone modulation with DSB.

AM

In the case of conventional AM, the average power is obtained by summing the powers for the three components in (4-55).

$$P = \frac{A^2}{2} + \frac{1}{2}\left(\frac{mA}{2}\right)^2 + \frac{1}{2}\left(\frac{mA}{2}\right)^2 \qquad (4\text{-}71)$$

$$= \frac{A^2}{2}\left(1 + \frac{m^2}{2}\right)$$

It is convenient in the case of conventional AM to define the unmodulated carrier power P_c as

$$P_c = \frac{A^2}{2} \qquad (4\text{-}72)$$

The unmodulated carrier power P_c is the power level specified in commercial broadcasting. The average power can then be expressed as

$$P = P_c\left(1 + \frac{m^2}{2}\right) \qquad (4\text{-}73)$$

The peak of the envelope of the composite signal is $(1 + m)A$, and peaks occur when $\omega_m t = 2k\pi$ (k an integer). The peak envelope power is

$$P_p = \frac{(1 + m)^2 A^2}{2} = P_c(1 + m)^2 \qquad (4\text{-}74)$$

The ratio of peak envelope power to average power for a tone modulated conventional AM signal is

$$\frac{P_p}{P} = \frac{(1 + m)^2}{1 + m^2/2}$$ (4-75)

The preceding quantities are tabulated in Table 4-1 with the AM data shown for several values of m. It is interesting to observe the manner in which the peak power varies with the modulation factor. For small values of the modulation factor, the average power increases only slightly, but the peak power increases substantially. For 100% tone modulation, *the average power is 1.5 times the unmodulated carrier power, and the peak power is four times the unmodulated carrier power.* The power levels of most AM transmitters are specified in terms of the unmodulated carrier power. Thus, a 1000-W commercial broadcasting transmitter will generate 1500 W of average power and 4000 W of peak power under conditions of 100% tone modulation. These facts must be considered in designing the RF power-amplifier stage and the antenna system.

Another point of interest with the AM signal is the relationship of the sideband power to the total power. For $m = 1$, the average power in one sideband P_{1sb} is

$$P_{1sb} = \frac{1}{2}\left(\frac{A}{2}\right)^2 = \frac{A^2}{8} = \frac{P_c}{4}$$ (4-76)

The ratio of total average power to power in one sideband for $m = 1$ is

$$\frac{P}{P_{1sb}} = \frac{1.5 P_c}{P_c/4} = 6$$ (4-77)

Note that (4-76) and (4-77) apply only to 100% tone modulation.

In completing this section, it should be emphasized once again that the power relationships developed here apply in exact form only to *single-tone*

Table 4-1. Normalized Power Relationships for Single-Tone Modulation.

	P	P_p	P_p/P
DSB	$\dfrac{A^2}{4}$	$\dfrac{A^2}{2}$	2
SSB	$\dfrac{A^2}{2}$	$\dfrac{A^2}{2}$	1
AM	$\left(1 + \dfrac{m^2}{2}\right) P_c$	$(1 + m)^2 P_c$	$\dfrac{(1 + m)^2}{1 + \dfrac{m^2}{2}}$
AM, $m = 1$	$1.5 P_c$	$4 P_c$	2.667
AM, $m = 0.5$	$1.125 P_c$	$2.25 P_c$	2

modulation (except, of course, for the two-tone test relationships). These results are very useful in analyzing and specifying AM systems, but they are not necessarily the same as would be obtained with complex modulating signals. Indeed, the power relationships for complex modulating signals depend on the type of modulating signal and can be described only in statistical terms for the general case.

Although a 1-Ω resistor was assumed in the computation of the normalized power relationships, all power *ratios* such as P_p/P apply for any resistance levels since the resistances cancel in the division process. Thus, an equation such as (4-73), for example, applies to any AM transmitter with tone modulation.

Example 4-5

A certain commercial AM transmitter has an unmodulated carrier power level of 1 kW, and the antenna presents a resistive load of $R_a = 50 \ \Omega$ to the transmitter at the operating frequency. Determine the total average power, peak power and power contained in each sideband for single-tone modulation corresponding to (a) 50% modulation and (b) 100% modulation. Determine the antenna input rms voltage V_a and current I_a corresponding to (c) no modulation, (d) 50% modulation, and (e) 100% modulation.

Solution

(a) Since the carrier power is $P_c = 1000$ W, the results of Eqs. (4-73) and (4-74) may be readily employed to determine the total average power and the peak power. The power contained in one sideband in each case is half the difference between the total average power and the carrier power of 1000 W. The calculations are summarized as follows for $m = 0.5$:

$$P = 1000 \left[1 + \frac{(0.5)^2}{2} \right] = 1125 \text{ W} \tag{4-78}$$

$$P_p = 1000 (1 + 0.5)^2 = 2250 \text{ W} \tag{4-79}$$

$$P_{1sb} = \frac{1125 - 1000}{2} = 62.5 \text{ W} \tag{4-80}$$

(b) The computations in part (a) are repeated for $m = 1$.

$$P = 1000 \left[1 + \frac{(1)^2}{2} \right] = 1500 \text{ W} \tag{4-81}$$

$$P_p = 1000 (1 + 1)^2 = 4000 \text{ W} \tag{4-82}$$

$$P_{1sb} = \frac{1500 - 1000}{2} = 250 \text{ W} \tag{4-83}$$

Note that the increases in each of the power values are *not* a linear function of *m*.

(c) For a given average power level P, the rms current, voltage, and antenna resistance must satisfy the standard AC power relationships:

$$P = I_a^2 R_a \tag{4-84}$$

$$P = \frac{V_a^2}{R_a} \tag{4-85}$$

$$V_a = R_a I_a \tag{4-86}$$

If the antenna had presented a complex load (i.e., both real and imaginary parts) to the transmitter, more general ac power relationships would be required.

With no modulation, we have from (4-84)

$$I_a = \sqrt{\frac{P}{R_a}} = \sqrt{\frac{1000}{50}} = 4.472A \tag{4-87}$$

From (4-86), the voltage is

$$V_a = 4.472 \times 50 = 223.6 \text{ V} \tag{4-88}$$

(d) The steps given in part (c) are repeated for $P = 1125$ W, corresponding to 50% modulation:

$$I_a = \sqrt{\frac{1125}{50}} = 4.743A \tag{4-89}$$

$$V_a = 4.743 \times 50 = 237.2 \text{ V} \tag{4-90}$$

(e) For 100% modulation, $P = 1500$ W and the current and voltage are

$$I_a = \sqrt{\frac{1500}{50}} = 5.477A \tag{4-91}$$

$$V_a = 5.477 \times 50 = 273.9 \text{ V} \tag{4-92}$$

4-9 SWITCHING MODULATORS

In this section, we will explore a special class of modulator circuits, denoted as *switching modulators,* that are widely used in communications systems of all types and deserve special consideration. These circuits depend on abrupt switching states to generate the various sidebands required and are most often implemented with diodes, although at very low frequencies (e.g., control system applications) they can even be implemented with mechanical switches. These circuits are capable of modulation, demodulation, and frequency translation, and depending on the nature of the actual components, they can be used over an extremely wide frequency range from near dc to the microwave range.

One major reason for this special consideration is that switching modulators generate many undesired components along with the desired components, and filtering is usually required. Thus, it is necessary for the user to understand and predict the nature of these undesired components so that proper and adequate filtering can be employed. Fortunately, it is possible to accurately predict the spurious components, and our attention will be directed toward that goal in this section.

Switching modulators can be classified as either (1) *single-balanced* or (2) *double-balanced.* One common example of a single-balanced modulator is the *bridge modulator,* which will be used as the basis for the explanation that follows. A bridge modulator connected as a *shunt switch* is shown in Fig. 4-22a. For the circuit to function properly, it is necessary that the magnitude A of the carrier be large compared with the maximum magnitude of the modulating signal $x(t)$. Typically, A is ten or more times as large. In this case, the conduction states of the diodes are controlled almost completely by the

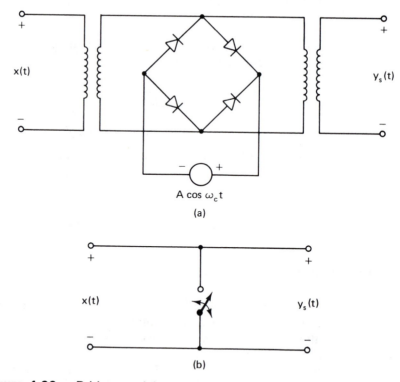

Figure 4.22. Bridge modulator connected as a shunt switch. (Signal and carrier internal resistances are not shown.)

carrier and are not affected significantly by the modulating signal. When the actual carrier polarity is aligned exactly with the reference polarity shown on the figure, the four diodes are reverse biased, and the signal $x(t)$ appears at the output. The carrier itself does not appear at the output in the ideal case because the voltage drops across adjacent diodes are equal in magnitude and opposite in polarity and thus cancel.

When the actual carrier polarity is opposite to the reference polarity shown on the figure, the four diodes are forward biased, and a short appears across the line. Thus, the signal $x(t)$ is shorted to ground. Due to the size of the carrier, the states of the diodes change abruptly when the carrier voltage crosses zero.

The net result of the circuit is the equivalent of a shunt switch across the line, as shown in Fig. 4-22b. During one half-cycle of the carrier frequency, the switch is open, and during the other half-cycle, the switch is closed.

We will now perform an important mathematical analysis of this circuit in order to show what can be achieved in the form of modulation. Consider the various waveforms shown in Fig. 4-23. An arbitrary modulating signal is shown in Fig. 4-23a, and an assumed two-sided amplitude spectrum is shown in Fig. 4-23b. The wave form $y_s(t)$ at the output of the bridge modulator is shown in Fig. 4-23e. Observe that $y_s(t)$ is zero half of the time and an exact replica of the input for the other half. Mathematically, this waveform can be expressed as the product of $x(t)$ and a pulse switching function $p_1(t)$ whose form is shown in Fig. 4-23c. Thus,

$$y_s(t) = x(t)p_1(t) \qquad (4\text{-}93)$$

The spectrum of $p_1(t)$ is, of course, that of a pulse train having a 50% duty cycle and is shown in Fig. 4-23d. The function $p_1(t)$ can be expressed in exponential Fourier series form as

$$p_1(t) = \sum_{-\infty}^{\infty} \overline{P}_{1n}\, e^{jn\omega_c t} \qquad (4\text{-}94)$$

where the amplitude spectrum of $\overline{P}_{1n}$ is

$$P_{1n} = \frac{1}{2}\frac{\sin(n\pi/2)}{n\pi/2} \qquad (4\text{-}95)$$

(The reader can refer back to Chapter 2 for a review of periodic pulse trains and their spectra if necessary.)

Substituting (4-94) into (4-93), we have

$$y_s(t) = x(t) \sum_{-\infty}^{\infty} \overline{P}_{1n} e^{jn\omega_c t} \qquad (4\text{-}96)$$

$$= \sum_{-\infty}^{\infty} \overline{P}_{1n} x(t) e^{jn\omega_c t}$$

166

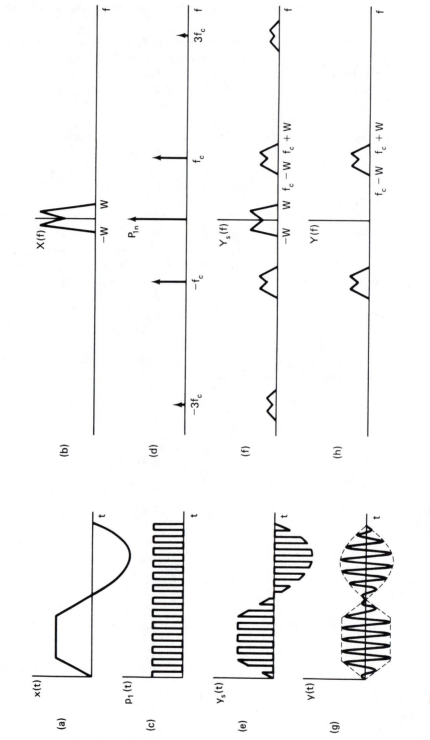

Figure 4-23. Time-domain and frequency-domain forms encountered in a single-balanced switching modulator.

Fourier transformation of both sides of (4-96) yields

$$\overline{Y}_s(f) = \sum_{-\infty}^{\infty} \overline{P}_{1n} \overline{X}(f - nf_c) \qquad (4\text{-}97)$$

This result indicates that an infinite number of spectral functions is obtained. Each function represents the form of the original spectrum shifted to and centered at one of the harmonic components of $p_1(t)$ and multiplied by the appropriate coefficient $\overline{P}_{1n}$. The form of the overall amplitude spectrum is shown in Fig. 4-23f.

Observe that, if a DSB signal centered at f_c is desired, a bandpass filter centered at f_c and having a bandwidth $2W$ is required. The result of such a filtering operation is shown in Fig. 4-23g, and the corresponding amplitude spectrum is shown in Fig. 4-23h. These final results are seen to be identical to those that would be obtained from an ideal balanced modulator as discussed earlier in the chapter. Thus, one approach to implementing a balanced modulator is to periodically gate the modulating signal off and on at the carrier rate and filter out the desired component.

An interesting point to observe is that each of the frequencies contained in the switching waveform also carries the message signal in a DSB form. Indeed, one way of interpreting the result of a switching modulator is to consider that the modulating signal modulates *each* component of the square wave. Thus, by proper filtering, it would be theoretically possible to produce a modulated signal at any one of the harmonics of the carrier frequency.

The preceding development began with the use of a bridge modulator connected as a shunt switch. It is also possible to achieve the same effect with the bridge modulator connected as a *series switch,* and consideration of this circuit will be left as an exercise for the reader (Prob. 4-23).

The concept of the *double-balanced* modulator will now be investigated. The particular circuit that will be used in this development is called a *ring modulator* and is shown in two different layouts in Fig. 4-24a and b. (Both forms are shown because the circuit appears in equipment schematic diagrams both ways, and it may not be obvious that the two forms are exactly the same circuit.)

In the explanation that follows, reference will be made to the layout of Fig. 4-24a. As in the earlier example, the carrier level is large compared to the modulating signal so that switching action is abrupt and dependent almost completely on the carrier. Assume that initially the actual polarity of the carrier is the same as the assumed reference polarity on the figure. In this case, the two parallel diodes will both be forward biased, while the crossed diodes will be reverse biased. Carrier components around the circuit are equal in magnitude and opposite in phase and thus cancel as far as the input-output signal path is concerned. The resulting action of the circuit can be simplified during this portion of the cycle to the simple form shown in Fig. 4-24c for the signal path.

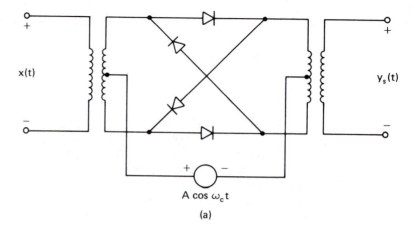

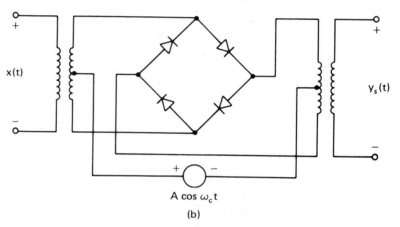

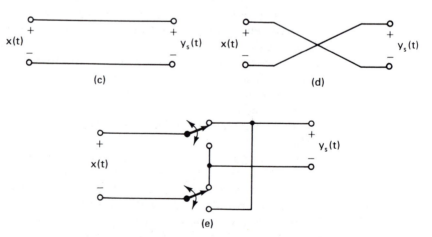

Figure 4-24. (a) Ring modulator shown two ways [in (a) and (b)] and its equivalent switching forms [(c), (d), and (e)]. (Signal and carrier internal resistances are not shown.)

Assume now that the actual polarity of the carrier is opposite to the assumed polarity. The parallel diodes are now reverse biased, but the crossed diodes are forward biased. This action can be represented by the simple form shown in Fig. 4-24d for the signal path.

The combined action of the two preceding operations can be illustrated by the switching process shown in Fig. 4-24e. On one-half of the carrier cycle, the input is connected directly to the output, while on the other half-cycle, the input is reversed in sign and applied to the output.

This circuit can be analyzed mathematically in the same manner as for the single-balanced modulator. However, there are some important differences in the results, as will be seen shortly. First, refer to Fig. 4-25 in which a series of time-domain and spectral forms are shown. An arbitrary modulating signal is shown in Fig. 4-25a. In this case, the switched waveform $y_s(t)$ takes the form of Fig. 4-25e in which the signal alternately appears first without change and then with an abrupt signal reversal. This signal can be considered as the product of the original signal $x(t)$ and a pulse switching waveform $p_2(t)$, whose form is shown in Fig. 4-25c. This function possesses a Fourier series representation of the form

$$p_2(t) = \sum_{-\infty}^{\infty} \overline{P}_{2n} e^{jn\omega_c t} \tag{4-98}$$

in which the amplitude spectrum of $\overline{P}_{2n}$ is

$$P_{2n} = \frac{\sin(n\pi/2)}{n\pi/2}, \qquad \text{for } n \neq 0$$
$$= 0, \qquad \text{for } n = 0 \tag{4-99}$$

The form of this spectrum is illustrated in Fig. 4-25d for a few components.

There are two significant differences between the spectrum of $p_2(t)$ for the double-balanced modulator and $p_1(t)$ for the single-balanced modulator: (1) There is no dc component in $p_2(t)$. (2) All the other components in $p_2(t)$ are twice as large as the corresponding components in $p_1(t)$.

Observing the composite spectrum shown in Fig. 4-25f, the first property results in a complete absence of the original spectrum in $Y_s(f)$. The second property results in a larger output for the filtered DSB signal obtained from a double-balanced modulator as compared with a single-balanced modulator.

A DSB signal obtained from filtering $y_s(t)$ and its spectrum are shown in Fig. 4-25g and h. Note that the relative magnitude of the signal in this case is larger than for the single-balanced modulator case shown in Fig. 4-23, as is to be expected.

Some interesting practical results will now be given for predicting the output frequency components of switching modulators. Assume that the baseband modulating signal frequencies range from dc to W hertz, and

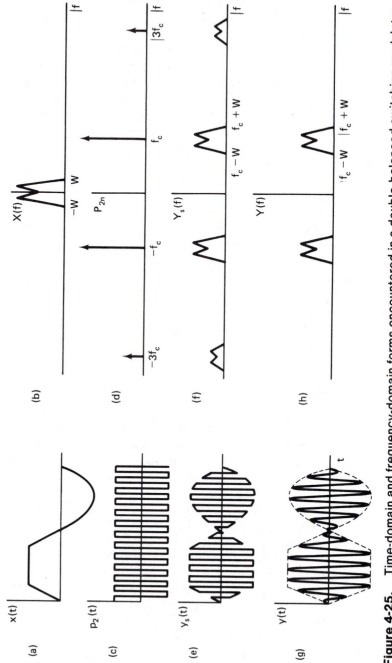

Figure 4-25. Time-domain and frequency-domain forms encountered in a double-balanced switching modulator.

assume that the switching frequency is f_c hertz. The ranges of components are summarized as follows:

Single-Balanced Modulator

dc to W (original signal spectrum)

$f_c - W$ to $f_c + W$

$3f_c - W$ to $3f_c + W$

$5f_c - W$ to $5f_c + W$

etc.

Double-Balanced Modulator

$f_c - W$ to $f_c + W$

$3f_c - W$ to $3f_c + W$

$5f_c - W$ to $5f_c + W$

etc.

Example 4-6

Consider again the multitone modulating signal of Ex. 4-1.

$$x(t) = 10 \cos 2\pi \times 10^3\, t + 8 \cos 4\pi \times 10^3\, t + 6 \cos 8\pi \times 10^3\, t \quad (4\text{-}100)$$

Assume that the signal is applied to the input of a switching modulator having a carrier frequency of 100 kHz. On a positive frequency basis, list all frequencies below 600 kHz appearing in the output before filtering for (a) a single-balanced modulator and (b) a double-balanced modulator.

Solution

(a) Reviewing the work of this section, the switching modulator output is equivalent to the sum of the components of the switching waveform, each separately modulated by the input signal. For the single-balanced modulator having a carrier (or switching) frequency of 100 kHz, the components of the square wave are at the following frequencies: dc, 100, 300, 500, 700 kHz, and so on. The signal components have frequencies of 1, 2, and 4 kHz. Since the switching function for the single-balanced modulator has a dc component, the frequencies of the modulating signal appear in the output. In addition, the sum and difference frequencies formed from the combination of the switching frequency components and the three input components appear in the output. A list of all the frequencies below 600 kHz in the output follows:

1, 2, 4 kHz (input components)

96, 98, 99, 101, 102, 104 kHz

296, 298, 299, 301, 302, 304 kHz

496, 498, 499, 501, 502, 504 kHz, etc.

(b) For a double-balanced modulator, there is no dc component in the switching function, so the input modulating signal does not appear in the output. Thus, the list of frequencies for the double-balanced case is the same as for the single-balanced case except that the input components of 1, 2, and 4 kHz should be omitted from the list. Of course, the magnitudes of the components for the double-balanced case are larger by a factor of 2 than for the single balanced case.

For both the single- and double-balanced modulators, the magnitudes of the various components follow a $\sin x/x$ function envelope and become smaller with increasing frequency. While a precise mathematical development could be made in this example, we have chosen to illustrate how some relatively simple arithmetic calculations can be made to predict the major properties of a more complex spectrum.

PROBLEMS

4-1 A carrier with frequency of 500 kHz is modulated in an ideal balanced modulator by the signal $x(t) = 6 \cos 4\pi \times 10^3 t + 4 \cos 10\pi \times 10^3 t$.

(a) Using only simple arithmetic, list the frequencies (in hertz) appearing in the DSB output of the modulator.

(b) Develop a formal expression for the DSB output $y(t)$. Assume a cosine function with a normalized amplitude of unity for the carrier.

(c) Plot the one-sided spectra for both the input and output of the modulator.

4-2 Repeat part (b) of Prob. 4-1 if the carrier is assumed to be a sine function.

4-3 Consider the system described in Prob. 4-1, and assume that an SSB signal is to be generated.

(a) For USB, list the frequencies present in the SSB output and write an equation for $y(t)$.

(b) Repeat part (a) for LSB.

4-4 Assume that the DSB signal generated in Prob. 4-1 is applied directly to a product detector having a local oscillator carrier of the form $c(t) = \cos (10^6 \pi t + \theta)$. After multiplication, the signal is applied to a low-pass filter having a cutoff frequency just above 5 kHz.

(a) For $\theta = 0$, show that the output $x_d(t)$ of the low-pass filter is directly proportional to the modulating signal $x(t)$.

(b) For $\theta = 90°$, show that $x_d(t) = 0$.

4-5   Assume that the USB SSB signal generated in Prob. 4-3a is applied directly to the product detector of Prob. 4-4.

    (a)   For $\theta = 0$, show that the output $x_d(t)$ is directly proportional to $x(t)$.

    (b)   For $\theta \neq 0$, show that the effect is to shift the phase in each of the components of $x(t)$, but that each of the magnitudes remains the same as in part (a).

4-6   Repeat Prob. 4-5 for the LSB signal of Prob. 4-3b.

4-7   Consider the phase-shift method of SSB discussed in Ex. 4-3 and the corresponding block diagram in Fig. 4-10.

    (a)   Show that if $\hat{y}_1(t)$ is added to $y_1(t)$ in the summer, the output will be an LSB SSB signal.

    (b)   Show that if the phase shift of the carrier is changed to $+90°$ while the phase shift of the modulating signal remains at $-90°$, the output will be an LSB SSB signal. For this part, $y_1(t)$ is subtracted from $y_1(t)$ as indicated on the figure. (This result holds true if the phase shift of either the carrier or the intelligence, but not both, is changed to $+90°$.)

4-8   A symmetrical squarewave with no dc component and a frequency of 1 kHz is applied to an ideal balanced modulator having a sinusoidal carrier with a frequency of 4 kHz. Sketch the output over the range of two cycles of the square wave. You may assume that a square-wave cycle begins at the same instant of time as a carrier cycle begins.

4-9   A signal $x(t)$ is band-limited from dc to W hertz. It is applied to a DSB modulator. Find the maximum value of f_c expressed in terms of W such that the bandwidth of the DSB signal is no less than 1% of f_c.

4-10   Repeat Prob. 4-9 for a SSB output signal.

4-11   The following two signals are applied as the two inputs of an ideal balanced modulator:

$$x_1(t) = A_1 \cos 20\pi t + A_2 \cos 60\pi t$$
$$x_2(t) = B_1 \cos 2000\pi t + B_2 \cos 4000\pi t$$

List all positive frequencies (in hertz) present in the output.

4-12   A symmetrical square wave with no dc component and a frequency of 1 kHz is passed through a low-pass filter with a cutoff frequency of 10 kHz. The band-limited signal then modulates a 500-kHz carrier in an ideal balanced modulator. List all positive frequencies (in hertz) in the DSB output.

4-13   Assume that the carrier $c(t)$ used in a certain balanced modulator contains an undesirable second harmonic term; that is,

$$c(t) = \cos \omega_c t + \alpha \cos 2\omega_c t$$

For an arbitrary baseband modulating signal $x(t)$, determine expressions for the DSB output, $y(t)$, and its amplitude spectrum $Y(f)$. Sketch the forms for $X(f)$ and $Y(f)$.

4-14   If the carrier frequency f_c in a balanced modulator is too low relative to the highest baseband frequency W, the spectrum of the DSB signal may overlap the baseband spectrum and make exact recovery of the original signal impossible. Using a spectral diagram, determine an equality for the carrier frequency f_c that will ensure that spectral overlap does not exist.

4-15   Consider a received DSB signal of the form of Eq. (4-28). Assume that the receiver product detector oscillator has an error in both frequency and phase as given by Eq. (4-36). Show that the detected output is of the form of Eq. (4-37).

4-16   Consider a received SSB signal of the form of Eq. (4-32) based on a discrete-spectrum message signal of the form of Eq. (4-7). Assume that the receiver product detector oscillator has an error in both frequency and phase as given by Eq. (4-36). Show that the detected output is of the form of Eq. (4-38).

4-17   Show that the system of Fig. 4-16 produces an AM signal.

4-18   Consider the multitone signal of Prob. 4-1, and redefine it as $x_1(t)$. The signal is to be used to modulate a 100-kHz carrier with conventional AM.

    (a)   Define the normalized form of the modulating signal $x(t)$.
    (b)   Assuming a final carrier amplitude of 100, determine an expression for the modulated signal $y(t)$ in terms of m.
    (c)   Sketch the one-sided spectra for $m = 1$.
    (d)   Repeat part (c) for m = 0.5.

4-19   Consider the particular AM signal of Ex. 4-4 with the resulting form for 100% modulation as given by Eq. (4-48). Assume that $y(t)$ represents the voltage in volts at the terminals of an antenna whose input impedance is 50 Ω resistive. Determine the (a) average total power, (b) peak power, (c) unmodulated carrier power, and (d) power contained in each sideband. (e) Determine the antenna rms current I_a.

4-20   A certain AM transmitter has an unmodulated RF carrier power of 1 kW. Determine the total power, the power in each sideband, and the peak power for each of the following modulation percentages using single-tone modulation: (a) 25%, (b) 50%, (c) 100%.

4-21   The input impedance at the base of a certain 10-kW commercial broadcasting station antenna is 50 Ω resistive. For single-tone modulation, the rms ammeter at the base of the antenna reads 16 A. Determine the percentage of modulation. (*Note:* The power level as specified for commercial AM stations is the unmodulated carrier power.)

4-22   Assume that the multitone signal of Prob. 4-1 is applied to the input of a switching modulator having a carrier frequency of 500 kHz. On a positive frequency basis, list all frequencies below 3 MHz appearing in the output before filtering for (a) a single balanced modulator and (b) a double balanced modulator.

4-23 Devise a circuit in which the bridge modulator is connected as a series switch.

4-24 By comparing Eqs. (4-95) and (4-99) for a given signal level, what is the decibel difference in levels between the desired output signals (about f_c) for the single and double balanced modulators?

4-25 One classical approach to achieving a balanced modulator is illustrated by the circuit shown in Fig. P4-25. Assume that the two transistors have identical characteristics and are biased in a nonlinear region such that the collector signal current i_c for either is related to the base-to-emitter voltage v_{be} by

$$i_c = a_1 v_{be} + a_2 v_{be}^2$$

The untuned voltage on the secondary of the output transformer is then

$$v_0 = K(i_{c1} - i_{c2})$$

(a) Obtain an expression for the output v_0 before filtering, and demonstrate that one component of this output corresponds to a balanced modulator operation.

(b) By assuming an arbitrary modulating signal for $v_1(t)$ and a sinusoidal carrier for $v_2(t)$, sketch the spectrum for $V_0(f)$ and discuss the filtering requirements for DSB or SSB generation. Is this a single balanced or a double balanced modulator?

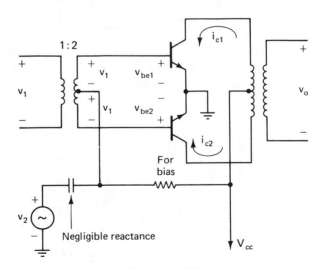

Figure P4-25

4-26 The system shown in Fig. P4-26 is a simplified *speech scrambler* used to ensure communication privacy and to foil wiretapping. By sketching the spectrum at each stage, demonstrate that the output spectrum is reversed in frequency. How would you unscramble the signal? (From A. B. Carlson, *Communications Systems,* 2nd Ed., McGraw-Hill, Book Co., New York, 1975). Assume that the filters are ideal.

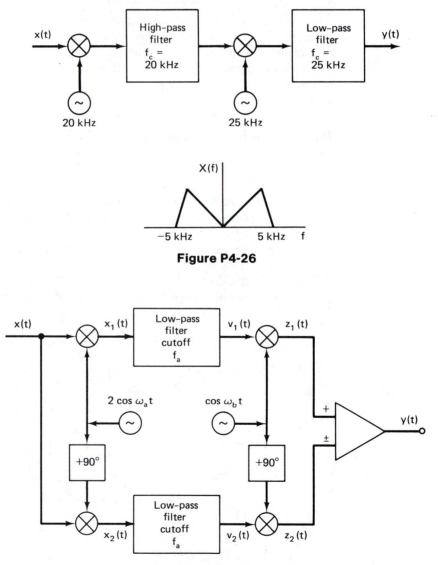

Figure P4-26

Figure P4-27

4-27 The system shown in Fig. P4-27 represents the Weaver method (also called the "third method") for generating an SSB signal. Assume that the modulating signal is a discrete-spectrum signal of the form of Eq. 4-7. The first oscillator frequency is $f_a = W/2$, where $W = Nf_1$ is the highest frequency in the baseband spectrum. The second oscillator frequency f_b can be in the RF range. Solve for the various signals in the system, that is, $x_1(t)$, $x_2(t)$, $v_1(t)$, $v_2(t)$, $z_1(t)$, and $z_2(t)$. Assume that the filters are ideal.

(a) Show that if $y(t) = z_1(t) + z_2(t)$, the output $y(t)$ is a USB SSB signal referred to an effective reference carrier frequency $f_b - W/2$.

(b) Show that if $y(t) = z_1(t) - z_2(t)$, the output $y(t)$ is an LSB SSB signal referred to an effective reference carrier frequency $f_b + W/2$.

4-28 The system shown in Fig. P4-28 represents the signal-processing strategy of a certain doppler radar velocity measurement system. The transmitter radiates a directional signal at frequency f_c, from which a return is received from the target. However, the return signal is at a frequency $f_c + f_d$, where f_d is the doppler shift. The doppler shift is directly proportional to the velocity component in the direction of the beam; that is,

$$f_d = \frac{2f_c v}{c} \cos \theta$$

where v is the velocity of the target and c is the speed of light. Analyze the signals passing through the system, and demonstrate that the dc output Y is a quantity whose magnitude is proportional to the velocity and whose sign determines the direction of the velocity (i.e, either toward the receiver or away from the receiver). The two low-pass filters are identical and have cutoff frequencies between f_d and f_c.

4-29 In some non-linear modulation operations, square-law devices are used. A square-law device has an output v which is related to the input u by

$$v = au^2 \qquad (1)$$

where a is a constant. Certain types of diodes exhibit a square-law characteristic over a portion of their operating region. In general, a square-law device changes completely the nature of a given signal, and for conventional linear operations, such an operation would be intolerable. However, certain modulation, detection, and mixing operations can be achieved with such a device if the operation is carefully controlled. Some possible applications will be demonstrated in Probs. 4-30, 4-31, and 4-32.

In this problem, the change of the bandwidth resulting from passing a baseband signal through a square-law device will be demonstrated with a specific example. Consider a baseband signal containing a dc component, a fundamental with frequency f_1, and components up to the third harmonic as

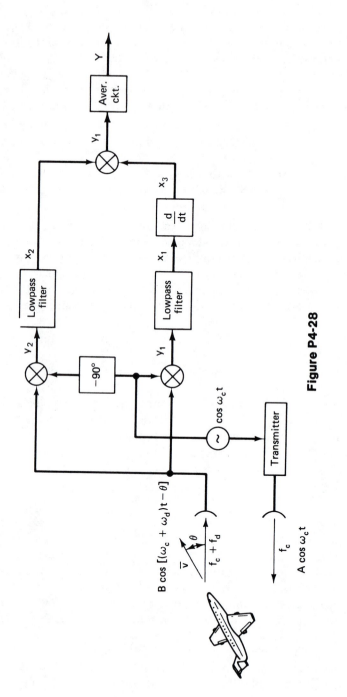

Figure P4-28

follows:

$$u = A_0 + A_1 \cos \omega_1 t + A_2 \cos 2\omega_1 t + A_3 \cos 3\omega_1 t \qquad (2)$$

Assume that this signal is applied as the input to the square-law device defined in (1). Expand the output into single frequency terms, and demonstrate that the highest cyclic frequency contained in v is $6f_1$, meaning that the bandwidth is doubled. This result can be extended to the general case as follows: *If a baseband signal with bandwidth from dc to W is applied to the input of a square-law device, the output signal has a bandwidth from dc to 2W.*

4-30   A square-law device can be used to generate a DSB signal as will be demonstrated in this problem. Consider a device whose input-output characteristic is of the form of (1) in Prob. 4-29. Assume a modulating signal $x(t)$ with baseband bandwidth W and a carrier with frequency f_c, where $f_c \gg W$. Assume that the modulating signal and the carrier term are first added as follows:

$$u = A \cos \omega_c t + x(t)$$

This signal is then applied as the input to the square-law device.

(a)   Determine the output v and list the frequency range of each of the terms.

(b)   Assume that v is passed through a filter having a passband from $f_c - W$ to $f_c + W$. Show that the filter output is a DSB signal. (Hint: refer to Prob. 4-29 for help with dealing with $x^2(t)$.)

4-31   The combination of a linear and a square-law characteristic can be used to generate an AM signal as will be demonstrated in this problem. This method is rarely used in commercial systems due to the lack of preciseness in controlling the modulation parameters, but it is often used as a "quick" method for generating AM products in routine tests where accuracy is not required. Consider a device whose output v is related to the input u by

$$v = a_1 u + a_2 u^2$$

where a_1 and a_2 are constants. Assume a *normalized* modulating signal $x(t)$ with bandwidth W. A carrier with frequency f_c, where $f_c \gg W$, is added to the modulation as follows:

$$u = A \cos \omega_c t + B x(t)$$

(a)   Determine the output v and list the frequency range of each of the terms.

(b)   Assume that the signal v is passed through a filter having a passband from $f_c - W$ to $f_c + W$. Show that the filter output is an AM signal having a modulation index

$$m = \frac{2a_2 B}{a_1}$$

4-32 It is possible, with some distortion, to detect an AM signal under limited conditions with a square-law device as will be demonstrated in this problem. Consider an AM signal of the form:

$$y = A [1 + m x(t)] \cos \omega_c t$$

where $x(t)$ is assumed to be a *normalized* modulating signal. Assume that the AM signal is applied as the input to the square-law device defined in (1) of Prob. 4-29. (a) Determine the output v, and list the frequency range of each of the terms. Note that a portion of the spectrum of $x^2(t)$ occupies the same frequency range as $x(t)$ and prevents an *exact* recovery of the modulating signal. (b) Assume that the dc component is of no value and can be removed. Assume that the signal v is passed through a filter having a passband from just above dc to W. Show that *if $m \ll 1$,* the detected output is *nearly* proportional to the intelligence $x(t)$. (However, a small distortion term will always be present.)

Angle-Modulation Methods

5

5-0 INTRODUCTION

The various forms of amplitude modulation were developed in some detail in Chapter 4. The emphasis in this chapter will be directed toward *angle-modulation* methods. With each form of amplitude modulation previously considered, it was observed that the amplitude of the composite signal was varied in some manner in accordance with the information signal. In contrast, angle-modulation signals usually have constant amplitudes, but the angle or argument of the carrier function is varied in accordance with the intelligence. The various forms of amplitude and angle modulation are often referred to collectively as *analog-modulation* methods.

The two forms of angle modulation that have received the widest emphasis in practical systems are (1) phase modulation and (2) frequency modulation. Both forms will be discussed in this chapter.

5-1 ANGLE MODULATION

Consider the function defined by

$$y(t) = A \cos \left[\omega_c t + \phi_i(t) \right] \tag{5-1}$$

For $\phi_i(t) = 0$, this equation describes an ordinary sinusoid having an amplitude A and a radian frequency ω_c. However, assume now that $\phi_i(t)$ is made to vary in some way as a controlled function of a given message signal. Except for certain special cases, the function will no longer appear as a simple sinusoid. The waveform will still oscillate between the limits of $+A$ and $-A$, but the manner and rate in which the oscillations occur are very dependent on the phase $\phi_i(t)$. This idea is illustrated in Fig. 5-1 for some arbitrary phase function.

This concept is the basis for *angle modulation,* which for the present discussion can be defined as a process in which the angle of a sinusoidal reference function is varied in accordance with a modulating signal. However,

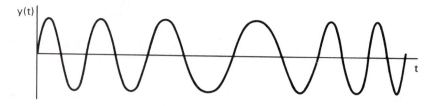

Figure 5-1. Illustration of an arbitrary angle-modulated "sinusoid."

it will be observed later that the function to which the modulation is initially applied need not be a sinusoid at all, but in most applications the final signal after processing is based on a sinusoidal reference, so that form will be assumed in the development that follows.

The two forms of angle modulation of primary interest are *phase modulation (PM)* and *frequency modulation (FM)*. The general forms of the composite signals differ only in the relationship of the angle variation to the modulating signal. In fact, it is usually not possible to tell by a simple inspection process whether a given angle-modulated signal is PM or FM. Furthermore, many modulation systems employ a combination of both PM and FM. For these reasons, it is neither necessary nor desirable to study PM and FM as completely separate concepts since many of the results are common to both.

We will now define several angular terms that will be used frequently in the developments that follow.

$$\phi_i(t) = \text{instantaneous } \textit{signal} \text{ phase angle}$$
$$\Phi_i(t) = \text{instantaneous } \textit{total} \text{ phase angle} = \omega_c t + \phi_i(t)$$
$$\omega_i(t) = \text{instantaneous } \textit{signal} \text{ radian frequency}$$
$$\Omega_i(t) = \text{instantaneous } \textit{total} \text{ radian frequency} = \omega_c + \omega_i(t)$$

Observe that for both phase and frequency there is both a *signal* function and a *total* function. The signal function in each case will be the portion that represents the direct effect of the modulating signal. The total function has an additional term, which represents the higher-frequency carrier reference. The carrier reference term ensures that the total spectrum will be shifted upward in frequency and centered at f_c for transmission purposes, as will be seen later. Note that the carrier reference term appears as $\omega_c t$ in the phase function, but appears as ω_c in the radian frequency function.

The relationships between the signal phase and frequency functions are defined as follows:

$$\omega_i(t) = \frac{d\phi_i(t)}{dt} \tag{5-2}$$

and

$$\phi_i(t) = \int_0^t \omega_i(t) \, dt \tag{5-3}$$

where it has been assumed for convenience that the initial phase angle is zero. The relationships between the total phase and frequency functions are similar and are given by

$$\Omega_i(t) = \frac{d\Phi_i(t)}{dt} \tag{5-4}$$

and

$$\Phi_i(t) = \int_0^t \Omega_i(t) \, dt \tag{5-5}$$

where again the initial phase has been assumed to be zero. Note that the total frequency and phase are related to each other in the same manner as the signal frequency and phase are related to each other.

A composite angle-modulated signal may now be expressed in terms of the preceding definitions as

$$y(t) = A \cos \Phi_i(t) \tag{5-6a}$$

$$= A \cos \left[\omega_c t + \phi_i(t)\right] \tag{5-6b}$$

The instantaneous total frequency corresponding to this function may be determined by application of (5-4), and the instantaneous signal radian frequency may be determined from (5-2). Incidentally, in final practice form, it is usually the cyclic frequency in hertz that is of direct interest, but rather than confuse the issue now with too many separate variables, we will continue working with radian frequency in many of the immediate steps that follow. When desirable, the instantaneous signal cyclic frequency will be expressed simply as $f_i(t) = \omega_i(t)/2\pi$.

The appropriateness of these definitions may be verified by considering the special example of a constant frequency sinusoid. In this case, the signal phase $\phi_i(t)$ may be assumed to be zero, and $\Phi_i(t) = \omega_c t$. Application of (5-4) yields for the instantaneous frequency $\Omega_i(t) = \omega_c$, which is a constant value. This result is certainly in total agreement with our previous concept of frequency. In the more general case, however, the definition of instantaneous frequency may or may not produce a result that is immediately intuitively obvious. After all, our earlier concept of "frequency" was based on the notion that the frequency of a sinusoid is fixed and does not change with time. Now we are facing a new concept entirely, one that defines a frequency as a time-varying quantity, and the results may be difficult to grasp at first.

We will now investigate the manner in which PM and FM are generated from an arbitrary modulating signal. As was done for conventional AM in Chapter 4, it is convenient to consider the modulating signal normalized to a maximum magnitude of unity. Let

$x(t)$ = normalized modulating signal (maximum magnitude = 1)

$\Delta\phi$ = maximum phase deviation in radians (referred to carrier reference phase $\omega_c t$)

$\Delta\omega$ = maximum radian frequency deviation in radians per second (referred to carrier reference frequency ω_c)

Δf = maximum cyclic frequency deviation in hertz = $\Delta\omega/2\pi$

Phase Modulation (PM)

With PM, the signal phase $\phi_i(t)$ is made to be proportional to the modulating signal.

$$\phi_i(t) = \Delta\phi\, x(t) \tag{5-7}$$

The total phase is obtained by adding the carrier phase to (5-7), and the resulting composite PM function is

$$y(t) = A \cos\left[\omega_c t + \Delta\phi\, x(t)\right] \tag{5-8}$$

The instantaneous radian frequency for the PM signal is

$$\Omega_i(t) = \omega_c + \Delta\phi\, \frac{dx(t)}{dt} \tag{5-9}$$

For the PM signal, the instantaneous signal phase is seen to be directly proportional to the modulating signal, but the instantaneous signal radian frequency is directly proportional to the derivative of the modulating signal.

Frequency Modulation (FM)

With FM, the instantaneous signal radian frequency $\omega_i(t)$ is made to be proportional to the modulating signal.

$$\omega_i(t) = \Delta\omega\, x(t) \tag{5-10}$$

The instantaneous total radian frequency $\Omega_i(t)$ is obtained by adding the carrier frequency ω_c to (5-10).

$$\Omega_i(t) = \omega_c + \Delta\omega\, x(t) \tag{5-11}$$

The instantaneous total phase $\Phi_i(t)$ corresponding to this function is

$$\Phi_i(t) = \int_0^t \Omega_i(t)\, dt = \omega_c t + \Delta\omega \int_0^t x(t)\, dt \tag{5-12}$$

where the initial phase angle has been assumed to be zero. The composite FM signal is

$$y(t) = A \cos\left[\omega_c t + \Delta\omega \int_0^t x(t)\, dt\right] \tag{5-13}$$

The mathematical forms for PM and FM have been developed in the preceding few paragraphs, and a comparison of (5-8) and (5-13) reveals the differences. In both cases, the phase and the frequency will change in some manner as the modulating signal changes. However, a direct relationship for phase exists with PM and a direct relationship for frequency exists for FM with respect to the modulating signal. Because of the derivative/integral relationships between PM and FM, it is possible to use some of the modulator and detector circuits designed for PM with FM, and vice versa.

Consider the PM modulator block shown in Fig. 5-2a. The circuit produces an output in which the signal phase is directly proportional to the modulating signal at the input. However, if the signal is integrated first, comparison of Eqs. (5-8) and (5-13) leads to the conclusion that the output will actually be an FM signal with respect to the modulating signal. Next consider the FM modulator block shown in Fig. 5-2b. This circuit produces an output in which the signal frequency is proportional to the modulating signal at the input. If the signal is differentiated first, the output will actually be a PM signal with respect to the modulating signal.

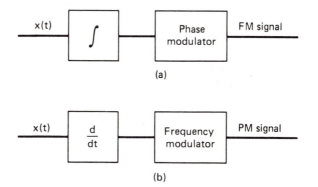

(a)

(b)

Figure 5-2. (a) A PM modulator may be used to generate an FM signal if the modulating signal is integrated first. (b) An FM modulator may be used to generate a PM signal if the modulating signal is differentiated first.

Consider now the PM detector shown in Fig. 5-3a. This circuit produces an output voltage proportional to the signal phase of the modulated signal at the input. The circuit may be used to properly demodulate an FM signal if a differentiator circuit processes the signal at the output of the detector. Finally, the FM detector circuit shown in Fig. 5-3b may be used to properly demodulate a PM signal if an integrator circuit processes the signal at the output of the detector.

A final important point for consideration in this section is the question of the power contained in an angle-modulated signal. Assuming an amplitude A, the normalized power generated by the unmodulated carrier in a reference 1-Ω resistor is clearly $P = A^2/2$. However, does this power change when modulation is applied? We recognize that the signal still oscillates over the same amplitude range as before, so one might intuitively suspect that the

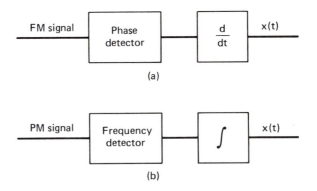

Figure 5-3. (a) A PM detector may be used to detect an FM signal if the detected signal is differentiated. (b) An FM detector may be used to detect a PM signal if the detected signal is integrated.

average power remains the same. This suspicion is indeed confirmed, and it can be shown by a rigorous development that the average power remains the same *provided* the average is taken over a sufficiently long interval. This restriction is made because it is possible to define certain modulating signals in which a short-term average of the total power might lead to a different result. Since average power is usually defined in a manner to ensure a long-term trend, this point need not cause concern in a practical sense. To summarize, the long-term average power in an angle-modulated signal remains constant with or without modulation and is $P = A^2/2$ in a 1-Ω reference. This point often results in a simpler approach to the design for the power-amplifier stages in an angle-modulated transmitter as compared with one for AM.

Example 5-1

A certain low-frequency function generator can be angle modulated with various wave forms. Based on a 2-Hz sinusoid before modulation is applied, the modulated output signal under certain conditions is given by

$$y(t) = 10 \cos (4\pi t + \pi t^2) \qquad (5\text{-}14)$$

Considering that the 2-Hz reference represents the carrier frequency, determine (a) instantaneous total phase, (b) instantaneous signal phase, (c) instantaneous total frequency (radian and cyclic), and (d) instantaneous signal frequency (radian and cyclic). (e) Sketch the forms of the instantaneous signal phase and frequency.

Solution

(a) From direct inspection of (5-14) and a comparison with (5-8), the instantaneous total phase is

$$\Phi_i(t) = 4\pi t + \pi t^2 \qquad (5\text{-}15)$$

(b) The carrier term is $4\pi t$ since this term represents a cyclic frequency of $4\pi/2\pi = 2$ Hz. The instantaneous signal phase is thus

$$\phi_i(t) = \pi t^2 \tag{5-16}$$

Since this function grows without bound as t increases, it is not possible to define a peak frequency deviation $\Delta\phi$.

(c) The instantaneous total radian frequency is

$$\Omega_i(t) = \frac{d\Phi_i(t)}{dt} = 4\pi + 2\pi t \tag{5-17}$$

The corresponding instantaneous total cyclic frequency in hertz is

$$\frac{\Omega_i(t)}{2\pi} = 2 + t \tag{5-18}$$

(d) The instantaneous signal radian frequency is

$$\omega_i(t) = 2\pi t \tag{5-19}$$

The corresponding instantaneous signal cyclic frequency is

$$f_i(t) = \frac{\omega_i(t)}{2\pi} = t \tag{5-20}$$

(e) The instantaneous signal phase and frequency functions, as given by (5-16) and (5-20), respectively, are shown in Fig. 5-4. It is impossible to state whether the given angle-modulated signal is PM or FM unless the form of the modulating signal is specified. The reader should verify that the signal could be called a PM signal if the modulating signal were known to be a parabolic waveform, and the signal could be called an FM signal if the modulating signal were known to be a straight-line (ramp) waveform.

5-2   SINGLE-TONE PHASE AND FREQUENCY MODULATION

As a matter of important practical interest, we will now consider the forms for both PM and FM signals when the modulating signal is a single frequency tone. Assume that the modulating signal is a unit amplitude sinusoid of the form

$$x(t) = \cos \omega_m t \tag{5-21}$$

Phase Modulation

The signal phase $\phi_i(t)$ will be directly proportional to $x(t)$ and will have the form

$$\phi_i(t) = \Delta\phi \cos \omega_m t \tag{5-22}$$

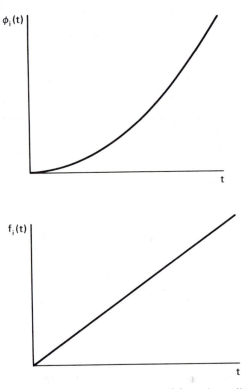

Figure 5-4. Instantaneous signal phase $\phi_i(t)$ and cyclic frequency $f_i(t)$ of signal of Example 5-1.

The total phase will be

$$\Phi_i(t) = \omega_c t + \Delta\phi \cos \omega_m t \qquad (5\text{-}23)$$

The composite signal $y(t)$ for tone-modulated PM will thus be

$$y(t) = A \cos (\omega_c t + \Delta\phi \cos \omega_m t) \qquad (5\text{-}24)$$

The instantaneous signal frequency corresponding to the PM signal can be readily determined to be

$$\omega_i(t) = \frac{d\phi_i(t)}{dt} = -\omega_m \Delta\phi \sin \omega_m t \qquad (5\text{-}25)$$

Observe that the peak frequency deviation of the PM signal is $\omega_m \Delta\phi$ (in radians per second) or $f_m \Delta\phi$ (in hertz), which means that the peak frequency deviation for PM is directly proportional to the modulating frequency.

Frequency Modulation

The signal frequency $\omega_i(t)$ will be directly proportional to $x(t)$ and will thus have the form

$$\omega_i(t) = \Delta\omega \cos \omega_m t \tag{5-26}$$

This results in an instantaneous signal phase of

$$\phi_i(t) = \int_0^t \omega_i(t) \, dt = \frac{\Delta\omega}{\omega_m} \sin \omega_m t \tag{5-27}$$

where an initial zero phase angle has been assumed. The total phase is

$$\Phi_i(t) = \omega_c t + \frac{\Delta\omega}{\omega_m} \sin \omega_m t \tag{5-28}$$

where again the initial phase angle has been set to zero. The composite modulated signal is then given by

$$y(t) = A \cos \left(\omega_c t + \frac{\Delta\omega}{\omega_m} \sin \omega_m t \right) \tag{5-29}$$

An important parameter β, which is called the *modulation index*, will be defined as

$$\beta = \frac{\Delta\omega}{\omega_m} = \frac{\Delta f}{f_m} \tag{5-30}$$

Substituting β in (5-29), the single-tone modulated FM signal can be expressed as

$$y(t) = A \cos \left(\omega_c t + \beta \sin \omega_m t \right) \tag{5-31}$$

As we will see in the next section, β is a very important parameter in determining the transmission bandwidth. Observe that the modulation index is the ratio of the *maximum frequency deviation* to the *particular modulating frequency*. With pure FM, the instantaneous frequency deviation $\Delta\omega$ (or Δf) is a constant at all frequencies, so β varies inversely with frequency; that is, it has its largest value at the lowest frequency and its smallest value at the highest frequency.

Some interesting and useful comparisons between PM and FM can be made from these results for tone modulation. Assume that $\Delta\phi$ is fixed for some reference PM modulator, and assume that $\Delta\omega$ is fixed for a similar reference FM modulator. Assume that a unit sinusoidal modulating signal is applied to each modulator, and consider the results as the frequency of the sinusoid changes. In each case we will look at the *peak* (or amplitude) of the phase and frequency deviation as a function of the frequency of the modulating sinusoid.

From Eq. (5-22), the peak value of the phase deviation for the PM modulator does not change with frequency, and so the result is as shown in Fig. 5-5a. However, from Eq. (5-25), the peak frequency deviation for the PM modulator is a linearly increasing function of frequency as shown in Fig. 5-5b. The minus sign in (5-25) is ignored since the magnitude of the sinusoid is still considered positive.

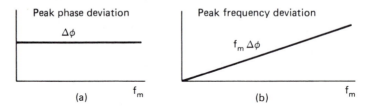

Figure 5-5. (a) Peak phase deviation, and (b) peak frequency deviation as a function of modulating frequency for a PM modulator.

From Eq. (5-26), the peak value of the frequency deviation for the FM modulator does not change with frequency, and this result is shown in Fig. 5-6a. However, from Eq. (5-27), the peak phase deviation for the FM modulator is a decreasing function of frequency as shown in Fig. 5-6b.

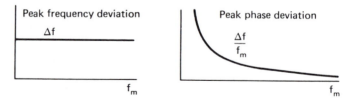

Figure 5-6. (a) Peak frequency deviator, and (b) peak phase deviation as a function of modulating frequency for an FM modulator.

The forms of the actual PM and FM signals are illustrated in Fig. 5-7 along with the modulating sinusoid. Observe the close similarities between the two modulated functions, but note that similar portions of a given cycle occur at different times.

One final point concerning the forms of tone-modulated PM and FM will be made. A comparison of (5-24) for PM and (5-29) for FM might lead one to the erroneous conclusion that a cosine function always appears in the argument for the PM signal, and a sine function always appears in the argument for an FM signal. However, this pattern developed only because the assumed modulating tone was a cosine function. When the modulating signal is a sine function, the opposite pattern occurs, and verification of this point will be left as an exercise for the reader (Prob. 5-10). If an arbitrary angle-modulated signal having tone modulation is given, it may not be

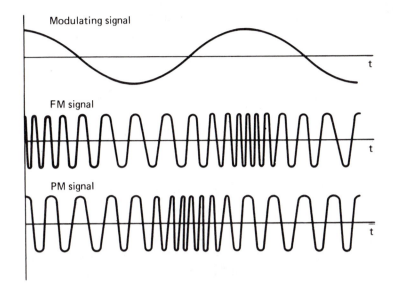

Figure 5-7. Sinusoidal modulating signal and the corresponding forms for the FM and PM signals.

possible to identify whether the signal is PM or FM without further information.

Example 5-2

A certain FM tone-modulated signal is given by

$$y(t) = 100 \cos (2\pi \times 10^8 \, t + 20 \sin 2\pi \times 10^3 \, t) \qquad (5\text{-}32)$$

Determine the (a) unmodulated carrier frequency in hertz, (b) modulating frequency in hertz, (c) modulation index β, and (d) maximum frequency deviation Δf in hertz. (e) If the signal is a voltage in volts, determine the average power dissipated in a 50-Ω resistive load.

Solution

(a) In this part and most of the parts that follow, the given function can be compared with (5-29) and (5-31). From the carrier phase term $2\pi \times 10^8 \, t$, the unmodulated carrier frequency is $f_c = 10^8$ Hz = 100 MHz.

(b) The radian modulating frequency is identified from the sine function in the argument as $\omega_m = 2\pi \times 10^3$ rad/s. Hence, $f_m = 1$ kHz.

(c) The modulation index is immediately identified as $\beta = 20$.

(d) Since $\beta = \Delta f / f_m$, and both β and f_m are known, the frequency deviation is determined to be $\Delta f = \beta f_m = 20 \times 1$ kHz = 20 kHz.

(e) Assuming that the signal is a voltage, the average power dissipated in a 50-Ω resistor is $P = (100)^2/(2 \times 50) = 100$ W. As explained earlier, this power is the same as the unmodulated carrier power.

Example 5-3

Assuming that the function given in Ex. 5-2, that is, Eq. (5-32), is a PM signal, determine (a) maximum phase deviation and (b) the functional form of the normalized modulating signal.

Solution

(a) The mathematical form of the argument is not quite the same as that of (5-24) due to a difference in the phase of the modulating signal, but, as already explained, this is an arbitrary choice. The maximum phase deviation is readily seen to be $\Delta\phi = 20$ rad. Note that the maximum phase deviation for the PM tone-modulated signal has the same numerical value as β when the function was considered to be an FM signal. In general, β for FM and $\Delta\phi$ for PM have similar properties at a *given frequency*. However, if the frequency were changed, but the amplitude of the modulating signal remained constant, the value 20 would remain constant if the signal were PM and would change if the signal were FM. (Why?)

(b) The functional form of the normalized modulating signal is readily determined as

$$x(t) = \sin 2\pi \times 10^3 \, t \qquad (5\text{-}33)$$

Example 5-4

A certain 100-MHz sinusoidal carrier is to be frequency modulated by a 2-kHz tone. The maximum frequency deviation is to be adjusted to ± 10 kHz. (a) Write an expression for the composite FM signal assuming cosine functions for both the modulating signal and the carrier. (b) Assume now that the carrier is to be phase modulated by the same tone, and the maximum phase deviation is to be adjusted to ± 3 rad. Write an expression for the composite PM signal making the same assumptions as in part (a).

Solution

(a) Reviewing the form of the tone-modulated FM signal, it is observed that a cosine modulating signal results in a sine term in the argument due to the integration of frequency to obtain phase. The modulation index is $\beta = \Delta f/f_m = 10$ kHz/2 kHz $= 5$. Assuming an arbitrary amplitude A for the signal, we have

$$y(t) = A \cos (2\pi \times 10^8 \, t + 5 \sin 4\pi \times 10^3 \, t) \qquad (5\text{-}34)$$

(b) For the PM signal, the phase variation in the argument has the same form as the modulating signal, and the peak signal phase deviation is 3 rad. Hence,

$$y(t) = A \cos (2\pi \times 10^8 t + 3 \cos 4\pi \times 10^3 t) \qquad (5\text{-}35)$$

5-3   SPECTRUM OF TONE-MODULATED FM SIGNAL

So far in this chapter, angle modulation has been considered from a somewhat general point of view, and equal emphasis has been given to both PM and FM. In actual practice, FM has some advantages that make it more desirable for most (but not all) applications than PM. (The advantages of FM will be partially considered later in this chapter, but a full development will be delayed until its performance in the presence of noise can be considered in Chapters 10 and 11.)

Instead of continuing with the dual developments providing complete results for both PM and FM, the developments in this section and in portions of later sections will focus specifically on FM signals. However, in most cases, the results can be adapted to PM, as will be shown where appropriate.

The primary objective in this section is to investigate the form of the spectrum of a tone-modulated FM signal and to use this result as a basis for determining the bandwidth requirements for complex modulating signals. With the various forms of AM, it was possible in most cases to derive complete expressions for the spectra even when the modulating signals were complex. Unfortunately, this is not the case with FM. It is possible to derive meaningful closed-form expressions of FM spectra only for a few special modulating waveforms. For this reason, much emphasis has been directed in FM analysis to the problem of estimating the transmission bandwidth required for complex waveforms by using approximate bounds determined from simpler waveforms. Fortunately, appropriate estimates have been developed, and the success of the many operating FM systems certainly indicates the validity of this approach.

The analysis begins with the form of a single-tone modulated signal as given by Eq. (5-31), which is repeated here for convenience.

$$y(t) = A \cos (\omega_c t + \beta \sin \omega_m t) \qquad (5\text{-}36)$$

The problem confronting us is how to determine the spectrum of the signal of (5-36). An attempt at evaluating the frequency content of the signal using a basic definition of spectrum is a most difficult chore indeed! Fortunately, certain tabulated series expansions may be used in an indirect manner, as will be seen shortly. The first step is to expand $y(t)$ using the standard trigonometric identity for the cosine of the sum of two angles. This yields

$$y(t) = A \cos \omega_c t \cos (\beta \sin \omega_m t) - A \sin \omega_c t \sin (\beta \sin \omega_m t) \qquad (5\text{-}37)$$

The next step involves using appropriate closed-form expansions appearing in advanced mathematics texts to simplify the sinusoidal terms with sinusoidal arguments. These expansions make use of Bessel functions of the first kind, whose values are well tabulated in the literature. The specific expansions involved are the following:

$$\cos(\beta \sin \omega_m t) = J_0(\beta) + \sum_{\substack{n=2 \\ n \text{ even}}}^{\infty} 2J_n(\beta) \cos n\omega_m t \qquad (5\text{-}38)$$

$$\sin(\beta \sin \omega_m t) = \sum_{\substack{n=1 \\ n \text{ odd}}}^{\infty} 2J_n(\beta) \sin n\omega_m t \qquad (5\text{-}39)$$

The function $J_n(\beta)$ represents a Bessel function of the first kind, of order n and argument β. Note that for a particular value of n and a particular value of β, the quantity $J_n(\beta)$ is a single value, so the presence of the Bessel functions should not obscure the trigonometric forms of the expansions in (5-38) and 5-39).

When the expansions in (5-38) and (5-39) are substituted in (5-37), two series involving sinusoidal product terms are obtained. In each series, the terms will each involve the product of a sinusoid with frequency ω_m and a sinusoid with frequency ω_c. The identities provided at the beginning of Chapter 4, that is, Eqs. (4-1) and (4-2), may be used to simplify the expansions. The reader is invited to show that the result obtained after simplification is

$$\frac{y(t)}{A} = J_0(\beta) \cos \omega_c t - J_1(\beta) \cos(\omega_c - \omega_m)t + J_1(\beta) \cos(\omega_c + \omega_m)t \qquad (5\text{-}40)$$

$$+ J_2(\beta) \cos(\omega_c - 2\omega_m)t + J_2(\beta) \cos(\omega_c + 2\omega_m)t$$

$$- J_3(\beta) \cos(\omega_c - 3\omega_m)t + J_3(\beta) \cos(\omega_c + 3\omega_m)t + \ldots$$

The form of (5-40) suggests that there are an infinite number of components in the spectrum. If one accepted this thought without reason, it could lead to the obviously incorrect conclusion that an infinite bandwidth would be required to transmit an FM signal. The fact is that for a given β the various terms in the expansion diminish rapidly beyond a certain point in the series so that the spectrum of (5-40) can be truncated to a finite practical bandwidth, as will be seen shortly.

The form of the spectrum as deduced from (5-40) is quite interesting and leads to some important properties. A component having the frequency f_c appears at the center of the spectrum. This frequency is, of course, the same as the frequency of the unmodulated sinusoid, so this term is referred to as the carrier for obvious reasons. All the other terms are collectively referred to as sidebands, and, as previously noted, there are theoretically an infinite number of such components. Spectral frequencies greater than f_c represent the upper sideband components, and frequencies less than f_c represent the lower sideband components. All components are displaced apart by a fixed frequency increment equal to the modulating frequency f_m. The *magnitudes* of components an equal frequency increment on either side of the carrier are equal; that is, the magnitudes of spectral components have even symmetry about f_c. However, the signs of components an odd integer multiple of f_m on one side of the carrier are the opposite of those components on the other side. A representative spectrum is shown in Fig. 5-8. The components having negative signs are actually inverted in this figure to reinforce the preceding discussion. In subsequent spectral diagrams, only the magnitudes will be shown.

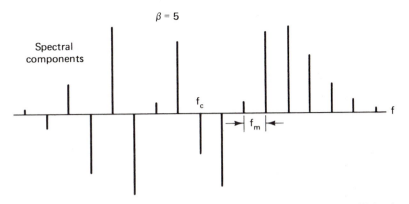

Figure 5-8. Representative spectrum of tone-modulated FM signal exhibiting signs of components.

A point that may be puzzling to the reader is the fact that the Fourier expansion of the FM signal results in a series of *fixed-frequency* sinusoids, which seems to be contradictory to the basic assumption that the FM signal has a frequency that is *varying* with time. This paradox is resolved by recognizing that two separate concepts of frequency are at work here. The concept of *instantaneous frequency* is a dynamic parameter relating the variation of an angular function with a modulating signal, and the resulting composite signal is no longer a single frequency. On the other hand, the concept of *spectral frequency* permits us to determine those spectral components that comprise the signal in a Fourier sense, and each of the components is a steady-state sinusoid with a fixed frequency. If all the fixed-frequency components were added together in the form described by the Fourier expansion, the result would definitely be the FM signal in which the instantaneous frequency varied with time.

An important question for consideration at this point is that of determining the approximate transmission bandwidth required for an FM signal, since we clearly cannot employ an infinite bandwidth. Expressed differently, how many terms in the expansion of (5-40) are needed for reproduction such that the resulting signal will have negligible distortion? Since the components in the spectrum are spaced apart by f_m, we would intuitively expect the transmission bandwidth to be heavily influenced by f_m. However, from observing the form of the FM signal, we would expect the bandwidth possibly to be greater, and perhaps much greater, than the corresponding bandwidth for an AM signal with the same modulating signal. We might also suspect that the bandwidth will be related in some way to the modulation index.

The process of estimating the practical bandwidth required is subject to different interpretations and depends on what type of criterion is employed. The very curious reader who is willing to search through different textbooks may be surprised to find somewhat different formulas and estimation rules, with the result that different bandwidth values could be obtained for the same specified conditions. The differences lie in the nature of the criteria employed and the assumptions made in the approximations involved. In actual practice, enough leeway is usually allowed in the design process to accommodate reasonable differences in these estimates.

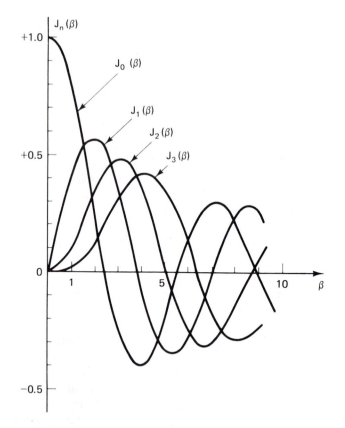

Figure 5-9. The Bessel functions $J_n(\beta)$ for $n = 0, 1, 2,$ and 3.

Most estimates for FM bandwidth computations have been developed from detailed studies of the properties of the Bessel sideband components as a function of modulation index. The graphical forms of the Bessel functions of several orders as a function of β are illustrated in Fig. 5-9. Since only limited accuracy can be obtained from the curves, several tabulated values are given in Table 5-1. Only values greater than 1% of the carrier level are shown, so it is possible to determine the number of sideband components in each case that have magnitudes greater than the 1% level. Incidentally, this particular choice is one that has been used as a criterion for bandwidth, but it is not as easy to work with as the one we will consider in the next section.

5-4 CARSON'S RULE FOR ESTIMATING FM AND PM BANDWIDTH

The particular procedure that will be employed in this book for estimating transmission bandwidth for FM (and also for PM) is known as *Carson's rule*. It is based on the criterion that the number of sidebands selected should be the minimum number that will result in transmission of no less than 98% of the total power. This choice may seem arbitrary, but the particular criterion results in a rather simple bandwidth formula. In practice, Carson's rule has proved to be quite adequate for initial estimates, and it can be modified if necessary.

Table 5-1. Selected Values of the Bessel Functions.[a]

n			$J_n(\beta)$			
	$\beta = 0.1$	$\beta = 0.5$	$\beta = 1$	$\beta = 2$	$\beta = 5$	$\beta = 10$
0	0.9975	0.9385	0.7652	0.2239	−0.1776	−0.2459
1	0.0499	0.2423	0.4401	0.5767	−0.3276	0.04347
2		0.03125	0.1149	0.3528	0.04657	0.2546
3			0.01956	0.1289	0.3648	0.05838
4				0.03400	0.3912	−0.2196
5					0.2611	−0.2341
6					0.1310	−0.01446
7					0.05338	0.2167
8					0.01841	0.3179
9						0.2919
10						0.2075
11						0.1231
12						0.06337
13						0.02897
14						0.01196

[a]For a given β, only components greater than 1% of the peak carrier level are shown.

The basis for Carson's rule is that if β assumes integer values, it can be verified by computation that the number of sidebands on either side of the carrier required for transmission of no less than 98% of the power is always $\beta + 1$ sidebands. For example, if $\beta = 3$, at least 98% of the power is transmitted if 4 sidebands on either side of the carrier are selected, resulting in the need to transmit 8 sidebands (plus the carrier). The formula is then extended to noninteger values of β, and so Carson's rule for transmission bandwidth B_T can be stated as

$$B_T = 2(1 + \beta)f_m \qquad (5\text{-}41)$$

An alternative, and equally useful, form for Carson's rule is obtained by expanding (5-41) and recognizing that $\beta f_m = \Delta f$. This form reads

$$B_T = 2(\Delta f + f_m) \qquad (5\text{-}42)$$

Both forms will be used freely throughout the book.

Two limiting cases of bandwidth will next be considered. The first is called *narrowband FM (NBFM)* and is the range for very small β. In this range, Carson's rule from (5-41) reduces to

$$B_T \simeq 2f_m \qquad (5\text{-}43)$$

This value of transmission bandwidth is the same as for AM and indicates that only one sideband on either side of the carrier is significant. The exact boundary for NBFM is a matter of interpretation, but we will somewhat arbitrarily select the range $\beta \le 0.25$ for defining NBFM. One might question whether (5-43) is a good approximation for (5-41) at the upper range ($\beta = 0.25$). However, this is more of a limitation of Carson's rule than of the approximation between (5-41) and (5-43). In later computations, the simplified form of (5-43) will be used when $\beta \le 0.25$.

Wideband FM (WBFM) can be defined as that range of β in which the transmission bandwidth is greater than the corresponding bandwidth for AM with the same modulating frequency. Having defined $\beta \le 0.25$ as the region for NBFM, one might infer that the equality $\beta > 0.25$ could be used to define WBFM. However, the primary advantages that can be achieved through WBFM do not appear at the lower end of this range, so in practice reference to WBFM usually implies a value of β typically *much larger than 0.25*.

As β increases to a very large value, Carson's rule from (5-42) reduces to the approximate value

$$B_T \simeq 2\Delta f = 2\beta f_m \qquad (5\text{-}44)$$

In this "very wideband" range, all significant spectral components lie in the frequency range governed by the instantaneous frequency deviation (i.e., $\pm \Delta f$). At lower values of β, some significant components appear at frequencies outside the range of the instantaneous frequency deviation. The exact value at which the approximation of (5-44) can be applied is somewhat arbitrary, but the practice will be made in this book that this limiting form will be used whenever $\beta \geq 100$.

Some of the other bandwidth criteria appearing in the literature tend to produce bandwidths somewhat greater than the values predicted by Carson's rule. Some of these methods involve graphical results and are not as easy to interpret and apply as Carson's rule. In actual practice, some additional leeway will usually be required for worst-case conditions, and actual system adjustments will be made. Considering these facts, Carson's rule is widely used for its simplicity and ease of interpretation, and it will serve as the standard method for this book.

To summarize this section up to this point, Carson's rule as given by either (5-41) or (5-42) may be used to estimate the bandwidth required for transmission of a tone-modulated FM signal. The bandwidth is centered at the carrier frequency f_c. The approximate relationship of (5-43) can be employed for NBFM (i.e., $\beta \leq 0.25$). Conversely, when β is very large (e.g., 100 or more), the approximate relationship of (5-44) can be employed. However, these two limiting cases merely represent conveniences, and one would still be reasonable to apply the basic formula to all cases.

The variation of the spectra with different modulation parameters will be illustrated with some spectral plots. Consider the magnitude plots shown in Fig. 5-10 in the vicinity of f_c. The modulating frequency is fixed at $f_m = 1$ kHz in all these plots, so the spacing between components is fixed at 1 kHz for all cases. In Fig. 5-10a, $\Delta f = 100$ Hz and $\beta = 100/1000 = 0.1$, so the result is NBFM. Only two sideband components are significant, and $B_T = 2$ kHz. In Fig. 5-10b, $\Delta f = 1$ kHz and $\beta = 1000/1000 = 1$. Two sideband components on either side of the carrier are significant and $B_T = 4$ kHz. In Fig. 5-10c, $\Delta f = 10$ kHz and $\beta = 10/1 = 10$. In this case, it is assumed that there are $10 + 1 = 11$ significant components on either side of the carrier, and $B_T = 22$ kHz. Although not shown, if Δf were increased to, say, 100 kHz, in which case $\beta = 100/1 = 100$, the bandwidth would increase to approximately $B_T = 200$ kHz.

The plots shown in Fig. 5-11 illustrate the effect of holding Δf constant and changing f_m. In all these plots, the frequency deviation is fixed at $\Delta f = 10$ kHz. In Fig. 5-11a, the modulating frequency is $f_m = 40$ kHz, and $\beta = 10/40 = 0.25$. The result is NBFM and $B_T = 80$ kHz. In Fig. 5-11b, the modulating frequency is reduced to $f_m = 10$ kHz, and $\beta = 10/10 = 1$. The bandwidth is $B_T = 40$ kHz. In Fig. 5-11c, the modulating frequency is assumed to be reduced all the way to 100 Hz, and $\beta = 10,000/100 = 100$. The bandwidth is very nearly $B_T = 20$ kHz, and all significant components lie in the range governed by the ± 10-kHz frequency swing. In view of the large number of components in the frequency spectrum, a rectangular block has been shown on the figure to define the proper range.

The general estimates that have been developed were based on single-tone modulation, so a fundamental question is how to determine the bandwidth

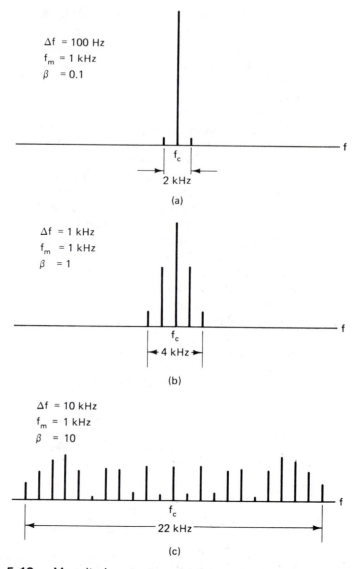

Figure 5-10. Magnitude spectra of tone-modulated FM signal with f_m fixed for different values of Δf.

required for more complex modulating signals. The FM spectra resulting from most modulating signals are quite complex and will not be considered here. Fortunately, the results for the tone-modulated signal can be used to predict approximate bandwidth limits for general modulating signals.

The strategy involved is to select the set of conditions for the complex modulating signal that would produce the largest transmission bandwidth

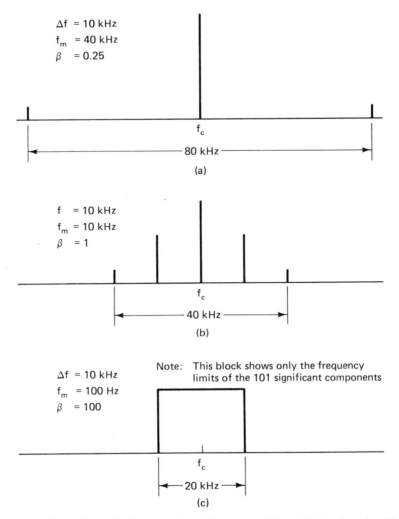

Figure 5-11. Magnitude spectra of tone-modulated FM signal with Δf fixed for different values of f_m.

when viewed from a single-tone modulation basis. Since it is assumed that we are dealing with the basic form of FM at this time, the frequency deviation Δf does not vary with frequency. With this condition, the highest modulating frequency determines the bandwidth. (This point will be demonstrated in detail in Ex. 5-5.)

A particular parameter of interest when dealing with complex modulating signals is the *deviation ratio D*. This quantity is defined as

$$D = \text{minimum value of } \beta = \frac{\Delta f}{W} \qquad (5\text{-}45)$$

where W represents the highest possible modulating frequency (baseband bandwidth). If the deviation ratio and the highest modulating frequency are specified, the appropriate bandwidth may be determined from Carson's rule. The parameter D is treated the same as β in the formulas, and W is treated the same as f_m. Observe that the minimum value of β determines the bandwidth for the case when Δf is fixed (pure FM) since this value occurs at the highest frequency.

A final point of interest concerns the application of the preceding results to a PM signal. The development of a Fourier expansion for a tone-modulated PM signal follows the same general approach employed with the FM signal. The results differ only in minor sign and phase details and in the fact that $\Delta \phi$ appears in place of β in the Bessel function arguments. The magnitudes of the spectral terms have exactly the same forms as for FM with $J_n(\Delta \phi)$ replacing $J_n(\beta)$. Thus, the bandwidth rules as developed for FM apply directly to PM tone modulation with β for FM being replaced by $\Delta \phi$ for PM.

For PM with a complex modulating signal, it is noted that $\Delta \phi$, unlike β, remains constant with frequency, so the parameter D has no significance for PM. Instead, the maximum bandwidth is determined from the specified $\Delta \phi$ and the highest modulating frequency W. Carson's rule is then applied with $\Delta \phi$ treated the same for PM as β was used for FM.

Example 5-5

The commercial FM broadcasting system existing in the United States employs the following parameters for monaural transmission as specified by the Federal Communications Commission (FCC): maximum frequency deviation = ± 75 kHz and highest modulating frequency = 15 kHz. (a) For single-tone frequency modulation and a fixed deviation of ± 75 kHz, determine the approximate transmission bandwidth for each of the following frequencies: 25 Hz, 75 Hz, 750 Hz, 1.5 kHz, 5 kHz, 10 kHz, 15 kHz. (b) Determine the approximate transmission bandwidth required when the FM transmitter is modulated by a complex audio signal whose highest frequency is 15 kHz.

Solution

(a) For each of the specific modulating frequencies, the value of $\beta = \Delta f / f_m$ should be determined, and the appropriate bandwidth formula can then be used. The results are tabulated in Table 5-2, and some of the trends will be briefly discussed. For relatively low modulating frequencies, the values of β are quite large and fall into the upper range of WBFM. Thus, for the first three entries in the table, the bandwidth is $B_T = 2\Delta f = 150$ kHz. Eventually, β drops to the lower range of WBFM where $\beta + 1$ sidebands on either side of

Table 5-2. Data for Example 5-5

f_m	β	B_T (kHz)
25 Hz	3000	150
75 Hz	1000	150
750 Hz	100	150
1.5 kHz	50	153
5 kHz	15	160
10 kHz	7.5	170
15 kHz	$D = 5$	180

the carrier are considered significant, and the remaining results were determined on that basis. Note that none of the results falls in the NBFM range.

(b) It is readily observed from Table 5-2 that the highest modulating frequency results in the largest bandwidth, as expected. Note that the deviation ratio is $D = \Delta f / W = 75$ kHz/15 kHz = 5, and this value of β results in the largest bandwidth. The transmission bandwidth for the composite modulating signal is approximately $B_T = 180$ kHz. Commercial FM channels are spaced 200 kHz apart, which provides some additional guard band and a transition region between channels. In addition, two stations in the same geographical area are not normally assigned to adjacent channels, so this serves as an additional safeguard.

The preceding analysis was chosen to illustrate the use of the approximations for estimating bandwidth required, and the results are as accurate as the approximations permit for the data given. In actual practice, however, commercial FM systems employ preemphasis to the modulating frequency to improve the signal-to-noise ratio. (This concept will be explained in Chapters 10 and 11.) The result is that Δf does not remain constant over the entire modulating frequency range as assumed here, and so commercial FM is not really "pure FM" by the basic definition. More will be said about this later, but the results of this problem would be directly applicable to commercial monaural FM without preemphasis, and the overall results do not differ significantly from those obtained with preemphasis.

One more property concerning FM bandwidth will be noted before leaving this example. Observe from Table 5-2 that there is not as much of a variation in bandwidth as one might expect for such a wide variation in the signal frequency. Indeed, while the modulating frequency varies over a very wide range from 25 Hz to 15 kHz, the transmission bandwidth varies only over a range from 150 to 180 kHz. In fact, in the lower part of the modulating frequency range from about 25 to 750 Hz or so, the bandwidth is virtually constant. These results are shown graphically as curve (a) in Fig. 5-12. A smooth curve has been extrapolated between points. The property of reasonably constant bandwidth as a function of modulation frequency is generally

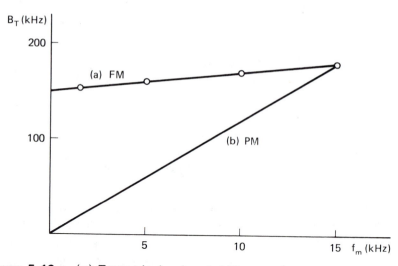

Figure 5-12. (a) Transmission bandwidth as a function of modulating frequency for FM, and (b) for PM, using conditions given in Examples 5-5 and 5-6.

true with *wideband FM systems* (but not with narrowband FM) and is one of the features that contrasts FM from PM, as will be illustrated in Ex. 5-6.

Example 5-6

As a problem of hypothetical interest, *suppose* commercial monaural FM had been established as a pure *PM* system. Assume that the maximum deviation of ±75 kHz and the highest modulating frequency of 15 kHz still applies. Determine the bandwidth as a function of the modulating frequency.

Solution

From a review of Eq. (5-25) and the discussion of Sec. 5-2, it is recalled that the maximum frequency deviation for tone-modulated PM is a linear function of the frequency. Since the maximum frequency deviation is ±75 kHz and the highest modulating frequency is 15 kHz, then $\Delta f/f_m$ = 75 kHz/15 kHz = 5. Thus $\Delta\phi$ in this example is the same as D in Ex. 5-5. However, in this example, $\Delta\phi$ will remain at a value of 5, while Δf will change with frequency.

Since $\Delta\phi$ remains at 5, the lower range rule for wideband FM is applicable and there are about 6 significant sidebands on either side of the carrier. The bandwidth B_T is approximately

$$B_T \approx 12f_m \qquad (5\text{-}46)$$

This result shows that the bandwidth for PM is *directly proportional to the modulating frequency*. This is in contrast to wideband FM in which the bandwidth is reasonably constant. The results for PM are shown graphically as curve (b) in Fig. 5-12. Note that, because of the same maximum frequency deviation, the curves for FM and PM coincide at the highest modulating frequency.

5-5 PHASOR REPRESENTATION OF MODULATED SIGNALS

The familiar phasor diagram of steady-state ac circuit analysis may be adapted in a special way for analyzing certain types of modulated waveforms. This approach leads to a graphical interpretation of the properties of modulated waveforms and is particularly useful in comparing the details of different modulation schemes.

In classical steady-state ac circuit analysis, the phasor approach is most often used in cases where all sources in the circuit are sinusoidal and are of the same frequency. By representing voltages and currents as complex exponentials, the exponential function appears as a factor in all terms and may be removed. The resulting voltages and currents in the circuit are then phasors whose positions are "frozen" or fixed in relationship to one another, and the important properties of amplitude and phase for the various components can then be determined by algebraic manipulations using complex numbers.

Consider now the case of a modulated waveform whose form can be represented as a series of sine and/or cosine components. [For example, review the form of a tone-modulated FM signal as given by Eq. (5-40).] From our knowledge of ac circuit theory, we could select any single component at a particular frequency and represent it by a phasor. The process could be repeated at each of the different frequencies in the spectrum in order. However, at each step in the process, the phasor concept is being applied at a *single frequency* using this standard approach.

What we are looking for now is a means of expressing a *complete* modulated waveform having *different* frequencies in some sort of phasor form with all components appearing simultaneously. This can be achieved by selecting one of the frequency components as the reference for the phasor rotation and suppressing or removing the exponential factor corresponding to this term from all other terms in the series. This results in one term being fixed while all other components rotate either clockwise or counterclockwise with respect to the reference term.

The most common approach is to select the *carrier* component as the reference and to consider the rotation of all phasor components with respect to the carrier. The carrier exponential factor is removed from all terms, which leaves the carrier in a fixed or reference position. All phasor sidebands will then be rotating in either positive or negative directions at integer multiples of the modulating frequency.

The basis for this concept lies in the property that a modulated sinusoid containing amplitude and/or angle modulation about a carrier frequency f_c can be expressed in the form

$$y(t) = A(t) \cos [\omega_c t + \phi(t)] \qquad (5\text{-}47)$$

where $A(t)$ represents the time variation of the envelope and $\phi(t)$ represents the total phase angle variation about the reference carrier phase. It may not be obvious to the reader, but all modulated waveforms that we have encountered in the book can be represented in the form of (5-47) by suitable manipulation. Observe that this form includes the possibility of amplitude modulation, angle modulation, or both at the same time.

By Euler's formula, the cosine function can be represented as the real part of the complex exponential. The phasor form is obtained by replacing the cosine function by the complex exponential function, with the fact understood that the real part represents the actual variable of interest. Thus, we can express (5-47) as

$$y(t) = A(t) \cos [\omega_c t + \phi(t)] = \mathrm{Re} \left\{ A(t) \epsilon^{j[\omega_c t + \phi(t)]} \right\} \qquad (5\text{-}48)$$

where $\mathrm{Re} \{ \ \}$ indicates the "real part" of the quantity in brackets. The real-part designation is then dropped, and a complex time function $\bar{y}(t)$ is defined as

$$\bar{y}(t) = A(t) \epsilon^{j[\omega_c t + \phi(t)]} \qquad (5\text{-}49)$$

Complex algebraic operations can now be performed on this function, but it should be understood that it is actually the real part that is ultimately of interest.

The form of (5-49) can be rearranged as

$$\bar{y}(t) = A(t) \epsilon^{j\phi(t)} \epsilon^{j\omega_c t} \qquad (5\text{-}50)$$

Let

$$\bar{Y}(t) = A(t) \epsilon^{j\phi(t)} \qquad (5\text{-}51)$$

The quantity $\bar{Y}(t)$ represents a complex phasor in which the variation with respect to the carrier frequency ω_c is eliminated. In conventional single frequency ac analysis, this function would have constant amplitude and constant phase. However, in the case at hand, both the amplitude $A(t)$ and the phase $\phi(t)$ are, in general, time-varying functions, although they will normally be varying at a much slower rate than the carrier.

If the complex phasor $\bar{Y}(t)$ is given, the complex time function $\bar{y}(t)$ is readily found by multiplying $\bar{Y}(t)$ by the complex exponential representing the carrier frequency. This follows from (5-50) and (5-51), and we have

$$\bar{y}(t) = \overline{Y}(t)\epsilon^{j\omega_c t} \qquad (5\text{-}52)$$

Finally, the real time function $y(t)$ is found by taking the real part of (5-52).

If $y(t)$ is given directly in the form of (5-47), the phasor $\overline{Y}(t)$ can be readily determined by comparing the various terms with those in (5-51). On the other hand, if $y(t)$ is given in the form of a trigonometric series, it is necessary to perform certain operations on the terms to arrange them in the proper form. The following procedure may be used:

1. By various trigonometric identities, arrange all the terms as cosine functions. The identities of Eqs. (4-1), (4-2), and (4-3) and the diagram of Fig. 2-6 are often used for this purpose.
2. Replace each cosine term by a complex exponential carrying the same argument:

$$B \cos\left[(\omega_c + \omega_m)t + \theta\right] \Longrightarrow B\epsilon^{j[(\omega_c+\omega_m)t+\theta]}$$

The resulting function is $\bar{y}(t)$.
3. Each term can be arranged to have $\epsilon^{j\omega_c t}$ as a factor. Drop this factor, and the result is $\overline{Y}(t)$.
4. The complex phasor $\overline{Y}(t)$ can be represented graphically by displaying all the phasor components and combining them using standard phasor addition rules. However, except for the carrier component, which is fixed, all other phasor components will have angles that are varying with time. The resulting magnitude of the combination is $A(t)$, and the resulting angle is $\phi(t)$.

Example 5-7

A certain modulated waveform can be expressed as

$$y(t) = A \sin \omega_c t + \frac{mA}{2} \cos (\omega_c - \omega_m)t - \frac{mA}{2} \cos (\omega_c + \omega_m)t \quad (5\text{-}53)$$

Determine the complex phasor form of the signal including the magnitude $A(t)$ and the phase angle $\phi(t)$.

Solution

The first step is to express all components in terms of cosine functions. Since the two sideband components are already expressed in cosine forms, we need only change the carrier. Since the angles having $\omega_m t$ products will be automatically expressed in radians, the carrier phase shift will also be expressed in radians for consistency. We have

$$y(t) = A \cos \left(\omega_c t - \frac{\pi}{2} \right) + \frac{mA}{2} \cos (\omega_c - \omega_m)t - \frac{mA}{2} \cos (\omega_c + \omega_m)t \quad (5\text{-}54)$$

The three components can now be expressed as real parts of complex exponentials.

$$y(t) = \text{Re} \left[A\epsilon^{j[\omega_c t - (\pi/2)]} + \frac{mA}{2} \epsilon^{j\omega_c t - j\omega_m t} - \frac{mA}{2} \epsilon^{j\omega_c t + j\omega_m t} \right] \quad (5\text{-}55)$$

The complex function $\bar{y}(t)$ is formed by dropping the real-part designation.

$$\bar{y}(t) = A\epsilon^{j\omega_c t} \epsilon^{-j(\pi/2)} + \frac{mA}{2} \epsilon^{j\omega_c t} \epsilon^{-j\omega_m t} - \frac{mA}{2} \epsilon^{j\omega_c t} \epsilon^{j\omega_m t} \quad (5\text{-}56)$$

The phasor $\bar{Y}(t)$ is obtained from $\bar{y}(t)$ by suppressing the exponential carrier factor.

$$\bar{Y}(t) = A\epsilon^{-j(\pi/2)} + \frac{mA}{2} \epsilon^{-j\omega_m t} - \frac{mA}{2} \epsilon^{j\omega_m t} \quad (5\text{-}57\text{a})$$

$$= A\underline{/-(\pi/2)} + \frac{mA}{2} \underline{/-\omega_m t} - \frac{mA}{2} \underline{/\omega_m t} \quad (5\text{-}57\text{b})$$

In the expression of (5-57b), the standard phasor notation has been employed but with the angles understood to be in radians. If desired, the term $-(mA/2)$ $\underline{/\omega_m t}$ could be replaced by $(mA/2) \underline{/\omega_m t + \pi}$.

A phasor diagram corresponding to $\bar{Y}(t)$ is shown in Fig. 5-13a. Observe that the carrier is fixed at an angle of $-\pi/2$ radians, while the two sideband components each rotate about the carrier. The lower sideband is rotating with angular velocity ω_m in a clockwise (negative) direction, and its angular orientation with respect to the positive real axis at any time is $-\omega_m t$. The

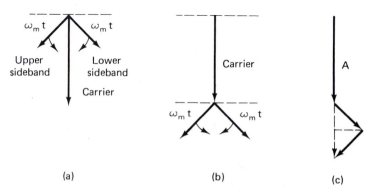

(a)   (b)   (c)

Figure 5-13. Phasor diagrams for modulated waveforms of Example 5-7.

upper sideband is rotating with angular velocity ω_m in a counterclockwise (positive) direction. However, due to the negative sign of the magnitude (or, equivalently, the additional π radians phase shift), its angular orientation with respect to the negative real axis at any time is $+\omega_m t$.

The three components may be readily added by the phasor diagram as shown two separate ways in Fig. 5-13b and c. Note that due to the symmetry involved the components perpendicular to the carrier cancel, leaving a net phasor having the same direction as the carrier. The magnitude of this phasor is the carrier magnitude plus the sum of two components each having a value $(mA/2) \sin \omega_m t$. The result is

$$A(t) = A + 2\left(\frac{mA}{2}\right) \sin \omega_m t \qquad (5\text{-}58)$$

$$= A(1 + m \sin \omega_m t)$$

The net angle $\phi(t)$ in this example remains constant and is

$$\phi(t) = -\frac{\pi}{2} \qquad (5\text{-}59)$$

The result in this example may be recognized as a tone-modulated AM wave form. The function given in (5-53) is not quite in the same form as developed in Chapter 4 owing to a phase shift in both the carrier and the modulating signal. In fact, when (5-58) and (5-59) are combined, and the carrier term is restored, the result is

$$y(t) = A(1 + m \sin \omega_m t) \cos\left(\omega_c t - \frac{\pi}{2}\right) \qquad (5\text{-}60)$$

$$= A(1 + m \sin \omega_m t) \sin \omega_c t$$

This result could have been obtained by applying appropriate trigonometric identities to (5-53). However, the intent of the example was to illustrate the application of the complex phasor concept rather than to immediately simplify the given function.

This example involved a function having only amplitude modulation and no angle modulation. In Ex. 5-8 at the end of the next section, a function having both amplitude and angle modulation will be considered. However, it is best to wait until the reader has observed some of the phasor differences between AM and NBFM in the next section before encountering this more general case.

5-6 NARROWBAND FREQUENCY MODULATION

The concept of narrowband frequency modulation (NBFM) was introduced in Sec. 5-4 along with other forms. Some very interesting and practical results can be deduced by applying the phasor form developed in Sec. 5-5 to the NBFM signal. While NBFM has only limited usefulness for direct RF communications, a detailed study of the process is important for two reasons: (1) Many commercial transmitters designed for WBFM actually start with an NBFM signal and by means of frequency multiplication (to be studied in Sec. 5-7), convert the signal into a suitable wideband form at the output of the transmitter. (2) By comparing the phasor forms of NBFM and AM, some useful and interesting interpretations can be made.

Consider again the equation of a tone-modulated signal as developed earlier in the chapter.

$$y(t) = A \cos (\omega_c t + \beta \sin \omega_m t) \qquad (5\text{-}61)$$

By the basic trigonometric identity for the cosine of the sum of two angles, (5-61) may be readily expanded as

$$y(t) = A \cos \omega_c t \cos (\beta \sin \omega_m t) - A \sin \omega_c t \sin (\beta \sin \omega_m t) \qquad (5\text{-}62)$$

We will now restrict (5-62) to the narrowband case, which requires that β be very small. Specifically, we will assume that $\beta \ll \pi/2$ for the development at hand. In this case, $\cos (\beta \sin \omega_m t) \approx 1$ and $\sin (\beta \sin \omega_m t) \approx \beta \sin \omega_m t$. The resulting function will be designated as $y_{nb}(t)$, and it can be expressed as

$$y_{nb}(t) = A \cos \omega_c t - A\beta \sin \omega_m t \sin \omega_c t \qquad (5\text{-}63)$$

This function can be expanded into the form

$$y_{nb}(t) = A \cos \omega_c t - \frac{A\beta}{2} \cos (\omega_c - \omega_m)t + \frac{A\beta}{2} \cos (\omega_c + \omega_m)t \qquad (5\text{-}64)$$

As expected for NBFM, the signal consists of a carrier plus two adjacent sidebands, each of which is located f_m hertz from the carrier. The reader may observe that (5-64) appears remarkably similar to the expression for an AM tone-modulated signal as encountered in Chapter 4. However, there is a significant difference, as will be seen shortly. This difference is best seen through the use of the complex phasor concept developed in the last section.

We will begin the phasor development with a tone-modulated AM signal as developed in Chapter 4. Using $y_{AM}(t)$ to represent this function, we have

$$y_{AM}(t) = A \cos \omega_c t + \frac{mA}{2} \cos (\omega_c - \omega_m)t + \frac{mA}{2} \cos (\omega_c + \omega_m)t \qquad (5\text{-}65)$$

Using the concepts of the last section, the AM signal can be represented as a complex phasor $\overline{Y}_{AM}(t)$ of the form

$$\overline{Y}_{AM}(t) = A + \frac{mA}{2} \epsilon^{-j\omega_m t} + \frac{mA}{2} \epsilon^{j\omega_m t} \qquad (5\text{-}66a)$$

$$= A\underline{/0°} + \frac{mA}{2}\underline{/-\omega_m t} + \frac{mA}{2}\underline{/\omega_m t} \qquad (5\text{-}66b)$$

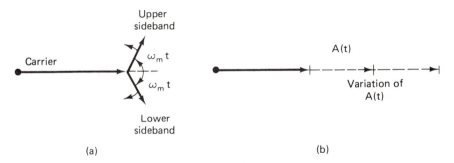

Figure 5-14. Modulated phasor form for AM tone-modulated signal.

The phasor representation of (5-66) is shown in Fig. 5-14a. The exponential phasor term with the negative argument is rotating in the negative (clockwise) direction, and the term with the positive argument is rotating in the positive (counterclockwise) direction. An important property to observe is that for any angle $\omega_m t$ the projections of the two rotating phasors in the direction perpendicular to the carrier are equal in magnitude and opposite in sign. Hence, these components cancel. However, the components parallel to the direction of the carrier are always in the same direction and thus add. This means that the resultant on the phasor diagram will always retain the same angular orientation, but the magnitude will vary above and below the carrier level, as illustrated by the locus shown in Fig. 5-14b.

The resulting amplitude function is determined to be

$$A(t) = A + 2\left(\frac{mA}{2}\cos\omega_m t\right) = A(1 + m\cos\omega_m t) \qquad (5\text{-}67)$$

The phase function is simply

$$\phi(t) = 0 \qquad (5\text{-}68)$$

as expected for an AM signal.

Next we will inspect the corresponding situation for NBFM. The function of (5-64) can be expressed as a complex phasor $\overline{Y}_{nb}(t)$ in the form

$$\overline{Y}_{nb}(t) = A - \frac{A\beta}{2}\epsilon^{-j\omega_m t} + \frac{A\beta}{2}\epsilon^{j\omega_m t} \qquad (5\text{-}69a)$$

$$= A\underline{/0°} + \frac{A\beta}{2}\underline{/-\omega_m t + \pi} + \frac{A\beta}{2}\underline{/\omega_m t} \qquad (5\text{-}69b)$$

This function is shown in Fig. 5-15a. Observe that the phasor rotating in the negative direction has an additional phase shift of π radians. For any angle $\omega_m t$, the projections of the two rotating phasors in the direction parallel to the carrier are equal in magnitude and opposite in sign. Hence, these components cancel. However, components in the direction perpendicular to the carrier are

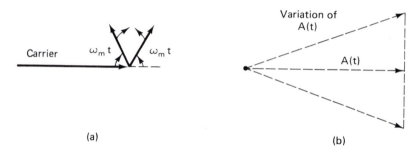

(a)   (b)

Figure 5-15.   Modulated phasor form for NBFM signal.

always in the same direction and add. The result will follow the locus shown in Fig. 5-15b. The resulting phasor appears to "swing" back and forth about the reference carrier position.

Observe that there will be some variation in the amplitude, but for small β, the amplitude variation should be insignificant. Specifically, from the phasor diagram, the reader is invited to show (Prob. 5-21) that the amplitude of the phasor is

$$A(t) = A \sqrt{1 + \beta^2 \sin^2 \omega_m t} \tag{5-70}$$
$$\approx A, \quad \text{for small } \beta$$

On the other hand, the reader is invited to show (Prob. 5-22) that the phase function is

$$\phi(t) = \tan^{-1} (\beta \sin \omega_m t) \tag{5-71}$$
$$\approx \beta \sin \omega_m t, \quad \text{for small } \beta$$

From this analysis, it is clear that the major difference between AM and NBFM lies in the relative phase of the sideband components with respect to the carrier. Indeed, an AM signal can be converted to a NBFM signal by changing the phase of the carrier with respect to the sidebands, as will be discussed in Sec. 5-8.

Example 5-8

Consider a function consisting of a carrier and *one* sideband of the form

$$y(t) = A \cos \omega_c t + B \cos (\omega_c + \omega_m)t \tag{5-72}$$

Show that this signal contains *both amplitude* and *angle* modulation, and determine the amplitude and angle functions.

Solution

Both terms in (5-72) are cosine functions, so the phasor form $\overline{Y}(t)$ is readily determined as

$$\overline{Y}(t) = A + B\epsilon^{j\omega_m t} \tag{5-73a}$$

$$= A\underline{/0^\circ} + B\underline{/\omega_m t} \tag{5-73b}$$

A phasor diagram is shown in Fig. 5-16. It is necessary to resolve the rotating phasor into two components, one in the direction of the carrier, and one in the direction perpendicular to the carrier. The resulting amplitude is then determined as

$$A(t) = \sqrt{(A + B \cos \omega_m t)^2 + B^2 \sin^2 \omega_m t} \tag{5-74}$$

$$= \sqrt{A^2 + B^2 + 2AB \cos \omega_m t}$$

The phase function is

$$\phi(t) = \tan^{-1}\left[\frac{B \sin \omega_m t}{A + B \cos \omega_m t}\right] \tag{5-75}$$

From the results of (5-74) and (5-75), it can be observed that both the amplitude and phase angle functions vary in some way with the modulating signal, so the signal contains both amplitude and angle modulation. However, the functional relationships for arbitrary A and B are somewhat complex, so it is unclear at this point whether or not a signal could be recovered from either the amplitude or the angle without distortion. The reader is invited to show (Prob. 5-23) that, if certain inequalities are met, the function may display usable modulation forms.

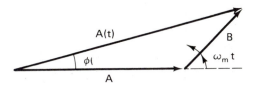

Figure 5-16. Phasor diagram for Example 5-8.

5-7 FREQUENCY TRANSLATION AND MULTIPLICATION

In this section, we will explore the concepts of *frequency translation* and *frequency multiplication* and their effects on modulated waveforms. Although our focus in this chapter is on angle modulation, it is convenient to observe at the same time how these operations affect other forms of modulation.

Frequency translation can be viewed as a process in which the spectrum of a given signal is shifted to a different frequency range, but in which the shape of the spectrum remains unchanged in the shift. Such a process is illustrated on a one-sided spectral basis in Fig. 5-17. An arbitrary signal spectrum is shown in Fig. 5-17a. If the signal is shifted to a higher frequency range, as shown in Fig. 5-17b, the process is referred to as an *up-conversion*. If the signal is shifted to a lower frequency range, as shown in Fig. 5-17c, the process is called a *down-conversion*.

With some thought, we recognize that the generation of an SSB signal as discussed in Chapter 4 is a special type of up-conversion, and its detection is a special type of down-conversion. The special nature in this case is the fact that the frequency conversions themselves were viewed as modulation and demodulation operations. On the other hand, the more general approach that we are pursuing now is that frequency conversion can be applied to virtually any type of signal containing an arbitrary form of modulation.

Frequency conversion is frequently necessary at both the transmitter and at the receiver because signals are often easier to generate and detect at different frequency ranges than are appropriate for RF transmission. Thus, a

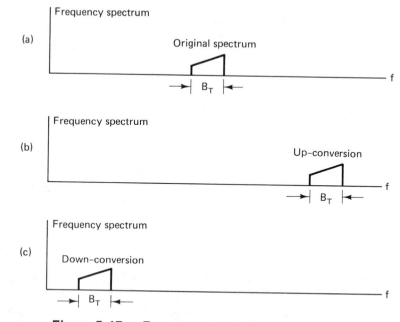

Figure 5-17. Frequency translation or conversion.

transmitter might generate a modulation signal at a frequency well below the RF transmission range, but by means of one or more successive frequency shifts, the signal could be translated to any arbitrary RF range. At the receiver, the signal could in turn be down-converted in one or more stages to a range in which proper detection is more easily implemented. It will also be shown here and in later work that proper filtering at both ends of the system can often be better achieved through frequency translation.

Frequency translation can be achieved through a balanced modulator followed by a filter to remove the unwanted sideband and other spurious components. As a matter of accepted terminology, balanced modulators designed specifically for frequency-translation purposes are usually called *mixers*. Many practical mixers utilize the switching modulator principle, as discussed in Sec. 4-9, so that many harmonic components are generated along with the two sidebands. This makes it imperative that good filtering be employed in mixer systems to remove all undesired components.

Although our primary interest in this chapter is on angle modulation, some of the results to be developed apply equally well to other forms of modulation, so we will assume a function of the form

$$y_1(t) = A(t) \cos \left[\omega_{c1} t + \phi(t) \right] \tag{5-76}$$

where $A(t)$ represents some arbitrary amplitude modulation function, $\phi(t)$ represents some arbitrary angle modulation function, and ω_{c1} represents an initial carrier radian frequency.

Even though many practical mixers employ switching modulator circuits that generate many spectral components, we will assume an ideal multiplier in the development that follows. The process that we wish to illustrate can be done much easier with this idealized assumption, and once understood on this basis the effects of the other harmonics could then be added. Thus, assume that the signal of (5-76) is mixed (multiplied) with a sinusoid of radian frequency ω_0 in an ideal mixer, as shown in Fig. 5-18. The input amplitude spectrum $Y_1(f)$ is shown in a *simplified form* in Fig. 5-19a, with an assumed carrier component at the middle of the spectrum. The mixing component at ω_0 is also shown.

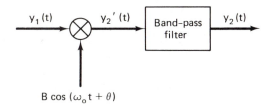

Figure 5-18. Frequency translation or mixing system.

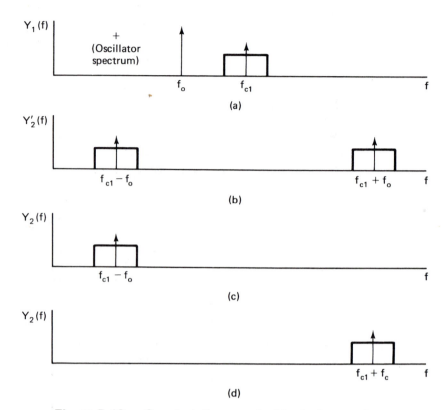

Figure 5-19. Spectral diagrams for ideal mixer systems.

The output $y_2'(t)$ of the mixer before filtering can be expressed as

$$y_2'(t) = BA(t) \cos[\omega_{c1}t + \phi(t)] \cos(\omega_0 t + \theta) \qquad (5\text{-}77a)$$

$$= \frac{BA(t) \cos[(\omega_{c1} - \omega_0)t + \phi(t) - \theta]}{2}$$

$$+ \frac{BA(t) \cos[(\omega_{c1} + \omega_0)t + \phi(t) + \theta]}{2} \qquad (5\text{-}77b)$$

For convenience in discussion, assume for the moment that $\omega_{c1} > \omega_0$. Inspection of the two terms in (5-77b) reveals that two separate sideband terms are obtained at the output of the ideal mixer, a result which is certainly not new to us. The spectrum $Y_2'(f)$ is illustrated in Fig. 5-19b. The center radian frequency of the upper sideband is $\omega_{c1} + \omega_0$, and the center radian frequency of the lower sideband is $\omega_{c1} - \omega_0$.

Observe that *both the amplitude modulation function A(t) and the angle modulation function $\phi(t)$ are preserved in the translation process*. Of course, the resulting sidebands are multiplied by an arbitrary constant $(B/2)$, and an arbitrary constant phase shift has been added to the phase function. However, these quantities do not change the original form of the intelligence, which has been preserved in the translation process.

As far as the phase-shift term is concerned, some systems require strict phase and frequency coherency during frequency shifts, which means that the value of θ would be controlled. However, for the discussion here this point need not be considered.

For $\omega_{c1} > \omega_0$, the first component in (5-77b) has a lower carrier frequency than the input signal, and the second component has a higher carrier frequency than the input signal. Thus, depending on whether a down-conversion or an up-conversion is desired, a filter tuned to the appropriate sideband is used, and the output of the filter $y_2(t)$, as shown in Fig. 5-18, represents the desired frequency translated signal. The two possible amplitude spectra for $Y_2(f)$ are illustrated in Fig. 5-19c and d. The bandwidth of the filter must be sufficient to pass all the spectral components of the input modulated signal, and at the same time the filter must reject the unwanted sideband completely.

Some comments concerning the filtering requirements will be made. As long as the difference between the highest frequency component in the lower sideband is well below the lowest frequency component in the upper sideband, there is a reasonable guard band between the sidebands, and filtering is fairly straightforward. However, if the guard band is very small, it is quite difficult to properly filter the desired signal. In fact, there are certain combinations of frequencies and bandwidths such that portions of the two sidebands actually overlap, making the desired objective impossible to achieve. Some of these difficulties will be pursued in later sections of the book, so we will not attempt to completely analyze the situation now. The best procedure that can be stated at this time is to draw a spectral diagram each time a mixing operation is encountered and to carefully label all frequencies and bandwidths on the diagram. If the separation between sidebands is very small, or if there is spectral overlap, the required filtering may be very difficult or impossible to achieve. Some trial and error is often involved in the design of frequency translation systems.

To summarize, assuming that the two sidebands at the output of a mixer can be properly separated by filtering, the result of a frequency translation operation on a signal of the form of (5-76) can be expressed as

$$y_2(t) = KA(t) \cos [\omega_{c2} t + \phi(t) \pm \theta] \qquad (5\text{-}78)$$

where K and θ are constants and $\omega_{c2} = \omega_{c1} \pm \omega_0$ depending on whether up-conversion or down-conversion is desired. Both amplitude and angle information are maintained.

For convenience, we have assumed so far that $\omega_{c1} > \omega_0$. However, this inequality need not be satisfied at all. If $\omega_0 > \omega_{c1}$, it is possible to generate a down-converted sideband and an up-converted sideband, as we have shown for the other inequality, or it is possible to generate two up-converted sidebands. (See Prob. 5-32.)

Next we will consider the effect of *frequency multiplication*. Frequency multiplication is a process in which the center frequency of a given signal is multiplied by some integer constant so that the new signal appears at a higher frequency. However, in contrast to frequency translation, the process of frequency multiplication may alter the form of the modulated intelligence.

Circuits having nonlinear input-output characteristics generate various harmonic components of an input signal, with the amplitudes of the harmonic components being a function of the type of nonlinearity. Frequency multiplication is obtained by applying the input signal to a nonlinear circuit that has a strong component at the particular harmonic frequency desired. A filter is then used to pass the particular harmonic and to reject all other components.

Although there are many types of nonlinear circuits that will generate various types of harmonics, we will assume for the purpose of this development an ideal square-law characteristic. The results will provide us with sufficient insight to generalize the concept without attempting a more formidable (and unwieldy) derivation.

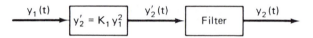

Figure 5-20. Frequency doubler using a square-law characteristic.

Assume a device whose input-output characteristics are given by $y_2' = K_1 y_1^2$, as shown in Fig. 5-20. The signal of (5-76) will be assumed as the input. The output $y_2'(t)$ of the square-law device before filtering will be of the form

$$y_2'(t) = K_1 \{A(t) \cos [\omega_{c1} t + \phi(t)]\}^2 \tag{5-79a}$$

$$= \frac{K_1}{2} A^2(t) + \frac{K_1}{2} A^2(t) \cos [2\omega_{c1} t + 2\phi(t)] \tag{5-79b}$$

The first term in (5-79b) is the square of the amplitude modulation multiplier function. In most cases of practical interest, the frequency components of $A(t)$, and hence those of $A^2(t)$, are much lower in frequency than the carrier frequency ω_{c1}. Thus, the first term can be readily filtered out, as illustrated in Fig. 5-20, and the output $y_2(t)$ can be expressed as

$$y_2(t) = KA^2(t) \cos [\omega_{c2} t + 2\phi(t)] \tag{5-80}$$

where $\omega_{c2} = 2\omega_{c1}$ and $K = K_1/2$. Thus, the center frequency has been multiplied by a factor of 2 when the circuit is a square-law device.

The effects on the amplitude and angle functions need to be considered. First, consider the amplitude factor. The original function of (5-76) contained the amplitude factor $A(t)$. However, the new function contains the amplitude squared function $A^2(t)$. In general, this represents a distortion of the envelope. Later in the text it will be shown that this operation is acceptable for certain types of signals in which only the presence of the signal or the presence of energy needs to be detected. However, for general analog amplitude intelligence, the resulting distortion is usually intolerable. Thus, *frequency multiplication results in a distortion of amplitude modulation data and is usable only in certain special cases.*

Now consider the effect on the angle modulation. The original function contained the angle function $\phi(t)$, while the frequency multiplied signal contains the function $2\phi(t)$. The form is preserved exactly, but the function is simply multiplied by a factor of 2. This means that the effective phase and frequency deviation at any time have been doubled. Since the modulating signal (and hence the maximum modulating frequency) has not changed, the modulation index has been multiplied by 2.

To generalize this concept, the process of frequency multiplication may be applied to an angle modulated signal without introducing distortion to the intelligence. *When the center frequency is multiplied by a constant N, the frequency and phase deviation and the modulation index are all multiplied by the same constant N.*

In general, assume that the input-output characteristic of some arbitrary nonlinear device can be represented by a power series of the form

$$y_2 = K_0 + K_1 y + K_2 y^2 + K_3 y^3 + \cdots \qquad (5\text{-}81)$$

It can be shown that a harmonic at frequency $N\omega_1$ is produced by the $K_N y^N$ term. Thus, a circuit designed to be a frequency tripler should have as large a value of K_3 as possible. This would accentuate the third harmonic component and make it easier to separate.

In practice, the constants in (5-81) tend to diminish with larger N so that it becomes increasingly difficult to use higher-order products in a single step. For that reason, many frequency multiplication systems employ a cascade of lower-order multipliers in order to achieve a higher-order frequency multiplication. For example, a cascade of one doubler, two triplers, and one quadrupler will produce an effective frequency multiplication of 72.

By combining frequency multiplication and frequency translation, it is possible to generate angle modulated signals having a wide range of center frequencies and frequency deviations (or modulation indexes). This concept will be pursued at some length in the next section.

To summarize some major parts of this section as applied to angle modulated signals, both frequency translation and frequency multiplication can be successfully applied to angle modulation, and the intelligence is preserved in both cases. In frequency translation, the carrier frequency is shifted, but the frequency and phase deviation and modulation index remain unchanged. However, in frequency multiplication, the frequency and phase deviation and the modulation index are multiplied by the same integer as the carrier frequency.

Example 5-9

A certain FM signal has a center frequency of 1 MHz and a bandwidth of 100 kHz. It is desired to shift the center frequency to 400 MHz while retaining the exact form and sense of the spectrum. To ease the filtering requirements, a somewhat arbitrary but reasonable constraint is imposed on the output of each mixer to the effect that all sideband spectral components to be eliminated should be displaced by no less than 5% from the adjacent band edge of the desired sideband. Plan a tentative design for an appropriate frequency-translation system, specify mixing oscillator frequencies, and sketch the various spectra.

Solution

At the outset, it is worth pointing out that, while the signal is specified to be FM, the results would apply equally well to most other forms of modulated signals. This is true because the exact form of the spectrum must be preserved, and thus only frequency translation (and not multiplication) can be employed.

A problem such as this has a number of possible solutions, and some trial and error is usually involved. First, we will determine if it would be practical to make the desired frequency shift in one step. A simplified form of the given spectrum with an assumed carrier is shown in Fig. 5-21a. In this part only, both positive and negative frequencies are shown. A and B are used to indicate the original lower and upper portions, respectively, of the spectrum about 1 MHz. (Observe that A and B appear in reverse order at negative frequencies.) The final spectrum must have the same sense in this particular system, so A and B must be in the same order about 400 MHz as at 1 MHz. (In some frequency-translation systems, the spectrum may be inverted in sense without causing any difficulty. (Review, for example, SSB generation for the case where the lower sideband is selected.) If we attempted to shift the spectrum all the way to 400 MHz at one step, a mixing sinusoid with a frequency of 399 MHz would be required. (A frequency of 401 MHz would also shift the center frequency to 400 MHz, but the reader is invited to show that the component about 400 MHz, which arises from the translation of the original negative frequency components, would be inverted in sense.) The two sidebands at the output of the mixer corresponding to a mixing frequency of 399 MHz are shown in Fig. 5-21b. The lowest portion of the desired sideband has a frequency of about 399.95 MHz, and the upper portion of the sideband to be removed has a frequency of about 398.05 MHz. The second frequency is slightly greater than 0.995 times the first frequency or less than 0.5%

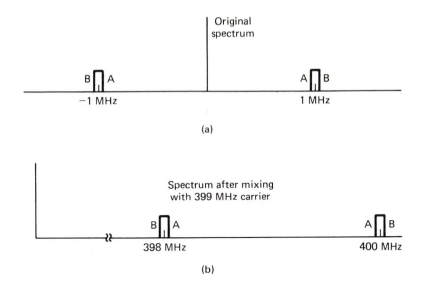

Figure 5-21. Original spectrum and up-converted spectrum after mixing with 399-MHz carrier in Example 5-9.

removed from the desired sideband. Adding to this is the fact that in any practical mixer, a small component of the mixing oscillator (399 MHz in this case) appears at the output due to imperfections in the balancing process, and this component would be less than 0.24% removed from the desired sideband. Thus, the filtering requirements in this case cannot be achieved with the original constraints as stated, so the possibility of translation in one step is eliminated. (The reader should not interpret this as an ironclad statement that it cannot be done in one step. If a sufficiently complex filter were implemented, it *might* be possible to achieve the desired result in one step. However, most real-life problems involve some trade-offs between the quantity of components and the ease of implementation per component, so a one-step solution would be quite difficult in this case.)

We will now consider the possibility of two steps of frequency translation. Since the center frequency must eventually be translated by a factor of 400, a reasonable approach is to shift by a factor of about 20 or so in each step. This process is illustrated in Fig. 5-22. The two-stage mixing process is illustrated in Fig. 5-22a, and the spectra at the output of the first and second mixers are shown in Fig. 5-22b and c. The first mixing oscillator is chosen to have a frequency of 19 MHz. The lowest portion of the desired sideband has a frequency of about 19.95 MHz, and the upper portion of the sideband to be removed has a frequency of about 18.05 MHz. The second frequency is slightly less than 0.905 times the first frequency or about 9.5% removed from the desired sideband. The filtering constraint as stated in the problem can thus be achieved. The mixing frequency at 19 MHz is displaced from the

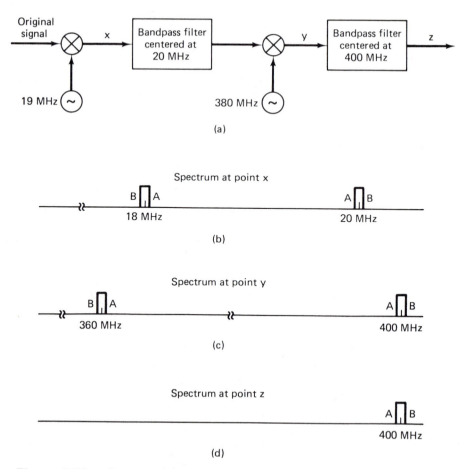

Figure 5-22. Proposed frequency translation system for Example 5-9 and spectra at different points.

desired sideband by slightly less than 5%, but since any component appearing at the output at that frequency will be very small anyway (typically 40 dB down from the sideband levels), it should not cause any special filtering problem.

The second mixer employs a mixing frequency of 380 MHz. The lowest portion of the desired sideband has a frequency of about 399.95 MHz, and the upper portion of the sideband to be removed has a frequency of about 360.05 MHz. The second frequency is about 0.9 times the first frequency or about 10% removed from the desired sideband, so the filtering constraint is again satisfied. The mixing frequency of 380 MHz is displaced from the desired sideband by slightly less than 5%, but for the same reason as before, this should not pose any serious problem. This system will be proposed as a possible solution, and the final filtered spectrum is illustrated in Fig. 5-22d.

As discussed in Chapter 4, if the mixers employ switching modulator concepts, double-sided modulated components centered at harmonics of the mixing frequency would appear at the output of each mixer. However, these products all fall well outside of the desired frequency range of each mixer output in this problem and should be easily eliminated with the band-pass filters employed in each case. However, in some design situations this is not the case, so one should always be alert to the possibility that such components could be troublesome.

Although the original sense of the spectrum had to be preserved at the output in this problem, it can be shown that, if the spectrum is inverted *twice* through the system, the original sense is restored. An alternative solution to this problem in which this concept appears is provided as an exercise for the reader (Prob. 5-27).

5-8   FM MODULATOR CIRCUITS

In this section, a survey of some representative techniques that may be used in circuits designed for generating FM and/or PM signals will be made. Although the technology used in implementing such circuits is changing constantly, there are some basic strategies that continue to be employed. The focus will be on these general strategies rather than on detailed circuit diagrams.

While the term *FM modulator* is perfectly proper to use for any circuit designed for generating an FM signal, the term *voltage-controlled oscillator* (VCO) is widely used in various segments of the industry for such circuits. In addition, the term *voltage-to-frequency (V/F) converter* is also used. The first two terms will be used in this book as the need arises.

Any circuit used for generating an angle modulated signal should produce a steady-state output at a fixed frequency when no modulating signal is present. When the data signal is applied, the frequency or the phase should change in accordance with the modulating signal. As was discussed earlier in the chapter, it is possible to use an FM generator to produce PM, and vice versa, so in accordance with the widest usage, primary attention will be focused momentarily on FM modulators.

For an ideal FM modulator or VCO, the frequency deviation Δf should be directly proportional to the magnitude of the modulating signal, as shown in the ideal characteristic of Fig. 5-23. (On the curve, Δf refers to any particular deviation rather than the maximum deviation as earlier notation established.) If the Δf versus input signal magnitude characteristic is not a straight line, the resulting FM signal will be distorted with respect to the intelligence, so care must be exercised in designing a circuit to produce a characteristic as close to a straight line as possible.

For wideband FM systems with large frequency deviations, it is often difficult to produce a straight-line characteristic over a wide frequency range.

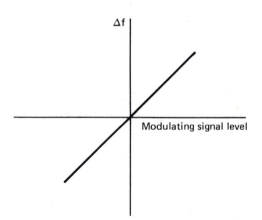

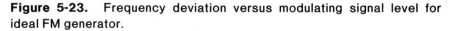

Figure 5-23. Frequency deviation versus modulating signal level for ideal FM generator.

One general strategy widely employed is to first generate the modulated FM signal at a much lower center frequency and with a much smaller deviation. By successive frequency multiplications coupled with frequency translations when appropriate, the center frequency and the deviation can both be adjusted to the final required range, as discussed in Sec. 5-7.

In some cases, particularly in telemetry data systems, all the modulator circuits up to the point where the FM or PM signal is generated are directly coupled. This permits very low frequency or even dc data to be transmitted in the system, whereas it is usually more difficult to transmit such data in many AM-type systems. The ability to process dc and very low frequency data is one of the significant advantages of FM. However, in commercial FM broadcasting, it is not necessary to transmit dc, so the various modulator circuits can be ac coupled in that case.

One widely used approach for FM modulators is the *variable parameter* concept. This concept has as its basis the fact that the frequency of many oscillator circuits depends on an *LC* product (or in some cases an *RC* product). If the value of one of the critical parameters can be varied in accordance with an applied signal, the instantaneous frequency can be varied. Depending on the circuit, it is possible to vary either the capacitance, the inductance, or the resistance. For the purpose of the present discussion, we will assume an *LC* oscillator and consider the capacitance as the variable parameter.

Consider the *LC* resonant circuit shown in Fig. 5-24, and assume that its resonant frequency controls the frequency of an oscillator, which is not shown. (There are a number of standard *LC* oscillator circuits that could be used.) Let L_0 and C_0 represent the fixed values of inductance and capacitance, and let ΔC represent a change in capacitance. The cyclic resonant

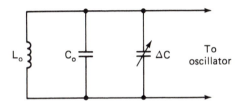

Figure 5-24. Resonant circuit used to tune oscillator in which the frequency deviation is a function of the change in capacitance.

frequency f for the circuit is

$$f = \frac{1}{2\pi \sqrt{L_0(C_0 + \Delta C)}} = \frac{1}{2\pi \sqrt{L_0 C_0}}\left(1 + \frac{\Delta C}{C_0}\right)^{-1/2} \tag{5-82}$$

Assume the change in capacitance is small compared with the fixed value; that is, $\Delta C/C_0 \ll 1$. The square root in (5-82) may then be approximated by the first two terms in the binomial expansion, and we have

$$f \approx \frac{1}{2\pi \sqrt{L_0 C_0}}\left(1 - \frac{\Delta C}{2C_0}\right) \tag{5-83}$$

In general, we desire that $f = f_0 + \Delta f$. We see then that

$$f_0 = \frac{1}{2\pi \sqrt{L_0 C_0}} \tag{5-84}$$

$$\frac{\Delta f}{f_0} = \frac{-\Delta C}{2C_0} \tag{5-85}$$

These results indicate that the change in frequency is directly proportional to the change in capacitance provided that the change in capacitance is small compared with the fixed capacitance. The negative sign indicates that an increase in capacitance produces a decrease in frequency, as would be expected.

We have thus seen that the instantaneous frequency of an LC oscillator can be made to change approximately linearly with a change in capacitance, provided that the change is small. The next part of the problem is to find a way to vary the capacitance with the modulating signal. This can be achieved either (1) indirectly or (2) directly. The indirect approach is the oldest and will be discussed first because of its widespread usage.

Indirect Variation

Indirect variation of a capacitance is achieved by connecting a capacitance across an amplifier whose gain changes as the level of the modulating signal changes. A simplified illustration of this concept is illustrated in Fig. 5-25.

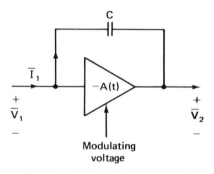

Figure 5-25. Illustration of how effective capacitance can be varied by varying gain.

The capacitor is connected between the input and output terminals of an inverting amplifier whose voltage gain $A(t)$ is varied in some manner as a function of the modulating signal. If the amplifier is assumed to have a very high input impedance, the total input phasor current $\bar{I}_1$ is given by

$$\bar{I}_1 = j\omega C(\bar{V}_1 - \bar{V}_2) \tag{5-86}$$

The output voltage $\bar{V}_2$ is related to the input voltage $\bar{V}_1$ by

$$\bar{V}_2 = -A(t)\bar{V}_1 \tag{5-87}$$

Substitution of (5-87) in (5-86) yields

$$\bar{I}_1 = j\omega C[1 + A(t)]\bar{V}_1 \tag{5-88}$$

The current $\bar{I}_1$ is seen to lead the input voltage by 90°, so it may be considered as a purely capacitive current. This result suggests that (5-88) may be expressed as

$$\bar{I}_1 = j\omega C_{eq}\bar{V}_1 \tag{5-89}$$

where C_{eq} may be recognized as

$$C_{eq} = [1 + A(t)]C \tag{5-90}$$

The effective capacitance is seen to be enlarged by the process, and it is a function of $A(t)$. If $A(t)$ is made to vary linearly about some nominal gain, the resulting capacitance variation about its nominal value will also vary linearly.

The reader may recognize that the basic capacitance multiplier technique here is the classical *Miller effect* in which a capacitance between input and output terminals of an inverting amplifier is effectively increased.

Gain variation with signal level is achieved by biasing the active device in a nonlinear portion of its characteristic curves. Care must be taken in the design to ensure that the gain variation is linear with the modulating signal. This usually restricts the fractional frequency deviation that can be achieved through this concept, although frequency multiplication can be employed, as previously noted.

The preceding steps have been chosen to illustrate for the reader the concepts involved rather than to provide a specific circuit. The most common circuit used through the years to obtain an indirect variation of capacitance (or inductance, too, for that matter) is the *reactance modulator*. A guided exercise in analyzing the details of one form of that circuit will be left as a problem for the interested reader (Prob. 5-34).

Direct Variation

Direct variation of the capacitance with modulating signal is possible with a voltage variable capacitive diode. This device uses the property that a reverse-biased PN junction diode displays a capacitance that is a function of the applied voltage. The capacitance $C(v)$ as a function of the voltage is a nonlinear function of the form

$$C(v) = \frac{C_{j0}}{\left(1 + \dfrac{v}{V_{bi}}\right)^{n}} \qquad (5\text{-}91)$$

where C_{j0} is the junction capacitance with no voltage applied, v is the magnitude of the reverse voltage across the diode, and V_{bi} is called the "built-in potential" (typically 0.8 V for silicon and 0.4 V for germanium.) The value n depends on the type of junction and ranges between about $\frac{1}{2}$ to $\frac{1}{3}$. A possible circuit is shown in Fig. 5-26.

While it is not obvious at all from (5-91), it can be shown that for small changes in v relative to V_{bi} the quantity $C(v)$ can be made to change approximately linearly with the signal voltage.

At relatively low frequencies such as encountered in subcarrier generators for multiplex systems, *RC* relaxation oscillators may be frequency modulated

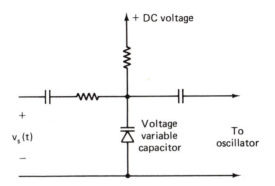

Figure 5-26.   FM generator using voltage-variable capacitive diode.

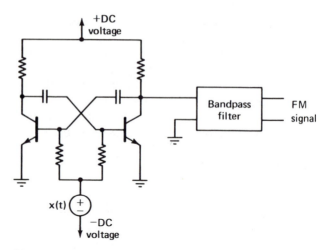

Figure 5-27. Voltage-controlled oscillator using an astable multivibrator.

by varying the trigger point in the cycle. Many of the integrated circuit timers and function generators available as VCO chips employ this concept. Since the internal operation of these chips would necessitate a more detailed explanation than desired here, the concept will be illustrated with the discrete component astable multivibrator shown in Fig. 5-27. The signal voltage either adds to or subtracts from the base return voltage. This varies the initial level at the base of one transistor as it is cut off by the negative step from the collector of the other transistor. The time between cutoff and conduction (which results when the base voltage climbs to zero) is then a function of the signal voltage. By careful design, a reasonably wide deviation can be achieved with good linearity. The initial output is a frequency-modulated square wave, but an appropriate filter at the output converts the FM square wave to an FM sine wave. The filter must be capable of passing all the sidebands about the fundamental but must reject all the harmonics and their associated sidebands.

One additional circuit that we will discuss deserves special recognition due to its historic significance. This is the Armstrong method, named after E. H. Armstrong, a major contributor to the development of FM. Consider the basic FM equation

$$y(t) = A \cos \left[\omega_c t + \phi_i(t) \right] \qquad (5\text{-}92)$$

as developed earlier in the chapter with

$$\phi_i(t) = \Delta \omega \int_0^t x(t)\, dt \qquad (5\text{-}93)$$

The expression of (5-92) can be expanded to yield

$$y(t) = A \cos \omega_c t \cos \phi(t) - A \sin \omega_c t \sin \phi(t) \qquad (5\text{-}94)$$

Assume narrowband FM, which requires that $\phi(t)$ be very small. In this case $\cos \phi(t) \simeq 1$ and $\sin \phi(t) \simeq \phi(t)$. The result of (5-94) may then be approximated by

$$y(t) \simeq A \cos \omega_c t - A \phi(t) \sin \omega_c t \qquad (5\text{-}95)$$

A circuit for realizing (5-95) is shown in Fig. 5-28. The modulating signal $x(t)$ is integrated in accordance with (5-93) and applied to a balanced multiplier along with a carrier $\sin \omega_c t$. The carrier is then shifted 90° and combined with the output of the multiplier. The result is an NBFM signal. A wideband FM signal is obtained by successive frequency multiplications of the NBFM signal. A typical system employing this concept will be explored in Ex. 5-10.

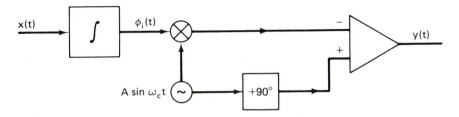

Figure 5-28. Generation of NBFM signal using Armstrong method.

Example 5-10

A certain commercial FM transmitter is to be designed to operate at a center frequency of 100 MHz and with a deviation of ±75 kHz. The modulating signal frequency range is from 50 Hz to 15 kHz. The Armstrong system is to be employed in the design. The center frequency of the basic modulator is selected to be at 100 kHz, and tests on the modulator indicate that a maximum phase deviation of $\Delta\phi_{max} = 0.5$ rad can be tolerated in the modulator itself. Formulate a possible block-diagram design layout for the transmitter with all frequency multiplication restricted to the use of doublers, triplers, and quadruplers. Prepare a table listing the center frequency, frequency deviation, deviation ratio, and transmission bandwidth at the output of each successive stage.

Solution

There are a number of possible solutions to this problem, so the particular solution to be discussed should not be interpreted as the only one or even necessarily the best solution. It is representative and was obtained after some trial and error.

First, we must determine the maximum frequency deviation at the output of the modulator itself. Since the modulator must produce FM at the output, the modulating signal is first integrated, and the value $\Delta\phi$ for the modulator is interpreted as β. We are given that $\Delta\phi_{max} = 0.5$, so this means that $\beta_{max} = 0.5$. Since $\beta = \Delta f / f_m$ and since Δf is fixed for FM, then $\beta_{max} = \Delta f / f_m(min)$, where $f_m(min)$ is the lowest modulating frequency. This deduction tells us that the *maximum* phase deviation occurs at the *lowest* modulating frequency. Thus, $\Delta f = \beta_{max} \times f_m(min) = 0.5 \times 50 = 25$ Hz. We thus conclude that the maximum frequency deviation at the output of the modulator is 25 Hz.

The final stage of the transmitter must have a maximum deviation of 75 kHz, so it is necessary to multiply the center frequency by $75 \times 10^3/25 = 3000$ in order to achieve the desired output frequency deviation. On the other hand, the center frequency need only be multiplied by a factor of 100 MHz/100 kHz = 1000. This means a down-conversion will be required somewhere in the multiplier chain so that we do not "overshoot" the desired output center frequency. One could suggest the possibility of doing all the multiplication first and then down-conversion, but this would involve a down-conversion from near 300 MHz to 100 MHz. This process could certainly be achieved, but 300 MHz is at the edge of the UHF frequency range, and implementation would be somewhat easier at a lower frequency. Consequently, the best choice is to make a down-conversion somewhere along the multiplier chain well below the final RF frequency.

We next select a combination of multipliers to achieve the required multiplication ratio. Some trial and error is involved, and it can be shown that a multiplication of exactly 3000 cannot be achieved using only doublers, triplers, and quadruplers. However, this need not be a cause for alarm, because we can always exceed the ratio slightly and then, by reducing the drive slightly to the modulator circuit, accomplish the desired result. Specifically, we find the combination of five quadruplers and one tripler produces a multiplication of $4^5 \times 3 = 3072$. (Of course, each quadrupler could be replaced by two doublers if desired, but we will assume quadruplers for convenience in the analysis.)

The actual block diagram of the system is shown in Fig. 5-29, and a tabular presentation of the data requested in the problem statement is presented in Table 5-3. The various letters in the left column of the table correspond to particular points shown on the block diagram. Since the multiplication ratio is slightly greater than needed, Δf at the modulator output is reduced from 25 Hz to about 24.41 Hz. This value represents the approximate Δf that, when multiplied by 3072, will yield a maximum deviation of 75 kHz at the output. The various values have been slightly rounded in the table for convenience. When the transmitter is constructed, the magnitude of the modulating signal will have to be carefully adjusted so that the maximum possible value of the input signal produces a frequency deviation at the output of 75 kHz.

The value D at each point in the table is the deviation ratio defined by $D = \Delta f / W = \Delta f / 15 \times 10^3$. This parameter, of course, is important in determining the transmission bandwidth, which in turn is used in designing the filters that

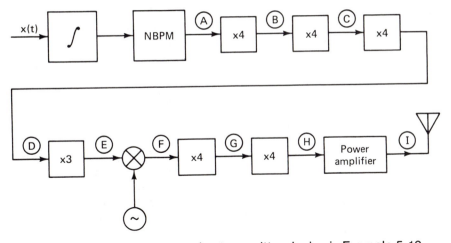

Figure 5-29.   Data concerning transmitter design in Example 5-10.

appear at the different multiplier outputs. Filters are necessary to eliminate spurious spectral components arising from the nonlinear multiplication operations. However, the filters are understood to be contained within the functional operations on the figure and are not shown.

The frequency converter is placed at a point in which the center frequency is 19.2 MHz. This position is somewhat arbitrary, but reasonable. The mixing frequency is selected by first determining what the output frequency must be. Since an additional multiplication of $4 \times 4 = 16$ remains, the desired output center frequency of the mixer must be 100 MHz/16 = 6.25 MHz. The mixing oscillator must then shift the incoming signal from 19.2 MHz down to 6.25 MHz. A mixing frequency of 12.95 MHz is one possible solution. The reader should verify that the output of the mixer before filtering will also contain an FM signal centered at 32.15 MHz, but this sideband is readily removed by a filter.

Table 5-3.   Data Concerning Transmitter Design in Ex. 5-10.

Point	f_c	Δf	D	B_T
A	100 kHz	24.41 Hz	1.63×10^{-3}	30 kHz
B	400 kHz	97.66 Hz	6.51×10^{-3}	30 kHz
C	1.6 MHz	390.6 Hz	0.026	30 kHz
D	6.4 MHz	1.562 kHz	0.104	30 kHz
E	19.2 MHz	4.688 kHz	0.313	39.38 kHz
F	6.25 MHz	4.688 kHz	0.313	39.38 kHz
G	25 MHz	18.75 kHz	1.25	67.5 kHz
H	100 MHz	75 kHz	5	180 kHz
I	100 MHz	75 kHz	5	180 kHz

The reader is invited to verify the correctness of the different values given in the table. Note that the signal remains as an NBFM signal through the first four stages, so the simplified bandwidth approximation $B_T \approx 2W$ is used for those stages. Beginning with point E, Carson's rule is used for determining the bandwidth.

5-9 FM DETECTION CIRCUITS

In this section, we will investigate some of the concepts that can be employed in circuits used for detecting FM and/or PM signals. As in the last section, the discussion will concentrate primarily on FM signals. FM detector circuits are often referred to as *FM discriminators*.

The ideal FM detector is a circuit that produces zero output voltage when the frequency of the received signal is constant and equal to the carrier frequency, but in which the voltage changes linearly with changes in the input frequency. The ideal form of the output voltage versus input frequency deviation characteristic is shown in Fig. 5-30. Any actual detector characteristic must approximate this straight-line characteristic very closely if distortion is to be minimized.

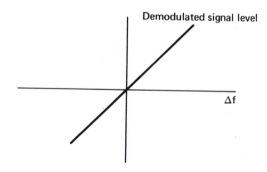
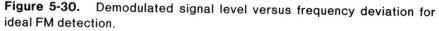

Figure 5-30. Demodulated signal level versus frequency deviation for ideal FM detection.

In the case of the FM modulator, we have seen that it is possible to generate the FM signal at a much lower deviation than required and then to multiply the smaller deviation to the correct value by the use of frequency multipliers. No simple reverse process exists in the case of frequency detection. Thus, the FM detector must be capable of operating over the entire frequency deviation range of the received signal. Of course, the received signal may be down-converted in frequency so that detection may be achieved about a more convenient center-frequency range than the RF transmission frequency.

An important general observation about FM signals is that the intelligence is always a function of the frequency (rate of zero crossings) only and is not a function of the amplitude. This allows various nonlinear operations to be performed on the amplitude without affecting the modulation. The most common operation of this type is *amplitude limiting,* which is used in many receivers to remove some of the amplitude noise, which can create various undesirable effects in the signal processing.

The input-output characteristic of an ideal limiter is shown in Fig. 5-31a, and a simplified implementation approach using diodes is shown in Fig. 5-31b. In practice, some of the standard FM discriminator circuits accomplish a degree of limiting along with the detection, although a separate limiting stage may be used if desired.

The effect of the limiter on a representative noisy signal is illustrated in Fig. 5-31c and d. Observe that the output has some of the properties of a square wave, but the information, which is contained in the zero crossings, is not affected.

Although excess amplitude noise can be eliminated or reduced by the limiting process, the FM signal is still disturbed by noise that has already affected the locations of zero crossings before reception. This noise cannot be eliminated since its presence is imbedded in the instantaneous zero crossings

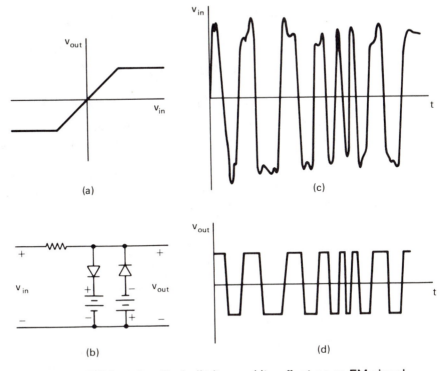

Figure 5-31. Amplitude limiter and its effect on an FM signal.

Figure 5-32. Theoretical concept of FM demodulation by differentiation and envelope detection.

and cannot be distinguished from the signal itself. However, as we will see in Chapters 10 and 11, the detected signal may be enhanced with respect to this noise provided that certain conditions are met.

A number of widely used FM detector circuits employ the concept of *slope detection*. This approach is based on the fact that certain circuits display a steep voltage versus frequency magnitude characteristic over a region that is equal to or can be approximated by a straight line.

To develop an analytical concept for such a circuit, consider the block diagram shown in Fig. 5-32. The block on the left is assumed to be an ideal differentiating circuit whose output $y_2(t)$ is related to the input $y_1(t)$ by

$$y_2(t) = \frac{dy_1(t)}{dt} \tag{5-96}$$

Assume that the input is an FM signal of the form

$$y_1(t) = A \cos \left[\omega_c t + \Delta \omega \int x(t) \, dt \right] \tag{5-97}$$

The output of the differentiator is

$$y_2(t) = -A \left[\omega_c + \Delta \omega \, x(t) \right] \sin \left[\omega_c t + \Delta \omega \int x(t) \, dt \right] \tag{5-98}$$

The function $y_2(t)$ contains an envelope amplitude multiplicative factor containing a constant plus a term proportional to the intelligence $x(t)$.

Assume that $\omega_c + \Delta \omega \, x(t) > 0$, which is normally the case, and assume also that the frequency range of the sinusoidal factor is much higher than that of the modulating signal. With these assumptions, the multiplicative factor may be extracted by an envelope detector, and the output of the envelope detector contains the intelligence. Letting $y_3(t)$ represent this quantity, it will be of the form

$$y_3(t) = k_1 + k_2 x(t) \tag{5-99}$$

The result is seen to contain the desired signal $x(t)$ plus a dc component. If we momentarily assume that $x(t)$ contains no dc or very low frequency data, a simple high-pass filter could be used to eliminate the dc component, and thus the output would be proportional to $x(t)$.

Although this circuit provides a sound theoretical approach for many FM detector circuits, a conventional differentiating circuit is seldom used. A conventional wideband differentiating circuit tends to accentuate high-frequency noise due to the increasing gain of the circuit as the frequency increases. Furthermore, with typical values of the operating parameters in FM systems, the dc level in (5-99) turns out to be very large compared with the signal level, and since the dc level must be removed, this would eliminate the possibility of dc data transmission. It was mentioned in the last section that direct dc data transmission was possible with FM, and this circuit would defeat that objective.

These problems can be overcome by several possible balanced arrangements that simultaneously achieve the following objectives: (1) The dc unwanted level is canceled at the center frequency by using two separate circuits connected in a special way. (2) The differentiating characteristic (increasing voltage versus frequency) is preserved and linearized only over the range of the input signal frequency deviation. This straight-line characteristic over a band-pass region may be closely approximated by resonance curves of tuned circuits.

One possible circuit using these principles is the *balanced discriminator* shown in Fig. 5-33. Consider the two halves of the secondary winding along with the capacitors as two separate parallel resonant circuits. Referring to Fig. 5-34, assume that the upper resonant circuit is tuned to $f_c + f_a$ where f_c is the carrier frequency, and assume that the lower one is tuned to $f_c - f_a$. The quantity f_a is a carefully selected increment on each side of f_c. The two tuned circuits are assumed to be identical except for the difference in the center frequencies as shown. The diodes in conjunction with the two RC circuits perform an envelope detection on the two signal components at the two respective ends of the transformer. The output voltage is the difference between these two separate voltages.

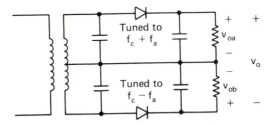

Figure 5-33. Balanced discriminator circuit.

Observe that the dc levels cancel provided that both circuits and the transformer are perfectly balanced. This eliminates one of the problems associated with the basic differentiating scheme. The high-frequency noise problem associated with a conventional differentiator is alleviated by the fact

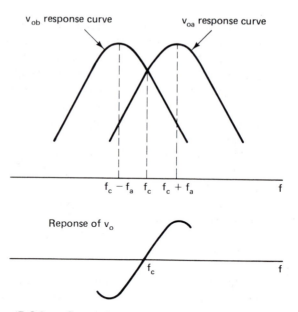

Figure 5-34. Operation of balanced discriminator circuit.

that the linear voltage variation with frequency is true only over a limited frequency range, as can be seen from Fig. 5-34. Over this range, the circuit is acting in a sense like a differentiator, but it is strictly bandlimited.

The preceding qualitative discussion will now be developed in a more analytical form. Assume that over a limited range of the tuned circuit magnitude curves in the vicinity of $f_c \pm f_a$, the variation is linear with frequency. In this case the upper voltage v_{0a} is approximately

$$v_{0a} \simeq A + Kf \qquad (5\text{-}100)$$

The lower voltage v_{0b} is then seen to be

$$v_{0b} \simeq A - Kf \qquad (5\text{-}101)$$

The output voltage v_0 is

$$v_0 = v_{0a} - v_{0b} = 2Kf = K_d f \qquad (5\text{-}102)$$

where K_d is a resulting discriminator constant in volts per hertz.

The form of the discriminator output voltage versus frequency given in (5-102) will be assumed in later developments irrespective of the type of discriminator. Furthermore, in spite of our aversion to the ideal differentiating circuit, it is sometimes convenient to assume an ideal differentiating characteristic in analytical calculations involving FM systems.

Two other examples of circuits that make use of the slope-detection property are the *Foster Seely discriminator* and the *ratio detector*, which are

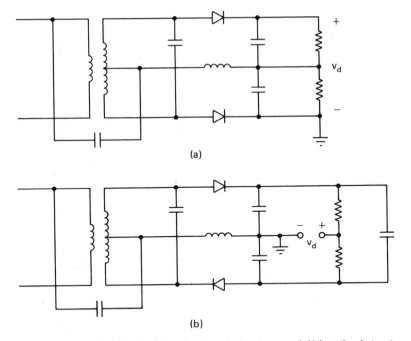

(a)

(b)

Figure 5-35. (a) Foster Seely discriminator, and (b) ratio detector.

illustrated in Fig. 5-35. Detailed discussions of these circuits are given in various communications electronics texts and will not be considered here.

A completely different approach to FM detection is the pulse averaging discriminator (PAD) whose block diagram is shown in Fig. 5-36. The FM signal is applied to a circuit that generates a trigger pulse for every positive-going zero crossing. (The circuit could also be implemented with negative-going zero crossings.) The triggers are used to initiate a fixed pulse from a monostable or one-shot multivibrator. The pulse width T_m of the multivibrator is chosen to be one-half of the period corresponding to the center frequency of the unmodulated carrier f_c; that is,

$$T_m = \frac{1}{2f_c} \tag{5-103}$$

Assume that in the rest state of the multivibrator the output voltage is $-A$, and during the duration of the pulse, it is $+A$. The various waveforms

Figure 5-36. Block diagram of pulse averaging discriminator.

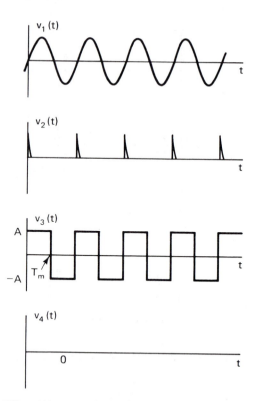

Figure 5-37. Waveforms in PAD when input frequency = f_c.

associated with this operation are shown in Figs. 5-37 through 5-39. The case when the input frequency if f_c is illustrated in Fig. 5-37. In this case, the voltage at the output of the multivibrator is a balanced square wave, so the net area in a cycle is zero. Thus, the output of the averaging circuit (low-pass filter) is zero.

Assume now that the instantaneous frequency changes abruptly to a value below f_c, as shown in Fig. 5-38. The pulse width T_m continues to be the same, but the distance between successive trigger pulses increases, and the negative part of the square-wave cycle is longer than the positive part. The net area is thus negative, and the output of the filter is a negative dc value.

Finally, assume that the frequency changes abruptly to a value higher than f_c, as shown in Fig. 5-39. In this case, the positive part of the square wave is longer than the negative part, and the net area is positive. Thus, the output of the filter is a positive dc value.

Operation of the circuit has been illustrated with abrupt step changes in the instantaneous input frequency. These results should demonstrate to the reader that dc data transmission is possible with FM, since the two cases of Figs. 5-38 and 5-39 correspond to dc modulating signals with negative and

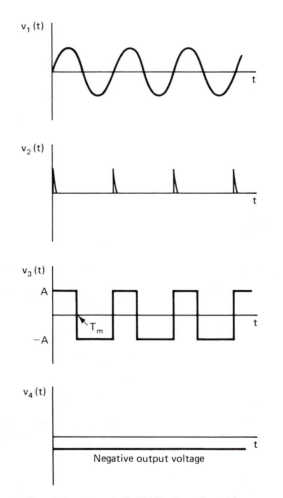

Figure 5-38. Waveforms in PAD when input frequency $< f_c$.

positive values, respectively. With time-varying data, the output of the filter will continually change in accordance with the instantaneous input frequency if the circuit is designed properly. The output filter must be chosen to have a flat passband amplitude characteristic over the baseband modulating signal frequency range, but it must reject frequency components in the range near f_c and higher.

No modern treatment of FM detection would be complete without consideration of the *phase-locked loop* (PLL) discriminator. There are so many facets of PLL analysis and operation that several complete books have been written on the subject! The availability of low-cost chips containing complete PLL circuits has increased the usage in many and varied applications.

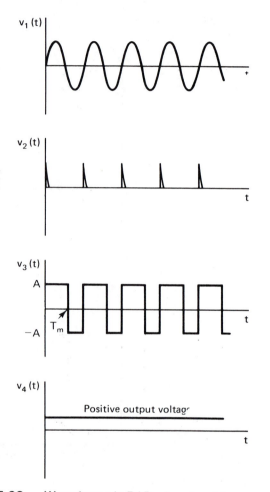

Figure 5-39. Waveforms in PAD when input frequency $> f_c$.

Possible applications for the PLL include discriminators, tracking filters, frequency multipliers, and frequency synthesizers.

We will concentrate here on a qualitative explanation of the PLL used as a frequency discriminator for FM detection. A block diagram of the PLL arranged for this application is shown in Fig. 5-40. The complete system includes a phase comparator, a VCO, and a loop filter. Most PLL chips require one or more external resistors and capacitors to be connected in the loop filter to establish the proper time constants for the particular application.

The phase comparator is characterized by an output voltage proportional to the difference in phase between the incoming signal on the left and the

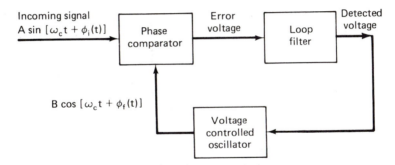

Figure 5-40. Block diagram of a phase-locked loop (PLL) frequency discriminator.

feedback signal. The VCO produces an output signal whose frequency deviation about the center frequency is proportional to its input voltage, which actually represents the detected output voltage $v_0(t)$. The loop filter helps to establish the proper transient response and filtering as required for complete transient and steady-state operation.

In simplest terms, the feedback signal mixes with the incoming signal to force the net instantaneous signal frequency at the output of the phase comparator to be zero. This means that the instantaneous frequency of the feedback signal must be the same as that of the input signal. Since the frequency deviation at the output of the VCO is directly proportional to $v_0(t)$, this loop action forces $v_0(t)$ to be a direct replica of the intelligence signal. In effect, the loop forces the VCO in the PLL to duplicate the action of the VCO back at the transmitter, so the voltage $v_0(t)$ is forced to duplicate the modulating voltage back at the transmitter.

The preceding discussion, though simplified, should provide at least an intuitive approach to basic understanding of the PLL. A full analysis of this system requires the use of feedback control theory. There are some aspects of the system that are nonlinear, but many useful analysis and design properties of the PLL can be determined with the aid of a linearized Laplace transform model.

For the benefit of readers familiar with s-plane analysis using Laplace transforms, one form of this linearized model is shown in Fig. 5-41. The constant K_p is the phase comparator constant in volts per radian, K_v is the VCO constant expressed in radians per second per volt, and $H_L(s)$ is the transfer function of the loop filter. The quantities $\Omega_i(s)$, $\Phi_i(s)$, $\Phi_e(s)$, $\Phi_f(s)$, and $V_0(s)$ represent the Laplace transforms, respectively, of the instantaneous input signal radian frequency $\omega_i(t)$, the instantaneous input signal phase $\phi_i(t)$, the instantaneous phase error $\phi_e(t)$ at the output of the phase detector, the instantaneous signal feedback phase $\phi_f(t)$, and the detected output voltage $v_0(t)$.

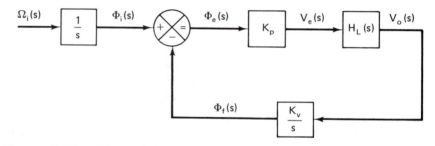

Figure 5-41. Form of the linearized Laplace transform model of the phase-locked loop frequency discriminator.

5-10 SUPERHETERODYNE RECEIVERS

The vast majority of receivers used in modern communications systems employ a concept known as the *superheterodyne* principle. This concept will be discussed in this section, and some representative block diagrams of superheterodyne receivers will be shown.

It should be noted at the outset that superheterodyne receivers are used with virtually all types of modulation systems. The development of this concept has been included in this chapter on FM only because this is a good point at which to bring together the techniques of frequency translation and filtering as used in the superheterodyne receiver.

First, some background discussion is required. A communications receiver has the complex functions of separating the desired signal from the large number of electromagnetic signals present at the input, amplifying the signal to an appropriate level, performing the demodulation or detection operation, and providing a baseband output in an acceptable form. This last function could range from providing audio power to a speaker in the case of a standard radio to transferring digital data to a computer memory system in the case of a complex telemetry data system. Several parameters used in describing the quality of various receiver functions will be introduced.

Sensitivity

Sensitivity is a measure of how well the receiver can respond to very weak signals. Theoretically, a receiver can be made to respond to an arbitrarily small signal by the addition of more amplifier stages. However, the real criterion is how well it can respond to small signals without masking the desired signal by the low-level noise introduced by additional stages. The major treatment of noise effects will be made in Chapter 8 and later, but it should be noted here that it is a difficult design task to produce a highly sensitive receiver that can still maintain a signal level well above the background noise level.

Sensitivity has been measured in several different ways. One common way is to specify the receiver input level in microvolts required to produce a certain specified signal to noise ratio at the detector output. A more recent standard is to specify the signal power in dBf (dB above 1 femtowatt, where 1 femtowatt $= 1$ fW $= 10^{-15}$ W) required to produce a certain specified signal-to-noise ratio at the detector output. The latter standard has the advantage that it is independent of the exact impedance level at the receiver input. With both of these standards, a *smaller* value for the sensitivity would indicate a *more sensitive* receiver. However, in qualitative references, it is natural to refer to a better receiver from a sensitivity point of view as one with greater sensitivity, so one has to be careful to imply the proper meaning in such a reference.

Selectivity

Selectivity is a measure of how well the receiver separates the desired signal from the undesired signals in adjacent frequency ranges. Receivers having higher selectivities have the capability of rejecting interfering components very close to the desired signal. Selectivity is often specified by two bandwidth parameters, one of which indicates the passband bandwidth and the other of which indicates the bandwidth at which the attenuation is some minimum stopband value. As we will see shortly, one significant advantage of the superheterodyne concept is that of providing a uniform selectivity over a wide band.

Consider now the task of separating the desired signal from other signals present in the case where the receiver must tune over a wide frequency range. Good selectivity demands a band-pass filter having a flat passband characteristic and a very pronounced attenuation rate in the stopband. Such filters are relatively straightforward to build for operation near a single frequency, but tuning such circuits over a very wide frequency range is something else entirely! Even with the single series or parallel resonant circuits, which represent some of the simplest band-pass structures, the ability to maintain a constant bandwidth over a wide frequency range is very difficult.

A parameter widely used in describing the relative selectivity of a filter is the *loaded Q* or overall circuit Q, which will be defined as

$$Q = \frac{\text{center frequency}}{\text{reference bandwidth}} \qquad (5\text{-}104)$$

This definition is based on the more fundamental Q property of series and parallel resonant circuits and represents an extension to band-pass structures of any type. The reference bandwidth is the frequency difference between adjacent points on either side of the center such that the response is down by the same value. In resonant circuits, the 3-dB bandwidth is usually employed, but other bandwidths are often used in precision filters (e.g., the 1-dB bandwidth).

One problem of tuning over a wide frequency range is the possible extreme range of Q that may be required. If the required Q is too low, there is difficulty in establishing proper stopband attenuation while simultaneously maintaining constant passband amplitude. Conversely, if the required Q is too high, the resulting filter is very sensitive to element value variations, and the range of component values may be very large. It is very difficult to establish an exact range for Q that is always feasible due to the many complex forms of technology available. Furthermore, the value of attainable Q is a function of the frequency range. For example, in the microwave range, a Q of several thousand can be attained with a cavity resonator, but such a value is virtually impossible in the lumped-parameter range. Realizing that there may be many exceptions to this pattern, the range $10 < Q < 100$ appears to be the most common approximate realistic working range for frequencies well below the microwave region.

A receiver in which all the tuning is performed directly at the required receiving frequency is called a *tuned radio frequency* (TRF) receiver. There is still a place for the TRF receiver, particularly in certain single-frequency receivers or in systems where a very limited amount of tuning is required. However, the TRF receiver has been largely superseded by the superheterodyne receiver, which will be considered next.

The block diagram of a single conversion superheterodyne receiver is shown in Fig. 5-42a and a corresponding spectral diagram is shown in Fig. 5-42b. Observe the break in the frequency scale since it is difficult to show on a single graph the typical frequencies involved to scale. In the discussion that follows, the reference will be shifted frequently back and forth from the block diagram to the spectral diagram.

The center frequencies of various portions of the receiver are indicated above the block diagram in Fig. 5-42a, and the corresponding bandwidths are indicated below the block diagram. The incoming signal is amplified first by an RF amplifier. The center frequency of this stage is f_c. A typical RF amplifier has a band-pass characteristic sufficiently broad to pass the highest bandwidth signal required. As we will see later, some selectively is desired in the RF stage, but it need not possess the sharp stopband attenuation rate eventually needed to reject adjacent interfering signals. Indeed, if it did, all the filtering could be performed in this stage! In some receivers, the RF stage has no band-pass characteristic at all, but best receiver design suggests some selectivity for reasons that will be clearer later. A typical broadband characteristic is shown over the top of the other components about f_c in Fig. 5-42b.

In the mixer stage, the incoming RF signal centered at f_c is mixed with a local oscillator (LO) sinusoid whose frequency is f_{LO}. The LO tuning circuit is coupled with the RF amplifier tuning circuit so that the frequency difference between f_c and f_{LO} is a constant, and this constant frequency is the intermediate frequency (IF), which will be denoted as f_{IF}. Coupling of the tuning is

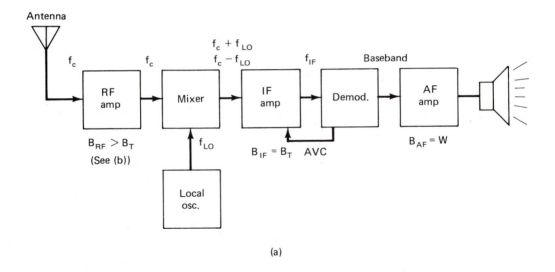

(a)

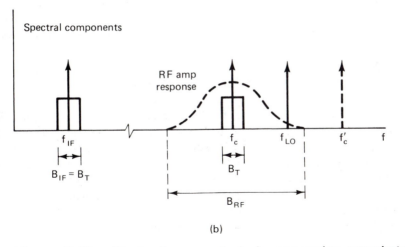

(b)

Figure 5-42. Block diagram of single conversion superheterodyne receiver and associated spectral components.

usually achieved by a two-section variable capacitance in which both values change as the tuning knob is turned.

The LO frequency may be higher than the RF frequency or it may be lower. When the LO frequency is higher, we have

$$f_{IF} = f_{LO} - f_c \qquad (5\text{-}105)$$

When the LO frequency is lower, the relationship is

$$f_{IF} = f_c - f_{LO} \qquad (5\text{-}106)$$

The spectral diagram assumes a higher LO frequency, and this assumption will be made in this development for convenience.

The mixer could be any balanced modulator type of circuit that performs an effective multiplication of the two signals, as we have observed in various applications throughout this chapter and in Chapter 4. Significant in the output of the mixer are the sum and difference frequency components centered at $f_{LO} + f_c$ and $f_{LO} - f_c$. Each of these components contains the original message sidebands (or sideband for SSB) about the center frequency. Depending on whether the mixer is a "perfect" multiplier, a switching modulator, or one of the other various nonlinear components that have been used for this operation, other products may appear in the output as well, although only the IF difference frequency component centered at $f_{IF} = f_{LO} - f_c$ is shown in Fig. 5-42b.

We will now focus on the major function of the IF amplifier, the establishment of the effective selectivity of the receiver. The frequency range of the IF amplifier is selected so that a very nearly rectangular band pass can be constructed. Once aligned, this band-pass characteristic will remain fixed and will not be tuned. The bandwidth B_{IF} of the IF amplifier is set at the required transmission bandwidth B_T, that is, $B_{IF} = B_T$. The mixer output component at $f_{LO} + f_c$ is easily rejected by the IF stage, and the component at $f_{LO} - f_c = f_{IF}$ falls right across the IF band-pass characteristic and is transmitted and amplified. Other undesirable products of the mixer will be rejected by the IF amplifier unless they happen to appear in the IF bandwidth.

The strategy of the superheterodyne receiver can now be deduced. Instead of attempting to tune across the RF spectrum with a highly selective variable center frequency filter, a fixed-frequency filter is established and optimized. The various signals are then shifted to the frequency range of the fixed IF filter by the mixer and LO and are then separated from other components in that frequency range. The selectivity of the receiver is then determined by the IF filter band-pass characteristic.

Following amplification and filtering in the IF amplifier, the signal is applied to the modulator or detector stage, in which it is converted to baseband form. The baseband signal is then amplified and processed in whatever form is required. The diagram in Fig. 5-42a indicates an audio frequency (AF) amplifier and a speaker, which would suggest the common household radio receiver situation.

One additional item in Fig. 5-42a is the *automatic volume control* (AVC) line. Signals appearing at receivers vary considerably in their levels. The AVC signal is a rectified bias voltage or current proportional to the level of the receiver signal. This voltage is applied to an earlier stage in the receiver in which the gain can be reduced by application of a dc voltage. Thus, strong stations cause a larger AVC voltage to appear, which acts to reduce the gain. A typical AVC circuit does not completely eliminate the variation of signal

strengths, but it does reduce the effective dynamic range between the strongest and the weakest signals.

We have seen the advantages of the superheterodyne receiver in establishing a fixed selectivity for a receiver as a function of the tuning, but what are the disadvantages? A particular problem called *image frequency interference* does suddenly appear. Refer again to the spectral diagram in Fig. 5-42b. Observe the component at f_c' shown as a dashed line. This component is called the *image frequency,* and the value of this frequency is

$$f_c' = f_{\text{LO}} + f_{\text{IF}} = f_c + 2 f_{\text{IF}} \qquad (5\text{-}107)$$

Suppose a signal at the image frequency is passed by the RF amplifier. It will then mix with the LO signal and will appear at f_{IF} along with the desired signal. Once it appears in the IF passband at the same frequency as the desired signal, there is very little hope of rejecting it.

The technique for minimizing image interference is to ensure that it does not get through the RF amplifier. This can be achieved by having a moderate degree of selectivity in that stage. Specifically, the frequency range from $f_c' - B_T/2$ to $f_c' + B_T/2$ represents the range that when mixed with f_{LO} would fall in the IF bandpass. Consequently, this range should be well into the stopband portion of the RF amplifier response curve, as illustrated in Fig. 5-42b.

As the IF frequency is increased, the moderate filtering on the RF amplifier stage is reduced and the corresponding image frequency is farther away from f_c. Conversely, however, it is often easier to design and implement an IF amplifier with high selectivity at a relative low frequency. The selection of an IF frequency is then an engineering compromise.

Because of these conflicting requirements of high IF selectivity at lower IF frequencies and ease of image rejection at high IF frequencies, complex receivers employing more than one level of conversion have been developed (e.g. *double-conversion* and *triple-conversion* superheterodyne receivers). A block diagram of a double-conversion receiver is shown in Fig. 5-43. The first mixer stage down-converts the signal to an IF frequency $f_{\text{IF}}^{(1)}$. This particular IF frequency is relatively high and eases the burden on the RF amplifier and thus helps to create a good image-rejection process. The second mixer stage down-converts the signal to a final IF frequency range $f_{\text{IF}}^{(2)}$, which is usually much lower than $f_{\text{IF}}^{(1)}$. This stage has been optimized to produce a very nearly rectangular band-pass characteristic, so the selectivity of the receiver is established at this point.

Throughout the analysis, we have assumed that $f_{\text{LO}} > f_c$. When $f_{\text{LO}} < f_c$, the same general principles apply except that the image component is at $f_{\text{LO}} - f_{\text{IF}} = f_c - 2 f_{\text{IF}}$. For either case, the image component is on the opposite side of the LO frequency for the desired signal by a frequency increment equal to f_{IF}.

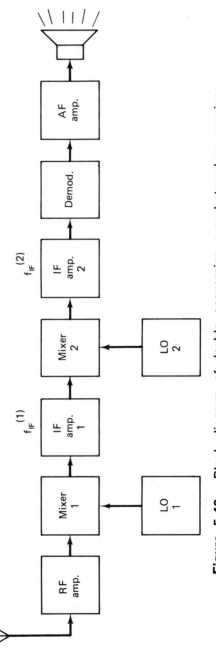

Figure 5-43. Block diagram of double conversion superheterodyne receiver.

Example 5-11

Most commercial FM receivers are single-conversion superheterodyne types and are capable of tuning over the FM broadcast band from approximately 88 to 108 MHz. The IF frequency in these receivers is usually 10.7 MHz. The bandwidth allocated by the FCC for each station is 200 kHz. (a) Suppose that very selective filtering were performed in the RF stage (i.e., the TRF receiver concept). Determine the range of approximate effective circuit Q that would be required. (b) Determine the approximate Q required for the IF filter in the superheterodyne receiver. (c) Determine the range of LO oscillator and image frequencies if the LO frequency is higher than the signal frequency. (d) Repeat part (c) if the LO frequency is lower than the signal frequency.

Solution

(**a**) For this part and for part (b), a bandwidth of 200 kHz will be used. This may or may not be the optimum bandwidth to use for the filter depending on the exact passband shape of a given filter and how sharply it attenuates in the stopband. However, it is sufficiently close for the purpose desired since approximate values were requested.

At the lower end of the FM band, the Q required would be

$$Q = \frac{88 \times 10^6}{200 \times 10^3} = 440 \tag{5-108}$$

At the high end of the FM band, the Q required would be

$$Q = \frac{108 \times 10^6}{200 \times 10^3} = 540 \tag{5-109}$$

Both of these Q values are rather difficult to obtain.

(**b**) At the IF frequency for the superheterdyne receiver, the required Q is

$$Q = \frac{10.7 \times 10^6}{200 \times 10^3} = 53.5 \tag{5-110}$$

which is much more reasonable and easier to implement. In addition, the IF amplifier is fixed, while the TRF tuning circuit would require maintaining a high Q over a 20-MHz tuning range.

(**c**) If the LO frequency is higher than the signal frequency, it must be 10.7 MHz higher at all frequencies. Thus, if the signal varies from 88 to 108 MHz, the LO oscillator must tune from 88 + 10.7 = 98.7 MHz to 108 + 10.7 = 118.7 MHz.

The image frequency is 10.7 MHz above the LO oscillator frequency, so at the low end of the dial, the image is 98.7 + 10.7 = 109.4 MHz, and at the high end of the dial, the image is 118.7 + 10.7 = 129.4 MHz.

The ranges involved are summarized as follows:

Signal frequency = 88 to 108 MHz
LO frequency = 98.7 to 118.7 MHz
Image frequency = 109.4 to 129.4 MHz

At a given frequency, the RF amplifier passband should be sufficiently selective to reject the corresponding image frequency, but since the image frequency is 21.4 MHz away, a much more moderate effective Q for the RF tuning circuit would suffice. The minimum Q of the RF amplifier would depend on the level of the image rejection desired and other factors, but a realistic design can be achieved with the conditions given.

It is interesting to note that the lowest possible image frequency (109.4 MHz) is above the high end of the FM band by a margin of 1.4 MHz. This eliminates the possibility of another FM station being the interfering image component, and this point was no doubt considered in the selection of the common IF frequency.

(d) The analysis proceeds in essentially the same fashion as in part (c) except that the LO frequency is always lower than the signal frequency by 10.7 MHz, and the image frequency is always lower than the LO frequency by another 10.7 MHz. The ranges involved are summarized as follows:

Signal frequency = 88 to 108 MHz
LO frequency = 77.3 to 97.3 MHz
Image frequency = 66.6 to 86.6 MHz

PROBLEMS

5-1 A certain laboratory function generator can be angle modulated with various waveforms. Based on a 100-Hz sinusoid before modulation is applied, the output signal with modulation is

$$y(t) = 20 \cos \left[200\pi t + 40\pi(1 - \epsilon^{-2t})\right]$$

Determine expressions for the following functions: (a) instantaneous total phase, (b) instantaneous signal phase, (c) instantaneous total frequency (radian and cyclic), and (d) instantaneous signal frequency (radian and cyclic). (e) Sketch the forms of the instantaneous signal phase and frequency.

5-2 If the angle modulated signal of Prob. 5-1 is known to be a PM signal, determine an expression for the normalized modulating signal $x(t)$.

5-3 If the angle modulated signal of Prob. 5-1 is known to be an FM signal, determine an expression for the normalized modulating signal $x(t)$.

5-4 A certain signal $x_1(t)$ over an interval of 5 s is described by the equation

$$x_1(t) = 2t, \qquad 0 \le t \le 5$$

This signal is to be used to angle modulate an oscillator whose output with no modulation applied is a 1-kHz sinusoid. Write an expression for the composite modulated signal $y(t)$ over the given interval if *phase modulation* is to be used with $\Delta\phi = 2$ rad. (*Hint:* It is convenient to first change $x_1(t)$ to a normalized modulating signal $x(t)$ having a maximum magnitude of unity.)

5-5 Repeat Prob. 5-4 if *frequency modulation* is to be used with $\Delta f = 100$ Hz.

5-6 A certain sweep frequency generator produces a near-sinusoidal output whose frequency changes linearly with time from f_1 at $t = 0$ to f_2 at $t = T$. Write expressions for (a) instantaneous total radian frequency and (b) instantaneous total phase. (There is no carrier reference in this case.) If the output has an amplitude of A, write an expression for the composite time signal.

5-7 The triangular waveform of Fig. P5-7 *phase modulates* a high-frequency carrier. Sketch the form of (a) the instantaneous signal phase and (b) the instantaneous signal cyclic frequency in hertz.

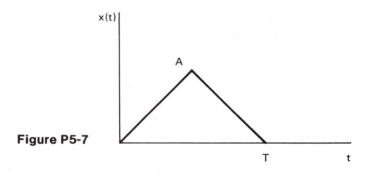

Figure P5-7

5-8 Repeat Prob. 5-7 if the triangular waveform *frequency* modulates a high-frequency carrier. In part (a), assume that $\phi_i(0) = 0$.

5-9 A certain FM tone-modulated signal is given by

$$y(t) = 20 \cos(6 \times 10^8\, t + 30 \sin 10^4\, t)$$

Determine (a) unmodulated carrier frequency in hertz, (b) modulating frequency in hertz, (c) modulation index β, and (d) maximum frequency deviation Δf in hertz.

5-10 In the text, the equations for PM and FM tone-modulated signals were developed with the assumption that the modulating signal was a unit amplitude cosine function. Develop the form of the PM and FM signals when the modulating signal is a sine function; that is, $x(t) = \sin \omega_m t$. You may still employ the cosine function for the carrier.

5-11 A procedure that is widely used for calibrating and measuring FM frequency deviation is based on the fact that the carrier vanishes for specific values of the modulation index with tone modulation. Assuming that the range of frequency deviation is known so that the specific zero on the Bessel curve can be identified, the modulating frequency is carefully adjusted until the carrier component vanishes. (The absence of the carrier is observed with a narrow-band filter or with a spectrum analyzer.) Let β_0 represent the value of β at which the carrier vanishes; that is, $J_0(\beta_0) = 0$. Determine an expression for Δf in terms of β_0 and the modulating frequency f_m.

5-12 A certain FM transmitter having a center frequency of 90 MHz is to be tested at different modulating frequencies using a single tone cosine function in each case. The maximum deviation is to be held at ±50 kHz throughout the test. Write an equation for the composite signal for each of the following modulating frequencies: (a) 100 Hz, (b) 1 kHz, (c) 10 kHz.

5-13 If the transmitter of Prob. 5-12 were a PM transmitter instead of an FM transmitter and if the maximum frequency deviation were the same as the FM transmitter at 10 kHz, what would be the maximum deviation at (a) 1 kHz and (b) 15 kHz?

5-14 A certain FM transmitter operates with a fixed maximum frequency deviation of ±20 kHz. Determine the approximate transmission bandwidth for a single-tone modulating signal at each of the following modulating frequencies: (a) 100 Hz, (b) 1 kHz, (c) 10 kHz, (d) 100 kHz.

5-15 A certain FM transmitter is modulated with a 1-kHz sinusoid. The frequency deviation is gradually increased while some tests are performed. Determine the approximate transmission bandwidth for each of the following frequency deviations: (a) ±100 Hz, (b) ±1 kHz, (c) ±10 kHz, (d) ±100 kHz.

5-16 A certain PM transmitter operates with a fixed maximum phase deviation of ±2 rad. Determine the approximate transmission bandwidth for a single-tone modulating signal at each of the following frequencies: (a) 100 Hz, (b) 1 kHz, (c) 10 kHz.

5-17 A certain PM transmitter is modulated with a 1-kHz sinusoid. The phase deviation is gradually increased while some tests are performed. Determine the approximate transmission bandwidth for each of the following phase deviations: (a) ±0.1 rad, (b) ±1 rad, (c) ±10 rad, (d) ±100 rad.

5-18 Using the data given in Table 5-1, verify that the total power remains constant as the modulation index changes by determining the power at each of the following values of β: (a) 0.1, (b) 0.5, (c) 1. (Very slight discrepancies are to be expected since the number of sidebands is finite in each case.)

5-19 Using the data provided in Table 5-1, verify the validity of the 98% power transmission criterion for Carson's rule by computing the percentage of total power contained in the bandwidth $2(\beta + 1)\Delta f$ for each of the following integer values of β: (a) $\beta = 1$, (b) $\beta = 2$, and (c) $\beta = 5$. (Remember that the total power is the same as that of an unmodulated sinusoid having the same amplitude as the FM signal.)

5-20 Using the bandwidth approximations, verify that the following statements are correct:

(a) In a narrowband FM system ($\beta \leq 0.25$), bandwidth increases linearly with the modulating frequency but is independent of changes in the frequency deviation as long as the modulating frequency is fixed.

(b) In a very wideband FM system (say $\beta \geq 100$), the bandwidth increases linearly with the frequency deviation but is independent of changes in the modulating frequency as long as the deviation is fixed.

(c) In a pure PM system ($\Delta\phi$ fixed), the bandwidth always increases linearly with the modulating frequency.

5-21 Verify Eq. (5-70).

5-22 Verify Eq. (5-71).

5-23 Consider the signal of Ex. 5-8 containing a carrier and one sideband, for which it was shown that both AM and FM were present.

(a) Show that if $B \ll A$, the amplitude function $A(t)$ resembles that of a conventional AM signal with a single tone modulation and $m = B/A$. [*Hint:* $\sqrt{1 + x} \approx 1 + (x/2)$ for $x \ll 1$]

(b) Show that for the same inequality in part (a), the phase angle function $\phi(t)$ resembles that of an angle modulated signal having a single tone modulation with $\Delta\phi$ (or β) $= B/A$.

5-24 A certain modulated signal can be expressed as

$$y(t) = -A \sin \omega_c t - \frac{kA}{2} \cos(\omega_c - \omega_m)t + \frac{kA}{2} \cos(\omega_c + \omega_m)t$$

Determine the complex phasor form of the signal including the magnitude $A(t)$ and the phase angle $\phi(t)$.

5-25 A certain signal is given by

$$y(t) = \cos \omega_c t + 0.2 \cos \omega_m t \sin \omega_c t$$

(a) Determine the complex phasor form of the signal and draw a phasor diagram.

(b) Determine expressions for $A(t)$ and $\phi(t)$.

5-26 A tone-modulated FM signal with $\beta = 1$ is passed through a narrow-band filter such that only the carrier and one pair of sidebands appear at the output. Neglect any phase shift produced by the filter.

(a) Show that the envelope of the output varies at a rate $2f_m$, where f_m is the modulating frequency.

(b) Determine the magnitude of the envelope.

5-27 In the frequency-translation system of Ex. 5-9, the mixing oscillator frequency was selected in each step such that the sense of the original spectrum was preserved all the way through. Redesign the system with the mixing oscillator frequencies such that the sense of the spectrum is reversed after the first translation, but restored in the second translation. The center frequency at the end of the first step can be 20 MHz as before. Sketch the spectral layout and label all frequencies as well as the senses of the spectra. Show that the constraints imposed on the system to ease filtering are still met.

5-28 A certain signal occupies the frequency range from 20 to 21 MHz. It is desired to shift this spectrum in one step so that it occupies the range from 150 to 151 MHz.

(a) Determine the mixing oscillator frequency required if the sense of the original spectrum must be preserved.

(b) Draw a spectral diagram and identify the major frequency components that must be eliminated by the filtering.

5-29 Suppose in Prob. 5-28 that the sense of the spectrum is not important. Repeat the analysis for the other possible solution.

5-30 A certain signal occupies the frequency range from 400 to 403 MHz. It is desired to shift this spectrum in one step so that it occupies the range from 30 to 33 MHz.

(a) Determine the mixing oscillator frequency required if the sense of the original spectrum must be preserved.

(b) Draw a spectral diagram and identify the major frequency components that must be eliminated by the filtering.

5-31 Suppose in Prob. 5-30 that the sense of the spectrum is not important. Repeat the analysis for the other possible solution.

5-32 In the development of Eqs. (5-77a) and (5-77b), it was assumed that $\omega_{c1} > \omega_0$. It is stated in the text that if $\omega_0 > \omega_{c1}$ it is possible to generate either a down-converted sideband and an up-converted sideband or two up-converted sidebands. Show that (a) if $\omega_{c1} < \omega_0 < 2\omega_{c1}$, one of the two sidebands is a down-converted component, and (b) if $\omega_0 > 2\omega_{c1}$, both of the two sidebands are up-converted components. (*Hint:* $\cos(-x) = \cos x$.)

5-33 Assume that the system of Ex. 5-10 is to be designed using three triplers and as many doublers as needed to complete the system. Formulate a

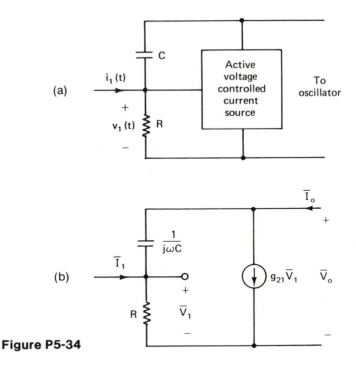

Figure P5-34

possible block-diagram design layout for the transmitter, and prepare a table similar to Table 5-3. The mixing frequency should be chosen to be less than 20 MHz. (It may be necessary to further reduce the deviation of the modulator.)

5-34 One of the most classical circuit forms for indirectly generating a variable reactive component is the reactance modulator. One form used for generating an effective variable capacitance is shown in Fig. P5-34a, and the equivalent circuit with assumed phasors is shown in Fig. P5-34b. In this form, the active device should function as a voltage-controlled current source. Furthermore, a nonlinear operating point is required such that the transconductance g_{21} varies with the input voltage. The input signal is assumed to be an ideal current source $i_1(t)$ or $\bar{I}_1$ in phasor form.

(a) Initially assume that $\bar{I}_1 = 0$. Show that

$$\bar{I}_0 = \frac{j\omega C(g_{21}R + 1)\bar{V}_0}{1 + j\omega RC}$$

(b) Assume that the frequency range is such that $\omega RC \ll 1$. Show that an effective capacitance C_0 appears across the output terminals whose value is

$$C_0 = (g_{21}R + 1)C$$

(c) Assume now that when the modulating signal current $i_1(t)$ is applied the gain is varied in direct proportion; that is, let

$$g_{21}(t) = g_0 + \alpha i_1(t)$$

where $g_{21}(t)$ is a time-varying gain, g_0 is the gain with no signal applied, and α is a constant. Show that the effective capacitance $C_0(t)$ is time varying and has the form

$$C_0(t) = (g_0 R + 1)C\,[1 + Ki_1(t)]$$

where K is a constant. This time-varying capacitance is connected across an LC oscillator circuit and provides the proper frequency modulation.

5-35 In this problem, the analysis of the phase comparator circuit shown in Fig. P5-35a will be made. The object of this circuit is to generate an output voltage that is a function of the phase difference between $\overline{V}_1$ and $\overline{V}_2$. Along with direct applications of this concept, the circuit also serves as the basis for the Foster-Seely discriminator. Assume that $\overline{V}_1 = A\underline{/0°}$ and $\overline{V}_2 = A\underline{/90° - \theta}$, where θ is any arbitrary phase shift shifted back from a 90° fixed reference. The time constants of the RC filters are chosen to be long compared with the

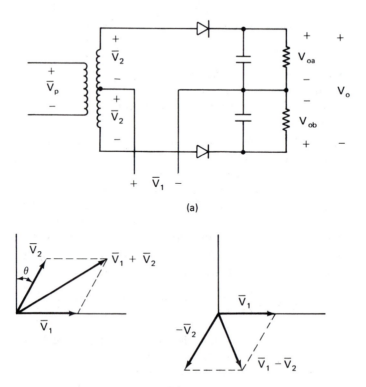

(a)

(b)

Figure P5-35

carrier frequency, but short compared with time changes of the phase angle. Assuming for the moment a constant value of θ, the dc voltages V_{0a} and V_{0b} are $V_{0a} = |\overline{V}_1 + \overline{V}_2|$ and $V_{0b} = |\overline{V}_1 - \overline{V}_2|$. The output voltage is $V_0 = V_{0a} - V_{0b}$. A phasor diagram depicting these results is shown in Fig. P5-35b. Show that the output voltage for small θ is approximately

$$V_0 \approx \sqrt{2}\, A \sin \theta \approx \sqrt{2}\, A\theta$$

[*Hint:* $\sqrt{1 + x} \approx 1 + (x/2)$ for $x \ll 1$.] This result indicates that the output voltage is a linear function of θ for small θ (i.e., $\sin \theta \approx \theta$ for $\theta \ll \pi/2$). Note, however, that a 90° reference phase difference is required.

5-36   Most commercial AM receivers are single conversion superheterodyne types and are capable of tuning over the AM broadcast band from approximately 540 to 1600 kHz. The IF frequency in these receivers is usually 455 kHz. The bandwidth allocated by the FCC for each station is 10 kHz.

   (a)   Suppose that a TRF receiver were used. Determine the range of approximate effective circuit Q that would be required.
   (b)   Determine the approximate circuit Q required for the IF filter in the superheterodyne receiver.
   (c)   Determine the range of LO oscillator and image frequencies if the LO oscillator is higher than the signal frequency.
   (d)   Repeat part (c) if the LO frequency is lower than the signal frequency.
   (e)   In actual practice, most AM receivers use an LO frequency higher than the signal frequency. Discuss the reason with respect to the design of the LO.

5-37   The system shown in Fig. P5-37 is used in some high frequency receivers as an "image rejection mixer." The input signal is simultaneously

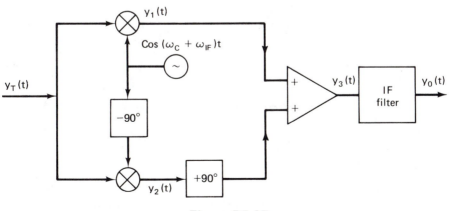

Figure P5-37

applied to two mixers for which the local oscillators are in phase quadrature. One of the mixer outputs is shifted in phase by 90° and summed with the other output. Assume that the total input signal $y_T(t)$ is of the form

$$y_T(t) = y(t) + y_i(t)$$

where $y(t) = A \cos \omega_m t \cos \omega_c t$ is the desired signal, which is assumed to be a tone modulated DSB signal for mathematical convenience. Let

$$y_i(t) = B \cos (\omega_c + 2\omega_{IF})t$$

be an undesired image component. The system shown assumes a local oscillator frequency higher than the input carrier frequency, but the concept can be applied to either case. Assuming ideal balanced modulator operations for the two mixers, determine $y_1(t)$, $y_2(t)$, and $y_3(t)$, and expand them to show the center frequencies of all terms. Recognizing that the IF filter will reject all high frequency terms, show that the output $y_0(t)$ is of the form

$$y_0(t) = C \cos \omega_m t \cos \omega_{IF}t$$

where C is a constant propertional to A. Note that the image component cancels out in the process. (The phase shift of a +90° in the lower path is assumed to apply over the DSB signal bandwidth centered at ω_{IF}.)

5-38 The form of the spectrum of a typical commercial TV channel is shown in Fig. P5-38. The video signal utilizes a *vestigial sideband*, which will be analyzed in detail in Prob. 5-39. The total channel bandwidth is 6 MHz. The audio signal is a conventional FM signal. The maximum frequency deviation is ±25 kHz. (a) Based on an assumed baseband bandwidth W = 15 kHz, calculate the approximate transmission bandwidth for the audio signal. (b) Assuming symmetry in the sidebands and an audio carrier 5.75 MHz above the lower bandedge, compute the approximate frequency limits of the audio spectrum as measured from the lower bandedge.

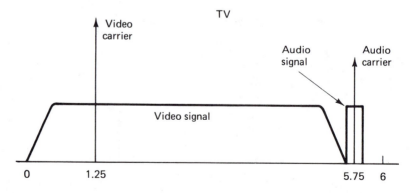

Frequency (MHz) above lower channel edge

Figure P5-38

5-39   Commercial television uses a form of AM called *vestigial sideband* (VSB) for the video signal. In this problem, some of the major properties of VSB will be investigated. VSB was postponed until Chapter 6 because it is best analyzed using the modulated phasor concept introduced in conjunction with FM.

VSB is a practical compromise between the bandwidth conservation of SSB and the detection simplicity of AM. The baseband bandwidth of typical video signals is about 4 MHz, so 8 MHz would be required for conventional AM. Since TV channels have a width of only 6 MHz and some space must be allocated for the audio signal, the VSB concept is utilized.

Refer to Fig. P5-39 for the discussion that follows. The video carrier is located 1.25 MHz above the lower channel edge. The video signal is generated as an AM signal but it is passed through a special vestigial filter before transmission. The entire upper sideband is maintained, but only a portion (a "vestige") of the lower sideband is transmitted. We will refer to portions of the video spectrum as "1" and "2" as follows:

Region 1: from about $f_c - 0.75$ MHz to $f_c + 0.75$ MHz where *both* sidebands are transmitted fully.

Region 2: from about $f_c + 0.75$ MHz to about $f_c + 4$ MHz where only one sideband is transmitted.

Consider a single component $\cos \omega_m t$ in the baseband signal, which would produce an AM signal $y_1(t)$ of the form

$$y_1(t) = A(1 + m \cos \omega_m t) \cos \omega_c t$$

$$= A \cos \omega_c t + \frac{mA}{2} \cos (\omega_c - \omega_m)t + \frac{mA}{2} \cos (\omega_c + \omega_m)t \qquad (1)$$

Assuming a normalized detection gain of unity, the output of a simple envelope detector after the dc component is eliminated would be

$$x_1(t) = mA \cos \omega_m t \qquad (2)$$

Observe that any baseband component from near dc to about 0.75 MHz would lie in region 1 and the detected signal would be of the form of (2).

Now assume a component $y_2(t)$ well above 0.75 MHz so that the lower sideband in (1) is eliminated.

   (a)   Write an expression for the modulated complex phasor function $\overline{Y_2}(t)$.

   (b)   Defining $\overline{Y_2}(t) = A(t) \underline{/\phi(t)}$, obtain an expression for $A(t)$.

   (c)   Obtain an expression for $\phi(t)$.

   (d)   Show that if $m \ll 1$, the envelope reduces to the approximate undistorted *form* of equation (2), but in which the detected value $x_2(t)$ has *half* the amplitude of the result of equation (2). [Hint: $(1 + a)^{1/2} \simeq 1 + a/2$ for $a \ll 1$.]

(e) Using ideal block filter characteristics, propose the form of a perfect filter that could be put in the receiver to compensate for the 2 to 1 amplitude ratio between components in regions 1 and 2.

Since the cost of a TV receiver having any filter close to the ideal determined in (e) would be increased markedly, the actual filter employed is based on the compromise characteristic shown in (b) of Fig. P5-39. Note that the relative amplitude for components displaced more than 0.75 MHz from

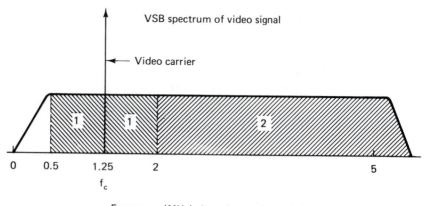

VSB spectrum of video signal

Video carrier

Frequency (MHz) above lower channel edge

(a)

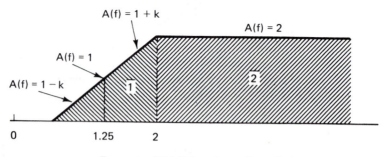

Amplitude response of receiver VSB filter

Frequency (MHz) above lower channel edge

(b)

Figure P5-39

the carrier (region 2 components) is 2. However the relative amplitude for components less that 0.75 MHz from the carrier (region 1 components) has odd symmetry about the carrier and can be expressed as $1 + k$ for $f > f_c$ and $1 - k$ for $f < f_c$.

(f) Show that any component in region 2 will have an approximate detected value given by equation (2).

(g) Show that a component in region 1 will have a detected value approximately the same as a component in region 2. (Assume $m \ll 1$.)

Pulse-Modulation and Multiplex Systems

6

6-0 INTRODUCTION

The first objective of this chapter is to develop the concepts of sampling and pulse modulation. The sampling theorem will be presented as the theoretical basis for all pulse-modulation systems. Several pulse-modulation methods will then be considered. The initial emphasis will be devoted to pulse amplitude modulation since it serves as a conceptual link between analog signals and virtually all forms of pulse and digital modulation. The concepts of pulse-width modulation and pulse-position modulation will also be presented.

The second objective of the chapter is to develop the basis for multiplexing in communications systems. Both time-division and frequency-division multiplexing will be considered. Some of the differences and the similarities between these two general categories will be discussed.

6-1 SAMPLING THEOREM

All modulated waveforms considered thus far in the book have been continuous-time signals or, in more casual terminology, analog modulated signals. We will now begin the study of the classes of modulation that involve pulse- and/or digital-modulation techniques. Pulse and digital modulated signals are characterized by a sequence of unique pulses or digital numbers, each of which represents the message signal at a particular value of time. The resulting signals are then defined only at discrete values of time, and portions of the signal are obviously "missed" by this process. However, it will be shown that if the signal is band-limited, and if a certain minimum sampling rate is met, the resulting signal can theoretically be completely reconstructed at the receiver.

The next chapter will be devoted exclusively to those methods usually designated as "digital" communications, while this chapter will deal with those that are more commonly designated as "pulse" communications. However, both general categories have the same theoretical basis, and that background will be established in this chapter.

The starting point for discussion of pulse- and digital-modulation techniques is the concept of a *sampled-data signal*. A sampled-data signal (also called a *sampled signal*) is one that consists of a regular sequence of encoded samples of a reference continuous-time or analog signal. There are a variety of ways in which the samples may be generated, encoded, transmitted, and decoded with significant resulting differences in the transmission bandwidth and performance characteristics.

Consider the arbitrary analog signal $x(t)$ shown in Fig. 6-1a. A particular form of a sampled-data signal $x_s(t)$ corresponding to $x(t)$ is shown in Fig. 6-1c. The sampled signal is obtained by observing $x(t)$ during short intervals of time of width τ seconds. The *sampling rate* is designated as f_s and $f_s = 1/T$, where T is the sampling period and is the time between the beginning of one sample and the beginning of the next sample. During the interval between samples, $x(t)$ is not observed at all. However, samples of other signals could be inserted in the "open space," as will be discussed later. However, the focus at the moment is directed toward the analysis of a single sampled signal, so the space will be considered unused at this time.

Note that since the sampling rate is a frequency it is perfectly proper to express a given sampling rate in hertz. One hertz is interpreted as one sample per second in this context. While hertz will be generally employed for sampling rates in this book, the term "samples per second" will be used when it adds clarity to a discussion.

The form of the signal generated by the process shown in Fig. 6-1 can be described as a *nonzero width, natural sampled signal*. The "non-zero width" description represents the fact that the sampling pulses occupy some width, however small, and this is always the case with actual samples of analog signals. However, the concept of impulse sampling, in which the samples are assumed to have zero width, is very useful in certain analytical developments and will be considered in the next section. The "natural sampled" description refers to the fact that the top of each sample in Fig. 6-1 follows the analog signal during the short interval τ. This form is the easiest to analyze mathematically and serves as a proper starting point for other forms. Observe that $x_s(t)$ can be generated from $x(t)$ by a switching arrangement in which $x(t)$ is switched periodically on for an interval τ and then turned off for the remainder of a period.

Irrespective of how the sampled signal is actually generated, it is convenient mathematically to express $x_s(t)$ as the product of $x(t)$ and the periodic

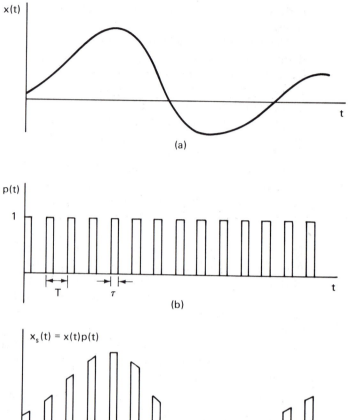

Figure 6-1. Development of sampled-data signal using nonzero width pulses and natural sampling.

pulse train $p(t)$ shown in Fig. 6-1b. Since the pulses assume only the two values 0 and 1, multiplication of $x(t)$ by $p(t)$ is equivalent to the sampling operation shown. Thus, we can write

$$x_s(t) = x(t)\ p(t) \tag{6-1}$$

The properties of the frequency spectrum of the sampled-data signal will now be investigated. We will assume a *baseband* signal for the purpose of this development, since the majority of signals for which sampling is applied are of that form. In Fig. 6-2a, an arbitrary baseband amplitude spectrum $X(f)$ is assumed for the analog signal. The spectrum of the periodic sampling pulse

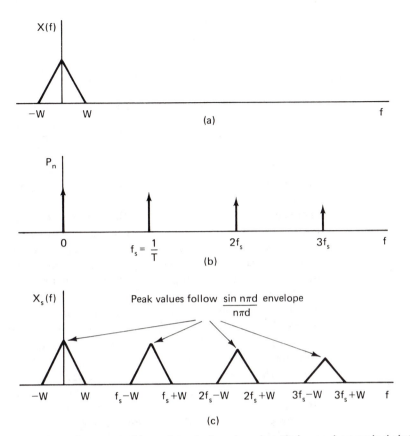

Figure 6-2. Spectra of baseband signal, pulse train, and sampled-data signal. (The three parts of this figure correspond to the three parts of Figure 6-1.)

train $p(t)$ is shown in Fig. 6-2b. The spectrum contains a dc component, a fundamental component at $f_s = 1/T$, and an infinite number of harmonics at integer multiples of the sampling frequency. Specifically, the reader can review appropriate sections of Chapter 2 to deduce that $p(t)$ can be expressed in a complex exponential Fourier series as

$$p(t) = \sum_{-\infty}^{\infty} \overline{P}_n \epsilon^{jn\omega_s t} \tag{6-2}$$

where the amplitude spectrum corresponding to $\overline{P}_n$ is

$$P_n = d \frac{\sin n\pi d}{n\pi d} \tag{6-3}$$

and $d = \tau/T$ is the duty cycle.

With the Fourier series form of (6-2) substituted in (6-1), the sampled signal can be expressed as

$$x_s(t) = x(t) \sum_{-\infty}^{\infty} \overline{P}_n \epsilon^{jn\omega_s t} = \sum_{-\infty}^{\infty} \overline{P}_n x(t) \epsilon^{jn\omega_s t} \qquad (6\text{-}4)$$

Fourier transformation of both sides of (6-4) yields

$$\overline{X}_s(f) = \sum_{-\infty}^{\infty} \overline{P}_n \overline{X}(f - nf_s) \qquad (6\text{-}5)$$

The form and relative magnitude pattern (but not necessarily actual magnitude levels) of $X_s(f)$ are illustrated in Fig. 6-2c. (Only a limited portion of the negative frequency range is shown.) It is observed that the spectrum of the sampled-data signal consists of the original baseband spectrum plus an infinite number of shifted or translated versions of the original spectrum. These various translated spectral components are shifted in frequency by increments equal to the sampling frequency and its harmonics. The magnitudes of the spectral components are multiplied by the P_n coefficients so that they diminish with frequency. However, for a very short duty cycle (i.e., $\tau \ll T$), the relative magnitudes of the components diminish very slowly so that the resulting spectrum can be quite broad in form. (The *actual* magnitudes are small for a short duty cycle, but the *relative* magnitudes remain nearly the same for a wide frequency range.)

From the nature of the spectrum of the sampled signal, it can be seen that the form of the original baseband spectrum is preserved in the component corresponding to $n = 0$ in (6-5). This component is multiplied by $\overline{P}_0$, which may be small compared with unity if $d \ll 1$, so the level may be reduced. Nevertheless, since the shape is preserved, the form of the spectrum of the original signal is maintained in the frequency range from dc to W *provided* that there is no overlap in the spectrum from any of the other components.

We will now explore this question of possible spectral overlap in some detail since it leads to one of the most important results in communications theory. We have assumed that the signal $x(t)$ is band-limited to W hertz. From Fig. 6-2c, it is observed that the lowest positive frequency corresponding to any of the shifted components is $f_s - W$. To be able to recover the original signal from the sampled signal, it is necessary that no portion of the spectrum of the first translated component overlap the original spectrum, which means $f_s - W \geq W$. This leads to the inequality

$$f_s \geq 2W \qquad (6\text{-}6)$$

Equation (6-6) is a statement of *Shannon's sampling theorem,* which is the foundation of all sampled-data, pulse, and digital signal and modulation systems. It states that a baseband signal must be uniformly sampled at a rate at least as high as twice the highest frequency in a spectrum in order to be recoverable by direct low-pass filtering. (There are special ways in which the strict requirements of the theorem can be relaxed for band-pass signals, but for the more common baseband signal case, the strict requirements will be assumed.) Although the "greater than or equal" statement of (6-6) has theoretical significance, in actual applications of the theorem, a sampling rate somewhat greater than the theoretical minimum is employed. The reason for this can be readily deduced from Fig. 6-2c by noting that, if $f_s = 2W$, there is no gap in the spectrum between the original spectrum and the first translated portion, and a perfect block filter would be required to separate the components. Thus, some frequency interval, called a *guard band,* should be provided in order that the first translated component can properly be rejected by a realistic filter. A typical example of actual sampling rates employed is that of commercial voice telephone sampling systems. Based on an assumed value of $W = 3.3$ kHz, a sampling rate $f_s = 8$ kHz is used while the theoretical minimum would be $f_s = 6.6$ kHz.

If the minimum sampling rate required as dictated by (6-6) is not met, some of the spectral components will overlap, as illustrated by Fig. 6-3. In this

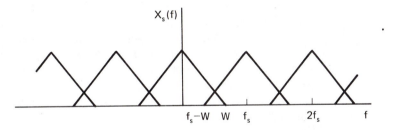

Figure 6-3. Illustration of aliasing resulting from inadequate sampling rate.

case components of the original spectrum appear at the same locations as other components and cannot be uniquely determined or separated. This process is called *aliasing,* and, as the name implies, spectral components may appear to be different than what they really are.

A convenient definition in sampling theory is the *folding frequency* f_0. It is given by

$$f_0 = \frac{f_s}{2} = \frac{1}{2T} \qquad (6\text{-}7)$$

The folding frequency is simply the highest theoretical frequency that can be processed by a sampled-data system with sampling rate f_s without any aliasing. The term *Nyquist rate* refers to the minimum theoretical sampling rate at which a given signal could be recovered and is $2W$ for a baseband signal with bandwidth W.

It is important in a sampled-data system that the minimum sampling rate be satisfied for *all* spectral components in a given signal, even though some may be out of the frequency range of interest. One might erroneously believe that, as long as the minimum sampling rate is met for all components of interest, the desired portion of the signal could be reconstructed. However, if there are components in the signal having a higher frequency range than the folding frequency, aliasing will occur and the desired portion can be obscured. This concept will be illustrated with a specific set of conditions in Ex. 6-2 at the end of this section. We will emphasize here that the minimum sampling rate given by (6-6) must be satisfied for *all* components within a given baseband spectrum being sampled, even though some portions of the spectrum may not be of interest.

Since it is not always possible to predict precisely the upper frequency limit of a given complex signal, a common practice employed in sampled-data baseband systems is to pass the analog signal through a low-pass filter (called an *aliasing filter*) before sampling. The cutoff frequency of the filter is chosen in accordance with the expected frequency range of the data being sampled, but is always less than or equal to the folding frequency. Consequently, if components higher than the folding frequency appear, they are suppressed by the filter, and aliasing is prevented or at least maintained to a tolerable level.

We will complete this section with some "philosophical notes" about the sampling process and some of the assumptions made. Some readers may be puzzled by the conclusion that the signal may be completely recovered from a set of narrow samples taken at discrete intervals. Referring back to Fig. 6-1, it is obvious that in the sampled signal we can observe the original analog signal for only a small fraction of the total time. Do we lose something by this process? Said differently, could something happen between samples that we might miss and therefore lose some of the desired information? The answer to this question lies at the core of the band-limited assumption made at the beginning. If the signal is truly band-limited to W hertz, nothing could happen in any interval of time less than or equal to $1/2W$ seconds that can't be predicted from the finite samples themselves. If new information appears in the interval that can't be reconstructed from the samples, the bandwidth assumed at the outset was not adequate. These points may be difficult to accept, but they can be rigorously established based on the assumed bandwidth limitation.

A deeper and closely related question concerns the assumption of a band-limited signal. Throughout the text and in all communications system design, it is necessary to assume the existence of band-limited signals. Yet there are more advanced theoretical developments that suggest that in order for a signal to be truly band-limited, it must exist for an infinite length of time. Since virtually all real signals exist for a finite length of time, the corresponding spectra theoretically are infinite in extent. In a practical sense, however, this need not cause any great concern. We have already encountered a number of spectral forms in the text that have theoretically occupied an infinite spectral range, but in which a finite practical bandwidth approximation could be made. Thus, in assuming a bandwidth W, we are implying that the spectral content above W is negligible for practical purposes.

One reason for pursuing this point now is that such spectral content above the folding frequency will result in some aliasing error when the signal is sampled. As noted earlier, a presampling alisasing filter can reduce these components to insignificant levels. However, the concept of restoring the analog signal to its original form without any change is dependent on the assumption of a truly band-limited signal, and since this assumption is "slightly shaky," one might conclude that there would be some distortion introduced to the signal, however slight it may be.

Returning to the "real-world" point of view, it can be said that the assumption of band-limited signals in virtually all applications is reasonable, and if the parameters of the system are properly chosen, an analog signal can be restored to a sufficiently high accuracy for most applications. The success of the many sampled-data systems in operation is certainly a testimony to that fact.

To summarize, a baseband signal bandlimited to W hertz can in theory be reconstructed from samples of the signal taken at a uniform rate equal to or greater than $2W$ samples/second. In practical terms, the rate is almost always selected to be somewhat greater than the theoretical minimum. Reconstruction of the signal can be achieved by passing the signal through a low-pass filter having good passband characteristics over the range from dc to W hertz, with high stopband attenuation present for frequencies in the range $f_s - W$ and higher. This process will be discussed in more detail later.

Example 6-1

A certain telemetry signal with a duration of 20 seconds is known to have spectral content from near dc to about 1 kHz but has negligible spectral content above that frequency. The signal is to be recorded, converted to digital format, and stored in memory for subsequent processing. To ease in reconstruction of the signal, the sampling rate is to be selected 25% greater than the theoretical minimum. What is the minimum number of samples of the signal that must be taken?

Solution

The theoretical minimum sampling rate would be $f_s = 2 \times 1$ kHz $= 2000$ samples/s. However, the actual sampling rate is chosen to be 25% greater, so $f_s = 1.25 \times 2000 = 2500$ samples/s. Since the signal has a duration of 20 s, the number of samples is $N = 2500$ samples/s $\times 20$ s $= 50,000$ samples. (Each sample would likely be represented by one word of digital memory.)

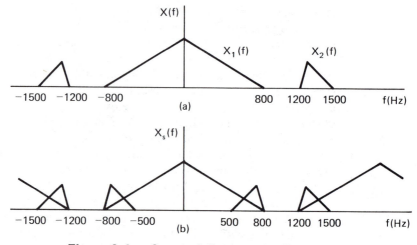

Figure 6-4. Spectral diagrams for Example 6-2.

Example 6-2

A certain signal designated as $x(t)$ is composed of two components $x_1(t)$ and $x_2(t)$ with $x(t) = x_1(t) + x_2(t)$. The portion of interest is $x_1(t)$, and its amplitude spectrum $X_1(f)$ is band-limited from dc to 800 Hz, as shown in Fig. 6-4a. The amplitude spectrum $X_2(f)$ of the second component occupies a spectrum from 1200 to 1500 Hz, as shown. It is necessary to sample the signal so that it can be encoded in a pulse-modulation form. An inexperienced (and naive) young engineer believes that since the component of interest extends only to 800 Hz a sampling rate greater than 1600 samples/s should suffice for reconstruction. To provide some guard band, he or she actually selects a sampling rate of 2000 samples/s. (a) Show by constructing a spectral diagram for the sampled signal that there is a fallacy in his (or her) reasoning and that it will not be possible to reconstruct the signal properly. (b) What steps can be taken to rectify the problem?

Solution

(a) Let $x_s(t)$ represent the sampled version of the analog signal $x(t)$. The amplitude spectrum $X_s(f)$ is obtained by sketching the form of the original spectrum plus successive shifts on the original spectrum as indicated by (6-5). In many cases, enough information about the form of the composite spectrum can be deduced from one or two shifted components. While the P_n factors serve as multiplicative constants for the various terms, the primary information needed here is frequency data, so it is not necessary to be too precise about the levels involved in this problem.

The forms of the baseband and sampled spectra over a reasonable frequency range are shown in Fig. 6-4b. Observe that the lower portion of the first translated component, which corresponds to the spectrum of $x_2(t)$, overlaps the upper 300-Hz range of the spectrum of $x_1(t)$. There is no way that components of $x_1(t)$ in this frequency range could be separated from the undesirable components of $x_2(t)$. Thus, aliasing of the spectrum has occurred.

This problem has illustrated that a sampling rate greater than twice the highest frequency of a desired signal is not sufficient if other components appear in the signal above the folding frequency.

(b) There are two approaches to solving the problem. The first is to increase the sampling rate to a value greater than twice the highest frequency (i.e., greater than $2 \times 1500 = 3000$ Hz). However, this solution would probably not be the best approach unless there is some reason to preserve $x_2(t)$, which seemed to be unimportant in the statement of the problem. A better solution probably would be to pass $x(t)$ through a presampling low-pass filter that has a fairly sharp cutoff frequency just above 800 Hz and in which the attenuation is quite high in the range from 1200 to 1500 Hz. The filter effectively eliminates $x_2(t)$, and then 2000 samples/s should be adequate.

6-2   IDEAL IMPULSE SAMPLING

The form of the sampled-data signal in the last section was derived on the assumption that each of the samples had a nonzero width τ. We will now consider the limiting case that results when the width τ is assumed to approach zero. In this case, the samples can be conveniently represented as a sequence of impulse functions, and the resulting signal is referred to as an *impulse sampled signal*.

While the analog samples of any real-life signal could never reach the extreme limit of zero width, the limiting concept serves two important functions: (1) If the widths of the actual pulse samples are very small compared with the various time constants of the system under consideration, the impulse function assumption is a good approximation and it leads to simplified analysis. (2) When a signal is sampled, converted from analog to digital form, and processed with digital circuits, it may be considered simply as a sequence of numbers occurring at specific points in time. A very convenient way of modeling such a signal for certain analytical purposes is through the impulse-sampling representation.

The form of the ideal impulse sampled signal is illustrated in Fig. 6-5. An arbitrary analog signal shown in Fig. 6-5a is sampled (or modulated) by the periodic impulse train shown in Fig. 6-5b. The resulting impulse sampled signal is shown in Fig. 6-5c.

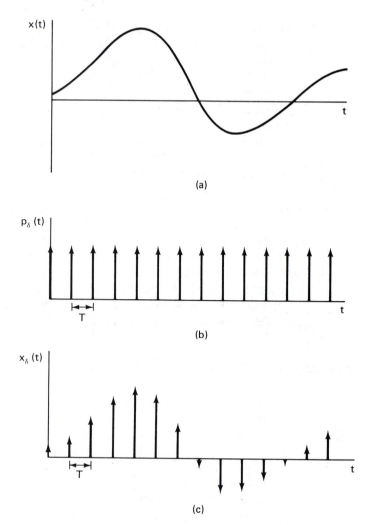

(a)

(b)

(c)

Figure 6-5. Development of sampled-data using ideal impulse sampling.

As a prelude to developing the form of the spectrum, consider the periodic impulse train shown in Fig. 6-5b. This function can be expressed as

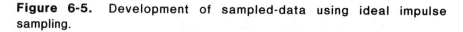

$$p_\delta(t) = \sum_{-\infty}^{\infty} \delta(t - nT) \qquad (6\text{-}8)$$

where $\delta(t\text{-}nT)$ is the mathematical symbol for an impulse occurring at $t = nT$.

Since the impulse train is periodic in time, it can be expanded in a complex exponential Fourier series of the form

$$p_\delta(t) = \sum_{-\infty}^{\infty} \overline{P}_{\delta n} \epsilon^{jn\omega_s t} \tag{6-9}$$

where $\overline{P}_{\delta n}$ represents the coefficients of the Fourier series expansion. Recall from Chapter 2 that the exponential series coefficients can be expressed in general as

$$\overline{P}_{\delta n} = \frac{1}{T} \int_{-(T/2)}^{T/2} p_\delta(t) \epsilon^{-jn\omega_s t} \, dt \tag{6-10}$$

In the interval of one cycle centered at $t = 0$, $p_\delta(t)$ contains only the single inpulse $\delta(t)$. Thus,* from the basic definition of the impulse function, $\overline{P}_{\delta n}$ is obtained as

$$\overline{P}_{\delta n} = \frac{1}{T} \int_{-(T/2)}^{T/2} \delta(t) \epsilon^{-jn\omega_s t} \, dt = \frac{1}{T} = f_s \tag{6-11}$$

From this result, it is deduced that the Fourier coefficients of the impulse train all have equal weight and there is no convergence. This is compatible with the nonperiodic single impulse in which the Fourier transform was developed in Chaper 2 and found to be simply a constant independent of frequency. For the periodic impulse train, the spectrum is also a constant but is now defined only at discrete multiples of the fundamental frequency.

The Fourier series for the periodic impulse train can then be expressed as

$$p_\delta(t) = \sum_{-\infty}^{\infty} \frac{1}{T} \epsilon^{jn\omega_s t} \tag{6-12}$$

The impulse sampled signal, which will be designated as $x_\delta(t)$, can be expressed in the form

$$x_\delta(t) = x(t) p_\delta(t) \tag{6-13}$$

Substitution of (6-12) in (6-13) results in

$$x_\delta(t) = x(t) \sum_{-\infty}^{\infty} \frac{1}{T} \epsilon^{jn\omega_s t} \tag{6-14}$$

or

$$x_\delta(t) = \frac{1}{T} \sum_{-\infty}^{\infty} x(t) \epsilon^{jn\omega_s t} \tag{6-15}$$

Fourier transformation of both sides of (6-15) yields

$$\overline{X}_\delta(f) = \frac{1}{T} \sum_{-\infty}^{\infty} \overline{X}(f - nf_s) \tag{6-16}$$

*Refer back to Eq. (2-73b) in Ex. 2-8.

The frequency-domain equivalents of the preceding operations are illustrated in Fig. 6-6. An assumed arbitary baseband amplitude spectrum for $x(t)$ is shown in Fig. 6-6a and the discrete spectrum of the periodic pulse train is shown in Fig. 6-6b. The form of the amplitude spectrum of the ideal impulse sampled signal is shown in Fig. 6-6c. The general form is similar to the form derived from the nonzero width natural sampling process shown in Fig. 6-2, and the basic sampling requirements developed in the last section apply here. However, a comparison of the figures and Eqs. (6-3), (6-5), and (6-16) reveals one difference between the spectral forms. The spectral components derived with nonzero pulse widths gradually diminish with frequency, and the magnitudes follow a $\sin x/x$ function envelope. However, components derived from impulse sampling are all of equal magnitude and do not diminish with frequency.

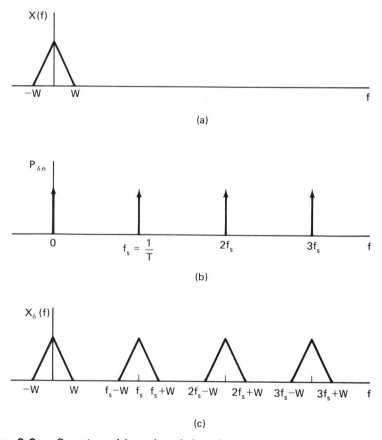

(a)

(b)

(c)

Figure 6-6. Spectra of baseband, impulse train, and impulse-sampled signal. (The three parts of this figure correspond to the three parts of Figure 6-5.)

An important deduction from this discussion is that the spectrum of an impulse sampled signal is a periodic function of frequency. The "period" in the frequency domain is equal to the sampling frequency f_s. Thus, ideal impulse sampling in the time domain leads to a periodic spectrum in the frequency domain. In simplest terms, when a signal is sampled in the time domain by a hypothetical uniform impulse train, the spectrum of the sampled signal is the original spectrum plus an infinite number of shifted versions of the original spectrum all having equal magnitude.

6-3 PULSE-AMPLITUDE MODULATION

So far in this chapter, we have concentrated primarily on establishing the basic properties of sampled-data signals in the time and frequency domains and the corresponding sampling requirements. We will now begin to consider some of the specific pulse-modulation methods as they apply to communications signal processing.

The first method for consideration is that of *pulse-amplitude modulation* (PAM). A PAM signal is a sampled-data signal consisting of a sequence of pulses in which the amplitude of each pulse is proportional to the analog signal at the corresponding sampling point. We immediately recognize that the form of the signal used in the basic development of the sampled-data concept in Sec. 6-1 fits this definition. Indeed, many PM signals are generated in the manner described in that section, and so the waveform that was shown in Fig. 6-1c is a good example of a natural sampled PAM signal. The PAM terminology was not used in Sec. 6-1 since the intent was to establish basic properties of sampled signals without getting involved in specific modulation terminology at that point.

Along with the natural sampled PAM signal, however, it is also necessary to consider a slightly different form, which frequently arises in both PAM systems and in conjunction with other pulse-modulation techniques. This function will be designated as a *flat-top PAM signal,* and an illustration of this form is shown in Fig. 6-7. The flat-top sampled PAM signal consists of purely rectangular pulses whose amplitudes represent the analog signal at *one* particular point in the sampling interval (usually the beginning). Thus, the pulses do not follow the analog signal during the intervals that the samples are being taken, but instead represent a single level of the signal during the interval.

It can be shown that the spectrum of the flat-top sampled signal contains some distortion with respect to the ideal natural sampled signal. However, if the widths of the pulses are very small compared to the sampling interval, the resulting distortion is usually negligible. Even when the pulse widths are substantial, it is possible to compensate for this distortion by an equalizing filter.

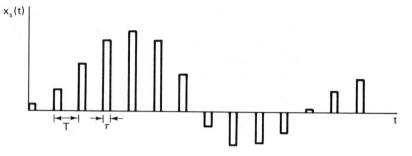

Figure 6-7. Form of flat-top sampled PAM signal.

A PAM signal may be used as a specific type of desired signal for pulse encoding and transmission. On the other hand, many systems utilizing other forms of pulse or digital modulation start with a PAM signal, and the signal is then converted to the particular desired form. Thus, a PAM signal may be the desired end goal in a given application, or it may simply be the means to an end.

In considering the properties of PAM signals, as well as other pulse-modulated methods to be considered, there is one important difference in the spectral form compared with the analog methods considered earlier in the book. All analog methods considered involved translating the modulating spectrum to a higher frequency, usually in the RF range, and eliminating all low-frequency components. However, the signal spectral forms shown in previous figures in this chapter indicate that the final PAM spectrum starts in the baseband range and may extend all the way down to dc if the message signal has a dc component. Thus, PAM is not suitable in its basic form for RF transmission.

The purpose of "modulation" as applied to PAM and the other pulse methods we will discuss is different than that of the various analog modulation methods previously considered. With the analog methods, the primary objective was to translate the spectrum to a frequency range in which electromagnetic energy transmission could be achieved. In pulse modulation, the primary objective of the modulation process is to replace the continuous signal by a sequence of discrete samples primarily for sharing the transmission system with other signals.

Some pulse-modulated waveforms are transmitted directly at baseband over wire links. However, when RF transmission is required, the PAM signal could be applied as a complex baseband modulating signal to a high-frequency transmitter, and the resulting spectrum would be shifted upward to the RF range. The resulting signal would have undergone two levels of modulation, one of which converted the continuous signal to a sequence of discrete samples, and the other of which shifted the spectrum of the sampled signal to the RF range to enhance transmission. More will be said about this after multiplexing is introduced in the next section.

A fundamental question with PAM concerns the bandwidth required to transmit the encoded signal. Referring back to Fig. 6-2, it is observed again that not only does the original spectrum appear from dc to W, but shifted versions of the spectrum appear about all frequencies that are integer multiples of the sampling rate. The peak values of these shifted components slowly reduce in magnitude as a function of the frequency since these peaks are proportional to the $\sin x/x$ function representing the P_n amplitude coefficients. We would intuitively expect that a reasonable number would have to be transmitted in order to maintain the integrity of the PAM signal.

A different and more convenient way of looking at the PAM signal is through the concept of pulse transmission, as discussed in Chapter 3. To preserve the quality of a PAM signal, the pulse amplitudes must be preserved, and they must not be allowed to smear excessively. On the other hand, perfect reproduction of the beginning and ending of the pulses has been found to require more bandwidth than desired for most of the types of applications for which PAM is used. Specifically, a "coarse" type of reproduction criteria has been found to be sufficient so that the baseband bandwidth B_T can be expressed as

$$B_T = \frac{K_1}{\tau} \qquad (6-17)$$

where τ is the width of each pulse sample.

The constant K_1 depends on the spacing between adjacent pulses, the level of acceptable adjacent pulse crosstalk, the sharpness of the cutoff rate, and other factors. In certain theoretical developments, the value $K_1 = 0.5$ is used. In this case, the result of (6-17) is in perfect agreement with the "coarse" pulse transmission criterion of Chapter 2. Certain idealized filter characteristics are theoretically capable of processing PAM signals with this minimum bandwidth, but most practical systems employ a greater bandwidth.

For a particular K_1 in (6-17), the transmission bandwidth becomes a specific frequency relative to the $\sin x/x$ envelope in Fig. 6-2. For example, if $K_1 = 1$, the bandwidth used includes all translated components up to the first zero crossing of the envelope at a frequency $1/\tau$, which is not shown in Fig. 6-2. For $\tau \ll T$, this may involve transmitting through the bandwidth a large number of the shifted spectral terms. For $K_1 = 0.5$, the bandwidth would include all shifted terms up through a frequency equal to half the first zero-crossing frequency.

The generation of a PAM signal may be achieved by gating on the analog signal periodically for a duration τ and turning it off for the remainder of each sampling period. Among other things, a balanced modulator may be used if the carrier waveform is a narrow pulse train. (See Prob. 6-6.) Many

PAM signals are formed as a portion of a time-division multiplexing system, and further discussion of PAM signal generation will be made in Sec. 6-4.

The reconstruction of a continuous signal from a PAM signal represents a rather interesting concept. Assume for this purpose a single PAM signal, and refer back to the spectral diagrams of Fig. 6-2. It is readily observed that the portion of the sampled spectrum from dc to W is exactly the same as the original spectrum. Therefore, if all the components above W are eliminated, the spectrum of the original unsampled signal will remain, and the signal will be restored. Thus, *the original analog signal is restored by passing a PAM signal through a suitable low-pass filter.* The filter should have a flat passband from dc to W, and it should display a sharp cutoff between W and $f_s - W$. We now see more clearly the necessity for maintaining some guard band between W and $f_s - W$. Without it, no real filter can eliminate the undesired translated components while passing the desired components without distortion.

The reader may be puzzled by the mystery of this process. In order to "fill-in" the time function between samples, we "dumped" portions of the frequency function. This seemingly contradictory process is a result of some of the subtleties of the sampling process and the associated spectral forms. When the original analog signal was defined at all values of time, the spectrum existed only over a narrow frequency range. When the signal was sampled at discrete values of time, a big uncertainty suddenly appeared over the time signal. The corresponding frequency function is required to be much broader in order to allow the time signal to exist only over a narrower range of time. However, when a return to the continuous time signal is desired, the higher-frequency components are no longer needed and can be eliminated. This type of process appears in a number of different areas of mathematics, physics, and engineering.

The actual level of the recovered signal may be quite small compared to the original signal due to the loss of energy resulting from filtering. Frequently, *holding circuits* are used in conjunction with the filter as a means of maintaining a reasonable level of energy in the filtered signal, as well as for easing a portion of the filtering requirements.

While there are different orders for holding circuits, we will focus on the *zero-order* form. The operation of a zero-order holding circuit is illustrated by the waveforms of Fig. 6-8. The waveform of Fig. 6-8a represents a single flat-top PAM signal for which restoration is desired. The output of a zero-order holding circuit is shown in Fig. 6-8b. Observe that a given pulse establishes an output proportional to the pulse level very quickly. This level is retained at a constant value until the next pulse arrives. The additional energy provided to the signal by this process is apparent by the increased area.

Observe that the resulting signal is a type of "staircase" approximation of the desired analog signal. As such, it contains less high-frequency content than the PAM signal, but it is certainly not a good representation of the

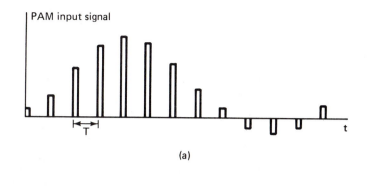

(a)

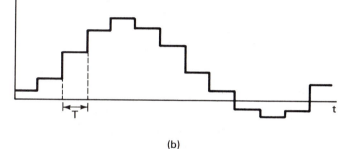

(b)

Figure 6-8. Operation of zero-order holding circuit.

original signal at this point. Additional low-pass filtering is required, and this filtering will serve to smooth the transitions between adjacent points.

Example 6-3

An analog signal $x(t)$ whose amplitude frequency spectrum is shown in Fig. 6-9a is sampled at a rate $f_s = 5$ kHz. The width of each sample is $\tau = 40$ μs. (a) Sketch the form of the sampled spectrum from dc to the first zero crossing of the pulse spectrum. (b) Compute the approximate transmission baseband transmission bandwidth for both $K_1 = 0.5$ and $K_1 = 1$ in Eq. (6-17).

Solution

(a) The basic concept to remember is that the spectrum of the sampled signal is obtained by reproducing the form of the spectrum of the unmodulated signal plus translated versions of the unmodulated spectrum at integer multiples of the sampling frequency. In this problem, we wish to explore the spectrum in somewhat more detail by observing the relative convergence of

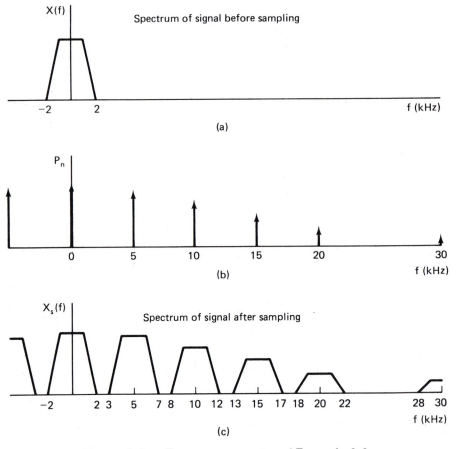

Figure 6-9. Frequency spectra of Example 6-3.

the spectral components. This should help in justifying the finite bandwidth rules.

First, we observe the spectrum of the pulse train, which may be considered as a multiplier for the continuous signal to yield the sampled signal. Since the sampling rate is $f_s = 5$ kHz, the time between successive pulses is $T = 1/f_s = 200$ μs. The duty cycle is thus $d = \tau/T = 40$ μs$/200$ μs $= 0.2$. Recalling the spectrum of a periodic pulse train from Chapter 2, the form for $d = 0.2$ is shown in Fig. 6-9b up to the vicinity of the first zero crossing, which occurs at a frequency $f = 1/\tau = 25$ kHz. Observe that the magnitudes of the terms reduce in accordance with the $\sin x/x$ function as the frequency increases.

The spectrum of the sampled signal is obtained by centering the spectrum of the unsampled signal about each of the pulse-train spectral frequencies and multiplying by the respective coefficient amplitudes. The amplitude scale of the pulse spectrum and the sampled signal spectrum are both amplified in order to enhance the illustration.

(b) For a choice $K_1 = 0.5$ in Eq. (6-17), the bandwidth required is

$$B_T = \frac{0.5}{\tau} = \frac{0.5}{40 \times 10^{-6}\,\text{s}} = 12.5 \text{ kHz} \qquad (6\text{-}18)$$

This is the minimum bandwidth used in system-level calculations and can be achieved only under idealized filter conditions. In many systems, a constant closer to $K_1 = 1$ is used. For this choice, the bandwidth would be

$$B_T = \frac{1}{\tau} = \frac{1}{40 \times 10^{-6}\,\text{s}} = 25 \text{ kHz} \qquad (6\text{-}19)$$

6-4   TIME-DIVISION MULTIPLEXING

We have seen that an analog signal can be represented by a series of discrete samples, each of which represents the analog signal at a particular sample point. The resulting signal can be reconstructed to within a tolerable deviation from the original signal by low-pass filtering if the sampling rate meets a certain specific minimum. The bandwidth of the resulting PAM signal has been shown to be much greater than that of the analog signal and is an inverse function of the width of the discrete samples used. The main question at this point is why should this be done? In other words, what can be gained by representing the signal as a series of short samples since the required transmission bandwidth has obviously been increased by this process?

The motivation for PAM (as well as other pulse-modulation forms) is a process called *time-division multiplexing* (TDM). To illustrate this process, consider a baseband transmission system in which a number of separate data signals must be transmitted over a single communications link. The link in this case could be a transmission line or cable between the source and the destination. The point is that there is only one transmission medium, and there are a number of different data sources.

The process is illustrated in Fig. 6-10. A *commutator* appears at the input and a *decommutator* appears at the output of the link. In virtually all

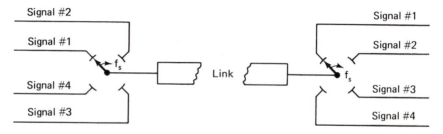

Figure 6-10. Concept of time-division multiplexing illustrated with mechanical commutator and decommutator.

systems, the commutation is performed electronically, but it is easier to illustrate the process mechanically at this point. At the input, the commutator sequentially samples each data signal in order. The decommutator at the receiving end is assumed to be synchronized with the one at the transmitter and routes the correct signal to its proper destination. Thus, as far as each individual output is concerned, the signal appears as a sampled PAM signal, and it can be reconstructed by proper filtering in the given data output. For the illustration shown, there is a short "dead space" between successive samples. The unused space is often desirable in order to allow some spreading of the pulses resulting from the finite bandwidth limitations of any realistic transmission system. Without some space, there may be a noticeable amount of *crosstalk* between corresponding channels. Each signal must be sampled at the minimum Nyquist rate for its particular bandwidth. For the moment, it is convenient to consider each channel as having an equal bandwidth, which is assumed to be W. Thus, the commutators must rotate at a rate no less than $2W$ revolutions per second; that is, each signal must be sampled at a rate no less than $2W$ times per second.

A typical layout of a TDM composite signal is shown in Fig. 6-11. While a variety of examples could be shown, this particular example has seven separate data signals all sampled at the same rate. A space equal to a sample pulse duration is maintained between any two successive pulses, and flat-top sampling is employed. A *frame* is one complete structured time interval in which each of the data pulses, plus any additional system information, is provided. Along with samples of the seven data signals, one synchronizing pulse is transmitted in each frame for this system. While a number of possible methods for synchronization can be used, in this system a dc bias level is added to all data signals so that they are always positive. The sync pulse is then transmitted as a negative pulse so that the receiver can readily recognize its presence. Each time the sync pulse appears, the receiver commutator is realigned with that of the transmitter by pulse selective circuits. Some of the hardware details of PAM systems will be considered later.

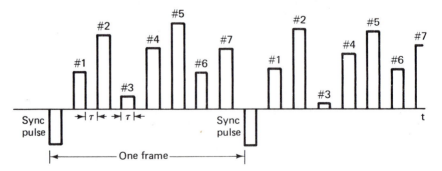

Figure 6-11. Representative PAM TDM signal format.

A lower bound for the minimum transmission bandwidth of a PAM multiplex system will now be developed. It will be assumed that there are k signals to be multiplexed, and each has a baseband bandwidth W. Several somewhat unrealistic limiting-case assumptions will now be made, but this is justified by the intent to develop a limiting lower bound for the bandwidth, rather than a final realistic estimate. These assumptions are (1) no spacing between successive pulses, (2) sampling rate equal to minimum Nyquist rate, and (3) no sync pulses. Finally, the lower-bound transmission bandwidth rule of $0.5/\tau$ will be used. The form of the assumed signal is shown in Fig. 6-12. The value T_f represents the *frame time*.

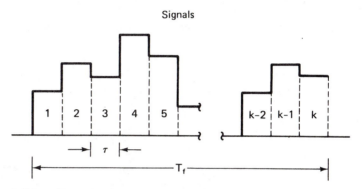

Figure 6-12. Form for hypothetical PAM signal used in developing lower bound for transmission bandwidth.

Each signal must be sampled no less than $2W$ times per second. Since T_f represents the time between samples, $T_f \leq 1/2W$. As previously stated, the minimum Nyquist rate will be selected, so

$$T_f = \frac{1}{2W} \tag{6-20}$$

A given frame contains k samples with each having a width τ, so $T_f = k\tau$. Thus,

$$\tau = \frac{T_f}{k} = \frac{1}{2kW} \tag{6-21}$$

The minimum baseband transmission bandwidth is then assumed to be

$$B_T = \frac{0.5}{\tau} = 0.5(2kW) = kW \tag{6-22}$$

The lower bound for the transmission bandwidth for this hypothetical case turns out to be simply the number of signals times the bandwidth per signal.

This result is interesting both from a philosophical point of view and a practical point of view. From a philosophical point of view, "we don't get something for nothing." As we increase the number of signals transmitted over a given channel, we have to increase the bandwidth accordingly, even though the sampling rate per channel may remain the same. The fact that the total available information increases as the sampling rate increases results in the need for a greater bandwidth to process the information. From a practical point of view, the result of (6-22) sets the lower bound for the transmission bandwidth, and the trade-off between the number of channels and the bandwidth per signal is readily deduced from that expression.

To summarize, it will be emphasized again that the assumptions made in developing these results are somewhat unrealistic and were made for the purpose of developing a limiting-case lower bound. In practice, spaces between pulses are often employed, the sampling rate is usually greater than the minimum Nyquist rate, a sync pulse is normally used, and the transmission bandwidth used may be selected to be greater than predicted by the $0.5/\tau$ rule. All these factors tend to increase the transmission bandwidth so that the actual bandwidth may be two or three times the theoretical minimum in some cases.

So far we have considered only cases in which all the data signals in the TDM system were assumed to have identical bandwidths so that it was logical to sample all signals at the same rate. However, suppose that the different channels have different bandwidths. One obvious solution is to use the signal with the highest frequency as the worst-case bound and to set a basic sampling rate for all the signals at a somewhat greater value than the Nyquist rate for the reference signal. This means, of course, that adequate or more than adequate sampling for all signals will automatically be met. However, this solution tends to be less attractive if there are wide differences in the upper frequency limits of the different signals. In this case, lower-frequency signals are being sampled at much higher rates than necessary, and some inefficiency in both hardware and bandwidth utilization results.

A more attractive solution in some cases is to devise a frame sequence which samples the higher-frequency signals more often than lower-frequency signals. While the timing and synchronizing of the signal may be more complex with such an arrangement, the overall benefits may be worth the extra complexity in many cases. Few general rules can be given for such a design as each one must be judged on its own merits. The critical requirement is that each signal must be sampled at a rate greater than its particular Nyquist rate. An example of a more complex sampling scheme for signals having widely different bandwidths will be considered in Sec. 7-5.

Generation of a PAM multiplexed signal is most easily implemented with a combination of analog switches and multiplexing circuits. A circuit diagram of a system to combine four channels is shown in Fig. 6-13. The switches are special P-channel JFET's, which are available as individual units or as a

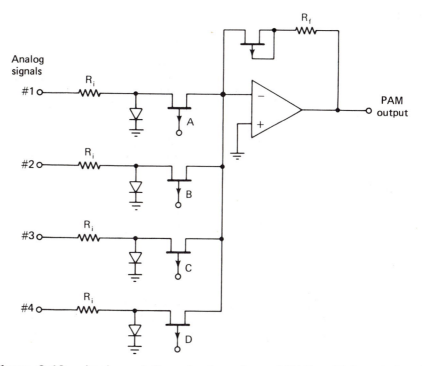

Figure 6-13.   Implementation of a four-channel PAM multiplexed signal using *P*-channel JFET switches.

combination of several on a single IC chip. In the latter form, an extra compensating JFET (shown at the top) is also included.

A given *P*-channel JFET acts as an open circuit when the gate voltage is more positive than a certain minimum voltage (related to the pinch-off voltage). Conversely, when the gate voltage is nearly zero, the JFET is turned on and acts like a small series resistance (typically less than 100 Ω). By connecting the gate terminals (designated as *A*, *B*, *C*, and *D*) to suitable counter-type circuits, the gates can be turned on in sequence in accordance with a master clock reference. The gate voltage levels of many analog switch circuits such as this are compatible with basic digital logic levels such as TTL, which minimizes the interfacing problems. For example, a TTL logic 1 might open the switch and a TTL logic 0 might close the switch.

The samples of the different signals are combined in the analog operational amplifier summing circuit. The compensating switch acts as a series resistance approximately equal to the series resistance of either of the series switches when they are on, and this compensates for the uncertainty in the gain level due to the switch resistance.

Example 6-4

Consider a PAM time-division multiplexing system with seven signals plus sync having a frame format of the form illustrated in Fig. 6-11. Assume that each signal has a baseband characteristic and is band-limited to $W = 1$ kHz. The sampling rate is selected to be 25% greater than the theoretical minimum Nyquist rate. (a) Using the $0.5/\tau$ rule, determine the approximate baseband bandwidth required for the composite PAM signal. (b) If the composite PAM signal is used to amplitude modulate a high-frequency RF carrier, determine the RF bandwidth required for the resulting high-frequency signal.

Solution

(a) The sampling rate f_s is to be set at 1.25 times the theoretical minimum so $f_s = 1.25 \times 2 \times 1000$ Hz $= 2.5$ kHz for each signal. The frame time T_f is

$$T_f = \frac{1}{2.5 \times 10^3} = 0.4 \text{ ms} \qquad (6\text{-}23)$$

The bandwidth is determined from the minimum pulse width. From Fig. 6-11, it is noted that there are 16 intervals of width τ in a given frame (7 data pulses, 1 sync pulse, and 8 open spaces). The pulse width τ is thus

$$\tau = \frac{T_f}{16} = \frac{0.4 \text{ ms}}{16} = 25 \text{ } \mu\text{s} \qquad (6\text{-}24)$$

The composite baseband bandwidth B_T is

$$B_T \approx \frac{0.5}{25 \times 10^{-6} \text{ s}} = 20 \text{ kHz} \qquad (6\text{-}25)$$

The reader is invited to show that the limiting-case lower-bound bandwidth for the seven signals based on the theoretical development in this section would be 7 kHz. The value of 20 kHz is much more realistic.

(b) If the composite PAM signal amplitude modulates a high-frequency carrier, the composite bandwidth is simply

$$B_T = 2 \times 20 \text{ kHz} = 40 \text{ kHz} \qquad (6\text{-}26)$$

6-5 PULSE-TIME MODULATION METHODS

With pulse-amplitude modulation, each pulse has a *fixed width,* but the pulse amplitude is proportional to the analog signal at each sample point. We will now consisder a class of pulse-modulation techniques in which each pulse has *fixed amplitude,* but in which some time characteristic of the pulse is used to represent the analog signal. These techniques are referred to collectively as *pulse-time modulation* methods. The two specific methods that will be considered are *pulse-width modulation* and *pulse-position modulation.* Each will be considered individually.

Pulse-width modulation (PWM) is a sampled-data process in which the *width* of each pulse is varied directly in accordance with the amplitude of the modulating signal at the particular sample point. Pulse-width modulation is also denoted by the title *pulse-duration modulation (PDM),* and both designations appear in the literature. This process is illustrated in Fig. 6-14. The assumed analog signal shown in Fig. 6-14a is converted by appropriate electronic circuits to the pulse train shown in Fib. 6-14b. Observe that the most positive peak corresponds to the widest pulse, and the most negative peak corresponds to the narrowest pulse.

Since there are obvious limits to both the maximum and minimum widths attainable, a PWM system must be carefully designed in accordance with the maximum and minimum signal levels expected. A certain nominal pulse width corresponds to zero signal level. Based on the maximum and minimum

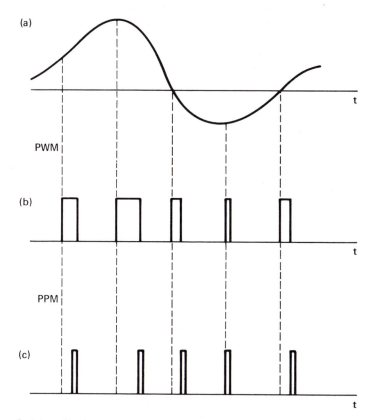

Figure 6-14. Forms for pulse width modulated (PWM) and pulse position modulation (PPM) signals.

signal levels expected, the sensitivity of the electronic circuitry is designed so that the maximum pulse width does not exceed the available time interval and so that the minimum pulse width remains greater than zero.

Pulse-position modulation (PPM) is a sampled-data process in which the *position* of each pulse is varied directly in accordance with the amplitude of the modulating signal at the particular sample point. This process is illustrated in Fig. 6-14c for the analog signal shown in Fig. 6-14a. In this case, the most positive peak corresponds to maximum shift from a reference point (the beginning in this case) in each pulse interval, while the most negative peak corresponds to no shift at all. As in the case of PWM, the system must be carefully designed to accommodate the range of the analog signal involved.

Both PWM and PPM techniques can be used in time-division multiplexing systems involving a number of separate signals. In each case a frame is divided into a number of separate slots according to the number of data signals involved, plus additional requirements such as sync pulses. Each slot is chosen sufficiently wide to allow the necessary pulse expansion as in PWM or the necessary pulse shifting as in PPM.

We will now explore the question of baseband bandwidth required for PWM and PPM transmission. The exact Fourier spectrum for either case is quite difficult to obtain mathematically, so that approach will not be pursued. However, some results for estimating the approximate bandwidth will be considered. Unlike the case of PAM, where only "rough" pulse reproduction was sufficient, both PWM and PPM require much more accurate pulse fidelity. This fact can be deduced intuitively by recognizing that any significant smearing or broadening of a pulse could obscure either the width or the location of the pulse. Thus, accurate or "fine" pulse reproduction is required.

The approximate bandwidth for nearly exact pulse reproduction was shown in Chapter 2 to be inversely proportional to the rise time of the pulse. Letting t_r represent the rise time, the proportionality constant for baseband transmission was conveniently rounded to 0.5 so that $B_T \approx 0.5/t_r$. This rule will be applied to the bandwidth computation for both PWM and PPM. Using this approach, the rise time allowed for the encoded pulses must be either known, estimated, or at least specified in conjunction with a PWM or PPM system design.

Since the rise time is normally quite short compared with the pulse width, the transmission bandwidth for either PWM or PPM is usually quite large compared with the corresponding bandwidth for PAM. As the rise time is reduced, the pulse width or the pulse position can be determined with a finer degree of accuracy (provided the noise level is below a certain minimum level). However, the trade-off that must be made to obtain the higher accuracy is increased transmission bandwidth.

We will now consider some approximate analogies between the pulse-modulation methods that have been encountered in this chaper and certain of the analog methods considered earlier in the text. In some ways, PAM has properties that compare with certain properties of the AM forms. (No particular AM method will be delineated since there is no one-to-one correspondence with a particular method.) In both cases, the information is encoded as amplitude functions, and the transmission bandwidths tend to be in the same general range for comparable baseband bandwidths.

The pulse-time modulation methods compare in some ways with angle-modulation methods. The analogies are not exact but by "stretching the point" one might say that PWM compares with FM and that PPM compares with PM. It turns out that the increased bandwidth required for PWM or PPM is comparable to the increased bandwidth required with FM or PM for the same modulating signal bandwidth.

While there are some classical discrete-component circuits that have been used for generating PWM and PPM signals, these circuits have been largely superseded by the use of integrated circuit timer modules. A number of IC timing circuits are available, and many of these permit pulse width or position modulated pulses to be obtained by suitable external connections. The interested reader is referred to the various integrated circuits applications manuals for further details.

Example 6-5

A certain PWM system is characterized by five channels each having a baseband bandwidth equal to 1 kHz. The sampling rate is to be selected to be twice the theoretical minimum rate. System specifications dictate that each pulse must be reproduced to a sufficient accuracy that the rise time cannot exceed 1% of the time allocated to the given channel sample. Determine the approximate minimum transmission bandwidth.

Solution

The minimum sampling rate would be 2×1 kHz = 2 kHz, but the actual sampling rate is specified to be twice the theoretical minimum, so $f_s = 2 \times 2$ kHz = 4 kHz. The frame time is thus

$$T_f = \frac{1}{4 \text{ kHz}} = 0.25 \text{ ms} \qquad (6\text{-}27)$$

The frame is divided into five slots for the five individual channels so the time T_0 devoted to each channel is equal to

$$T_0 = \frac{T_f}{5} = \frac{0.25 \text{ ms}}{5} = 50 \text{ } \mu\text{s} \qquad (6\text{-}28)$$

Each actual pulse will vary in width during a 50-μs interval in accordance with the magnitude of the modulating signal. The width may never exceed 50

μs, and there is a minimum width, which would be based on the rise-time specification. The actual variation of the pulse width would be adjusted according to the circuitry used to vary from the minimum width selected to a maximum width not exceeding 50 μs.

The maximum rise time is 1% of T_0 or

$$t_r = 0.01 \times 50 \ \mu s = 0.5 \ \mu s \qquad (6\text{-}29)$$

The approximate transmission bandwidth B_T is then

$$B_T \simeq \frac{0.5}{t_r} = \frac{0.5}{0.5 \times 10^{-6}} = 1 \text{ MHz} \qquad (6\text{-}30)$$

6-6 FREQUENCY-DIVISION MULTIPLEXING

In Sec. 6-5, the concept of time-division multiplexing was developed on the basis of time sharing a channel, with pulse samples of the various signals being multiplexed. We will now momentarily return to the "analog world" to discuss the other broad class of multiplexing systems.

Frequency-division multiplexing (FDM) is a concept in which a number of different data signals are transmitted together through the use of different subcarrier frequencies. In TDM all signals use the *same frequencies* but operate at *different times,* while in FDM, all signals operate at the *same time* with *different frequencies.*

A block diagram illustrating the form of an FDM transmitter system is shown in Fig. 6-15. Each data signal is applied to a separate modulator, and each modulator uses a separate *subcarrier* frequency. The various subcarrier frequencies are relatively low and would not normally be suited for RF transmission. The modulated subcarriers are all combined into a linear summing circuit, and this composite signal modulates the RF transmitter. The resulting RF signal can then be transmitted by electromagnetic means. The modulating signal for the RF transmitter in this case does not represent a single data source, but it represents a special combination of several different sources, each of which is a modulated signal.

The general form of the composite baseband spectrum at the output of the summing circuit is shown in Fig. 6-16. The line in the middle of each channel is used in this figure to illustrate possible carrier frequencies, which are designated as $f_{c1}, f_{c2}, \ldots, f_{ck}$. Depending on the type of modulation used in the subcarrier modulators, the subcarrier spectral lines may or may not appear in the actual spectrum. Each channel in this particular illustration is assumed to occupy the same bandwidth, but this is not a general requirement of FDM systems.

Observe that the subcarriers are chosen so that there is no spectral overlap between the different channels. Indeed, it is necessary that some *guard band* be provided between successive channels, as is shown in this illustration. The

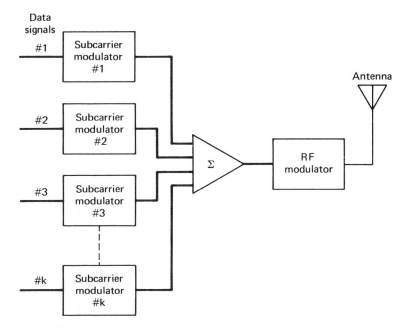

Figure 6-15. Block diagram of a frequency-division multiplexing transmitter system.

guard band is necessary in order that the separate channels can be separated at the receiver with practical filters, as will be discussed shortly. Note that portions of two separate channel spectra appear between two successvie subcarrier frequencies. Thus, each FDM system must be carefully designed with strict attention paid to minimum channel spacing and guard band between channels. Channels spaced too closely will suffer from crosstalk resulting from spectral spillover. Recall that TDM also suffered from the possibility of crosstalk, but in that case it was a result of signals spaced too closely in *time,* while for FDM, crosstalk results from signals spaced too closely in *frequency.*

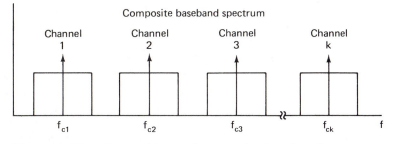

Figure 6-16. General form of composite baseband spectrum.

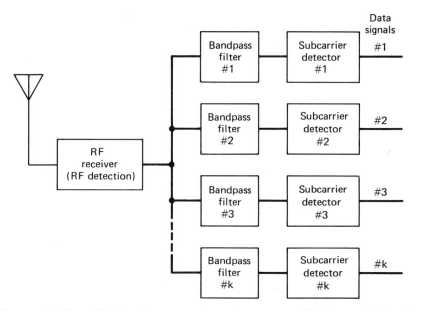

Figure 6-17. Block diagram of a frequency-division multiplexing receiver system.

A block diagram illustrating the form of an FDM receiver system is shown in Fig. 6-17. The front portion of the system is an RF receiver having sufficient bandwidth to accommodate the usual relatively wideband multiplexed FDM. The RF signal is detected in accordance with standard techniques, and the output of the receiver block on the figure represents the composite baseband signal, whose spectrum has the form of Fig. 6-16. The composite baseband signal is applied to a bank of band-pass filters. There is one filter for each subcarrier frequency, and the center of the passband for a given filter corresponds to the center of the corresponding spectrum. The bandwidth of each filter must be carefully selected to pass the spectrum for the given channel, but it must provide a significant level of attenuation at the band edges of the two corresponding adjacent channels.

A complex design trade-off results from the spacing between channels versus the filter complexity required at the receiver. As the space between channels is decreased, the RF transmission bandwidth is reduced. However, the required filter complexity at the receiver increases so some reasonable compromise is required.

So far no mention has been made of the form of analog modulation used. Indeed, any of the various analog methods, such as SSB, DSB, AM, and FM, could be used for either the subcarrier modulation or the RF modulation. Thus, there are many combinations that could be used, and a number of different systems have been implemented over the years. In practice, however, there are a few types that have been used more than others owing to their superior performance.

Before discussing any specific systems, some standard terminology will now be introduced. The term X/Y is widely employed, with X referring to the form of the subcarrier modulation and Y referring to the RF carrier modulation. Thus, a system designated as AM/FM would employ AM for the subcarrier modulators and FM for the RF modulator.

One system that has been used in some applications is SSB/SSB. This system is of interest at the outset because it requires the minimum transmission bandwidth of any FDM system for a given set of baseband bandwidths. A lower bound for the minimum transmission bandwidth of an FDM SSB/SSB system will now be developed. It will be assumed that there are k signals to be multiplexed, and each has baseband bandwidth W. The unrealistic assumption of no guard band between channels will be made for this particular development, and the spectrum of the baseband composite signal will thus have the form shown in Fig. 6-18. Observe in this hypothetical layout that channel 1 could represent an actual baseband signal without translation, while the other channels would require frequency translation. Each translated channel could represent the lower sideband if the carrier had been located at the upper end of the given channel or the upper sideband if the carrier had been located at the lower end of the channel.

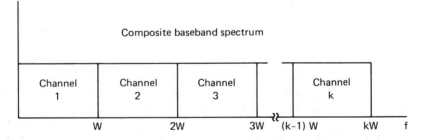

Figure 6-18.   Form for hypothetical baseband composite spectrum of SSB/SSB system used in developing lower bound for transmission bandwidth.

The composite baseband spectrum extends from dc to kW as shown in the figure. Since the RF bandwidth of an SSB signal is the same as the bandwidth of the modulating signal, we readily deduce that

$$B_T = kW \qquad\qquad (6\text{-}31)$$

Referring back to Eq. (6-22) momentarily, a rather important conclusion can be inferred. Under very idealized conditions, the minimum transmission bandwidth of an FDM system is the same as that of a TDM system. In both cases, the transmission bandwidth is the bandwidth per signal times the number of signals.

An interesting observation is the different way in which the time and frequency variables are shared in TDM and FDM. In TDM, the signals occupy different time spaces, but they share the same frequency spaces. In FDM, the signals share the same time spaces (all signals are on all the time), but they occupy different frequency spaces. Yet the total transmission bandwidths for the two separate cases are the same.

It will be emphasized again that the assumptions made in developing this FDM lower bound, like the assumptions made in developing the TDM lower bound, are unrealistic. In practice, guard bands would be required between all channels, so the actual RF bandwidth would be much greater than the hypothetical lower bound.

Probably the most widely employed FDM method is FM/FM. This technique is used extensively in the aerospace telemetry field. It offers the significant advantage of FM in being relatively immune to noise under certain conditions. (The capability of FM to enhance the signal-to-noise ratio will be developed and discussed in Chapters 10 and 11). The use of FM also allows the transmission of very slowly varying (i.e., "dc") data, which is very difficult to process with most of the AM methods.

Because of the widespread use of FM/FM systems, some standardization has been achieved in the industry. Standards have been developed by the Inter-Range Instrumentation Group (IRIG), an organization with representation from the various military branches. One can purchase VCOs and discriminators operating at standard IRIG frequencies from the various telemetry firms.

One set of IRIG standards that will be used for discussion will be that of the Proportional-Bandwidth Subcarrier Channels, for which some of the pertinent data extracted from the standards is given in Table 6-1. The table actually shows two possible options, ±7.5% deviation and ±15% deviation. (The deviation is the maximum percentage deviation measured from the subcarrier center frequency.) The "proportional" designation refers to the fact that the bandwidth allocated to channels increases with the center frequency of the subcarrier channels. This is a natural sort of allocation and tends to result in more equal implementation complexity for the different channel components. There are also Constant-Bandwidth Subcarrier Channel standards available from IRIG in which the bandwidth allocated per channel is the same.

Example 6-6

Consider an FM/FM FDM system consisting of the IRIG proportional bandwidth channels 1 to 4 of Table 6-1 operated with a deviation ratio $D = 5$ for each channel. (a) Develop a spectral layout for the composite baseband signal indicating the approximate spectral limits of each channel. (b)

Table 6-1. Inter-Range Instrumentation Group (IRIG) Proportional-Bandwidth Telemetry Channels ($D = 5$)

IRIG Channel Number	Channel Center Freq. (Hz)	Deviation (%)	Frequency Deviation Limits (Hz)		Data Cutoff Freq. (Hz)
1	400	±7.5	370	430	6
2	560	±7.5	518	602	8
3	730	±7.5	675	785	11
4	960	±7.5	888	1,032	14
5	1,300	±7.5	1,202	1,398	20
6	1,700	±7.5	1,572	1,828	25
7	2,300	±7.5	2,127	2,473	35
8	3,000	±7.5	2,775	3,225	45
9	3,900	±7.5	3,607	4,193	59
10	5,400	±7.5	4,995	5,805	81
11	7,350	±7.5	6,799	7,901	110
12	10,500	±7.5	9,712	11,288	160
13	14,500	±7.5	13,412	15,588	220
14	22,000	±7.5	20,350	23,650	330
15	30,000	±7.5	27,750	32,250	450
16	40,000	±7.5	37,000	43,000	600
17	52,500	±7.5	48,562	56,438	790
18	70,000	±7.5	64,750	75,250	1050
19	93,000	±7.5	86,025	99,975	1395
20	124,000	±7.5	114,700	133,300	1860
21	165,000	±7.5	152,625	177,375	2475
A	22,000	±15	18,700	25,300	660
B	30,000	±15	25,500	34,500	900
C	40,000	±15	34,000	46,000	1200
D	52,500	±15	44,625	60,375	1575
E	70,000	±15	59,500	80,500	2100
F	93,000	±15	79,050	106,950	2790
G	124,000	±15	105,400	142,600	3720
H	165,000	±15	140,250	189,750	4950

Compute the RF transmission bandwidth if the deviation ratio of the RF transmitter is also 5.

Solution

(a) To simplify the organization of the data developed, Table 6-2 has been prepared. The process involved will be illustrated with channel 1. The maximum deviation of ±7.5% corresponds to an actual deviation of ±30 Hz;

Table 6-2. Data of Example 6-6.

Channel	Modulating Frequency	Frequency Deviation	Bandwidth	Hertz		
				Lower Frequency	Center Frequency	Upper Frequency
1	6	±30	72	364	400	436
2	8	±42	100	510	560	610
3	11	±55	132	664	730	796
4	14	±72	172	874	960	1046

that is, the instantaneous frequency varies from 370 up to 430 Hz as indicated in Table 6-1. For $D = 5$, the baseband bandwidth (highest modulating frequency) for this channel is $W = \Delta f/D = 30 \text{ Hz}/5 = 6$ Hz. Using Carson's rule, the bandwidth B_1 for channel 1 is

$$B_1 = 2(\Delta f + W) = 2(30 + 6) = 72 \text{ Hz} \qquad (6\text{-}32)$$

It is assumed that the spectrum is symmetrical about the center frequency of 400 Hz. The lower limit of the spectrum is thus $400 - 72/2 = 364$ Hz, and the upper limit is $400 + 72/2 = 436$ Hz. These results are summarized in the first line of Table 6-2. The reader is invited to verify the values for the other channels.

The form of the composite spectrum is shown in Fig. 6-19. Observe that the horizontal scale begins just below the lower limit of channel 1, so the open area from dc to that frequency is not shown. All the channel spectra are shown here in continuous "block" form for ease of illustration, with the carrier assumed in the middle. The actual spectra for realistic modulating signals will usually show a significant rolloff rate near the band edges.

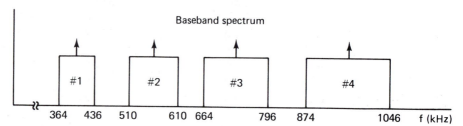

Figure 6-19. Composite baseband spectrum for FM/FM system of Example 6-6.

(b) The composite modulating signal consisting of the four channels has the highest frequency at 1046 Hz. Since $D = 5$, the RF frequency deviation is $\Delta f = 1046 \times 5 = 5230$ Hz. The RF transmission bandwidth B_T is then determined as

$$B_T = 2(5230 + 1046) = 12{,}552 \text{ Hz} = 12.552 \text{ kHz} \qquad (6\text{-}33)$$

PROBLEMS

6-1 A certain analog signal has a duration of 1 min. The spectral content ranges from near dc to 500 Hz. The signal is to be sampled, converted to digital format, and stored in memory for subsequent processing.

    (a)   Determine the theoretical minimum number of samples that must be taken if eventual reconstruction of the analog signal is desired.

    (b)   Repeat part (a) if the sampling rate is selected to be twice the theoretical minimum.

6-2 A certain analog signal is bandlimited from near dc to 2 kHz. The signal is to be sampled at a rate such that the lowest possible frequency of the first translated component is 50% greater than the highest possible frequency of the baseband spectrum to ease filtering in signal reconstruction. Sketch the form of the spectrum of the sampled signal and determine minimum sampling rate.

6-3 A certain sampled-data system operates at a sampling rate of 10,000 samples/s.

    (a)   Determine the theoretical highest frequency in a baseband signal such that the signal could be reconstructed from samples taken by the system.

    (b)   Repeat part (a) if a guardband of 4 kHz is to be established between the upper range of the baseband signal and the lower range of the first translated component.

6-4 A certain analog system is assumed to be band-limited from dc to 1 kHz. A rectangular form for the spectrum may be assumed for convenience. The signal is sampled at a rate $f_s = 4$ kHz. Sketch the general form of the spectrum of the sampled signal if the pulse width of the sampling pulses is 50 μs.

6-5 For the system of Prob. 6-4, determine the approximate transmission bandwidth using Eq. (6-17) for both $K_1 = 0.5$ and $K_1 = 1$.

6-6 Using the mathematical representation of a PAM signal given by Eq. (6-1) and illustrated in Fig. 6-1, propose a technique for generating a PAM signal using a balanced modulator (multiplier). Sketch the required form of the carrier waveform.

6-7 For the PAM TDM system of Ex. 6-4, repeat all calculations if the sampling rate is increased to twice the theoretical minimum Nyquist rate.

6-8 A certain PAM TDM system has six signals plus synchronization with a frame format as shown in Fig. P6-8. The sync pulse occupies an interval 2τ, while each signal sample has a width τ. The receiver detects the wider sync pulse once per frame and realigns the timing sequence accordingly. Note that an interval τ is also inserted between adjacent samples.

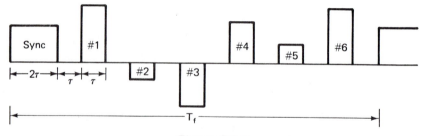

Figure P6-8

(a) Using the $0.5/\tau$ rule, determine the approximate baseband bandwidth required for the composite PAM signal if each channel has a 1-kHz bandwidth.

(b) If the composite PAM signal is used to amplitude modulate a high-frequency RF carrier, determine the RF bandwidth required for the resulting high-frequency signal.

(c) Repeat part (b) if FM is used for the RF modulation. Assume a deviation ratio of 5.

6-9 For the PWM TDM system of Ex. 6-5, determine the approximate transmission bandwidth if the sampling rate is reduced to the point where it is 25% above the theoretical minimum.

6-10 For the PWM TDM system of Ex. 6-5, determine the approximate transmission bandwidth if the rise time specification is changed so that it cannot exceed 0.2% of the time allocated to the given channel. All other specifications are the same as in Ex. 6-5.

6-11 An SSB/SSB FDM system is to be implemented using the following design strategy. Channel 1 will be retained directly at baseband. A guard band equal to 25% of the bandwidth of channel 1 will be maintained between the upper edge of channel 1 and the lower edge of channel 2. A guard band equal to 25% of the bandwidth of channel 2 will be maintained between the upper edge of channel 2 and the lower edge of channel 3. And so on. Draw a spectral diagram for the composite baseband spectrum, label various frequencies, and compute the total transmission bandwidth if the system contains four data signals, each having a baseband bandwidth of 4 kHz.

6-12 Repeat Prob. 6-11 if the system contains four data signals with assignments and baseband bandwidths as follows:

Channel 1: $W = 4$ kHz
        2: $W = 6$ kHz
        3: $W = 12$ kHz
        4: $W = 15$ kHz

Digital
7 Communications

7-0 INTRODUCTION

The major emphasis in this chaper will be directed toward the development, operation, and general comparison of different types of digital communications systems. Digital communications systems have increased significantly in utilization and complexity as digital hardware and software have become more readily available. Evidence indicates that the utilization of digital communications systems and techniques will increase rapidly in the years ahead.

The basic concept of pulse-code modulation, which is the foundation for digital encoding processes, will be explored in detail in this chapter. Various data formats and conversion techniques will be discussed. Modulation methods used for encoding and transmitting digital data by RF processes will be surveyed. Along with the widespread binary encoding methods, some consideration of quadriphase systems, in which four levels are used, will be explored. An introduction to the rapidly evolving area of data communications will be given.

7-1 PULSE-CODE MODULATION

The general category of digital communications systems includes a number of different types of encoding, transmission, and decoding techniques, many of which will be discussed in this chapter. However, most are derived either directly or indirectly from the basic concept of *pulse-code modulation* (PCM). The concept of PCM is illustrated by the block diagram shown in Fig. 7-1. The elements in the block are identified in a form to illustrate to the reader the processes involved, rather than to show an actual implementation. In practice, several of these operations may occur in the same circuit.

Assume an analog signal $x(t)$, band-limited from near dc to W hertz, which is to be converted to PCM form. The signal is first filtered by an aliasing analog filter whose function is to remove any superfluous frequency

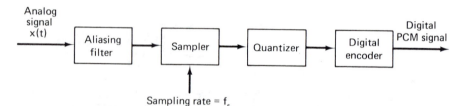

Figure 7-1. Block diagram illustrating the steps involved in generating a digital PCM signal. (In actual practice, several of the steps may be accomplished in the same circuit).

components above W that might appear at the input and be shifted into the data band by the aliasing effect. The signal is then sampled at a rate $f_s > 2W$. At this point in the block diagram, the signal has essentially the same characteristics as a sampled-data PAM signal discussed in the preceding chapter.

A process unique to all digital-modulation methods is now performed. Each sampled-data pulse is converted into (or replaced by) one of a finite number of possible values. This process is called *quantization,* and the quantizer block in the diagram of Fig. 7-1 is assumed to accomplish this purpose. Following the quantization process, each of the standard levels generated is encoded into the proper form for transmission or subsequent signal processing. Each of the composite encoded samples appearing at the output of the encoder is called a PCM *word.* Thus, the infinite number of possible amplitude levels of the sampled signal is converted to a finite number of possible PCM words.

The majority of PCM systems use the binary number system as a basis for encoding the digital PCM words. Each word then consists of a combination of only two digits: a 0 and a 1. This means that the transmission channel and the receiver circuitry need only be capable of recognizing two possible signal conditions. This situation results in some significant advantages, as we will see shortly. Since most PCM systems employ binary digits in the encoding process, the convention will be followed in this chapter that binary encoding will be assumed in various discussions unless otherwise stated. Systems employing encoding with more than two levels will be considered in Sec. 7-8.

Let n represent the number of *bi*nary dig*its* (bits) used in each word, and let m represent the number of possible distinct words that can be generated (alphabet size). We have

$$m = 2^n \qquad (7\text{-}1)$$

The number of possible words is very limited for only a few bits, but as the number of bits increases, the number of possible words increases exponentially. For example, the use of 12-bit words will result in 4096 possible levels at which a signal could be encoded. Thus, while PCM requires all signal levels to be encoded or quantized to a finite number of standard levels, the degree of accuracy in this process can be made as close as desired in theory. (In practice, there are limits to the ultimate accuracy attainable, but the same is true of analog systems as well.)

If the number of possible words m is known, the number of bits is given by

$$n = \log_2 m \qquad (7\text{-}2)$$

where n must be rounded upward to the next largest integer whenever the result is not an integer. For example, if 100 levels were desired, 7 bits would have to be used, which would actually provide the capability of 128 levels. For this reason, system designers usually specify the number of possible PCM levels directly as an integer power of 2.

The end result of the basic sampling, quantization, and encoding process in a binary PCM system then is a sequence of binary words, each having n bits. The value of each binary word represents in some predetermined way the amplitude of the analog signal at the point of sampling. Basically, what we have described up to this point is an analog-to-digital (A/D) conversion, which is used in a number of different types of data-processing systems other than those intended specifically for communications. Depending on the system requirements and other factors, some of the readily available A/D converters could be used for the PCM transmitter encoding process.

Skipping for the moment all the detailed steps between the encoding process and the receiver, an inverse decoding process at the receiver must be employed to convert the received digital words into a usable analog form, if the desired output is an analog signal. This process is actually a form of digital-to-analog (D/A) conversion, and some of the available D/A converters could be used for the receiver decoding purpose.

Before an erroneous impression is developed, it should be pointed out that PCM generation, transmission, and detection is much more than interfacing an A/D converter with a signal source and interfacing a D/A converter with the signal destination hardware. The intent here was to point out simply that the initial generation and final detection of the signals are essentially identical to the processes used in many other data systems, and some of the readily available hardware may be used for this purpose in many cases. As we will see, however, there are many other considerations that must be made for PCM systems between the point at which the initial signal is generated and the point at which the final analog signal is reconstructed.

A final qualitative point will be made before concluding this section. With most forms of communications systems, any degradation of the transmitted signal resulting from noise, distortion, fading, and the like, will adversely affect the quality of the received signal. With PCM systems, *as long as the signal level is sufficiently large that it is possible to tell the difference between the two possible signal levels transmitted, the signal can theoretically be reconstructed to the accuracy inherent in the sampling quantization process.* As we will see later in the book, errors may occur because noise may cause a transmitted 1 to appear as a 0, and vice versa. The point is, however, that a signal may undergo a significant amount of degradation and still be perfectly recognizable in terms of the two possible levels transmitted. Thus, as long as the two transmitted states can be recognized at the receiver, the signal can be "cleaned up," that is, replaced by a locally generated noise-free digital counterpart of the transmitted signal, and reconverted to analog form. This is one of the outstanding advantages of PCM systems.

7-2 BASIC PCM ENCODING AND QUANTIZATION

In this section, we will concentrate on the manner in which a PCM signal is encoded in its basic form and the corresponding quantization characteristics relating analog levels to the resulting digital levels. By "basic" encoding, we refer to the direct process of representing numbers in the conventional or *natural binary* number system form, as well as conversions or interpretations of those numbers in normalized or fractional forms. The encoding or conversion to special data forms will be considered in later sections.

Table 7-1. Natural Binary Numbers, Decimal Values, and Unipolar and Bipolar Offset Values for Four Bits

Natural Binary Number	Decimal Value[a]	Unipolar Normalized Decimal Value[b]	Bipolar Offset Normalized Decimal Value[b]
1111	15	$^{15}\!/_{16} = 0.9375$	$^{7}\!/_{8} = 0.875$
1110	14	$^{14}\!/_{16} = 0.875$	$^{6}\!/_{8} = 0.75$
1101	13	$^{13}\!/_{16} = 0.8125$	$^{5}\!/_{8} = 0.625$
1100	12	$^{12}\!/_{16} = 0.75$	$^{4}\!/_{8} = 0.5$
1011	11	$^{11}\!/_{16} = 0.6875$	$^{3}\!/_{8} = 0.375$
1010	10	$^{10}\!/_{16} = 0.625$	$^{2}\!/_{8} = 0.25$
1001	9	$^{9}\!/_{16} = 0.5625$	$^{1}\!/_{8} = 0.125$
1000	8	$^{8}\!/_{16} = 0.5$	0
0111	7	$^{7}\!/_{16} = 0.4375$	$-^{1}\!/_{8} = -0.125$
0110	6	$^{6}\!/_{16} = 0.375$	$-^{2}\!/_{8} = -0.25$
0101	5	$^{5}\!/_{16} = 0.3125$	$-^{3}\!/_{8} = -0.375$
0100	4	$^{4}\!/_{16} = 0.25$	$-^{4}\!/_{8} = -0.5$
0011	3	$^{3}\!/_{16} = 0.1875$	$-^{5}\!/_{8} = -0.625$
0010	2	$^{2}\!/_{16} = 0.125$	$-^{6}\!/_{8} = -0.75$
0001	1	$^{1}\!/_{16} = 0.0625$	$-^{7}\!/_{8} = -0.875$
0000	0	0	$-^{8}\!/_{8} = -1$

[a]Decimal value for integer representation of binary number.
[b]Decimal value with binary point understood on left-hand side of binary number.

For the discussion in this section, $n = 4$ bits will be used for illustration. This choice results in $m = 2^4 = 16$ words, which is sufficiently large that the general trend can be deduced and yet is sufficiently small that the results can be readily shown in graphical and tabular form.

The possible 16 natural binary words attainable with 4 bits and their corresponding decimal values are shown in the two left-hand columns of Table 7-1. (The two right-hand columns will be explained later.) Note that the smallest value is at the bottom of the table, and the largest value is at the top. For each binary number, the bit farthest to the left is called the *most significant bit* (MSB), and the bit farthest to the right is called *least significant bit* (LSB).

Depending on the exact circuitry involved, the dynamic range of the signal, and other factors, the 16 possible levels could be made to correspond to particular values of the analog signal as desired. However, some standardization has been achieved with many of the commercially available A/D and D/A converters, and the discussion here will concentrate on some of these standard forms. Typically, A/D and D/A converters are designed to operate with maximum input analog voltage levels of 2.5, 5, or 10 V (either positive or both positive and negative), although means are often provided for changing to other values if desired.

Because of these different voltage levels and the widely different decimal values of the binary number system as the number of bits is changed, it is frequently desirable to *normalize* the levels of both the analog signal and the digital words so that the maximum magnitudes of both forms have (or at least approach) unity. The normalized input analog voltage is defined as

$$\begin{array}{l} \text{normalized input} \\ \text{analog voltage} \end{array} = \frac{\text{actual input analog voltage}}{\text{full-scale voltage of A/D converter}} \quad (7\text{-}3)$$

where the full-scale voltage of the A/D converter is typically 2.5, 5, or 10 V. At the D/A converter in the receiver, the output voltage is

$$\begin{pmatrix} \text{actual output} \\ \text{analog voltage} \end{pmatrix} = \begin{pmatrix} \text{normalized value} \\ \text{of digital word} \end{pmatrix} \times \begin{pmatrix} \text{full-scale voltage} \\ \text{of D/A converter} \end{pmatrix} \quad (7\text{-}4)$$

Most of the subsequent discussions in this section will be based on the normalized forms of both the analog and digital values.

Normalization of the values of the binary words in Table 7-1 to a range less than 1 is achieved by adding a binary point to the left of the values as given in the table. However, to simplify the notation, the binary point will usually be omitted, but it will be understood in all discussions in which the normalized form is assumed.

We will now consider the manner in which the normalized analog level can be related to the normalized digital level. While a number of possible strategies could be used, the two most common forms employed in A/D conversion are (1) the *unipolar* and (2) the *bipolar offset representation*. Each will be considered separately.

The *unipolar representation* is most appropriate when the analog signal $x(t)$ is always of one polarity (including zero). If the signal is negative, it can be inverted before sampling, so assume that the normalized range of the analog signal is $0 \leq x < 1$. Let X_u represent the unipolar normalized quantized decimal representation of x following the A/D conversion. The 16 possible values of X_u for the case of 4 bits are shown in the third column of Table 7-1. Note that 0000 in binary corresponds to true decimal 0 and that the binary value 1000 corresponds to the exact decimal mid-range value 0.5. However, the upper value of binary 1111 does not actually reach decimal 1, but instead has the decimal value $\frac{15}{16} = 0.9375$.

Let ΔX_u represent the normalized step size, which represents on a decimal basis the difference between successive levels. This value is also the decimal value corresponding to 1 LSB and is in the general case

$$\Delta X_u = 2^{-n} \tag{7-5}$$

for the *unipolar* system. The largest normalized quantized decimal value attainable $X_u(\max)$ differs from unity by ΔX_u and is

$$X_u(\max) = 1 - 2^{-n} \tag{7-6}$$

for the *unipolar system*. Thus, unity can be approached to an arbitrarily close value, but it can never be completely reached. The resulting distribution about the midpoint is not completely symmetrical, but is has the advantages that the absolute zero levels of both number systems are identical, and the exact midpoint of the analog voltage corresponds to the binary value whose MSB is 1 and all other bits are 0.

The *bipolar offset representation* is most appropriate when the analog signal $x(t)$ has both polarities. Specifically, it assumes that the normalized range of the analog signal is $-1 \leq x < 1$. Let X_b represent the bipolar normalized quantized decimal representation of x following the A/D conversion. The 16 possible values of X_b for the case of 4 bits are shown in the fourth column of Table 7-1. Note that 0000 in binary corresponds exactly to the decimal value of -1, while the binary value 1000 corresponds to the exact decimal value of 0. However, the upper value of binary 1111 does not actually reach the decimal value 1 but instead has the decimal value $\frac{7}{8} = 0.875$.

Let ΔX_b represent the normalized step size or value corresponding to 1 LSB for the bipolar case. This value is

$$\Delta X_b = 2^{-n+1} \tag{7-7}$$

for the *bipolar offset* case, which is twice as large as for the unipolar case. The *percentage* resolution is the same as before, but since the peak-to-peak range of the input is twice as great, the difference between successive steps is twice as large. Depending on the application and the manner in which the signal is to be interpreted, this may or may not represent a difference in the ultimate accuracy of the results. The largest normalized quantized decimal value attainable $X_b(\max)$ is

$$X_b(\max) = 1 - 2^{-n+1} \qquad (7\text{-}8)$$

for the *bipolar offset* case.

Observe from the fourth column of Table 7-1 that for 4 bits there are 8 binary words corresponding to negative decimal quantized levels, 7 binary words corresponding to positive decimal levels, and 1 binary word (1000) corresponding to a decimal level of 0. In general, there are $m/2 = 2^{n-1}$ binary words corresponding to negative quantized decimal signal levels, $2^{n-1} - 1$ binary words corresponding to positive decimal values, and one word corresponding to decimal 0. This last level is always represented by a binary word whose MSB is 1 and whose other bits are 0.

As an additional point of interest, the two's-complement representation of a given binary number can be obtained by replacing the MSB in the bipolar representation by its logical complement. In fact, the bipolar offset representation is sometimes referred to as a "modified two's-complement" representation.

The quantized decimal values and their binary representations have now been defined for two possible forms. The exact manner in which the quantized decimal values represent different ranges of the analog signal is defined by means of the *quantization characteristic curve*. Both *rounding* and *truncation* strategies may be employed. In *rounding,* the sampled value of the analog signal is assigned to the *nearest* quantized level. In *truncation,* the sampled value is adjusted to the next lowest quantized level. For example, assume that two successive quantization levels correspond to 6.2 and 6.4 V, respectively. With rounding, a sample of value 6.29 V would be set at 6.2 V, and a sample of 6.31 would be set at 6.4 V. With truncation, both samples would be set to 6.2 V. As might be expected, the average error associated with truncation is greater than for rounding. Nevertheless, there are applications in which truncation is preferred.

For the development here, we will assume that rounding is employed in all A/D converters discussed. The quantization characteristic for an ideal 4-bit A/D converter employing rounding and a *unipolar encoding* scheme is shown in Fig. 7-2. The horizontal scale represents the decimal analog signal level on a normalized basis. [The actual signal level is a given normalized level multiplied by the full-scale voltage of the A/D converter. See (7-3).] The vertical scale represents the decimal representation of the equivalent quantized value on a normalized unipolar basis. The straight line on the graph represents the true analog characteristic in which the output would equal the input, which is, of course, not attainable due to the quantization effect.

Let us now investigate the manner in which the curve indicates the quantization effect. Whenever the normalized analog input $< \frac{1}{32}$, the quan-

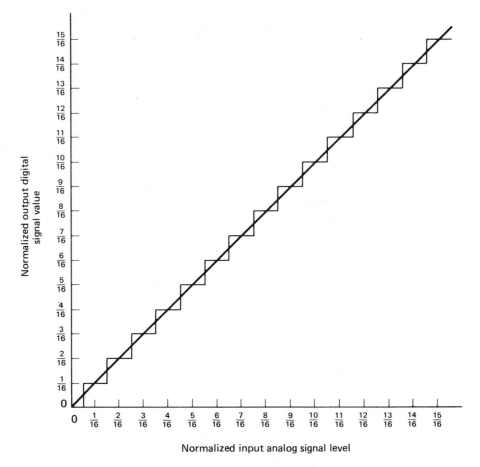

Figure 7-2. Unipolar quantization characteristic for a 4-bit A/D converter.

tized decimal value is 0, which is represented by binary 0000 (refer to Table 7-1). When the normalized analog input barely exceeds $\frac{1}{32}$, the quantized decimal value jumps to $\frac{1}{16}$, which is represented by binary 0001. This binary value remains the same until the decimal value of the normalized input just exceeds $\frac{3}{32}$, and so forth. At the other end of the scale, the largest binary value 1111 corresponds to the range of the analog input greater than $\frac{29}{32}$.

The vertical difference between the nonlinear quantization characteristic and the straight line represents the *quantization error*. Observe that for certain specific values there is no error, but the maximum error is $\frac{1}{32}$ (assuming that the maximum analog level is $\frac{31}{32}$). In general, for *unipolar* representation with quantizer *rounding,* the normalized resolution in the ideal case is

$$\text{normalized resolution} = \pm 2^{-(n+1)} \qquad (7\text{-}9)$$

The actual resolution in volts is

$$\text{actual resolution} = \pm 2^{-(n+1)} \times \text{full-scale voltage} \qquad (7\text{-}10)$$

The percentage resolution is

$$\text{percentage resolution} = \pm 2^{-(n+1)} \times 100\% \qquad (7\text{-}11)$$

The quantization characteristic of an ideal A/D converter employing rounding and a *bipolar offset* encoding scheme is shown in Fig. 7-3. The horizontal scale again represents the analog signal on a normalized basis, which in this case varies from -1 to $+1$. The curve is read in the same manner as in the preceding case except that the last column of Table 4-1 is now applicable.

Observe in this case that the maximum error is now $\frac{1}{16}$ on a normalized basis, a result of the fact that the peak-to-peak range of the normalized signal

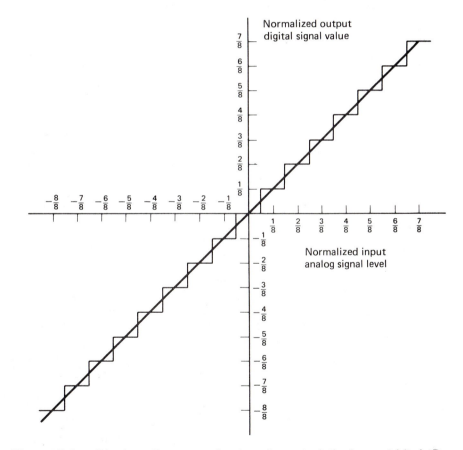

Figure 7-3. Bipolar offset quantization characteristic for a 4-bit A/D converter.

is now 2. In general, for the *bipolar offset* representation with quantizer rounding, the normalized resolution in the ideal case is

$$\text{normalized resolution} = \pm 2^{-n} \qquad (7\text{-}12)$$

The actual resolution in volts is

$$\text{actual resolution} = \pm 2^{-n} \times \text{full-scale voltage} \qquad (7\text{-}13)$$

The percentage resolution based on an assumed peak value of unity is

$$\text{percentage resolution} = \pm 2^{-n} \times 100\% \qquad (7\text{-}14)$$

[Some references use the *peak-to-peak* maximum value as a basis for the percentage resolution, and the result in that case is the same as for unipolar encoding, that is, Eq. (7-11).]

In casual usage, it is common to specify the resolution in bits. Thus, the system we have discussed might be specified as having a resolution of 4 bits, but the actual numerical resolution must be calculated by the appropriate formula.

Example 7-1

The normalized analog waveform shown as the smooth curve in Fig. 7-4 is to be converted to a PCM signal by a 4-bit A/D converter whose input-output normalized quantization characteristic is given by Fig. 7-3. Sampling will occur at $t = 0$ and at 1-ms intervals thereafter. Over the time interval shown, construct on the same scale the quantized form of the signal that is actually encoded and subsequently reconstructed at the receiver. In addition, list the corresponding binary words that are transmitted.

Solution

Since the analog signal is represented in normalized form, the values may be used directly on the quantization characteristic scales of Fig. 7-3. If the actual voltage level had been given instead, it would be convenient to divide the actual voltage level by the full-range voltage first so that the results could be used directly on the normalized scale. The quantized signal is constructed by reading the exact analog voltage at sampling points, noting which quantization level is nearest, and indicating the corresponding binary word using Table 7-1 if necessary.

At $t = 0$, the exact analog voltage is 0, so the quantized value is also 0. The corresponding binary bipolar offset value is 1000. At $t = 1$ ms, the analog voltage is closer to $\frac{1}{8}$ than 0, so the quantized level is $\frac{1}{8}$. The corresponding binary value is 1001. This process is continued for the entire duration of the

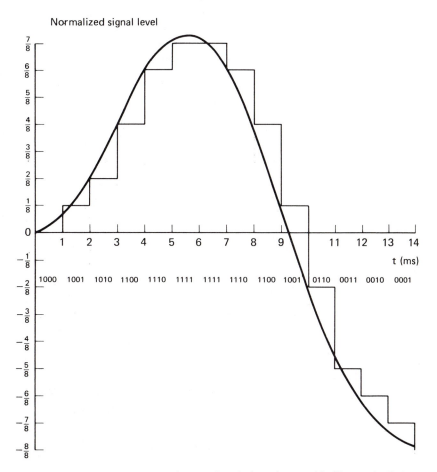

Figure 7-4.   Analog and quantized signals used in Example 7-1.

signal, and the results are summarized in Table 7-2 and partially in Fig. 7-4. The reader should find it instructive to verify the results given.

Example 7-2

A certain 8-bit A/D converter with a full-scale voltage of 10 V is to be employed in a binary PCM system. The input analog signal is adjusted to cover the range from zero to slightly under 10 V, and the converter is connected for unipolar encoding. Rounding is employed in the quantization scheme. Determine the following quantities: (a) normalized step size, (b) actual step size in volts, (c) normalized maximum quantized analog level, (d) actual maximum quantized level in volts, (e) normalized resolution, (f) actual resolution in volts and (g) percentage resolution.

Table 7-2. Data Supporting Example 7-1

Time (ms)	Closest Quantization Value	Binary Representation
0	0	1000
1	$\frac{1}{8}$	1001
2	$\frac{2}{8}$	1010
3	$\frac{4}{8}$	1100
4	$\frac{6}{8}$	1110
5	$\frac{7}{8}$	1111
6	$\frac{7}{8}$	1111
7	$\frac{6}{8}$	1110
8	$\frac{4}{8}$	1100
9	$\frac{1}{8}$	1001
10	$-\frac{2}{8}$	0110
11	$-\frac{5}{8}$	0011
12	$-\frac{6}{8}$	0010
13	$-\frac{7}{8}$	0001
14	$-\frac{8}{8}$	0000

Solution

The various quantities desired may be readily calculated from the relationships developed in this section. The results, with some practical rounding, are summarized in the following:

(a) The normalized unipolar step size is
$$\Delta X_u = 2^{-n} = 2^{-8} = 0.003906 \tag{7-15}$$

(b) Actual step size = $0.003906 \times 10 = 39.1$ mV $\tag{7-16}$

(c) The normalized quantized maximum analog level is
$$X_u(\max) = 1 - 2^{-n} = 1 - 2^{-8} = 0.9961 \tag{7-17}$$

(d) Actual maximum quantized voltage = $0.9961 \times 10 = 9.961$ V $\tag{7-18}$

(e) Normalized resolution = $\pm 2^{-(n+1)} = \pm 2^{-9} = \pm 0.001953$ $\tag{7-19}$

(f) Actual resolution = $\pm 0.001953 \times 10 = \pm 19.53$ mV $\tag{7-20}$

(g) Percentage resolution = $\pm 2^{-(n+1)} \times 100\% = \pm 2^{-9} \times 100\%$
$$= \pm 0.195\% \tag{7-21}$$

Example 7-3

The 8-bit A/D converter of Ex. 7-2 is to be changed so that it can be used with an analog signal having a range from -10 V to just under 10 V. Bipolar offset encoding is to be used. Repeat all the calculations of Ex. 7-2 for this case.

Solution

It is assumed that the full-scale voltage is again 10 V, but the peak-to-peak range is now 20 V. The various calculations with practical rounding of some of the results are summarized in the following:

(a) The normalized bipolar offset step size is

$$\Delta X_b = 2^{-(n-1)} = 2^{-7} = 0.007813 \qquad (7\text{-}22)$$

(b) Actual step size = $0.007813 \times 10 = 78.1$ mV $\qquad (7\text{-}23)$

(c) The normalized quantized maximum analog level is

$$\Delta X_b(\text{max}) = 1 - 2^{-(n-1)} = 1 - 2^{-7} = 0.9922 \qquad (7\text{-}24)$$

(d) Actual maximum quantized voltage = $0.9922 \times 10 = 9.922$ V $\qquad (7\text{-}25)$

(e) Normalized resolution = $\pm 2^{-n} = \pm 2^{-8} = \pm 0.003906 \qquad (7\text{-}26)$

(f) Actual resolution = $\pm 0.003906 \times 10 = \pm 39.1$ mV $\qquad (7\text{-}27)$

(g) Percentage resolution = $\pm 2^{-n} \times 100\% = \pm 2^{-8} \times 100\%$

$$= \pm 0.391\% \qquad (7\text{-}28)$$

7-3 COMPANDING

A special problem occurs with PCM systems in which the expected dynamic range of the signal is very large. To avoid saturation and subsequent signal clipping, the level of the signal must be adjusted so that its peak value never exceeds the full-scale signal level of the A/D converter. The resulting effect is that during intervals in which the signal level is very low the percent of quantization error increases significantly.

To illustrate this effect, assume that a given A/D converter has a full-scale voltage of 10 V, and assume that the actual resolution is about ± 4 mV, corresponding to a step size of 8 mV. During an interval in which the signal voltage is close to 10 V, the peak quantization error is in the neighborhood of (4 mV/10 V) $\times$ 100% = 0.04%. However, assume that over some interval of time, the signal level hovers around 10 mV, corresponding to a dynamic range of about 60 dB or so for the signal. In this time interval, the peak quantization error is in the vicinity of (4 mV/10 mV) $\times$ 100% = 40%! The result may be intolerable for many applications. The severity of this problem depends on the expected dynamic range of the signal and the number of bits used in the encoding process. In theory, a sufficient number of bits could be added to decrease the peak quantization error to a more tolerable level, but this is an inefficient and often impractical process, especially if the dynamic range of the signal is quite large.

There are two ways in which this problem can be partially alleviated, the first of which will be mentioned as a useful conceptual idea, and the second of which is the procedure most often used in practice. The first method is to employ nonuniform step increments in the A/D converter. At low signal levels, the step sizes would be very small, thus permitting a relatively high percentage of resolution. As the signal level increases, the step size can be increased, thus providing a more optimum distribution of the available number of steps. The ideal situation is one in which the step size is proportional to the signal level; that is, the step increment would follow an exponential taper. At the receiver, the taper of the D/A converter quantization curve would be required to have the inverse characteristic; that is, the steps would be larger at smaller signal amplitudes and smaller at larger amplitudes.

The second method, which accomplishes the same purpose with less difficulty, is to compress the dynamic range of the analog signal before sampling and to employ a conventional A/D converter with a uniform step size. At the receiver, the analog signal is first reconstructed with a conventional D/A converter, and the dynamic range of the analog signal is then expanded. The nonlinear process at the transmitter is called signal *compression,* and the process at the receiver is called *expansion.* The combined process is referred to as signal *companding.*

The input-output form of a typical compression characteristic curve is illustrated by Fig. 7-5a. Observe that at low input signal levels, the output level is increased considerably, thus "lifting" the signal to a higher level and thereby reducing the percent of quantization error. Of course, higher signals must be "pushed down," so to speak, to keep them in the proper range.

The input-output form of a typical expansion curve is shown in Fig. 7-5b. It is necessary to design the compression and expansion curves together for a particular system so that the two curves exactly offset each other. If this condition is not met, the resulting signal will obviously be distorted. The compression and expansion curves can be designed with either diode analog nonlinear shaping circuits or with a controlled nonlinear characteristic such as a logarithmic amplifier.

7-4 BASEBAND ENCODING FORMS

So far, all the binary PCM signal forms considered have employed natural binary encoding. While some systems transmit the natural binary words directly, other systems convert the natural binary words to special formats prior to actual transmission or high-frequency modulation. These other forms have advantages in particular situations in terms of ease of data processing, bandwidth requirements, synchronization, and other factors. Several of the most common baseband encoding forms will be discussed in this section.

The encoding schemes discussed in this section are primarily those that are used in establishing basic data formats and should not be confused with error detection and correction encoding, which is a different matter altogether. In

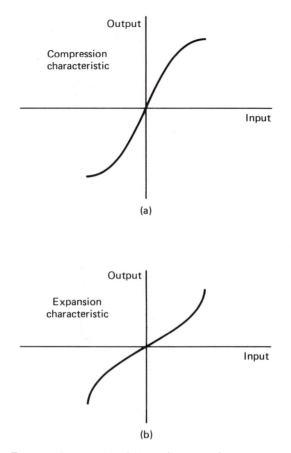

Figure 7-5. Forms of compression and expansion curves used in PCM systems to reduce relative quantization error effect.

fact, error detection encoding could be added to the formats discussed in this section.

It is worth mentioning at the outset that the designations for some of these encoding schemes are rather awkward sounding and have evolved over a period of time. Nevertheless, the terms are widely used and will be stated as accepted in the field. In the forms employed, a binary *one* is often designated as a *mark,* and a binary *zero* is designated as a *space.*

To illustrate some of the methods involved, refer to Fig. 7-6. A binary pattern of 1 0 1 1 0 0 0 1 1 0 1 is shown at the top, and some of the different forms in which this bit stream can be encoded are shown below. For all the waveforms given, no absolute levels are indicated, since this would depend on the particular voltage or current levels employed. Indeed, some data systems employ "negative logic" in which all the waveforms shown would be inverted

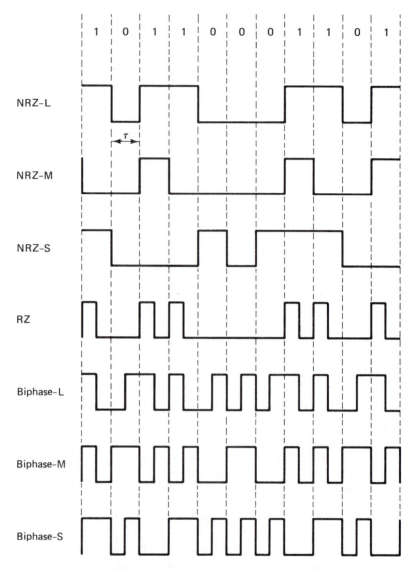

Figure 7-6. Different baseband encoding formats.

in level from those given. It will suffice to say that there are always two levels, one of which is a binary one or mark and the other of which is a binary zero or space, but the levels of both may be established quite arbitrarily.

In the discussion that follows, let τ represent the width in seconds of a data bit to be encoded. The actual encoded pulse may have a width τ or some fraction of τ depending on the particular scheme employed. Each of the individual waveforms will now be discussed in order.

NRZ-L

This designation refers to *non-return-to-zero level*. A 1 is always represented by one fixed level of width τ, and a 0 is always represented by the other level, also of width τ. This form is thus equivalent to a conventional binary representation, and each word would be a natural binary value if the signal to be encoded is expressed in natural binary form. Most of the other forms to be discussed are obtained by first generating an NRZ-L signal and converting it to one of the other forms by appropriate circuitry. For the data given in Fig. 7-6, the upper level is 1 and the lower level is a 0, as shown.

NRZ-M

This designation refers to *non-return-to-zero mark*. If a 1 is to be transmitted, there is always a change at the beginning of the bit interval. However, if a 0 is to be transmitted, there is no change. The reader can verify that this scheme is satisfied for the data shown in Fig. 7-6. The signal level just prior to the first data bit would be a 1 for the case shown. It will be seen later that this initial level is usually arbitrary.

NRZ-S

This designation refers to *non-return-to-zero space*. If a 0 is to be transmitted, there is always a change at the beginning of the bit interval. However, if a 1 is to be transmitted, there is no change. As in NRZ-M, the beginning level is arbitrary.

RZ

This designation refers to *return-to-zero*. If a 0 is to be transmitted, the signal remains at the 0 level (lower level in this case) for the entire bit interval. However, if a 1 is to be transmitted, a pulse of width $\tau/2$ is inserted in a designated part of the bit interval.

There is another form of RZ (not shown) that essentially utilizes three levels: no transmission, a positive pulse, and a negative pulse. The positive and negative pulses represent the 1 and 0 levels, and these pulses each have a width equal to $\tau/2$. Thus, there is always a "rest" period following the transmission of each data pulse.

Biphase-L

This designation refers to *biphase level*. Within each data bit interval, there are always two states, each of width $\tau/2$. If the data bit is a 0, the logical sequence 01 is inserted. Conversely, if the data bit is a 1, the sequence 10 is inserted. This encoding scheme is also referred to as a *split-phase* or *Manchester* code.

Biphase-M

This designation refers to *biphase mark*. A transition always occurs at the beginning of a bit interval. If the data bit is a 1, a second transition occurs at a time $\tau/2$ later. If the data bit is 0, no further transition occurs until the beginning of the next bit interval. Referring to Fig. 7-6, the level just prior to the beginning of the data stream was arbitrarily selected.

Biphase-S

This designation refers to *biphase space*. A transition always occurs at the beginning of a bit interval. If the data bit is a 0, a second transition occurs at a time $\tau/2$ later. If the data bit is a 1, no further transition occurs until the beginning of the next bit interval. As in the case of biphase-M, the initial level was arbitrarly selected.

It is not practical to attempt to discuss the various criteria that led to the evolution of all these as well as other encoding forms not shown, since some of the reasons are somewhat arbitrary and difficult to determine. However, a few of the general significant properties of some of these formats will be mentioned, and others will be considered later. Two particular comparison areas are the relative bandwidth requirements and the relative ease of synchronization.

The RZ and NRZ forms require the transmission of a dc component in the spectrum. This can be deduced by noting that for either form there is at least one possible transmission condition in which only one level would be transmitted for a long time (e.g., long string of 1's in NRZ-S). Unless the total transmission system has a response down to dc, this level cannot be maintained. In contrast, the biphase forms always have at least one transition per bit interval, so there is no need to retain the dc level. This property may be advantageous in signal-processing circuitry since ac coupling may be employed, or transformers could be used (e.g., as in telephone lines).

Another advantage of the biphase forms is that at least one orderly transition occurs per data-bit interval, and this property may be used to derive a synchronized clock reference at the receiver. In fact, the biphase forms may be generated by combining a data signal in an NRZ form with a reference clock so that the clock is actually transmitted with the signal.

While the preceding arguments would tend to favor the biphase data forms, there is, however, a distinct disadvantage in terms of the total required bandwidth. For the NRZ forms, the shortest pulse width is always equal to the data-bit interval width τ. In contrast, the biphase forms exhibit pulses that have a width $\tau/2$. Since the transmission bandwidth of a pulse is inversely proportional to the pulse width, the required transmission bandwidth for the biphase forms is twice the bandwidth for the NRZ forms if all other factors are equal. Incidentally, the RZ form requires both a dc response as well as a bandwidth comparable to that of the biphase forms.

The NRZ-L and NRZ-M formats are used in the encoding and detection of differentially coherent phase-shift keying (PSK) data. This concept will be explored in some depth in Sec. 7-7.

Irrespective of the particular data format employed in the encoding process, it is convenient to classify the final PCM signal, from the viewpoint of the relative levels used in the actual pulses, as either a *unipolar* signal or a *bipolar* signal. A *unipolar* signal is one in which one of the two possible levels corresponds to an actual zero voltage or current level (i.e., nothing transmitted), and the other level is either a positive or negative value as established in the design process. A *bipolar* signal employs a positive voltage or current as one level and a negative voltage or current as the other level.

These terms as applied to the final signals have nothing to do whatsoever with the terms "unipolar representation" and "bipolar offset representation" as used in the A/D encoding process, so one must be careful about the terminology to avoid confusion. As we will see later, unipolar and bipolar level characterizations as defined here are useful in the generation of RF modulated signals and in the determination of relative transmitted energy and power.

7-5 TIME-DIVISION MULTIPLEXING OF PCM SIGNALS

As in the case of other sampled-data signals discussed in Chapter 6, PCM signals from a number of different information sources may be combined together and transmitted over a common channel by means of *time-division multiplexing* (TDM). This process is illustrated for PCM in Fig. 7-7. In this particular simplified system, each of the individual data channels is sampled at the same rate, and a given word representing the encoded value of the particular sample is transmitted between the point of sampling that channel and the next channel in the sequence. As in the case of pulse modulation, there are numerous variations on this basic scheme that can be devised when different data sources have different bandwidths.

To ensure that the electronic commutation at the receiver is exactly in step with that at the source, some means of synchronization must be employed. One way this is frequently achieved is by means of a special synchronization word that is inserted at the beginning of a frame. The synchronized word is chosen to have a different pattern (or value) than other possible words that could be transmitted. At the receiver, all incoming words are sensed by digital circuitry, which establishes the beginning of a new frame when the special sync word is received.

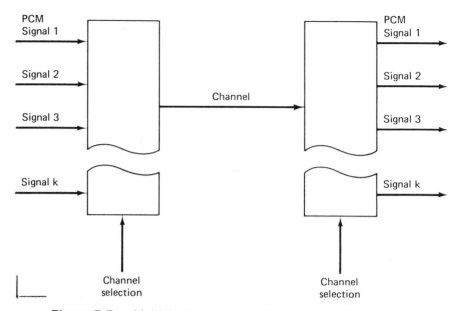

Figure 7-7. Multiplexing and demultiplexing of PCM signals.

Because of the fact that several pulses will have to be used to represent one data word with PCM, the effective baseband bandwidth of a TDM PCM signal is much greater than for a TDM PAM signal. In Chapter 6, an approximate lower bound for the bandwidth was developed with some idealized assumptions employed in the process. A similar development will now be performed for PCM.

A PCM TDM system in which there are k signals to be multiplexed will be considered, and each signal will be assumed to have a baseband bandwidth W. The same several somewhat unrealistic limiting-case assumptions made in Chapter 6 will be made again; they are (1) no spacing between successive pulses, (2) sampling rate set at minimum Nyquist rate, and (3) no sync information. An additional assumption made in this development is that the

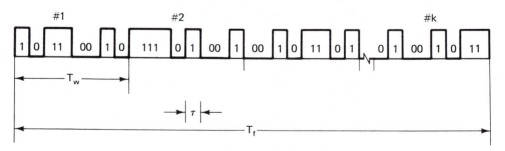

Figure 7-8. Frame format used in developing lower bound for transmission bandwidth of TDM PCM system.

actual bit pulses occupy the full width of the bit interval. As we saw in Sec. 7-4, some PCM encoded forms employ pulses shorter than the actual bit interval, and such systems would generally require greater bandwidths. Finally, the transmission bandwidth rule $B_T = 0.5/\tau$ will be used.

Referring to Fig. 7-8, there are three time intervals of significance. The *frame time* T_f is the total interval in which all signals are sampled. At the minimum Nyquist rate, T_f must satisfy

$$T_f = \frac{1}{2W} \qquad (7\text{-}29)$$

The *word time* T_w represents the duration of one particular encoded sample and is

$$T_w = \frac{T_f}{k} = \frac{1}{2kW} \qquad (7\text{-}30)$$

A given word is assumed to be represented by n bits, each of width τ. Thus,

$$\tau = \frac{T_w}{n} = \frac{1}{2nkW} \qquad (7\text{-}31)$$

The minimum baseband transmission bandwidth for the PCM TDM signal is then assumed to be

$$B_T = \frac{0.5}{\tau} = 0.5(2nkW) = knW \qquad (7\text{-}32a)$$

$$= kW \log_2 m \qquad (7\text{-}32b)$$

The lower bound for the transmission bandwidth in this hypothetical case turns out to be simply the product of the number of signals times the bandwidth per signal times the number of bits per word. Referring back to (6-22), it is noted that the minimum transmission bandwidth for PCM is n times the bandwidth required for PAM with all other factors equal. This particular point could also be deduced from the fact that the width of the transmitted pulses must be $1/n$ times the width required in PAM.

The second form of the bandwidth expression as given by (7-32b) displays a rather important result that should be emphasized. The inherent error in the PCM quantization process is inversely proportional to the number of levels m. As we will see later, this error can be viewed in much the same way as noise. Thus, the signal-to-noise ratio for the PCM signal increases rapidly with increasing m. However, the required transmission bandwidth increases as the logarithm of the number of levels, and this increase is generally very moderate. This suggests the fact that a moderate increase in transmission bandwidth can result in a significant increase in the signal-to-noise ratio. For example, in changing from an 8-bit system to a 9-bit system, the transmission bandwidth is increased only by about 12.5%, but the quantization error is reduced by 50%. There are other factors that must be considered in this process, but this example represents a special case of a rather general trade-off that appears in a number of complex communications systems. The concept is that with proper encoding and/or modulation, improved signal-to-noise ratio may be achieved at the expense of higher bandwidth, but the percentage of improvement is greater than the percentage of increase in bandwidth.

We wish to emphasize again that the assumptions made in deriving (7-32a) and (7-32b) represent limiting-case assumptions and are not intended to be workable formulas for a realistic system. The actual bandwidth when all realistic factors are considered could be much higher than these results predict. However, all the trends suggested by these results are true, and the implications of these simple formulas are very useful for comparing systems and for obtaining initial minimum estimates for bandwidth requirements.

Many multiplexing systems require the transmission of signals with widely different bandwidths. For example, the signals required to be transmitted from space vehicles may include such diverse data as the outputs of slowly varying instruments, voice data, and much higher-frequency video data. If all such signals were multiplexed in a simple frame structure, it would be necessary to sample all signals at a rate governed by the signal with the highest bandwidth. The result would be a very inefficient bandwidth utilization process.

In situations where the bandwidths of different signals vary widely, a more complex frame structure may be devised whereby some signals are sampled more than once per frame (i.e., at a "super" rate), and some signals are sampled less than once per frame (i.e., at a "sub" rate). Each design of this nature must be developed individually based on the number of signals and their respective bandwidths. However, a representative example will be discussed to illustrate some of the general features.

Consider the three-commutator system shown in Fig. 7-9.* The 25-channel commutator shown on the right will be referred to as the *prime commutator,* and one complete cycle of this unit is called a *prime frame.* The primary commutator has 25 slots. Notice, however, that every fifth slot (i.e., numbers 2, 7, 12, 17, and 22) are all connected together. The process applied to this signal is called *supercommutation,* and this analog signal is being sampled at five times the prime frame rate.

The commutator in the middle is called a *subcommutator.* The sampling rate of the subcommutator is one-fifth the rate of the prime commutator. Thus, each time the prime commutator samples channel 23, the subcommutator has advanced to the next of its five possible positions. Each subcommutator channel then is being sampled at a rate equal to one-fifth the prime sampling rate.

*This example system appeared in *EMR Telemeter,* published by EMR Telemetry, Sarasota, Florida.

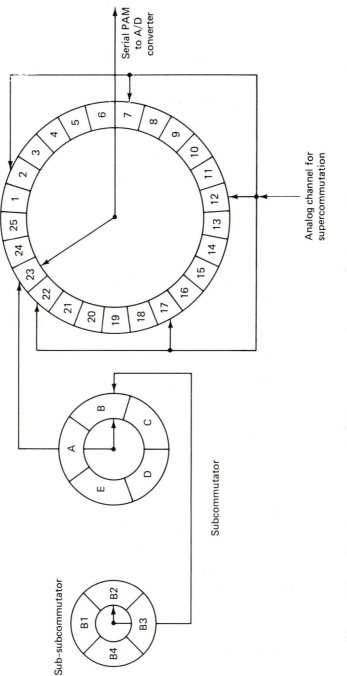

Figure 7-9. Sampling scheme involving supercommutation, subcommutation, and sub-subcommutation.

321

The commutator on the left is called the *sub-subcommutator*. The sampling rate of the sub-subcommutator is one-fourth the rate of the subcommutator or ($\frac{1}{5}$) $\times$ ($\frac{1}{4}$) = $\frac{1}{20}$ the rate of the prime commutator. Each time the subcommutator samples channel B, the sub-subcommutator has advanced to the next of its four possible positions. Each sub-subcommutator channel then is being sampled at a rate equal to one-twentieth the prime sampling rate.

The ratio of the bandwidth permitted in the highest-frequency data channel to the bandwidth permitted in the lowest-frequency channel is the ratio of the sampling rate of the supercommutated channel to the sampling rate of the sub-subcommutated channel and is $5/(\frac{1}{20})$ = 100. Note that it takes 20 complete cycles of the prime frame before all the signals are sampled at least once. This complete "supercycle" is called a *data field*.

Example 7-4

Consider a PCM TDM system in which 15 signals are to be processed. Each of the signals has a baseband bandwidth W = 1 kHz. The sampling rate is to be 25% greater than the theoretical minimum, and 8 bits are to be used in each word. Conventional NRZ-L encoding will be used, and an additional 8-bit sync word will be placed in each frame. Using the $0.5/\tau$ rule, determine the approximate minimum baseband bandwidth required.

Solution

To determine the bandwidth, it is necessary to determine the width of the shortest possible pulse. The sampling rate is

$$f_s = 1.25 \times 2W = 1.25 \times 2 \times 1 \text{ kHz} = 2.5 \text{ kHz} \tag{7-33}$$

The frame time is

$$T_f = \frac{1}{f_s} = \frac{1}{2.5 \times 10^3} = 0.4 \text{ ms} \tag{7-34}$$

The word time is

$$T_w = \frac{T_f}{k} = \frac{0.4 \text{ ms}}{16} = 25 \text{ } \mu s \tag{7-35}$$

where the value 16 represents the 15 data words plus the sync word for each frame. The bit interval is

$$\tau = \frac{T_w}{n} = \frac{25}{8} = 3.125 \text{ } \mu s \tag{7-36}$$

The approximate transmission bandwidth is

$$B_T = \frac{0.5}{\tau} = \frac{0.5}{3.125 \times 10^{-6}} = 160 \text{ kHz} \tag{7-37}$$

This value is somewhat larger than the idealized minimum suggested by (7-32), since more realistic assumptions were made in this case.

7-6 RF DIGITAL MODULATION METHODS

We have previously investigated the means by which a PCM signal could be generated and encoded in a number of possible formats. All the resulting encoded signals have baseband spectral forms so they can only be transmitted directly over wire links (or possibly links utilizing fiber optics). In this section, we will investigate some of the means by which the baseband digital PCM signals can be translated to radio-frequency ranges so that electromagnetic wave propagation can be utilized.

One could argue that once the PCM signal is generated it could be viewed in the same manner as any other data signal, and perhaps any of the various analog modulation systems considered earlier in the book could be used. In one sense this is true, and some systems have been implemented directly from this point of view. On the other hand, there are certain strategies for RF modulation that have been developed specifically with the transmission of digital data as the primary objective, and more optimum results can often be achieved by applying such methods. Some of the most common methods for RF modulation of digital data will be discussed in this section.

A few general (and perhaps philosophical) remarks about the relationship between digital and analog concepts will now be made. Many problems in modern communications involve concepts and hardware that have both digital and analog aspects, and it is sometimes difficult to separate the two areas. The various digital RF modulation techniques are good examples of these overlapping areas between the "digital world" and the "analog world." While the data itself may be digital in form, there is no straightforward way to transmit purely digital data by electromagnetic means. Our whole concept of spectral compatibility is based on sinusoidal signals, which are obviously analog in form. To transmit data that is spectrally compatible with other communications areas, the analog point of view must be considered for this purpose. Thus, RF modulation of digital data is neither a purely digital process nor a purely analog process, but it is a process having some of the attributes of both areas.

The discussion in this section will focus on three widely employed RF digital modulation methods:

1. Amplitude-shift keying (ASK)
2. Frequency-shift keying (FSK)
3. Phase-shift keying (PSK)

These three methods can be discussed together because they all involve direct binary encoding relationships between the baseband digital data and the corresponding RF characteristics. Other methods involving differential encoding and nonbinary systems will be discussed in later sections.

Radio-frequency digital modulation systems can be classified as either *coherent* or *noncoherent*. *Coherent* methods require a reference carrier at the receiver having the exact frequency and phase of the transmitter carrier. This process is similar to the concept of DSB synchronous detection encountered in Chapter 4. *Noncoherent* methods, on the other hand, do not require a reference carrier and can be detected by other means (e.g., envelope detection in some cases). (These concepts should not be confused with data synchronization, which may be required for both systems. The point of focus here relates only to the detection of the RF signal, which translates the RF signal back to the baseband frequency range.)

As a general rule, coherent systems tend to provide better performance in the presence of noise if all other factors are the same. However, noncoherent systems generally are less complex in design and operation, so both types of systems have their merits.

The three methods previously mentioned will now be discussed individually. Reference can be made to Fig. 7-10 for the discussions that follow. An

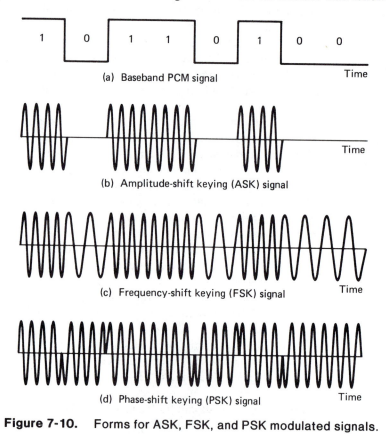

(a) Baseband PCM signal

(b) Amplitude-shift keying (ASK) signal

(c) Frequency-shift keying (FSK) signal

(d) Phase-shift keying (PSK) signal

Figure 7-10. Forms for ASK, FSK, and PSK modulated signals.

arbitrary baseband digital data stream is shown in Fig. 7-10a, and the different ways in which it modulates the RF carrier for the three separate forms are shown in Fig. 7-10b, c, and d. Bear in mind that in this figure it is not feasible to show more than a few RF cycles in a pulse width. Thus, the waveforms should be accepted for their illustrative and instructive values rather than for complete realism. In a practical system, there will usually be many (sometimes millions of) RF cycles in a pulse width.

Amplitude-Shift Keying

The process of *amplitude-shift keying* (ASK), which is sometimes referred to as *on-off keying* (OOK), is probably the simplest and most intuitive form of RF encoding. In this process a 1 is represented by turning the RF carrier on for the bit duration, and a 0 is represented by leaving the carrier off for the same interval. Thus, a fixed-frequency RF carrier is simply gated off and on in accordance with the bit value. The form of an ASK signal is illustrated in Fig. 7-10b.

Frequency-Shift Keying

The process of frequency-shift keying (FSK) involves the transmission of one or the other of two distinct RF frequencies. A 0 is represented by the transmission of a frequency f_0, and a 1 is represented by the transmission of a frequency f_1. The transmitter oscillator is thus required to switch back and forth between two separate frequencies at the data rate. The form of an FSK signal is illustrated in Fig. 7-10c.

Phase-Shift Keying

The process of phase-shift keying (PSK) utilizes a fixed-frequency sinusoid whose relative phase shift can be changed in abrupt steps. As we will see later, it is possible to utilize more than two possible phase shifts, but for the moment our attention is addressed to binary PSK. With binary PSK, the two relative phase shifts may be considered as 0° and 180°. The form of a binary PSK signal is illustrated in Fig. 7-10d.

Some of the possible implementation techniques that can be used in the generation and detection of these RF digital methods will now be discussed. An ASK signal can be generated by essentially the same process utilized in DSB analog systems, that is, a balanced modulator or multiplier. Consider the system shown in Fig. 7-11. A PCM signal having a unipolar level form is applied at the data input, and this signal is multiplied by the carrier. The carrier is thus multiplied by either a positive constant, which results in an RF burst, or by zero, which results in no transmission. Alternatively, the RF oscillator (or a buffer) can be turned on and off by the digital levels of the signal.

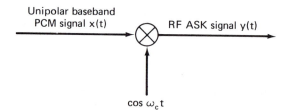

Figure 7-11. Generation of an ASK signal using a balanced modulator.

Let $x(t)$ represent the PCM data signal, and let f_c represent the cyclic carrier frequency. The form of the RF modulated signal $y(t)$ is

$$y(t) = x(t) \cos \omega_c t \qquad (7\text{-}38)$$

where a reference phase of $0°$ and unity magnitude have been assumed in the carrier for convenience. The reader can recognize that $y(t)$ is, in reality, simply a DSB signal whose properties were studied extensively in Chapter 4. As a result, the spectrum of the modulated signal contains two sidebands, each of which has the bandwidth of the baseband data stream $x(t)$.

(It is suggested that the reader pause briefly at this point to review again the concept of spectral frequency as discussed in Chapter 2. While only one sinusoidal frequency is transmitted, the actual spectral representation of the composite modulated signal depends on the data rate, and it contains two sideband regions about the carrier frequency.)

An ASK signal can be detected either noncoherently or coherently. *Noncoherent* detection can be achieved by the simple process of envelope detection, as illustrated by the block diagram of Fig. 7-12a. Following the

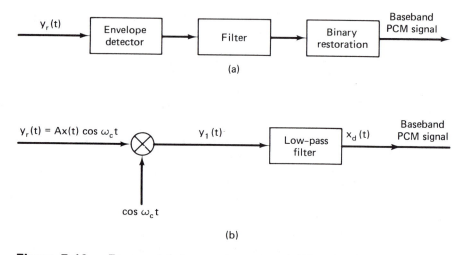

Figure 7-12. Forms of (a) noncoherent and (b) coherent detection of ASK signals.

actual extraction of the envelope, the signal is smoothed by a filter. The resulting signal at this point is no longer a simple binary two-state signal, since the finite bandwidth limitations of the channel have rounded the various pulses. By employing threshold circuits such as comparators, the imperfect pulses can be converted back to ideal forms. The block entitled "binary restoration" accomplishes this purpose, and the output is a reconstructured PCM replica of the transmitted signal (assuming no decision errors for bits).

The reader should note that, while envelope detection of DSB signal with *analog* modulated data was specifically prohibited in Chapter 4, envelope detection of digital data using an equivalent modulation process is perfectly permissible! Why?

Coherent detection of ASK data is also possible, and this approach is indicated by the block diagram of Fig. 7-12b. A coherent reference must be provided for this purpose. Assume that the received signal $y_r(t)$ is given by

$$y_r(t) = Ax(t) \cos \omega_c t \qquad (7\text{-}39)$$

The local coherent reference is given by $\cos \omega_c t$, and the output $y_1(t)$ of the coherent detector is

$$y_1(t) = Ax(t) \cos^2 \omega_c t \qquad (7\text{-}40a)$$

$$= \frac{A}{2}[x(t) + x(t) \cos 2\omega_c t] \qquad (7\text{-}40b)$$

The second term in the brackets of (7-40b) is a DSB signal centered at $2f_c$. Assuming that the carrier frequency is much greater than the bandwidth of the baseband PCM signal, this component is readily eliminated by the low-pass filter. The resulting detected signal $x_d(t)$ is then

$$x_d(t) = \frac{A}{2}x(t) \qquad (7\text{-}41)$$

which is the form of the baseband PCM data signal modified by a constant multiplier. The resulting digital data signal can then be converted back to analog form again.

An FSK signal can be generated by using the baseband data signal as the control signal for a voltage-controlled oscillator (VCO), as illustrated in Fig. 7-13. Since the baseband signal assumes only one of two values, the VCO generates only one of two possible sinusoidal frequencies. The level of the control voltage is adjusted in accordance with the VCO circuitry so that the

Figure 7-13. Generation of an FSK signal using a voltage-controlled oscillator.

two possible frequencies are produced at the proper levels of the PCM signal. It should be understood that while the FSK signal itself assumes only one of two frequencies, the spectrum is much more complex and contains many frequencies.

As in the case of the ASK signal, the FSK signal can be detected either noncoherently or coherently. *Noncoherent* detection can be achieved through the process illustrated in Fig. 7-14a. The signal is simultaneously applied to two band-pass filters, one tuned to f_1 and the other to f_0. When a 1 is transmitted, the output of the upper filter is maximum while the output of the lower filter should be rather low (assuming reasonably high signal-to-noise ratio and proper spacing between channels). When a 0 is transmitted, the opposite result occurs. The envelope detector in each path converts the corresponding RF pulses to baseband pulses. Finally, pulses from both paths are combined to yield a baseband data signal. The pulses from the lower path are subtracted from those in the upper path so that a bipolar baseband pulse stream is constructed. The resulting pulses are distorted as well as corrupted

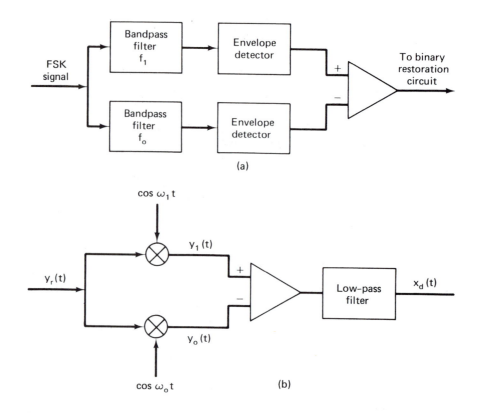

Figure 7-14. Forms of (a) noncoherent and (b) coherent detection of FSK signals.

by noise, so a binary restoration process should be used to generate a more ideal PCM data signal before D/A conversion.

Coherent detection of FSK signals is achieved by the process shown in Fig. 7-14b. This system requires local generated carriers having the same frequencies and phases as those used at the transmitter. For the purpose of illustration, assume that a 1 is transmitted, which means that the received signal is of the form

$$y_r(t) = A \cos \omega_1 t \tag{7-42}$$

Along the upper path, the output of the mixer is

$$y_1(t) = A \cos^2 \omega_1 t \tag{7-43a}$$

$$= \frac{A}{2}(1 + \cos 2\omega_1 t) \tag{7-43b}$$

The second term in (7-43b) is easily removed by the low-pass filter so that the remaining term at the output is

$$x_d(t) = \frac{A}{2} \tag{7-44}$$

which is simply a positive value indicating a 1. A similar development for the lower path would produce a negative constant in the output for the case when a 0 is transmitted.

A basic question concerns the level of *crosstalk,* that is, the effect produced in one channel when the signal corresponding to the other channel is transmitted. Assuming again that a 1 is transmitted, the output $y_2(t)$ of the lower channel is

$$y_2(t) = A \cos \omega_1 t \cos \omega_0 t \tag{7-45a}$$

$$= \frac{A}{2} \cos(\omega_1 - \omega_0)t + \frac{A}{2} \cos(\omega_1 + \omega_0)t \tag{7-45b}$$

The second term in (7-45b) is well above the passband of the low-pass filter and is readily removed. However, the first term has a frequency $f_1 - f_0$, which may be in the range of the low-pass filter. The problem can be circumvented by designing the low-pass filter to have a "notch," or at least very high attenuation, at the frequency $f_1 - f_0$. Thus, the crosstalk level can be managed by proper design of the signal-processing circuitry. The resulting detected signal $x_d(t)$ can then be applied to a circuit for proper binary restoration.

A PSK signal can be generated by the process shown in Fig. 7-15. Note that this is almost the same process used for generating ASK. However, one essential difference is that the PCM data stream has a *bipolar* form; that is, a 1 is represented by a voltage of one polarity and a 0 by a voltage of the opposite polarity. Thus, the same circuitry used for generating an ASK signal

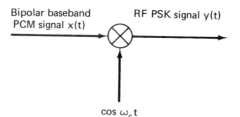

Figure 7-15. Generation of a PSK signal using a balanced modulator.

can be used for a PSK signal, but the level of the PCM modulating data stream is different for the two cases.

In the case of PSK then, the modulating signal $x(t)$ assumes one of two forms:

$$x(t) = +X \quad \text{or} \quad -X \tag{7-46}$$

The resulting modulated signal $y(t)$ can be expressed as

$$y(t) = x(t) \cos \omega_c t \tag{7-47}$$

which is, of course, a DSB signal again. Alternatively, $y(t)$ can be expressed as

$$y(t) = X \cos (\omega_c t + \theta) \tag{7-48}$$

where $\theta = 0°$ or $180°$.

While ASK can be detected either noncoherently or coherently, PSK can be detected only by a *coherent* process. (The reader should refer back to Fig. 7-10d to see why envelope detection will work for ASK, but not for PSK.) One particular process used for detecting PSK actually derives a phase coherent reference from the composite PCM signal, and this process will now be explored. Consider the system shown in Fig. 7-16, and assume that the received signal $y_r(t)$ is of the form

$$y_r(t) = Ax(t) \cos \omega_c t \tag{7-49}$$

The signal is first squared by a nonlinear circuit yielding

$$y_1(t) = A^2 x^2(t) \cos^2 \omega_c t \tag{7-50a}$$

$$= \frac{A^2}{2} x^2(t) (1 + \cos 2\omega_c t) \tag{7-50b}$$

The original modulating function $x(t)$ assumed both positive and negative values. However, the function $x^2(t) = X^2$ irrespective of whether $x(t) = +X$ or $-X$. This means that the second term in (7-50b) in the ideal case is simply

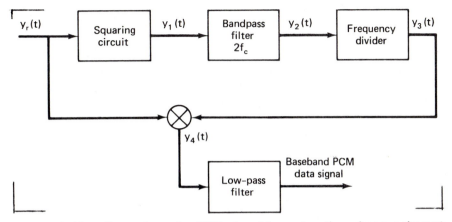

Figure 7-16. Detection of a PSK signal by extracting phase coherent reference from signal.

$(A^2 X^2/2) \cos 2\omega_c t$. Practically, there is some distortion and noise is present, so a band-pass filter (or phase-locked loop) having a very narrow passband and center frequency $2f_c$ is used to separate this component. The output y_2 of the narrow-band filter is then

$$y_2(t) = K_2 \cos 2\omega_c t \tag{7-51}$$

The frequency of this signal is twice the required frequency, so the frequency is divided by 2 to yield a signal

$$y_3(t) = K_3 \cos \omega_c t \tag{7-52}$$

The quantity y_3 is phase coherent with the reference carrier and is multiplied by the incoming signal to yield

$$\begin{aligned} y_4(t) &= y_3(t)y_r(t) \\ &= AK_3 x(t) \cos^2 \omega_c t \\ &= \frac{AK_3 x(t)}{2}(1 + \cos 2\omega_c t) \end{aligned} \tag{7-53}$$

The second term in the expansion of (7-53) is a DSB signal centered at $2f_c$. This component is readily removed by a low-pass filter, and the detected output is

$$x_d(t) = \frac{AK_3 x(t)}{2} = Kx(t) \tag{7-54}$$

where $K = AK_3/2$. The proper data signal is thus recovered.

7-7 DIFFERENTIAL PHASE-SHIFT KEYING

A special form of PSK offering certain advantages is *differentially-coherent phase-shift keying (DPSK)*. This signal form is obtained by first converting the basic NRZ-L form of the signal to either NRZ-M or NRZ-S before phase modulating the RF carrier. We will illustrate the process using an NRZ-S format and leave as some exercises for the reader to verify similar results for the NRZ-M format.

Consider the system shown in Fig. 7-17. The signal $x_1(t)$ is the input data stream, which is assumed to be encoded in NRZ-L format. The circuit preceding the balanced modulator consists of an EXCLUSIVE-NOR circuit and a 1-bit shift register (or delay element). Let τ represent the width of the bit interval. Observe that the output of the EXCLUSIVE-NOR circuit is delayed by 1 bit interval and applied as one of the inputs.

The truth table for the EXCLUSIVE-NOR circuit is shown below the circuit. Observe that the output is 1 when both inputs are the same, but it is 0 when the inputs are different. For illustration, assume the data stream $x_1(t)$ shown beneath the figure. We will arbitrarily assume a 1 for the output $x(t)$ of the EXCLUSIVE-NOR circuit 1-bit interval τ before the first data bit appeared. (The reader can verify in Prob. 7-16 that it does not make any difference as far as the final detected value is concerned.)

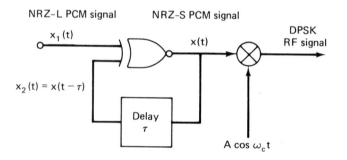

A	B	A⊙B
0	0	1
0	1	0
1	0	0
1	1	1

$x_1(t)$		1	0	1	1	0	1	0	0	1
$x(t)$	1	1	0	0	0	1	1	0	1	1
$x_2(t)$		1	1	0	0	0	1	1	0	1
Phase of $y(t)$	0°	0°	180°	180°	180°	0°	0°	180°	0°	0°

Figure 7-17. System used to generate differentially coherent phase-shift keying signal.

The initial 1 at the output is delayed 1 bit interval and appears at the input as $x(t - \tau)$ when the first data bit appears. It is compared with the first data bit, which is also a 1, and the resulting output is also 1. This value is delayed by τ and compared with the next input bit, which is 0, and this produces a 0 for that output. The tabulated values appearing below the figure should permit the reader to follow the pattern.

By carefully observing the relationship between the input data stream $x_1(t)$ and the modified data stream $x(t)$, it can be deduced that $x(t)$ represents an NRZ-S representation for the signal. (Refer back to Fig. 7-6 if necessary.) Thus, this circuit represents one way for converting NRZ-L data to NRZ-S data.

The NRZ-S signal can now be applied to the balanced modulator, and the resulting output is a form of a PSK signal. However, since the signal was differentially encoded, the more appropriate label DPSK is used. The last row in the tabulated data of Fig. 7-17 represents the relative phase of the reference sinusoidal carrier, which is assumed to be either 0° or 180°.

The process for generating the DPSK signal is interesting, but it offers no insight as to why such a seemingly complex encoding strategy is used. The answer lies in the detection process, which will now be discussed. Consider the block diagram shown in Fig. 7-18, and assume an ideal received signal $y_r(t)$ with no noise present. The received signal is delayed by the bit duration τ and applied as one input to an ideal multiplier, and the other input is the undelayed signal. The output $y_1(t)$ of the multiplier is given by

$$y_1(t) = [Ax(t) \cos \omega_c t] \cdot [Ax(t - \tau) \cos \omega_c(t - \tau)] \qquad (7\text{-}55a)$$
$$= A^2 x(t)x(t - \tau) \cos \omega_c t \cos \omega_c(t - \tau) \qquad (7\text{-}55b)$$

By using a standard trigonometric identity, (7-55b) can be expanded as

$$y_1(t) = \frac{A^2}{2} x(t)x(t - \tau) \cos \omega_c \tau + \frac{A^2}{2} x(t)x(t - \tau) \cos (2\omega_c t - \omega_c \tau) \qquad (7\text{-}56)$$

The second term in (7-56) is a DSB signal centered at $2f_c$ and is easily eliminated with a low-pass filter, assuming, of course, that the carrier frequency range is much higher than the highest frequency in the baseband spectrum.

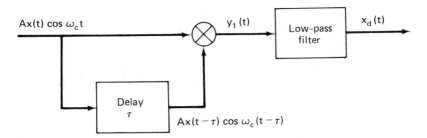

Figure 7-18. Detection system for DPSK signals.

The first term in (7-56) contains the constant factor $\cos \omega_c \tau$. Since the signal output should be as large as possible, an optimum design would call for $\omega_c \tau = 2\pi n$, where n is an integer. This result yields

$$2\pi f_c \tau = 2\pi n \tag{7-57a}$$

$$f_c = \frac{n}{\tau} \tag{7-57b}$$

or

$$\tau = \frac{n}{f_c} = nT_c \tag{7-57c}$$

where $T_c = 1/f_c$ is the carrier period. Stated in words, the bit interval τ should be selected so that it contains an *integer* number of cycles of the RF carrier. With this assumption, the detected output is

$$x_d(t) = \frac{A^2}{2} x(t)x(t - \tau) \tag{7-58}$$

At this point, we are still a little puzzled about the result since (7-58) does not look proper. However, let's see what this result means. Ignoring the $A^2/2$ constant, the product $x(t)x(t - \tau)$ is shown in Fig. 7-19. When the result is compared with the original data stream of Fig. 7-19a, we see that the product $x(t)x(t - \tau) = x_1(t)$ in bipolar form, a most fascinating result! We now see the primary advantage of the DPSK encoding process. The RF signal acts as its own synchronous reference for detected purposes, provided only that the reference be delayed by 1 bit interval.

The error rate with DPSK when noise is present is slightly higher than for conventional PSK under the most optimum conditions. It turns out that errors tend to occur in pairs of two, as will be illustrated in Fig. 7-20. The waveform in Fig. 7-20a is identical to the NRZ-S representation in Fig. 7-19 except for

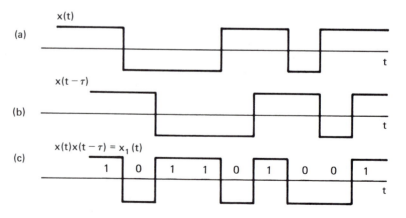

Figure 7-19. Waveforms pertaining to discussion of DPSK detection scheme.

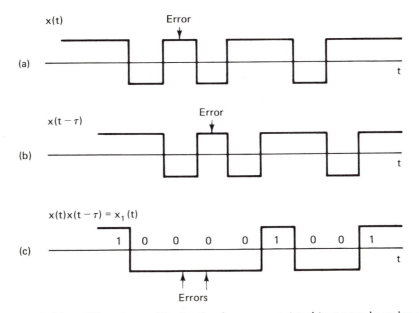

Figure 7-20. Waveforms illustrating how errors tend to occur in pairs of two when detecting DPSK signals.

the one bit error as shown. When the signal is delayed as shown in Fig. 7-19b, the same bit is obviously still incorrect. The resulting product shown in Fig.7-20c contains two bit errors. In spite of the increase in the number of errors, DPSK offers significant advantages in implementation and ease of detection, so it is a popular technique in PCM systems.

7-8 QUADRIPHASE SHIFT KEYING

All the PCM systems considered up to this point have utilized the binary number system for encoding; that is, only two levels are transmitted. However, it is possible to utilize more levels in the encoding process, and a number of systems have been developed with "m-ary" encoding, a term used to refer to the general process.

The primary advantage of utilizing more levels is that the true information rate can be increased significantly for a given bandwidth over the rate permitted for simple binary encoding. The disadvantage is that the receiver must be capable of resolving more than just two levels, so the detection process in general is more involved for m-ary systems. Indeed, the receiver complexity increases with the number of levels used.

Because of the relationship between m-ary level encoding and binary encoding, particularly in regard to implementation, most m-ary systems use a number of levels that can be expressed as an integer power of 2 (i.e., 4 levels, 8 levels, etc.). The most common is that of 4-level PCM, and our consideration will be limited to a discussion of that type of encoding.

The form of PCM encoding utilizing four levels is called *quaternary* encoding. The most common way in which quaternary encoding is converted to RF for transmission is through phase-shift keying, and the resulting process is called *quadriphase shift keying (QPSK)*. This section will be devoted to a discussion of the basic process of QPSK, and the results should give the reader an insight into the approach used for *m*-ary systems.

A QPSK signal is required to have four distinct states. This can be achieved by utilizing four separate reference phases for the transmitted signal. Refer to the relative phase diagram of Fig. 7-21. The four phase shifts and the corresponding cosine and sine references are shown on the diagram. The individual phase shifts differ by 90°; that is, they are all either in phase quadrature or 180° out of phase with each other.

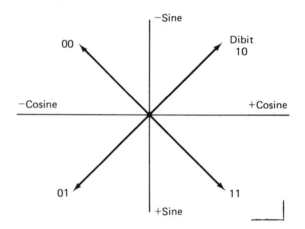

Figure 7-21. Relative phase sequence of four states (dibits) used in QPSK.

Each of the four possible states is referred to as a *dibit*. Since 2 bits would be required to represent four possible levels, each dibit represents the equivalent of 2 bits. Observe the 2-bit binary representations for the 4 dibits. These representations are meaningful in establishing a relationship between a binary representation and the corresponding dibit representation.

A means by which the RF signal can be generated is shown in Fig. 7-22. The input signal is a binary PCM representation of the data signal (e.g., output of an A/D converter). The signal is first applied to a 2-bit serial-to-parallel converter. Let τ represent the bit duration of the input binary signal. After parallel conversion of the first 2 bits, the first bit is applied to the upper path, and the second bit is applied to the lower path. The duration of these 2 bits can be 2τ since there is no need to change until two more binary bits have been received.

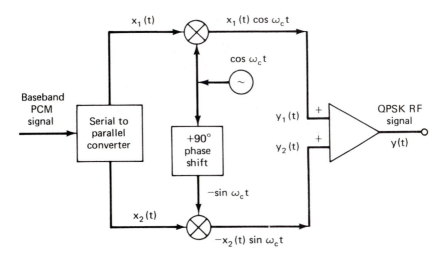

Figure 7-22. System used to generate QPSK signal.

The process of dividing $x(t)$ into two slower signals $x_1(t)$ and $x_2(t)$ is illustrated in Fig. 7-23. The *delay time* required to perform the serial-to-parallel conversion is not shown on the figure, so $x_1(t)$ and $x_2(t)$ would, in general, be delayed with respect to $x(t)$. The decimal numbers shown on the figure are used to label the bits for clarity. Observe that $x_1(t)$ assumes the

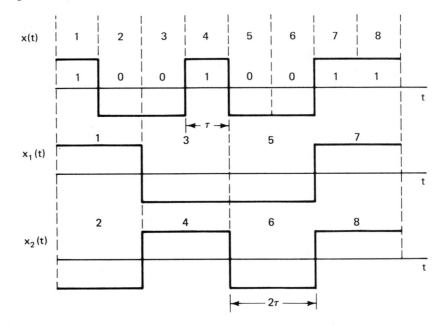

Figure 7-23. Waveforms illustrating channeling of successive bits in QPSK.

pattern of the odd bits, while $x_2(t)$ assumes the even bits. However, $x_1(t)$ and $x_2(t)$ each have bit periods of 2τ, so the transmission bandwidths of these signals should be half the bandwidth that would have been required if $x(t)$ had been transmitted directly.

Refer now back to Fig. 7-22. The signal $x_1(t)$ is multiplied by $\cos \omega_c t$, while the signal $x_2(t)$ is multiplied by $-\sin \omega_c t$. The results are denoted by $y_1(t)$ and $y_2(t)$, and they are added together to form the composite transmitted signal $y(t)$. Thus,

$$y(t) = y_1(t) + y_2(t) \tag{7-59a}$$
$$= x_1(t) \cos \omega_c t - x_2(t) \sin \omega_c t \tag{7-59b}$$

Both $y_1(t)$ and $y_2(t)$ are DSB signals centered at f_c, so their spectra overlap. Why then do they not interfere with each other? Stated differently, how can $y_1(t)$ and $y_2(t)$ be separated at the receiver when the frequency components occupy the same range? The answer lies in the fact that the two signals are in *phase quadrature;* that is, all spectral components of $y_2(t)$ are 90° out of phase with those of $y_1(t)$, and they can be separated by coherent detection, as we will see shortly.

The detection process for quadriphase is shown in Fig. 7-24. The signal is simultaneously applied to two channels. Assume that the form of the received signal $y_r(t)$ is

$$y_r(t) = Ax_1(t) \cos \omega_c t - Ax_2(t) \sin \omega_c t \tag{7-60}$$

A phase coherent reference $\cos \omega_c t$ is obtained for the lower channel. Let $y_1(t)$ and $y_2(t)$ represent the outputs of the upper and lower synchronous detectors, respectively. We have

$$y_1(t) = Ax_1(t) \cos^2 \omega_c t - Ax_2(t) \sin \omega_c t \cos \omega_c t \tag{7-61a}$$

$$= \frac{Ax_1(t)}{2} + \frac{Ax_1(t)}{2} \cos 2\omega_c t - \frac{Ax_2(t)}{2} \sin 2\omega_c t \tag{7-61b}$$

$$y_2(t) = -Ax_1(t) \cos \omega_c t \sin \omega_c t + Ax_2(t) \sin^2 \omega_c t \tag{7-62a}$$

$$= -\frac{Ax_1(t)}{2} \sin 2\omega_c t + \frac{Ax_2(t)}{2} - \frac{Ax_2(t)}{2} \cos 2\omega_c t \tag{7-62b}$$

Each of the preceding results contains two terms centered at $2f_c$ and one low-frequency data term. After appropriate low-pass filtering, we have for the data outputs $x_{d1}(t)$ and $x_{d2}(t)$

$$x_{d1}(t) = \frac{Ax_1(t)}{2} \tag{7-63}$$

$$x_{d2}(t) = \frac{Ax_2(t)}{2} \tag{7-64}$$

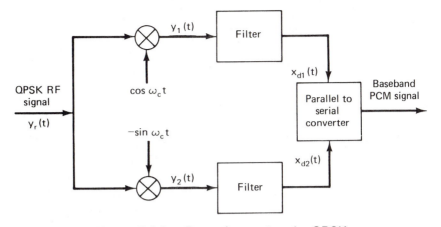

Figure 7-24. Detection system for QPSK.

Thus, the two separate individual data streams are obtained. By means of a parallel-to-serial conversion, the original data stream $x(t)$ is obtained.

7-9 DATA COMMUNICATIONS

The emphasis thus far in digital communications has been directed toward systems in which analog signals were first sampled and then converted into binary words whose values represented in some sense the levels of the corresponding analog samples. There is still another broad area of digital communications in which the individual words directly represent characters in a data-processing system. This includes data transmission between two separate computers, between a computer and a remote terminal, between a business terminal and a central accounting system, between a teletype unit and a video display unit, and many other possibilities. The key factor here is that both the input and output data employ digital codes corresponding to characters (e.g., letters of the alphabet and numbers), rather than samples of an analog signal.

While specific terminology to distinguish this form of digital communications from others has not been well defined, the term *data communications* seems to be about as good as any, so that label will be used in this book. Historically, some forms of data communications were in usage even before the advent of radio. The Morse code, in which letters and numbers are encoded by combinations of short pulses (dots) and long pulses (dashes), dates back to the nineteenth century. In this system, the accuracy of transmission and detection is greatly dependent on the skill of the human operators in generating and detecting the code words. Teletype transmission has also been in existence for many years and has been widely used in information systems.

A tremendous impetus to the field of data transmission has been generated by the digital computer field. As more and more computers have become associated with so many different aspects of modern life, there has been a subsequent increase in the utilization of data-transmission requirements. Furthermore, the advent of the microprocessor has resulted in so much potential for future applications, that none of us can imagine all the possibilities. Hence, the importance of data communications cannot be overemphasized because in a vast number of these potential applications, digital data will have to be shuffled back and forth.

A basic consideration in all digital data communications systems is that each of the possible characters must have a unique code word. A means must be provided at the data source for converting each possible character to its associated code word, and a similar means must be provided at the destination for converting the code word back to its proper character. Finally, some means must be provided for keeping the proper bits in synchronization so that the proper beginning and ends of words can be recognized.

Data transmission may be characterized as either *synchronous* or *asynchronous*. *Synchronous* transmission is achieved by means of a master clock or timed reference, which is either transmitted with the signal or available at both ends through some other means. *Asynchronous* (also called *start-stop*) data-transmission systems have the capability of initiating or terminating transmissions in a much more flexible manner. This is achieved by provided start and stop information with each data word, so it is essentially self-clocking. The advantage of asynchronous transmission is obvious for such applications as computer terminals and accounting systems. The disadvantage is that the actual data information rate is slower because of the necessity of providing self-clocking data with each digital word.

The most widely used encoding process in modern digital data systems is the *American Standard Code for Information Interchange,* hereafter referred to as the *ASCII* code. The ASCII code is basically a 7-bit code, but other bits are usually present in each word for reasons that will be discussed shortly. With 7 bits, the number of possible distinct words is $2^7 = 128$. This value is sufficient to encode all required letters (uppercase and lowercase), numerals, punctuation marks, and a number of special characters used in data transmission. In most applications, an additional parity check bit is added as a means of detecting a bit error. If odd parity is employed, the value of the parity bit is chosen to make the total numberof data 1's to be odd.

For asynchronous transmission, it is necessary to add additional start and stop bits. A start bit is the same length as each of the data bits. There is no limit to the maximum length of a stop bit, but the minimum length depends on the data rate and typically ranges from 1- to 2-bit intervals.

A complete summary of the ASCII code is provided in Appendix D. The 128 possible code words, the corresponding characters they represent, and the definitions are listed in this appendix. Many of the special characters employ terminology and concepts peculiar to data-processing procedures and will not be discussed here. Bit number 1 is normally transmitted first and is considered as the LSB, and bit number 7 is transmitted in the seventh position and is considered as the MSB. However, the terms LSB and MSB do not have the implications associated with numerical values, so these terms serve mainly to define the relative positions in this case. The numbers ranging from 0 to 7 in decimal along the top and 0 to 15 along the left represent the hexadecimal equivalent values of the binary words.

The manner in which a given character is encoded will now be illustrated in Fig. 7-25 for the uppercase letter W. The case of asynchronous transmission will be assumed. Prior to the beginning of the character, the signal is at the *mark* or 1 level. (The actual voltage or current associated with this level varies with the type of system.) The beginning of a new character is indicated by the *start* bit. The signal will always be at the mark level before the beginning of a new word, and the transition to the *space* level signals the beginning of a data word. Following the start bit the 7 bits representing the character W are now transmitted, with the LSB transmitted first and the MSB last. From Appendix D, the 7 bits arranged in the order of transmission from left to right are 1110101, but this corresponds to the usual form (MSB on left) for expressing the binary number of 1010111 as noted. This value could also be expressed as 57 in hexadecimal.

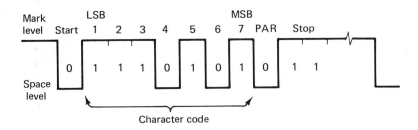

Figure 7-25. Asynchronous encoding of the letter W in ASCII form.

After the completion of the 7 data bits, an odd parity check bit is inserted, which in this particular example is a 0, since there was already an odd number of 1's (5). A *stop* signal is now transmitted. The stop signal is always at the mark level. As previously mentioned, it must have a minimum length, which in this example is assumed to be 2 bit intervals. The signal then remains at the mark level until the next start bit is received. This could represent any arbitrary length of time provided that the minimum stop element length has been satisfied. The reader can readily see that such asynchronous data transmission is appropriate for start-stop situations where the time between the generation of different characters could vary considerably (e.g., an individual operating a teletype terminal).

A number of older encoding processes used primarily in teletype communications employed 5 data bits. Some of these codes are still in use today, although the trend is toward ASCII compatibility in a large segment of the industry. Most of the 5-bit codes do not employ parity check bits, although start and stop elements are required for asynchronous systems. With 5 data bits, it is possible to represent only $2^5 = 32$ characters directly, which at first glance would seem to be inadequate since the number of required letters (26) plus the nine basic numbers totals 35. The situation is resolved by assigning a dual purpose to most of the characters in the same manner as a typewriter. All the 26 letters plus a few additional characters represent one of the two possible assignments, while all the numbers and additional characters constitute the other possible assignment (e.g., "carriage shift"). A special carriage-shift character precedes a data stream to indicate that, until the complementary character is received, all subsequent data will belong to one particular set. Since streams of letters and streams of numbers tend to be clustered separately in most messages, there are usually reasonable lengths of transmission in one mode before a change is required. In an extreme, but unlikely, situation in which every other character were a letter and the other were a number, the data rate would be slowed down considerably by the necessity to insert the carriage-shift character between every two data characters.

Two of the 5-bit codes are the Bell System *Baudot* code and the Western Union teletype code (several variations). There are also some codes that utilize 6 data bits (e.g., New York Stock Exchange code).

In describing the various data codes, it is convenient to define the *unit length* of the code. The *unit length* will be defined as the number of intervals, based on the minimum length of a given bit, that is required to completely

Table 7-3. Formats for Several Representative Codes Used in Data Communications

Intervals	Bell System Baudot TTY	Western Union Telex	10 Characters per Second ASCII (110 baud)	30 Characters per Second ASCII (300 baud)
Start bits	1	1	1	1
Data bits	5	5	7	7
Parity bits	None	None	1	1
Stop intervals	1.42	1.5	2	1
Total unit length	7.42	7.5	11	10

Each value refers to the number of minimum-width intervals used, and the unit length is the total number of these intervals per character.

represent a given character. This includes the data bits, the parity check bit (when present), the start bit, and the minimum length of the stop signal (in terms of the minimum number of bit intervals). For example, consider the form of ASCII indicated in Fig. 7-25. There are 7 data bits, 1 parity check bit, 1 start bit, and it was assumed that the minimum number of stop bits was 2. Thus, the unit length is $7 + 1 + 1 + 2 = 11$ units.

The unit lengths of several codes are compared in Table 7-3. The ASCII code transmitted at 10 characters per second corresponds to the example just considered, while the corresponding code for 30 characters per second has a unit length of 10 units (only 1 stop bit). As a point of interest, note that the stop signal for the Bell System Baudot code has a length of 1.42 units. This means that the stop signal is required to be 1.42 times the length of the other bits. This requirement was established to provide an adequate time for the mechanism in earlier teletype units to establish realignment following the completion of a character.

7-10 INFORMATION RATE AND SIGNALING RATE

Anyone who looks very closely at digital communications systems, particularly those used primarily for data transmission such as discussed in Sec. 7-9, will soon very likely encounter the terms "baud rate" and "bit rate." Based on this writer's personal experience, it is also very likely that these terms will be confusing, since they are used somewhat casually and are probably not fully understood by the majority of the people who use them. The problem is complicated by the fact that in many cases the baud rate and the bit rate are equal, and someone specifies one term when the other term is actually meant for the particular context.

Before commenting on the particular usage of the terms, we will back up slightly and consider some more basic and fundamental concepts. The first concept is that of *signaling speed,* and the second is *information rate.* Each will be discussed individually, and we will then relate the terms "baud rate" and "information rate" to these more general concepts.

The *signaling speed* is a measure of how fast individual elements could be (but not necessarily are) transmitted through the system. Specifically, it is defined mathematically as the reciprocal of the shortest element (in seconds) in the encoding scheme. Thus, in a system in which the duration of the shortest bit is 1 ms, the signaling rate would be 1000 elements per second. However, instead of saying "elements per second," the unit "baud" has come into usage and the corresponding signaling speed is referred to as the *baud rate.* Thus, if τ represents the shortest element in the code, the baud rate is

$$\text{baud rate} = \frac{1}{\tau} \qquad (7\text{-}65)$$

Note that in this definition nothing has been said about binary encoding. Indeed, this definition applies irrespective of the number of levels. Neither is continuous transmission required since the baud rate is based on the *shortest* element only. The baud rate for a given character length and structure would not be affected by whether the transmission is asynchronous or synchronous. Thus, baud rate indicates in a sense what a system could do in transmission rather than what it actually does. The baud rate also provides a direct means for predicting the bandwidth, since the required transmission bandwidth B_T is directly proportional to the reciprocal of the shortest element. Thus,

$$B_T \propto \text{baud rate} \qquad (7\text{-}66)$$

Next we will consider the concept of *information rate*. The theoretical basis of information is associated with the probability of occurrence, as was discussed in Chapter 1. Recall that, if an event were less likely to occur, more information was associated with a knowledge that the event had occurred. In this sense, then, the true information rate in a system is a function of the probability of occurrence of the different message outcomes.

In a digital communications system, only a finite number of possible words or characters can be transmitted. The occurrence of each word means something at the receiver whether it represents the level of a given sample of the signal or whether it represents an ASCII character in a data display system. From the information-theory point of view, the information is a function of the relative frequency of occurrence of the system.

Consider two hypothetical digital communications systems designated as *A* and *B*. In *A*, all the possible characters occur at about the same frequency or probability. In *B*, some of the characters occur at very widely separated intervals. Which of the two possible systems conveys the most information from an information-theory point of view? Obviously, the "rare" characters in *B* provide greater information peaks when they occur. However, in *A* the information rate is more uniform and consistent. It can be shown that system *A* provides a greater average information rate than *B*. Specifically, the average information rate in a digital communications system is achieved when all the possible characters have equal probabilities of occurrence.

At this point, the probabilistic point of view will be dropped in the discussion, and we will consider the concept of *maximum information rate*. The maximum information rate will be defined as the number of equivalent binary digits transmitted per second and is measured in bits per second (bits/s). The adjective "maximum" keeps us from violating the rules of information theory when the characters do not necessarily occur with equal probability. Thus, the maximum information would equal the true information rate only if the characters occurred with equal probability.

The term *equivalent binary digit* is one that needs some clarification. First, for binary systems, it is simply the actual number of bits, so its importance can be easily accepted. The problem becomes a bit more perplexing (a pun) when more than two levels are possible. In this case the number of equivalent binary digits corresponding to one character in the particular encoding scheme is the number of bits that would be required to encode the actual number of levels of the character in a binary representation. If the character has m possible levels, the number of binary digits required to encode the character is $\log_2 m$. Thus, if R characters per second are transmitted, the bit rate in bits per second is given by

$$\text{bit rate} = R \log_2 m \qquad (7\text{-}67)$$

where in most practical cases $\log_2 m$ is an integer number. For example, consider a quadraphase system in which 1000 dibits/s are transmitted. The bit rate is $1000 \times \log_2 4 = 1000 \times 2 = 2000$ bits/s. This number tells us that it would take 2000 bits/s to transmit the same amount of information in a conventional binary system. Observe that nothing was said about how long or how short any of the characters are; instead, the bit rate is determined by the actual number of events transmitted.

When are the baud rate and the bit rate equal? As a general rule, the two quantities are essentially the same under the following conditions: (1) Binary encoding is employed. (2) All elements used to encode characters are equal in width. (3) Synchronous transmission at a constant rate is employed. One can find isolated cases where the numerical values are the same even when these conditions are not met, but these conditions would indicate general equality. It should be indicated that parity bits are included as information bits, and start and stop signals are each weighted 1 bit. However, even if the stop signal occupies two regular pulse intervals, it is only counted at 1 bit since it indicates only one item of information.

Example 7-5

Consider the ASCII character in Fig. 7-25 with 2 stop bits assumed for each character. This particular format is used in certain systems designated as *110 baud*. Determine (a) the width of each bit interval, (b) the maximum number of characters that can be transmitted per second, and (c) the maximum information rate at maximum signaling speed.

Solution

(a) The baud rate is the reciprocal of the width τ of the shortest element, so

$$\tau = \frac{1}{\text{baud rate}} = \frac{1}{110} = 9.09 \text{ ms} \qquad (7\text{-}68)$$

(b) A given word contains 7 data bits, 1 start bit, 1 parity check bit, and 2 stop bits. The length is then $11 \times 9.09 = 100$ ms. The number of characters (words) that can be transmitted in 1 s is $1/100$ ms $= 10$ characters/s.

(c) At maximum signaling speed, the only difference between the baud rate and the information rate in this case is the fact that the 2 stop bits are counted only as 1 information bit. Thus, the number of information bits per character is 10, and since 10 characters are transmitted per second,

$$\text{information rate} = 100 \text{ bits/s} \qquad (7\text{-}69)$$

PROBLEMS

7-1   A certain 12-bit A/D converter with a full-scale voltage of 5 V is to be employed in a binary PCM system. The input analog signal is adjusted to cover the range from zero to slightly under 5 V, and the converter is connected for unipolar encoding. Rounding is employed in the quantization scheme. Determine the following quantities: (a) normalized step size, (b) actual step size in volts, (c) normalized maximum quantized analog level, (d) actual maximum quantized level in volts, (e) normalized resolution, (f) actual resolution in volts, and (g) percentage resolution.

7-2   The 12-bit A/D converter of Prob. 7-1 is to be changed so that it can be used with an analog signal having a range from -5 V to just under 5 V. Bipolar offset encoding is to be used. Repeat all the calculations of Prob. 7-1.

7-3   The level of a certain unipolar analog signal is adjusted to cover the range from zero to just under 10 V. It is to be converted to PCM with an A/D converter covering the range from 0 to 10 V. Determine the minimum number of bits required if the resolution must be within ± 5 mV.

7-4   Repeat Prob. 7-3 if the analog signal is bipolar and covers the range from -10 V to just under 10 V, and the A/D converter is connected for -10-V to 10-V operation.

7-5   For the unipolar signal of Prob. 7-3, assume that a new specification is given such that the resolution must be within $\pm 0.0125\%$. Determine the minimum number of bits.

7-6   Repeat Prob. 7-5 if the analog signal is bipolar and covers the range from -10 V to just under 10 V, and the A/D converter is connected for -10-V to 10-V operation. The percentage is still based on the peak value of 10 V.

7-7   In Prob. 7-6, suppose that the resolution of $\pm 0.0125\%$ were based on the percentage of the peak-to-peak voltage of 20 V. Determine the minimum number of bits.

7-8   The level appearing at the output of a certain transducer varies from 0 to just under 200 V. The signal is to be converted to PCM for telemetry transmission. This level exceeds that of most commercial A/D converters so that it is necessary to attenuate the signal before conversion and amplify it again at the receiver to reestablish the proper level. Accuracy of reproduction must be within ± 0.5 V. Determine the minimum number of bits required.

7-9 Design a circuit using one or more basic logic gates to convert an NRZ-L signal to an RZ signal. Assume standard TTL levels, and assume that a synchronous square wave having twice the data bit rate frequency is available. Show waveforms to verify the result.

7-10 Design a circuit using one or more basic logic gates to convert an NRZ-L signal to a biphase-L signal. Assume standard TTL levels, and assume that a synchronous square wave having twice the data bit rate frequency is available. Show waveforms to verify the result.

7-11 Consider the bit stream 01001110. Sketch the forms for each of the following data formats: (a) NRZ-L, (b) NRZ-M, (c) NRZ-S, (d) RZ, (e) biphase-L, (f) biphase-M, (g) biphase-S. On encoding schemes in which the initial level is arbitrary, assume a zero level just before the first bit is received.

7-12 Consider a PCM system in which 24 signals are to be processed. Each signal has a baseband bandwidth $W = 3$ kHz. The sampling rate is to be 33.3% higher than the theoretical minimum, and 8 bits are to be used in each word. Conventional NRZ-L encoding will be used, and an extra bit is added in each frame for sync.

    (a) Determine the bit rate.

    (b) Using the $0.5/\tau$ rule, determine the approximate minimum baseband bandwidth.

7-13 A certain TDM 12-bit PCM system must be designed to process six data channels. Channels 1 to 4 each have a baseband bandwidth $W = 1$ kHz, while channels 5 and 6 each have $W = 2$ kHz. To provide guard band, a sampling rate 25% above the theoretical minimum is to be employed.

    (a) Using only one commutator, devise a scheme whereby each of the channels is sampled at the proper rate uniformly.

    (b) Using the $0.5/\tau$ rule, determine the approximate minimum baseband bandwidth required.

7-14 Consider the complex sampling scheme illustrated in Fig. 7-9. Assume that the prime frame sampling rate is 2 kHz; that is, the prime commutator makes one "rotation" in 0.5 ms.

    (a) How many separate signals can be processed with the system as it is connected?

    (b) Assuming sampling at the Nyquist rate with no guard band provided, determine the maximum baseband data frequency for each of the available channels.

7-15 In this problem, you will make a brief study of PCM peak quantization error as a function of the required transmission bandwidth. The result should vividly display the exponential trade-off between improved accuracy (lower noise) and transmission bandwidth. Consider a PCM system with n bits and assume (1) minimum Nyquist rate, (2) no spacing between pulses, and (3) $0.5/\tau$ bandwidth rule. The peak quantization error will be defined as the magnitude of the percentage resolution for this purpose, and unipolar encoding will be assumed. Compute enough data to plot a curve of the percent of *peak quantization error* as a function of the ratio B_T/W, where B_T is the approximate baseband bandwidth. Use *semilog* paper with error on the log scale. While your result will be computed for integer values of n, extrapolate between points so that the nature of the relationship can be mostly clearly seen.

7-16 Consider the bit stream used in explaining the DPSK NRZ-S PCM signal in Figs. 7-17 and 7-19.

> (a) Develop a new table of entries for the case where the initial bit for $x(t)$ is assumed to be a 0.
>
> (b) By repeating the development of Fig. 7-19, verify that the original data stream is recovered.

7-17 Consider the bit stream used in the NRZ-S system of Figs. 7-17 and 7-19. Replace the EXCLUSIVE-NOR circuit by an EXCLUSIVE-OR circuit, and show that the resulting signal is an NRZ-M signal. Assume an initial 0 for $x(t)$.

7-18 For the NRZ-M system of Prob. 7-17, show that the detection scheme of Fig. 7-19 can be used to recover the original data stream provided that one extra operation is performed on the detected data.

7-19 Consider a data communications system employing the Bell System Baudot code for teletype communications. The character format for this code is listed in Table 7-3. Assume that the minimum width of a given interval in a character is $\tau = 22$ ms. Determine the (a) signaling speed (baud rate), (b) maximum number of characters transmitted per second, and (c) maximum information rate at maximum signaling speed.

7-20 The Baudot code under the condition of Prob. 7-19 is often said to have an approximate speed of "60 words per minute." If an average word is considered to be composed of five characters and there is a space equal to one character between successive words, justify the appropriateness of the description.

Introduction to Statistical Methods

8

8-0 INTRODUCTION

The primary objective of this chapter is to provide an introduction to those fundamentals of statistical analysis most important in communications system analysis. Statistical analysis is very useful in predicting the behavior of random signals encountered in actual systems. The properties of electrical noise, for example, can only be adequately described using statistical parameters.

The development begins with a detailed discussion of probability, which provides a basis for the description of statistical measures. The concept of the probability density function is then developed using a realistic type of example. Various statistical parameters are introduced and discussed. Major emphasis in this chapter is then directed toward the gaussian distribution. The most common forms of electrical noise can be represented by the gaussian function.

8-1 RANDOM SIGNALS

In Sec. 2-1, several classification schemes for signals were given. One particular classification involved the identification of a given signal as either a *deterministic signal* or a *random signal*. As a brief review, a deterministic signal was defined as one whose exact instantaneous behavior as a function of time was predictable, whether by an equation or by a point-by-point description (e.g., a graph). In contrast, a random signal is one whose exact behavior as a function of time cannot be predicted. An illustration of a typical random signal is shown in Fig. 8-1.

Virtually all the developments made thus far in the book have been made with deterministic signals. Such signals serve a very useful purpose in establishing operating bounds for communications systems, and their importance should not be minimized. However, most of the signals encountered in the actual operation of communications systems tend to be random in nature.

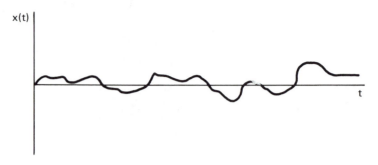

Figure 8-1.   A random signal.

Along with the uncertainty associated with a realistic desired signal, electrical noise is always present, and this noise tends to mask or obscure the signal. Noise is almost always a random function, and its instantaneous behavior can rarely be predicted. For short-range applications where plenty of power is available, one can simply overpower the noise by a "brute force" approach, which is usually the case, for example, in commercial broadcasting at short range. However, for long-range communications where power is at a premium, it is necessary to be able to predict how different communication systems behave in the presence of noise. Some of the modulation methods, for example, have the capability for enhancing the signal level with respect to the noise level. Statistical methods have also been developed for extracting signals from noisy backgrounds.

The term *random signal* will be used in this chapter to refer either to an actual communications signal of interest or to the associated electrical noise, both of which are random in nature. While the instantaneous values of random signals cannot be predicted exactly, there are many types of random signals encountered in communications systems that can be described by statistical representations. These statistical forms permit many of the important properties of the signals to be determined. For example, the mean value, the rms value, and the average power can often be determined directly from the statistical models.

To deal effectively with random signals on this basis, some of the basic concepts of probability and statistics must be understood. Most of this chapter will be concerned with the development of the concepts needed to deal with probabilistic and statistical representations for random signals.

It should be emphasized that the treatment here is intended only as an introduction to the concepts and their applications. Statistics is a complete discipline in itself, and hundreds of books have been written on the subject. Many specialists in the field have devoted their professional careers to the theory and applications of statistics. Our interest is to introduce a few important ideas as they relate to communications. Rigor will be sacrificed wherever a simpler intuitive approach is judged to be more appropriate.

8-2   PROBABILITY

Probability is associated with the very natural trend of a random event to follow a somewhat regular pattern if the process (often called the "experiment") leading to the event is repeated a sufficient number of times. For example, consider the process of flipping an unbiased coin. If the experiment is continually repeated over a large number of trials, we would expect to get about the same number of heads as tails. It is unlikely that the trend would appear in a small number of trials, but as the number of trials is continued, the expected trend would likely appear. Intuitively, we could say that the probability on a given trial of a head is 0.5 and the probability of a tail is 0.5.

Let us now describe in mathematical terms the concept of probability. Assume that a certain experiment has K possible outcomes (often called the "population") and that only one can occur as a result of performing the experiment. (The coin-flipping problem has two possible outcomes.) Let A_1 symbolically represent the condition that the first occurs; let A_2 symbolically represent the condition that the second event occurs; and so on. The probability that the first event occurs is designated as $P(A_1)$; the probability that the second event occurs is designated $P(A_2)$; and so on. Assume that a *large* number of trials n of the experiment are performed, which in the limit could be said to approach infinity. Assume that the first event occurs n_1 times, the second event n_2 times, and so on. The various probabilities are then defined as followed:

$$P(A_1) = \lim_{n \to \infty} \frac{n_1}{n} \tag{8-1}$$

$$P(A_2) = \lim_{n \to \infty} \frac{n_2}{n}$$

$$\vdots$$

$$P(A_K) = \lim_{n \to \infty} \frac{n_K}{n}$$

The question naturally arises as to how many times the experiment must be repeated before we can assume that the limits have been reached, since it is obviously impossible to repeat the experiment an infinite number of times. The question is a difficult one to answer and is a subject of fundamental importance in establishing the validity of statistical sampling. About all that will be said at this point is that a sufficiently large number of trials must be used to cause a natural trend of convergence in the values obtained from the ratios in (8-1). In general, the greater the number of independent trials in the experiment, the more confidence can be placed in the results.

Assume now that the probability values concerning the possible outcomes of a given experiment are known. Each value $P(A_k)$ is a number satisfying the condition

$$0 \le P(A_k) \le 1 \tag{8-2}$$

In general, the closer the value of $P(A_k)$ is to 1, the more likely the kth event will occur on a given trial, and the smaller the value of $P(A_k)$, the less likely the event will occur. On an average, the kth event will occur about every $1/P(A_k)$ trials. For example, if $P(A_1) = 0.1$, the first event in the population will occur about once every ten trials *on an average*.

If $P(A_k) = 1$, the kth event is certain; that is, it will always occur when the experiment is performed. Conversely, if $P(A_k) = 0$, the kth event will never occur. For example, consider a "two-headed" coin. Let H represent the event that a head occurs, and let T represent the event that a tail occurs. For this rather "loaded" case, it can be readily deduced that $P(H) = 1$ and $P(T) = 0$ for each trial.

For a population consisting of K possible outcomes, only one of which can occur for each trial of the experiment, it can be shown that

$$P(A_1) + P(A_2) + \cdots + P(A_K) = 1 \tag{8-3}$$

This equation indicates that, for every trial of the experiment, *one* (and only one) of the possible events *must* always happen.

Example 8-1

Assume that one card is to be randomly drawn from a standard 52-card deck. Using the relative frequency of occurrence as an intuitive basis, determine the probability that the card is (a) a red card, (b) a heart, (c) an ace, (d) the ace of spades, and (e) the ace of hearts or the ace of diamonds.

Solution

(a) Let R represent the condition that a red card is drawn. There are 52 cards of which 26 are red. Thus,

$$P(R) = {}^{26}\!/_{52} = \tfrac{1}{2} \tag{8-4}$$

(b) Let H represent the condition that a heart is drawn. Since there are 13 hearts,

$$P(H) = {}^{13}\!/_{52} = {}^{1}\!/_{4} \tag{8-5}$$

(c) Let A represent the condition that an ace is drawn. Since there are 4 aces,

$$P(A) = {}^{4}\!/_{52} = {}^{1}\!/_{13} \tag{8-6}$$

(d) Let AH represent the condition that the ace of hearts is drawn. There is only one ace of hearts so that

$$P(AH) = {}^{1}\!/_{52} \tag{8-7}$$

(e) The desired outcome can either be the ace of hearts or the ace of diamonds, so there are two possible favorable results. Letting E represent the condition that either desirable outcome occurs, we have

$$P(E) = {}^{2}\!/_{52} = {}^{1}\!/_{26} \tag{8-8}$$

8-3 MUTUAL EXCLUSIVENESS AND STATISTICAL INDEPENDENCE

In this section, the concepts of *mutual exclusiveness* and *statistical independence* will be treated. Although the two concepts are quite different, they are often confused with each other owing to certain similarities in their usage and form.

Recall from the last section that, when a term such as A_1 is used to represent a certain event occurring, this is done in a symbolic sense. Consequently, we can say that A_1 is "true" if the first event in a population occurs. This inference is the same as encountered in Boolean expressions. Thus, $A_1 = 1$ could represent a Boolean expression indicating that the first event has occurred. Likewise, if the first event does not occur, an appropriate Boolean expression would be $A_1 = 0$. Alternatively $\overline{A}_1 = 1$, where $\overline{A}_1$ is the complement of A ("not A"), states the same condition.

In the same sense then as combinational logic, we define a logical expression such as $A_1 + A_2$ as the condition that either A_1 *or* A_2 is true. Similarly, an expression such as A_1A_2 represents the condition that both A_1 *and* A_2 are true. These expressions need not be confused with ordinary algebraic expressions, since it is usually clear in the usage which meaning is implied.

From a probability point of view, the expression $P(A_1 + A_2)$ represents the probability that *either* A_1 *or* A_2 is true, that is, the probability that either the first or the second event has occurred. Similarly, the expression $P(A_1A_2)$

represents the probability that *both A_1 and A_2* are true, that is, the probability that both events have occurred.

We now define the concept of *mutual exclusiveness*. Two events are said to be *mutually exclusive* if

$$P(A_1 + A_2) = P(A_1) + P(A_2) \qquad (8\text{-}9)$$

Stated in words, two events are said to be mutually exclusive if the probability of *either* the first event *or* the second event occurring is the sum of the respective probabilities. Note that $A_1 + A_2$ on the left represents an "or" statement, while the sum on the right is a true algebraic sum.

The concept of mutual exclusiveness can be illustrated by Venn diagrams. In Fig. 8-2a there is no overlap between the area corresponding to event 1 and the area corresponding to event 2, so the events are mutually exclusive. However, in Fig. 8-2b there is a common overlapping area between the population corresponding to event 1 and that of event 2, so the two events are not mutually exclusive.

As examples, let A_1 represent the event that a person is in Norfolk, Virginia, at a given time and let A_2 represent the event that the same person is in New York City at the same time. These two events are mutually exclusive since a person obviously cannot be in both places simultaneously. As a second example, let A_1 represent the event that a person is in Norfolk, and let A_2 represent the event that a person is serving in the U.S. Navy. These two events are not mutually exclusive, since there are many persons in Norfolk who are also in the U.S. Navy.

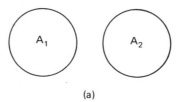

(a)

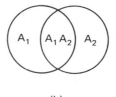

(b)

Figure 8-2. Venn diagrams for events that are (a) mutually exclusive and (b) not mutually exclusive.

If two events are not mutually exclusive, some portion of their respective populations overlap, so the probability of one or the other occurring would be weighted too heavily if the respective probabilities are simply added, as in (8-9). From the Venn diagram of Fig. 8-2b, it is seen that the area corresponding to $A_1 A_2$ appears both in the weighting for A_1 and the weighting for A_2 in the respective probabilities. Consequently, if $P(A_1)$ is added to $P(A_2)$, this common area has been added twice. To correct for this extra weighting, the term $P(A_1 A_2)$ is subtracted, and so the probability of either the first event *or* the second event occurring when the two events are not mutually exclusive is

$$P(A_1 + A_2) = P(A_1) + P(A_2) - P(A_1 A_2) \qquad (8\text{-}10)$$

Observe that this expression reduces to (8-9) when the two events are mutually exclusive since $P(A_1 A_2) = 0$ for that case.

We will now consider the concept of *statistical independence*. Two events are said to be statistically independent if

$$P(A_1 A_2) = P(A_1) P(A_2) \qquad (8\text{-}11)$$

Stated in words, two events are said to be statistically independent if the probability that *both* the first event *and* the second event occur is equal to the product of the respective probabilities. Note that $A_1 A_2$ on the left represents an "and" statement, while the product on the right is a normal algebraic multiplication. The expression $P(A_1 A_2)$ is called a *joint probability* of two events.

For events to be statistically independent there must be no cause-effect relationship between them; that is, the occurrence of either event should not in any way influence the occurrence of the other event. Some philosophers have argued that everything in the universe has some effect on everything else. If one accepts the broadness of this implication, possibly no two events could ever be truly statistically independent. However, in a practical sense, there are many events in which the assumption of statistical independence is reasonable.

As examples of statistical dependence and independence, let A_1 represent the event that the dc voltage in a transmitter power supply exceeds the maximum collector voltage rating for one of the transistors, and let A_2 represent the event that the corresponding transistor fails. These two events are not statistically independent, since the occurrence of the first event (higher power supply voltage) strongly affects the probable occurrence of the second event (transistor failure). On the other hand, if A_2 represents the event that a similar transistor fails in a completely separate transmitter operating from a different power supply, it is likely that the two events are statistically independent.

Statistical independence in the general case is not always so obvious. Indeed, there are sophisticated statistical tests that may be used to test for the degree of statistical dependency between different physical variables.

We next consider the question of evaluating a joint probability when the events are not statistically independent. This requires the use of a concept called *conditional probability*. The expression $P(A_2/A_1)$ is defined to mean "the probability that A_2 is true *given* that A_1 is true." Knowing that A_1 is true may influence the likelihood of A_2 being true so that conditional probability deals with events that may be statistically dependent. For example, the transistor in the example referred to earlier may normally have a low probability of failure under specified operating conditions. However, if we are *given* the fact that its collector supply voltage has exceeded the maximum rating, the conditional probability of failure may be quite high.

The Venn diagram of Fig. 8-2b may be used to infer a relationship for conditional probability. Assume that it is known that A_1 is true. This means that any possible outcome must be contained within the area representing A_1. Now what is the probability that A_2 is true if it is known that A_1 is true? The pertinent area is the common area for both A_1 and A_2; that is, $A_1 A_2$ must be true. The ratio of this area to the total area of A_1 provides a proper formulation for the conditional probability. Thus,

$$P(A_2/A_1) = \frac{P(A_1 A_2)}{P(A_1)} \tag{8-12}$$

The joint probability $P(A_1 A_2)$ may be determined from (8-12) as

$$P(A_1 A_2) = P(A_1) P(A_2/A_1) \tag{8-13}$$

This result can be used to determine joint probabilities when events are not statistically independent. In this case, it is necessary to first specify (1) the probability that one of the two events occurs and (2) the conditional probability that the second event occurs given that the first event has occurred.

If the two events are statistically independent, the fact that A_1 is true will in no way affect the outcome of A_2. In this case, the conditional probability of (8-12) reduces to the simple probability of A_2 being true; that is, $P(A_2/A_1) = P(A_2)$. Thus, (8-13) reduces back to the form of (8-11) for the case of statistical independence.

A few final descriptive remarks about mutual exclusiveness and statistical independence will be made. Most applications of the mutual exclusiveness concept involve probability evaluations in which it is necessary to determine if there are common elements of separate "population" groups representing different outcomes. Of particular concern is the case where one event or the other might occur in the probability evaluation. In contrast, statistical independence is primarily of interest when probabilities involving two or more events all occurring are to be evaluated and when it is desired to determine if the occurrence of any one of the events tends to influence the outcome of other events.

Example 8-2

A single card is to be drawn from a deck of cards. What is the probability that it will be *either* an ace *or* a king?

Solution

Let A represent the outcome of drawing an ace, and let K represent the outcome of drawing a king. The individual probabilities for a single draw are

$$P(A) = 4/52 = 1/13 \tag{8-14}$$

$$P(K) = 4/52 = 1/13 \tag{8-15}$$

The probability evaluation desired is $P(A + K)$. Whenever the probability of one event *or* another is desired, it is necessary to determine if the two events are mutually exclusive; that is, can a card simultaneously be an ace and a king? The answer is obviously no, so the two events are mutually exclusive. Thus,

$$P(A + K) = P(A) + P(K) = 1/13 + 1/13 = 2/13 \tag{8-16}$$

Example 8-3

A single card is to be drawn from a deck of cards. What is the probability that it will be *either* an ace *or* a red card?

Solution

Let R represent the outcome of drawing a red card. We have

$$P(R) = 26/52 = 1/2 \tag{8-17}$$

As in Ex. 8-2, the probability of drawing an ace is

$$P(A) = 1/13 \tag{8-18}$$

Again, we raise the question of mutual exclusiveness; that is, can a card simultaneously be an ace and a red card? The answer is yes in this case since there are two cards that belong to both groups. Thus, the events are not mutually exclusive in this case. The probability of a card being both an ace and a red card is

$$P(AR) = \frac{2}{52} = \frac{1}{26} \qquad (8\text{-}19)$$

From (8-10), we have

$$P(A + R) = \frac{1}{13} + \frac{1}{2} - \frac{1}{26} = \frac{7}{13} \qquad (8\text{-}20)$$

Example 8-4

Two cards are drawn in succession from a deck. The first card is replaced and the deck is reshuffled before the second card is drawn. What is the probability that the two cards will both be aces?

Solution

Let A_1 represent the condition that the first card is an ace, and let A_2 represent the same condition for the second card. Since the first card is replaced and the deck is reshuffled before the second card is drawn, there is no dependency between the two favorable outcomes and the individual probabilities are equal. Thus,

$$P(A_1) = \frac{1}{13} \qquad (8\text{-}21)$$

$$P(A_2) = \frac{1}{13} \qquad (8\text{-}22)$$

The desired outcome is that an ace is drawn on the first trial *and* an ace is drawn on the second trial, that is, $P(A_1 A_2)$. In view of the statistical independence,

$$P(A_1 A_2) = P(A_1)P(A_2) = (\tfrac{1}{13})^2 = \frac{1}{169} \qquad (8\text{-}23)$$

Example 8-5

Consider the same experiment and desired outcome as in Ex. 8-4, but assume that the first card is *not* replaced before drawing the second card.

Solution

The probability of drawing an ace on the first trial is the same as before; that is, $P(A_1) = \frac{1}{13}$. However, if an ace is drawn, there is one less ace in the deck and one fewer card. Thus, the occurrence of the first event will influence the occurrence of the second event, and so the two events are no longer statistically independent.

Assuming that an ace is obtained on the first draw, there are 51 cards remaining of which 3 are aces. Thus, the conditional probability that an ace is drawn on the second trial (A_2 is true), given that an ace was drawn on the first trial (A_1 is true), is

$$P(A_2/A_1) = \tfrac{3}{51} \qquad\qquad (8\text{-}24)$$

Using (8-13), we have

$$P(A_1 A_2) = P(A_1)P(A_2/A_1) = \tfrac{1}{13} \times \tfrac{3}{51} = \tfrac{1}{221} \qquad (8\text{-}25)$$

8-4 STATISTICAL FUNCTIONS

The major statistical functions of interest will be introduced through the discussion of a hypothetical measurement problem. While only a limited degree of accuracy will be assumed for the measurement, and the results will be assumed to be "nice and clean" for illustrative purposes, the process involved could represent a realistic type of measurement.

Consider the signal $x(t)$ shown in Fig. 8-3 for a short interval of time. This function represents a hypothetical random signal whose statistical properties are to be measured. After some lengthy observation of the signal on an oscilloscope, it is determined that the amplitude always varies from 0 V to slightly less than 8 V.

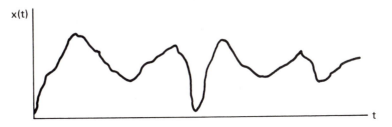

Figure 8-3. A short segment of a hypothetical random signal whose statistical behavior is to be determined.

To perform a statistical study of the signal, it is desired to determine what fraction of the time different amplitude levels of the signal will occur. The measurement system shown in Fig. 8-4 is implemented for this purpose. A 3-bit A/D converter connected for unipolar encoding with rounding is used to quantize successive independent samples of the signal into one of eight possible levels. However, instead of using one of the "standard" A/D voltage levels discussed in Chapter 7, the voltage level of the A/D converter is

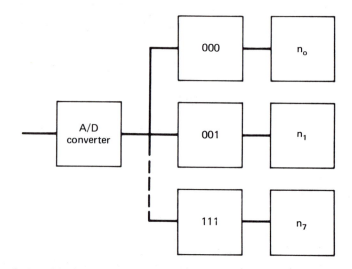

Figure 8-4. Measurement system used to study a hypothetical random signal over a period of time.

adjusted so that each step corresponds to 1 V. The assumed quantization characteristic is shown in Fig. 8-5.

Eight digital comparators are connected to the A/D converter, and the output of each comparator is connected to a counter. Each time a given binary word represents a true condition for a given comparator, the corresponding counter is advanced by one count. Thus, the value n_0 represents the number of times that the binary word 000 has appeared at the A/D converter output, n_1 represents the number of times 001 has appeared, and so on. Note, however, that the number at each counter output actually represents the number of

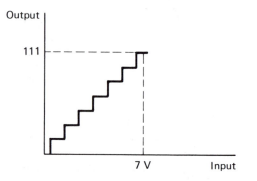

Figure 8-5. Quantization input-output characteristic of the A/D converter used in the measurement of the hypothetical signal.

analog samples falling within a given range of ± 0.5 V from the given midpoint.

For convenience, a discrete variable x_k will be defined to represent the eight possible levels in which the signal is quantized. Let $x_0 = 0$ V, $x_1 = 1$ V, $x_2 = 2$ V, ..., $x_7 = 7$ V. Thus x_k represents the midpoint of the corresponding analog quantization range, and it also represents the decimal value of the quantized digital word based on a step size of 1 V.

Assume that over some period of time 35,000 samples of the signal are taken. Assume that the distribution of the samples according to the defined x_k levels is as given in the first two columns of Table 8-1. (The third column will be discussed later.) Understand that the actual values of the analog signal could have been anywhere within a given range, but each x_k value represents only the discrete quantized value corresponding to the midpoint of that given range. Thus, we could say, for example, that 7500 samples fell within a range from 2.5 to 3.5 V, but it will be easier in subsequent references to indicate simply that there were 7500 samples of the 3-V quantized level.

The results of the measurement experiment will now be displayed graphically. Consider first the plot shown in Fig. 8-6, in which the number of times each discrete quantized value occurs is indicated. The exact number of times a given level occurs is of interest for some purposes, but in most applications the *exact* number of counts is not nearly as important as the *relative* number of counts for each level.

If we assume that the 35,000 samples taken is a sufficiently large number to produce a meaningful statistical pattern, each individual count can then be normalized or divided by the total number of samples. The results obtained are tabulated in the third column of Table 8-1 and are shown graphically in

Table 8-1.   Tabular Data Associated with the Hypothetical Measurement Experiment Shown in Figure 8-3

x_k (V)	Count	$f(x_k)$
0	500	0.0143
1	2,000	0.0571
2	4,000	0.1143
3	7,500	0.2143
4	10,000	0.2857
5	7,000	0.2000
6	3,000	0.0857
7	1,000	0.0286

$$\sum_{0}^{7} f(x_k) = 1$$

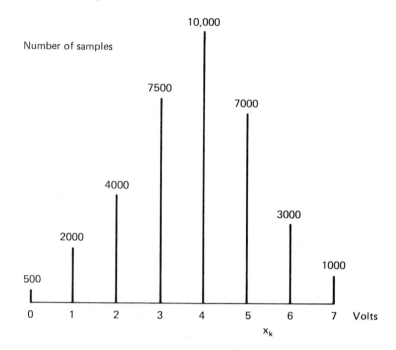

Figure 8-6. Results of the hypothetical measurement experiment displayed according to the exact number of counts for each level.

Fig. 8-7. Note that the shape of this graph is the same as that of Fig. 8-6, but the values are now all less than unity. In fact, from Table 8-1 it is noted that the sum of all the values in the third column is exactly unity.

The advantage of this normalized presentation is that the results are independent of the actual number of samples taken in establishing the distribution, provided, of course, that the number taken is sufficiently large to provide a meaningful description and provided that the statistical behavior of the signal does not change. These results can then be used as a form of statistical representation for the signal.

The plot of Fig. 8-7 provides a graphical display of the relative frequency of occurrence of different amplitude values of the quantized signal (or, more specifically, amplitude range bins for the analog signal). It is noted that the level of 4 V is the most common single quantized level, since its relative frequency of occurrence is the highest. Conversely, the level of 0 V is the least common value.

The function shown in Fig. 8-7 and listed in the third column of Table 8-1 is one form of a *probability density function* (often denoted simply as a PDF). A probability density function may be a function of one variable or more than one variable. It may be a function of a *discrete* variable or a *continuous*

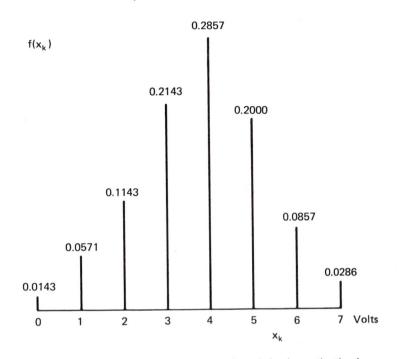

Figure 8-7. Normalized form of the results of the hypothetical measurement experiment. This form is now a *probability density function* of the random discrete variable, to be denoted simply as a discrete PDF.

variable. For convenience, the former will be referred to simply as a *discrete PDF,* and the latter will be referred to as a *continuous PDF.* A discrete PDF will be denoted by the form $f(x_k)$, and a continuous PDF will be denoted by the form $f(x)$. Subscripts on f will be added as required when there is more than one PDF to be considered. For example, in an analysis involving two continuous random variables $x(t)$ and $y(t)$, the corresponding PDFs will be denoted as $f_x(x)$ and $f_y(y)$.

The PDF that has been determined for our example is obviously a discrete PDF, and it is identified by the symbol $f(x_k)$, as shown in Fig. 8-7 and in Table 8-1. In this case, there are eight possible values that the discrete variable x_k may assume.

Some awkward but meaningful notation will now be introduced. Let X represent one random sample from the process (the particular quantized voltage corresponding to one reading in the present example). The expression $P(X = x_k)$ is defined to mean "the probability that a random sample of the experiment is equal to x_k." This probability is then related to the PDF by

$$P(X = x_k) = f(x_k) \tag{8-26}$$

Stated in words, the probability that one random sample of a process is equal to a particular discrete level is the value of the discrete PDF evaluated for that given level. Thus, a discrete PDF may be interpreted as a set of probability values, each of which predicts the probability that a random sample of the process will be the given value.

To illustrate this concept, assume that we wish to determine the probability that a single random sample of the given signal will correspond to a quantized voltage of $+2$ V. We have

$$P(X = 2) = f(2) = 0.1143 \qquad (8\text{-}27)$$

from either Table 8-1 or Fig. 8-7.

Since the signal in the hypothetical example was first sampled and quantized, all the measurements were assumed to be made directly on a discrete variable, and this was a natural way to introduce the reader to the concept of a discrete PDF. On the other hand, suppose we had desired to obtain a much higher resolution on the amplitude range bins for the given analog signal. We would obviously have to increase the number of bits in the A/D converter. For example, if a 10-bit A/D converter were used, 1024 level ranges could be theoretically established, and each sample could be resolved to an uncertainty of about ± 3.91 mV under ideal conditions. If the plot of the corresponding PDF were attempted, it would be extremely difficult to display the horizontal scale properly, since 1024 separate values would be required.

What is happening in this case is that the difference in successive quantization levels is becoming so small that the discrete variable x_k is beginning to approach a continuous variable x. Under these conditions, the mathematics, as well as the means by which the results are displayed, become quite clumsy to deal with on a discrete basis, so it is better to redefine the mathematical forms in continuous function form. The reader may recall from Chapter 2 that this is the same type of process that occurred when the Fourier transform was introduced as a limiting case of the Fourier series as the difference between successive frequencies was permitted to approach zero.

In obtaining an actual PDF from experimental data, it is often necessary to obtain the data through amplitude range bins, as we have described in our hypothetical experiment, so the initial result is a discrete PDF even though the process being studied may be continuous in form. If a continuous PDF is to be determined, it is necessary that a sufficient number of levels be employed so that the quantization uncertainty is within acceptable accuracy bounds. A subtle point is that, if the number of quantization levels is increased, the total number of samples taken must be increased, since it is necessary that a relatively large number of samples of *each level* be present before reasonable satistical certainty is assured.

We do not really know what the limiting true continuous PDF for the signal $x(t)$ in our example is like, but it could be determined quite closely by a more accurate measurement in accordance with the preceding discussion. The curve shown in Fig. 8-8 might be about what we would intuitively expect to obtain. Notice that even with only the eight levels used in our experiment, we could have obtained a crude estimate of the continuous PDF, but this may not be sufficient for highly accurate purposes.

Several points about the continuous PDF using Fig. 8-8 as a reference will now be made. The first point is that *all* values in the applicable voltage range are now possible rather than just the eight discrete values previously considered. However, for a true continuous PDF, the probability that a random sample is a single *exact* value is actually zero. Thus, while the peak of the curve of Fig. 8-8 occurs at about 4 V, the probability that a random sample of the signal is *exactly* 4 V is zero; that is, $P(X = 4) = 0$.

The preceding result may seem puzzling to the reader, but it is explained by realizing that there are theoretically an infinite number of possible values associated with a continuous random variable, so the probability of any single exact value occurring is zero. Instead, a continuous random variable must always be described by a possible range of values in order to be meaningful.

The term $P(x_1 < X < x_2)$ is interpreted to read "the probability that a random sample of the process lies between the limits of x_1 and x_2." This

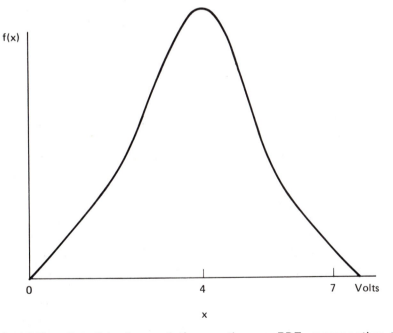

Figure 8-8. Possible form of the continuous PDF representing the analog signal in the hypothetical measurement experiment.

probability is determined from the continuous PDF $f(x)$ by the following integral:

$$P(x_1 < X < x_2) = \int_{x_1}^{x_2} f(x)\,dx \qquad (8\text{-}28)$$

Stated in words, the probability that a random sample of a continuous variable lies between the limits of x_1 and x_2 is the area under the curve of the PDF between x_1 and x_2. This concept is illustrated in Fig. 8-9.

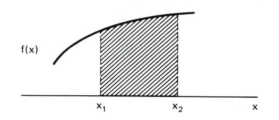

Figure 8-9. Probability of a random sample of a continuous variable being between two points is the area under the PDF curve between the points.

Since any random sample must lie somewhere between $-\infty$ and $+\infty$ with probability 1, the PDF must always be scaled to satisfy the following criterion:

$$\int_{-\infty}^{\infty} f(x)\,dx = 1 \qquad (8\text{-}29)$$

This result simply states that the net area under the curve of a continuous PDF must always be unity. If a PDF is constructed from empirical data, it is usually necessary to scale the vertical level so that (8-29) is satisfied.

Let us now return to the question that was raised earlier about the probability that a random sample assumed a given value. While the probability that a random sample of the signal being exactly 4 V is zero, we could, for example, evaluate $P(3.99 < X < 4.01)$ by determining the area under the curve of $f(x)$ over that range, and the result would not be zero. Thus, when working with a *continuous* PDF, it is always necessary that a range be specified for meaningful probability calculations. However, as observed earlier, an exact level of the variable may be meaningful for a *discrete* PDF.

The term *distribution* is often used in reference to a statistical representation of a random variable. Strictly speaking, there is a function called the *cumulative distribution function* that has a precise meaning in statistics. However, since that function will not be employed in this book, the term *distribution* will be used as a somewhat casual reference to probability density functions in accordance with popular convention.

The hypothetical signal used to develop the concept of the PDF in this section was actually a continuous variable, although the development began by representing it as a discrete variable through the process of A/D conversion. By allowing the quantization step size to become arbitrarily small, the form of the continuous PDF could eventually be closely approximated. There are, of course, variables that are strictly discrete in nature and would always be described by discrete PDFs. For example, an ideal square wave (assuming zero transition times between levels) will only have two possible levels. Logic levels in digital systems can only assume discrete levels and thus would be described statistically by discrete PDFs.

Example 8-6

A certain hypothetical random analog voltage $x(t)$ varies between the levels of 0 and 10 V at all times. Values closer to 10 V are increasingly more likely to occur than values near 0 V, and a sufficient amount of experimental data suggests that the continuous PDF may be represented by the function shown in Fig. 8-10. (a) Determine an equation for the PDF $f(x)$. (b) A single sample X of the signal is to be taken at some arbitrary time. Determine the following probabilities concerning the single sample X: (1) $P(X = 9)$, (2) $P(8.9 < X < 9.1)$, (3) $P(0 < X < 5)$, (4) $P(5 < X < 10)$, and (5) $P(0 < X < 10)$.

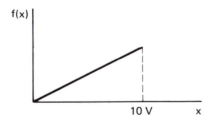

Figure 8-10.   Continuous PDF for Example 8-6.

Solution

(a) The PDF is a straight line through the origin over the range of interest, so

$$f(x) = Ax, \qquad 0 < x < 10 \tag{8-30}$$
$$= 0, \qquad \text{elsewhere}$$

where A is not yet known. The constant A is determined so that the net area under the PDF curve is unity. This evaluation follows:

$$\int_0^{10} f(x)\, dx = \int_0^{10} Ax\; dx = \frac{Ax^2}{2}\bigg|_0^{10} = 50A = 1 \qquad (8\text{-}31)$$

Thus,

$$A = \tfrac{1}{50} = 0.02 \qquad (8\text{-}32)$$

and

$$f(x) = \frac{x}{50}, \qquad 0 < x < 10 \qquad (8\text{-}33)$$

$$= 0, \qquad \text{elsewhere}$$

This example so far has illustrated a reasonably common occurrence in statistical analysis. The form of the PDF is known, but the level must be determined. The concept that the total area must be unity is the usual means by which the level of the PDF can be determined. Although the equation for $f(x)$ could readily be expressed in this example, there are many situations in which the PDF can only be plotted as a graph as determined from experimental data. In such cases, the area under the curve might have to be adjusted to unity by a numerical integration scheme.

(b) Now that the equation for $f(x)$ has been determined, the specific probability values requested are calculated as follows:

$$(1) \quad P(X = 9) = 0 \qquad (8\text{-}34)$$

This result was readily deduced from the earlier discussion indicating that the probability of a single *exact* value of a continuous random variable occurring is zero.

$$(2) \quad P(8.9 < X < 9.1) = \int_{8.9}^{9.1} \frac{x}{50}\, dx = \frac{x^2}{100}\bigg|_{8.9}^{9.1} = 0.036 \qquad (8\text{-}35)$$

In this case, a range for X about 9 V has been specified, and the probability is not zero. This result indicates that on an average about 36 of every 1000 samples of the signal fall in the range between 8.9 and 9.1 V.

Without showing the details, the reader may readily verify that (3) and (4) are

$$(3) \quad P(0 < X < 5) = 0.25 \qquad (8\text{-}36)$$

$$(4) \quad P(5 < X < 10) = 0.75 \qquad (8\text{-}37)$$

Finally, it should be obvious that

$$(5) \quad P(0 < X < 10) = 1 \qquad (8\text{-}38)$$

since all samples are assumed to exist over this range, and the net area is unity.

8-5 STATISTICAL AVERAGES

We have seen thus far that a random signal, whose exact behavior as a function of time cannot be predicted, can often be described on a probabilistic basis using a probability density function. The PDF provides a relative measure of the likelihood that the amplitude of the random signal lies within a given range or, in the case of a discrete value, assumes a given value.

In many communications applications of random signals, such quantities as the dc value or the rms value of the signal are of major importance. It turns out that these quantities may be determined directly from the probability density function in many cases. This section will be concerned with the evaluation of these quantities and other parameters of interest for both continuous and discrete variables.

To simplify the discussion that follows, a signal $x(t)$ (or x_k for the discrete case), which could represent either a voltage or a current, will be assumed. For power-calculation purposes, a normalized resistance of 1 Ω will be often assumed, as has been true throughout the text. This means that the instantaneous power $p(t)$ at any time is $x^2(t)$, independent of whether $x(t)$ is a voltage or a current when the 1-Ω reference is assumed.

We will first consider a *continuous PDF* $f(x)$. The *mean* value $\bar{x}$ of the signal is determined from the following equation:

$$\bar{x} = \int_{-\infty}^{\infty} x f(x)\, dx \qquad (8\text{-}39)$$

The limits indicate that the integral is evaluated over the entire range of the PDF.

The mean value $\bar{x}$ is the quantity that is most often referred to as the *average value*. However, in the terminology of statistics, the term "average" is more general and should not be used for a specific parameter. The term *mean value* is more correct and will be used here.

The second quantity of interest is the *mean-square value* $\overline{x^2}$, which is defined as

$$\overline{x^2} = \int_{-\infty}^{\infty} x^2 f(x)\, dx \qquad (8\text{-}40)$$

This quantity is the weighted accumulation of all the squared values of the random variable. The root-mean-square (rms) value x_{rms} is the positive square root of the mean-square value; that is,

$$x_{rms} = \sqrt{\overline{x^2}} = \sqrt{\int_{-\infty}^{\infty} x^2 f(x)\, dx} \qquad (8\text{-}41)$$

The next quantity of interest is the *variance,* which is denoted by σ^2, and it is defined as

$$\sigma^2 = \int_{-\infty}^{\infty} (x - \overline{x})^2 f(x)\, dx \qquad (8\text{-}42)$$

It will be left as an exercise for the reader (Prob. 8-29) to verify that σ^2 may be expressed in terms of the mean-square and mean values as follows:

$$\sigma^2 = \overline{x^2} - (\overline{x})^2 \qquad (8\text{-}43)$$

The form of (8-43) is often the easiest way to compute the variance.

The variance is the weighted accumulation of all the squared differences between the variable x and its mean value $\overline{x}$. Finally, the *standard deviation σ* is the square root of the variance: $\sigma = \sqrt{\sigma^2}$.

It can readily be seen from the preceding equations that if $\overline{x} = 0$ the mean-square value and the variance are equal: $\overline{x^2} = \sigma^2$. Likewise, the rms value and the standard deviation are equal: $x_{rms} = \sigma$ when $\overline{x} = 0$.

Some feeling for these quantities may be deduced from the curves of Fig. 8-11. In Fig. 8-11a and b, symmetrical distributions about midpoints are assumed. It can be shown that when a PDF is completely symmetrical about a given value of x this particular value is the mean value. On the other hand,

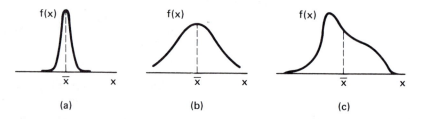

Figure 8-11. Curves used to illustrate certain properties of the statistical parameters.

the PDF shown in Fig. 8-11c is not symmetrical, so the mean value is not as obvious. However, the actual criterion is that the net area to the right of the mean be the same as the net area to the left of the mean. Observe that the mean value in this illustration is larger than the value corresponding to the peak of the PDF.

Looking again at Fig. 8-11a and b, it can be shown that the PDF of Fig. 8-11b has a larger variance than the PDF of Fig. 8-11a. The larger the variance, the more likely that random samples of the signal will be farther away from the mean value. Thus in Fig. 8-11a, the amplitude of the signal is always near $x = \overline{x}$, but in Fig. 8-11b the samples spread out considerably in range. Thus, variance is a measure of the relative spread of the possible values of the distribution about the mean value.

Next consider the case of a discrete PDF, which will be denoted as $f(x_k)$. The mean value $\overline{x}$ is determined from the summation

$$\overline{x} = \sum_k x_k f(x_k) \tag{8-44}$$

where the summation is performed over all possible values of x_k.

The mean-square value $\overline{x^2}$ is obtained from the summation

$$\overline{x^2} = \sum_k x_k^2 f(x_k) \tag{8-45}$$

The rms value x_{rms} is then given by

$$x_{rms} = \sqrt{\overline{x^2}} \tag{8-46}$$

The variance σ^2 is defined by the summation

$$\sigma^2 = \sum_k (x_k - \overline{x})^2 f(x_k) \tag{8-47}$$

As in the case of the continuous PDF, the variance may be expressed in the form

$$\sigma^2 = \overline{x^2} - (\overline{x})^2 \tag{8-48}$$

Finally, the standard deviation σ is simply the square root of the variance: $\sigma = \sqrt{\sigma^2}$.

The relative behavior of these different parameters for a discrete PDF corresponds to the same properties previously discussed for a continuous PDF. One seemingly peculiar point should be noted, however. A statistical parameter evaluated for a discrete PDF may not even correspond to one of the quantized values in the population. For example, the mean value as computed from the definition will often result in a value located between two adjacent quantization levels. However, the statistical parameter as evaluated from the definition is correct and need not represent any particular value of the population.

Some of the statistical definitions that have been discussed thus far sound very much like some of the waveform parameters obtained by time averaging as discussed in Chapter 2, for example, mean-square and rms values. A natural question then arises as to whether or not these quantities are, in fact, the same as those already familiar to us for deterministic signals. This question will be answered shortly, but several pertinent definitions need to be discussed.

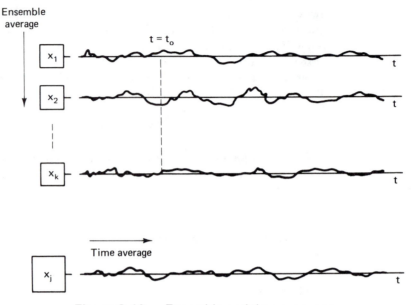

Figure 8-12. Ensemble and time averages.

First, we will introduce the concept of an *ensemble* average. Assume that we have available some arbitrary number of random noise generators, as illustrated in Fig. 8-12. Assume that the physical mechanisms that produce the random signals are similar in form but are physically separated in generating the actual noise voltages (or currents). The various signals then are assumed to be statistically independent.

Assume that at some time t_0 we can observe the output of each respective generator and obtain a set of values representing the respective amplitudes at that time. If the sample size is sufficiently large, the set of random samples obtained through this mechanism might be sufficient to deduce some important statistical properties of the signal. The process of determining the statistical properties by measuring the amplitude of different independent random generators at a particular value of time is called an *ensemble average*.

Assume now that attention is focused on *one* of the generators and that its behavior is closely observed over a period of time. Various averages could be measured or calculated, and "bin sorting" could be used to construct a PDF in the manner discussed in the last section. The process of determining the statistical properties by monitoring the behavior of a single generator over an interval of time is called a *time average*. A time average of a signal $x(t)$ will be represented by the notation $\langle x(t) \rangle$.

We now return to the question raised several paragraphs earlier, which will now be rephrased as follows: Is a particular ensemble average obtained from monitoring the amplitude of many sources at a single value of time equal to the corresponding time average obtained over a long period of time from a single generator? The answer in the most general case is *no*. However, for a variety of commonly occurring cases, the answer is a *"qualified" yes*.

An *ergodic process* is defined as one in which the ensemble averages are always equal to the corresponding time averages. How does one know if a given random process is ergodic? A realistic answer is that one does not often truly know. The means by which ergodicity is established for the general case is beyond the scope of most engineering books and will not be pursued here.

Before relieving the reader's possible concern at this point, one more definition will be given. A *stationary process* is one in which the statistics are independent of the time interval in which they are computed. Examples of a possible stationary process and a nonstationary process are illustrated in Fig. 8-13. Observe that the waveform of Fig. 8-13a *might* appear to maintain a reasonably similar pattern (one can never be sure from a simple observation) over the interval shown and is probably stationary, but the waveform of Fig. 8-13b slowly diverges as time passes as is not stationary.

An ergodic process is necessarily stationary, but a stationary process may or may not be ergodic.

Now let us return to the nagging question of the ergodic condition. Fortunately for us, the majority of waveforms that are encountered in communications system analysis and design are either ergodic (and also stationary) or sufficiently close over certain ranges that the ergodic assumption is often valid. In many cases, ergodicity has to be assumed because there is nothing else that can be done. One should be aware, however, that it is not always a valid assumption, and some widely employed results apply only when the process is ergodic.

Let us now assume that the random signals to be considered are members of ergodic (and stationary) processes. In this case, various statistical averages can be determined from either time or ensemble averages and the results are the same. In most practical cases, time averages are easier to determine than ensemble averages, since it may be difficult to obtain a sufficiently large

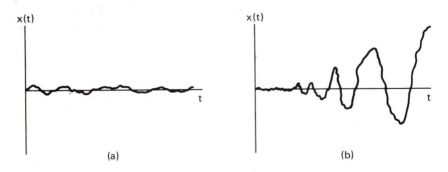

Figure 8-13.   (a) A possible stationary process and (b) a nonstationary process.

number of separate generators to make ensemble averages meaningful. Thus, the hypothetical experiment described in the last section, which was based on a single generator, could be used to infer general statistics for a random process of its type, provided that the process is ergodic.

For ergodic processes representing random signals, the various ensemble averages have particularly simple interpretations when related to the time averages. We start with:

$$\overline{x} = \langle x(t) \rangle = \text{dc value} \tag{8-49}$$

This statement indicates that the mean value computed from the PDF is identical to the time average, and the result is the dc value for the waveform. Thus, an ideal dc voltmeter will respond to the true mean value of a random signal from an ergodic process. (A practical dc voltmeter may exhibit fluctuations about the mean depending on the time constant of the meter.)

Next we have

$$\overline{x^2} = \langle x^2(t) \rangle = \text{total average power in 1-}\Omega\text{ reference} \tag{8-50}$$

This statement indicates that the mean-square value of the total signal computed from the PDF is the same as obtained from the corresponding time average, and this result is the average total power in a 1-Ω reference. The square root is the usual rms value; that is,

$$x_{\text{rms}} = \sqrt{\overline{x^2}} = \sqrt{\langle x^2 \rangle} = \text{total rms value} \tag{8-51}$$

Finally, we have

$$\sigma^2 = \overline{x^2} - (\overline{x})^2 = \langle x^2 \rangle - \langle x \rangle^2 = \text{ac power in a 1-}\Omega\text{ reference} \tag{8-52}$$

This statement indicates that the variance, which can now be computed from either an ensemble or a time average, is the ac power of the signal computed on a 1-Ω basis. The term "ac power" as used here refers to the fluctuation power about the mean. The square root is the ac rms value; that is,

$$\sigma = \sqrt{\sigma^2} = \text{ac rms value} \tag{8-53}$$

The result of (8-52) can also be expressed in the form

$$\text{ac power} = \text{total power} - \text{dc power} \tag{8-54}$$

where $(\overline{x})^2$ is interpreted as the "dc power" in a 1-Ω reference.

Example 8-7

Consider the random voltage $x(t)$ of Ex. 8-6, and assume that the signal is ergodic. Determine the following quantities for the signal: (a) dc value, (b) mean-square value, (c) total rms value, (d) variance, (e) ac rms value, and (f) average total power dissipated by signal in a 2-Ω resistor.

Solution

The PDF was developed in Ex. 8-6 and is repeated here for convenience.

$$f(x) = \frac{x}{50}, \qquad 0 < x < 10 \tag{8-55}$$
$$= 0, \qquad \text{elsewhere}$$

(a) The dc or mean value is

$$\overline{x} = \int_{-\infty}^{\infty} x f(x) \, dx = \int_0^{10} \frac{x^2}{50} \, dx = \left. \frac{x^3}{150} \right|_0^{10} = 6.667 \text{ V} \tag{8-56}$$

(b) The mean-square value is

$$\overline{x^2} = \int_{-\infty}^{\infty} x^2 f(x) \, dx = \int_0^{10} \frac{x^3}{50} \, dx = \left. \frac{x^4}{200} \right|_0^{10} = 50 \text{ V}^2 \tag{8-57}$$

(c) The total rms value is the square root of $\overline{x^2}$; that is,

$$x_{\text{rms}} = \sqrt{50} = 7.071 \text{ V} \tag{8-58}$$

(d) The variance is most easily determined from (8-43). We have

$$\sigma^2 = \overline{x^2} - (\overline{x})^2 = 50 - (6.667)^2 = 5.555 \text{ V}^2 \tag{8-59}$$

(e) The ac rms value is the same as the standard deviation and is

$$\sigma = \sqrt{5.555} = 2.357 \text{ V} \tag{8-60}$$

(f) Finally, the average power P dissipated in a 2-Ω resistor is

$$P = \frac{\overline{x^2}}{R} = \frac{\overline{x^2}}{2} = \frac{50}{2} = 25 \text{ W} \tag{8-61}$$

If the usual 1-Ω reference had been employed, the mean-square value and the power would have been identical.

Example 8-8

The *uniform* probability density function arises in situations in which all possible values of the random variable in the given range are equally likely to appear. (One application of the uniform PDF is in analyzing quantization noise in PCM, as will be seen in Chapters 10 and 11.) Consider the uniform PDF $f(x)$ representing an ergodic signal $x(t)$ shown in Fig. 8-14. Determine the mean, mean-square, variance, and rms value for the signal.

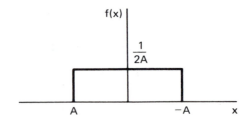

Figure 8-14. Uniform PDF used in Example 8-8.

Solution

The given PDF can be expressed as

$$f(x) = \frac{1}{2A}, \qquad -A < x < A$$
$$= 0, \qquad \text{elsewhere} \tag{8-62}$$

Observe that the level $1/2A$ is such that the net area is $(1/2A) \times (2A) = 1$, as required. If this level had not been given, it could have been readily determined from a knowledge of the range of the signal and the fact that the distribution is uniform.

The mean value can be readily determined from the symmetry of the distribution about $x = 0$, or one can derive it if desired. The result is

$$\overline{x} = 0 \tag{8-63}$$

Since the mean value is zero, the mean-square value and the variance are equal, and the result is determined from the following development:

$$\overline{x^2} = \sigma^2 = \int_{-A}^{A} x^2 f(x)\, dx = \int_{-A}^{A} \frac{x^2}{2A}\, dx = \frac{x^3}{6A}\Big]_{-A}^{A} = \frac{A^2}{3} \tag{8-64}$$

Again, since the mean value is zero, the total rms value and the standard deviation are equal; that is,

$$x_{\text{rms}} = \sigma = \frac{A}{\sqrt{3}} \tag{8-65}$$

This result indicates that the rms value of a uniformly distributed random zero-mean signal is $1/\sqrt{3}$ (or about 0.577) times the peak value. Contrast this with a deterministic sine wave in which the rms value is $1/\sqrt{2}$ times the peak value.

8-6 GAUSSIAN DISTRIBUTION

There are a variety of different statistical distributions that arise in the analysis and design of communications systems, some of which are used only in certain specialized applications. Rather than overwhelm the reader with a long list of probability density functions, the approach taken will be to introduce most of the important ones as they are needed in subsequent developments and problems. The exception to this rule is the *gaussian* distribution (also called the *normal* distribution), to which this entire section is devoted.

Without question, the *gaussian* probability density function represents the most important single type of distribution in applied statistics. Random physical quantities in a wide variety of different areas can be modeled by the gaussian PDF. In fact, a very basic result from statistical theory is the *central limit theorem,* which states that the distribution of the sum of a number of statistically independent random variables tends to approach a gaussian distribution in the limit irrespective of the forms of the individual distributions. Since there are so many physical processes in nature that depend on the sum of a number of independent contributions, the gaussian process appears quite frequently in modeling random phenomena.

The general shape of gaussian PDF is shown is shown in Fig. 8-15. The gaussian curve is sometimes referred to as "bell-shaped" curve, for obvious reasons. The mathematical form of the gaussian PDF is

$$f(x) = \frac{1}{\sqrt{2\pi\sigma^2}} \epsilon^{-(x - \bar{x})^2/2\sigma^2}, \qquad \text{for } -\infty < x < \infty \tag{8-66}$$

where $\bar{x}$ and σ^2 are immediately identified as the mean and variance in the PDF. If the defining equations for the mean and variance are applied to (8-66), the results would check with the values given. Thus, the advantage of

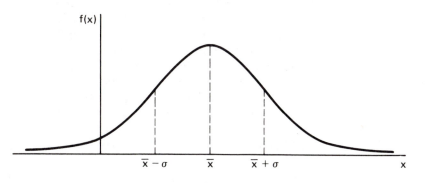

Figure 8-15. Form of the gaussian PDF.

the form of (8-66) is that the mean and variance can be immediately identified. For example, suppose we are given

$$f(x) = \frac{1}{\sqrt{200\pi}} \epsilon^{-(x-5)^2/200} \tag{8-67}$$

By comparing (8-67) with (8-66), it can be deduced that $\bar{x} = 5$ and $\sigma^2 = 100$ or $\sigma = 10$.

A comment about the infinite limits of the random variable will be made. In theory, the ideal mathematical form of the gaussian PDF indicates an infinite range on either side of the mean. In practice, the integral converges quite rapidly, and the area in the "tails" at a reasonable distance away from the mean becomes negligible. Consequently, the gaussian model often closely fits real-life signals that are obviously limited to a finite amplitude range. As is true with so many processes, the ideal mathematical form must be tempered by the real-life limitation of the actual system.

Consider now the problem of evaluating the area under the gaussian PDF, which is a necessary procedure for determining probability values associated with a gaussian random variable. Assume that it is desired to determine the probability that a random sample X of the signal be in the range $x_1 < X < x_2$. An expression for this probability is

$$P(x_1 < X < x_2) = \int_{x_1}^{x_2} \frac{1}{\sqrt{2\pi\sigma^2}} \epsilon^{-(x-\bar{x})^2/2\sigma^2} \, dx \tag{8-68}$$

The expression of (8-68) has associated with it both " bad news" and "good news." The bad news is that the integral involved cannot be expressed in closed form using standard functions, so a "brick wall" appears to have been reached. The good news is that certain integrals of this form have been evaluated by numerical methods over the years, and the results can be adapted to this problem.

Before pursuing that point, a general comment will be made. The gaussian PDF is treated in numerous books on statistics, communications, and many other subjects. Because of the necessity to utilize the results of numerical integration and the somewhat unwieldy nature of the mathematical development, several different forms for the evaluation have been developed. Indeed, it seems as if there are about as many variations on the form as there are books on the subject. All, of course, produce identical results, but the appearance of these different forms can be confusing. These comments were made to console the reader who starts searching through different books and finds different forms for the gaussian probability evaluation.

The important thing is not which method is used, but that the reader should find at least one method that he or she establishes enough proficiency with to permit the determination of areas under the gaussian curve when required. As a matter of interest, many programmable calculators are provided with programs that can be used to determine gaussian areas by direct numerical analysis. However, it is this author's opinion that the reader should first focus on establishing proficiency using tabular or graphical data before depending completely on a calculator or computer for this evaluation.

The particular approach used here will rely heavily on certain graphical presentations owing to the simplicity of usage. The results obviously cannot be read quite as accurately as tabular data, but the ease of extrapolation between points will permit the simplest means for evaluation. The results will be sufficiently accurate for the purposes at hand and for many of the problems encountered in communications systems analysis. Readers requiring more accuracy should consult one of the mathematical or statistical handbooks providing extensive tabular data or use one of the available computer or calculator programs.

The approach used here (and in most other places for that matter) is based on a normalized gaussian PDF $f(y)$ in which the variable y is assumed to have zero mean and unit variance; that is, $\bar{y} = 0$ and $\sigma^2 = 1$. This normalized PDF can then be expressed as

$$f(y) = \frac{1}{\sqrt{2\pi}} \epsilon^{(-y^2/2)}, \qquad \text{for } -\infty < y < \infty \tag{8-69}$$

A sketch of $f(y)$ is shown in Fig. 8-16.

The approach used will be to evaluate the area under the curve of $f(y)$ to the right of an argument y_1, as illustrated by the shaded area in Fig. 8-16. The function obtained from the integration will be designated as $Q(y_1)$, and it is defined as

$$Q(y_1) = \int_{y_1}^{\infty} f(y)\, dy = \int_{y_1}^{\infty} \frac{1}{\sqrt{2\pi}} \epsilon^{-(y^2/2)}\, dy \tag{8-70}$$

Once the results are accepted, there is no particular need to retain the variable y_1 in the form required to establish the integration. Instead, the

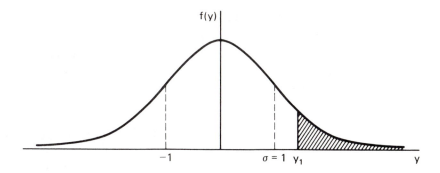

Figure 8-16. Form of the normalized gaussian PDF with shaded segment representing area used to obtain $Q(y)$.

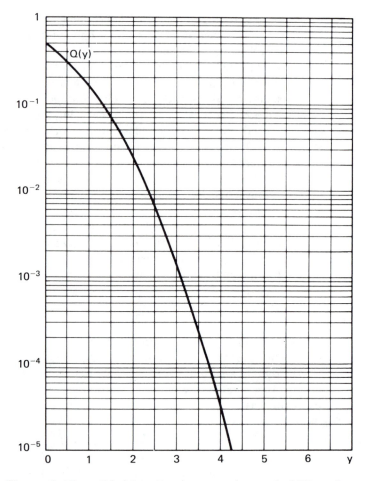

Figure 8-17. $Q(y)$ function for gaussian probability values.

subscript will be dropped and the variable y will be used to denote any particular point of interest.

The function $Q(y)$ is shown in Fig. 8-17 over the positive range of y. For the case of a negative argument $-y$, the value for the corresponding positive argument y is first read on the curve, and the following equation is used:

$$Q(-y) = 1 - Q(y) \qquad (8\text{-}71)$$

No special name will be attached to the curve of Fig. 8-17; instead it will be referred to simply as the $Q(y)$ function. This integration has been performed numerically, and the results can be obtained from mathematical tables.

Two particular limiting values of $Q(y)$ are $Q(-\infty) = 1$ and $Q(\infty) = 0$. In addition note that $Q(0) = 0.5$.

We will now consider the manner in which the $Q(y)$ function is used. Recall that the curve was derived from a gaussian PDF with zero mean and unit variance. Any actual variable x, corresponding to a PDF $f_x(x)$, must be transformed to one with zero mean and unit variance. This is achieved by the transformation

$$y = \frac{x - \bar{x}}{\sigma} \tag{8-72}$$

where $\bar{x}$ and σ are the mean and standard deviation corresponding to $f_x(x)$. To use the curves, all x values must be converted to corresponding y values by the transformation of (8-72). Conversely, if some y value is read off of either of the curves, the corresponding x value is determined from the inverse form of (8-72), which is

$$x = \sigma y + \bar{x} \tag{8-73}$$

Assume now that the area between $x = x_1$ and $x = x_2$ is to be determined: that is, an evaluation of $P(x_1 < X < x_2)$ is desired. This result can be expressed in terms of the $Q(y)$ function as follows:

$$P(x_1 < X < x_2) = Q\left(\frac{x_1 - \bar{x}}{\sigma}\right) - Q\left(\frac{x_2 - \bar{x}}{\sigma}\right) \tag{8-74}$$

The two arguments of the Q function in (8-74) correspond to the x values transformed to corresponding y values according to (8-72).

Example 8-9

A certain random voltage $x(t)$ has a gaussian distribution with a mean $\bar{x} = 10$ V and a standard deviation $\sigma = 5$ V. A single random sample X of the voltage is taken. Determine the following probabilities concerning the value (in volts) of the random sample: (a) $P(X > 20)$, (b) $P(X < 20)$, (c) $P(X < 0)$, (d) $P(0 < X < 20)$, (e) $P|X| > 20)$

Solution

All the probability calculations that follow make use of the $Q(y)$ function of Fig. 8-17 along with Eq. (8-74) for the specific calculation. In each case the actual PDF is transformed to a normalized PDF through the transformation of (8-72), which for this example is

$$y = \frac{x - 10}{5} \tag{8-75}$$

When $x = 20$, for example, $y = 2$. However, this transformation is automatically performed in the manipulation of (8-74) as it is given.

(a) $P(X > 20)$: The probability involved is the area to the right of $X = 20$, which corresponds to the area to the right of $y = 2$ for the normalized PDF. By definition, this is the value of the $Q(y)$ function evaluated for $y = 2$. Thus,

$$P(X > 20) = Q\left(\frac{20 - 10}{5}\right) = Q(2) = 0.0228* \qquad (8\text{-}76)$$

If one wishes always to use (8-74) exactly as given, the probability desired could be expressed as

$$
\begin{aligned}
P(20 < X < \infty) &= Q(2) - Q(\infty) \\
&= 0.0228 - 0 \qquad (8\text{-}77) \\
&= 0.0228
\end{aligned}
$$

as before, since $Q(\infty) = 0$.

(b) $P(X < 20)$: This probability is the area to the left of $X = 20$ (or $y = 2$ on a normalized basis). Since the area to the right of y is $Q(y)$, the area to the left is $1 - Q(y)$, so

$$P(X < 20) = 1 - Q\left(\frac{20 - 10}{5}\right) = 1 - Q(2) \qquad (8\text{-}78)$$

$$= 1 - 0.0228 = 0.9772$$

As in part (a), this result can be expressed explicitly in the form of (8-74) as follows:

$$
\begin{aligned}
P(-\infty < X < 20) &= Q(-\infty) - Q(2) \qquad (8\text{-}79) \\
&= 1 - 0.0228 \\
&= 0.9772
\end{aligned}
$$

as before, since $Q(-\infty) = 1$.

(c) $P(X < 0)$: In this case, (8-74) will be employed by expressing the desired probability as

$$P(X < 0) = P(-\infty < X < 0) = Q(-\infty) - Q\left(\frac{0 - 10}{5}\right) \qquad (8\text{-}80)$$

$$= 1 - Q(-2) = 1 - [1 - Q(2)] = Q(2) = 0.0228$$

in which case the use of (8-71) for the negative argument was used. As one develops insight into the form of the manipulations involved, some of the steps may often be omitted. For example, it turns out in this case that the value $x = 0$ corresponds to the normalized value $y = -2$, and the area involved is the

*The reader should not be disturbed if the curve can't be read to the accuracy provided in this development and in similar developments later. The author has access to extensive tabulated data!

area to the left of this point. However, due to the symmetry of the gaussian PDF, this area is the same as that to the right of $y = 2$, which has already been determined in part (a).

(d) $P(0 < X < 20)$: This calculation follows from (8-74) and some preceding results.

$$P(0 < X < 20) = Q\left(\frac{0 - 10}{5}\right) - Q\left(\frac{20 - 10}{5}\right) \tag{8-81}$$

$$= Q(-2) - Q(2)$$
$$= 0.9772 - 0.0228$$
$$= 0.9544$$

(e) $P(|X| > 20)$: This probability evaluation consists of the area to the right of $X = 20$ plus the area to the left of $X = -20$. This is equivalent to the area to right of $y = 2$ plus the area to the left of $y = -2$. The separate areas could be determined and added together to determine the total probability since the individual areas are mutually exclusive. Alternatively, the desired total area can be expressed in one equation as

$$P(|X| > 20) = 1 - P(-20 < X < 20) \tag{8-82}$$
$$= 1 - 0.9544 = 0.0456$$

in which case the result of part (d) was used.

PROBLEMS

8-1   A single card is drawn from a deck of cards. What is the probability that it will be (a) an ace of hearts, (b) a heart, (c) either a heart or a diamond, (d) either a red king or any queen, and (e) a face card?

8-2   A single card is drawn from a deck of cards. What is the probability that it will be (a) either a king or a face card, (b) either a king or a spade, and (c) either a heart or a face card?

8-3   Two cards are drawn in succession from a deck of cards. The first card is replaced and the deck is reshuffled before the second card is drawn. What is the probability that the two cards are (a) both kings, (b) an ace on the first draw and a king on the second draw, and (c) an ace and a king in either order?

8-4   Consider the same experiments and desired outcomes as in Prob. 8-3, but assume that the first card is not replaced before drawing the second card. Repeat the probability evaluations.

8-5   Three cards are drawn in succession from a deck of cards. After each draw, the card is replaced and the deck is reshuffled before the next card is drawn. What is the probability that the three cards are (a) an ace on the first draw, a king on the second draw, and a queen on the third draw, and (b) an ace, a king, and a queen in any order?

8-6 A certain hypothetical random analog voltage $x(t)$ varies between the levels of -2 V and $+6$ V at all times. Experimental data suggest that all voltage levels in the range are equally likely to occur.

    (a)   Determine an equation for the PDF $f(x)$.

    (b)   A random sample X of the voltage is taken. Determine the following probabilities concerning the single sample X: (1) $P(X = 2)$; (2) $P(1.95 < X < 2.05)$; (3) $P(X < 1)$; (4) $P(X < -1)$; (5) $P(X > 1)$.

8-7 Consider the random voltage $x(t)$ of Problem 8-6, and assume that the signal is ergodic. Determine the following quantities for the signal: (a) dc value, (b) mean-square value, (c) total rms value, (d) variance, (e) ac rms value, and (f) average total power dissipated in a 2-Ω resistor.

8-8 Assume that a certain quantized signal x_k can assume any one of four possible levels at a given time. The four levels are the integer values of voltage from 0 to 3 V. Experimental data taken over a period produce the following distribution for the levels:

Voltage x_k (V)	0	1	2	3
Number of samples	400	600	800	200

Define a discrete PDF for the signal and sketch it.

8-9 Consider the discrete random voltage of Prob. 8-8, and assume that the signal is ergodic. Determine the following quantities for the signal: (a) dc value, (b) mean-square value, (c) total rms value, (d) variance, (e) ac rms value, and (f) average total power dissipated in a 2-Ω resistor.

8-10 Consider the experiment of rolling two dice. Each die may show any one of the numbers 1 through 6 with equal probability. However, for the two dice, there are $(6)^2 = 36$ possible outcomes with net weights ranging from 2 through 12. (Some net weights may occur in several possible ways.) Let n represent the net weight of the outcome. Determine the PDF $f(n)$ for the experiment and sketch it.

8-11 A random unipolar binary signal produces zeros and ones with equal probability. A 0 appears as 0 V and 1 appears as E volts. Determine the PDF $f(x_k)$ and sketch it.

8-12 For the signal of Prob. 8-11, determine from the PDF the following quantities: (a) dc value, (b) mean-square value, (c) total rms value, (d) variance, and (e) ac rms value.

8-13 Repeat Prob. 8-11 for a bipolar binary signal (i.e., a 0 appears as $-E$ volts).

8-14 Repeat Prob. 8-12 for the assumptions of Prob. 8-13.

8-15 A random unipolar binary signal produces three times as many ones as zeros in a long data stream. A 0 appears as 0 V and a 1 appears as E volts. Determine the PDF $f(x_k)$ and sketch it.

8-16 For the signal of Prob. 8-15, determine from the PDF the following quantities: (a) dc value, (b) mean-square value, (c) total rms value, (d) variance, and (e) ac rms value.

8-17 Repeat Prob. 8-15 for a bipolar binary signal (i.e., a 0 appears as $-E$ volts).

8-18 Repeat Prob. 8-16 for the assumption of Prob. 8-17.

8-19 Consider an experiment having only two possible outcomes: A and $\overline{A}$. On a single trial of the experiment, the probability that A occurs (is true) is $P(A) = p$, which means that $P(\overline{A}) = 1 - p$. Consider then that m trials of the experiment are performed. The probability that A is true on exactly n of the trials ($n \leq m$) is given by the *binomial distribution*

$$f(n) = \frac{m!}{n!(m-n)!} p^n(1-p)^{m-n}$$

It can be shown that $\overline{n} = mp$ and $\sigma_n^2 = \overline{n}(1-p)$. To illustrate the use of this distribution, assume that a given binary signal is disturbed by such a large level of noise that errors in received bits occur about half the time. For three consecutive bits, calculate the following probabilities concerning the number of errors occurring in the bit stream: (a) no errors, (b) exactly one error, (c) exactly two errors, (d) exactly three errors.

8-20 Repeat Prob. 8-19 if the noise level is reduced such that errors in received bits occur about one-tenth the time.

8-21 Each of the possible conditions in parts (a) through (d) of Prob. 8-19 may be considered as mutually exclusive since a specific number of errors is implied in each case. Based on the data of that problem, determine the probability that there will be no more than one error in the three digits.

8-22 Repeat Prob. 8-21 for the lower noise level defined in Prob. 8-20.

8-23 The one-sided exponential PDF is given by

$$f(x) = \alpha \epsilon^{-\alpha x}, \quad 0 < x < \infty$$
$$= 0, \quad \text{for } x < 0$$

(a) Show that the net area under the curve is unity as required. Determine the (b) mean value, (c) mean-square value, and (d) standard deviation.

8-24 A certain random voltage $x(t)$ has a gaussian distribution with a mean $\overline{x} = 0$ V and a standard deviation $\sigma = 10$ V. A single random sample X of the voltage is taken. Determine the following probabilities concerning the value (in volts) of the random sample: (a) $P(X > 0)$, (b) $P(X > 10 \text{ V})$,

(c) $P(X < -10$ V$)$, (d) $P(-10 < X < 10)$, (e) $P(|X| > 10)$, (f) $P(X > 20)$, and (g) $P(|X| > 20)$.

8-25 Assume that the signal of Prob. 8-24 represents an undesirable noise signal and that it will trigger a "false alarm" in a circuit if the level exceeds 10 V. Suppose that the bandwidth of the circuit is such that two independent samples of the voltage have an "opportunity" to trigger the circuit in a certain time interval. Determine the probability of false alarm in the time interval.

8-26 A certain random voltage $x(t)$ has a gaussian distribution with a mean $\bar{x} = 12$ V and a standard deviation of $\sigma = 3$ V. A single random sample X of the voltage is taken. Determine the following probabilities concerning the value (in volts) of the random sample: (a) $P(X > 15)$, (b) $P(X < 9)$, (c) $P(X > 21)$, (d) $P(|X| > 21)$.

8-27 A certain random signal $x(t)$ has a gaussian distribution with a mean $\bar{x} = 0$ and a standard deviation σ. A random sample X is to be taken. A certain threshold level A_T is defined such that if $x(t)$ exceeds the threshold a "false alarm" will occur. Thus, $P(X > A_T) = p$ is to be studied. Determine the value of A_T in terms of σ for each of the following values of p: (a) $p = 0.5$, (b) $p = 0.1$, (c) $p = 10^{-2}$, (d) $p = 10^{-3}$, (e) $p = 10^{-4}$, (f) $p = 10^{-5}$.

8-28 Repeat Prob. 8-27 for the case where the signal has a mean value $\bar{x} = E$. The value of A_T in each case will now be expressed in terms of both σ and E.

8-29 Starting with the definition of σ^2 in Eq. (8-42), expand the integral, and verify the result of Eq. (8-43).

9 Noise

9-0 INTRODUCTION

The major objectives of this chapter are to provide an overview of electrical noise in general and to present techniques for analyzing thermal noise effects in particular. The effects of thermal noise occur in all communications systems, and other types of noise are often analyzed through equivalent thermal models.

The means by which noise effects are represented by either the concept of an equivalent noise temperature or by the concept of a noise figure will be presented. Techniques for determining the overall noise level for a receiver will be developed. The concepts discussed in this chapter will be used in the remainder of the book to predict overall system performance.

9-1 CLASSIFICATION OF NOISE

A most significant effect on the general operation and usable range of a given communications system is caused by the presence of noise, which disturbs or interferes with the desired signal and can render it completely unintelligible. There are a variety of different sources and types of electrical noise, and the different ways in which they have been categorized in the literature result in some ambiguities in providing a simple breakdown of the different types. However, on the broadest scale, noise can be classified as either *external* or *internal*.

External noise represents all the different categories of noise that arise outside of the communications system components and includes atmospheric noise, galactic noise, man-made noise, and interference from other communications sources. While it is possible to reduce the effects of external noise by filtering, amplitude limiting, changing the frequency of operation, and so on, nothing can be done within a communications system to completely eliminate these noise sources since they all arise external to the system. Each of the categories mentioned will be briefly discussed.

Atmospheric Noise

Atmospheric noise is produced mostly by lightning discharges in thunderstorms. It is usually the dominating external noise source in quiet locations at frequencies below about 20 MHz or so. However, the power spectrum of atmospheric noise decreases rapidly as the frequency increases so that this effect becomes relatively insignificant at frequencies well above this value. The level of atmospheric noise also decreases with increasing latitude on the surface of the globe, and it is particularly severe during the rainy season in regions near the equator.

Galactic Noise

Galactic noise is caused by disturbances originating outside the earth's atmosphere. The primary sources of galactic noise are the sun, background radiation along the galactic plane, and the many cosmic sources distributed along the galactic plane. The primary frequency range in which galactic noise is significant is from about 15 MHz to perhaps 500 MHz, and its power spectrum decreases with increasing frequency.

Man-Made Noise

Man-made noise is somewhat obvious from its title and consists of any source of electrical noise resulting from a man-made device or system. Among the chief offenders in this category are electric motors, automobile ignition systems, neon signs, and power lines. As one would likely suspect, the average level of man-made noise is significantly higher in urban areas than in suburban areas, and it is higher in suburban areas than in rural areas. This fact has led to the selection of certain remote rural areas for the locations of many of the satellite-tracking stations and radio astronomy observatories. The power spectrum of man-made noise decreases as the frequency increases, but the exact frequency range at which it becomes negligible is a function of its relative level. For example, in quite remote locations, the noise level from man-made sources will usually be below galactic noise in the frequency range above about 10 MHz or so.

One can debate as to whether or not *interference from other communication sources* should be classified as "noise." However, it produces many of the same interfering effects and can thus be classified as noise as far as the desired signal is concerned.

There are many sources of pollution of which most people are aware, for example, air pollution and water pollution. However, another form of pollution of major interest to the communications engineer or technologist, of which many people are not aware, is that of *spectral pollution*. Because of the large number of electromagnetic transmission sources, a large amount of spectral background radiation exists. The available frequency spectrum is another natural resource that is rapidly being depleted by the increasing utilization of so many different types of communications systems. As far as the discussion of noise sources is concerned, one must always be prepared to deal with interference from other sources operating at the same frequency at another location or, possibly, undesirable spectral emission at a different frequency than the frequency of operation. Two widely employed acronyms of interest in this context are RFI (radio-frequency interference) and EMC (electromagnetic compatibility).

The second broad category of noise is that of *internal* noise. Internal noise represents all the different categories of noise that arise inside the components of the communications system and includes thermal noise, shot noise, and flicker noise as the primary contributors. Although the term "internal" includes both the transmitter and receiver circuits, the region of primary concern is from the receiving antenna on through the receiver circuitry to the data output. It is in this portion of the system that internal noise is most troublesome in view of the typical small received signal levels. At the transmitter, sources of potential interfering noise such as power supply ripple, for example, can be virtually eliminated by good design, since the signal levels are quite large in comparison there.

Thermal Noise

Thermal noise is the result of the random motion of charged particles (usually electrons) in a conducting medium such as a resistor. Since all circuits necessarily contain resistive devices, thermal noise sources appear throughout virtually all electrical circuits. The power spectrum of thermal noise is quite wide and is essentially uniform over the RF spectrum of interest for most communications applications. Mathematical models for analyzing thermal noise are used either directly or indirectly for dealing with a variety of different types of noise. Consequently, much of the chapter is devoted to dealing with the development and application of thermal noise models, so we will postpone further consideration until the next section.

Shot Noise

Shot noise arises from the discrete nature of current flow in electronic devices such as transistors and tubes. For example, the electrons or holes crossing a semiconductor junction display a random variation of the time corresponding to the crossing, which in turn produces a random fluctuation of the current. The associated random variation in the current appears as a disturbance to the signal being processed by the device, and so the result is a form of noise. The power spectrum of shot noise is similar to that of thermal noise, and the two effects are usually lumped together for system analysis.

Flicker Noise

Flicker noise (also called $1/f$ noise) is a somewhat vaguely understood form of noise occurring in active devices such as transistors at very low frequencies. It is most significantly near dc and a few hertz and is usually negligible above about 1 kHz or so. Flicker noise is often a limiting factor for the minimum signal level that can be processed by a direct-coupled (dc) amplifier.

There are often other forms of noise that are peculiar to certain types of modulation systems; they will be discussed as the need arises. For example, we will see later that PCM systems exhibit an inherent noiselike uncertainty called *quantization noise,* and FM systems exhibit a form of noise in the receiver circuitry having a parabolic power spectrum.

9-2 THERMAL NOISE

This section, along with portions of the remaining sections of the chapter, will be devoted to a detailed discussion of the various properties of thermal noise. Not only is thermal noise of major importance in its own right, but many of the concepts utilized in modeling thermal noise are also used in studying other forms of noise.

As mentioned in the last section, thermal noise is the result of random fluctuations of the charge carriers in any conducting medium and is dependent on the temperature. As the physical temperature of the conductor increases, the random motion of the charge carriers is greater, and the thermal noise level in turn increases.

Consider a simple resistance of value R as shown in Fig. 9-1a. In basic circuit theory, the resistance is considered as a passive device with no energy present unless, of course, an external source is connected to it. However, that is the simplified point of view pertinent to lumped linear circuit theory. As it turns out, random motion of electrons inside the resistance results in the presence of a small amount of energy, which in turn can be represented as a small random voltage or current. For many applications, the desired signal levels are sufficiently large that they overshadow the small thermal noise

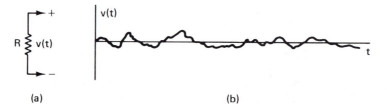

(a)   (b)

Figure 9-1.   (a) Simple resistance, and (b) small thermal random noise voltage appearing across it.

sources, but in communications applications where received signal levels are often down in the microvolt range, these thermal voltages may be quite significant and, in many cases, can completely mask the signal.

The actual noise voltage $v(t)$ appearing across a resistance is a random function, as illustrated in Fig. 9-1b, and it is assumed to be an ergodic function. Three general properties of this thermal noise will be stated at the outset before the various details are discussed:

1. The instantaneous amplitude of the noise voltage can be represented by a gaussian probability density function $f(v)$ with a mean $\bar{v} = 0$. The form of the variance will be discussed later.

2. The power spectrum of the noise (before filtering) is essentially constant for a given set of conditions over a very wide frequency range encompassing virtually all the frequency range used in conventional RF communications. Several variations of the power spectrum will be discussed throughout this chapter.

3. If gaussian distributed wideband noise is passed through a linear filter, the filtered output is also gaussian distributed. In general, this property is not true for all types of distributions, but a gaussian distribution remains as a gaussian distribution after linear filtering.

Because of the presence of a broad spectrum and the corresponding analogy with white light, thermal noise is frequently referred to as *white noise*.

Over the very wide frequency range in which the power spectrum is constant, the thermal noise appearing across the resistor is a function of the resistance, the absolute temperature, and the bandwidth over which the thermal noise appears. The bandwidth concept needs some further clarification, since the simple model shown in Fig. 9-1 shows no bandwidth-limiting parameters. Of course, all circuits are bandwidth limited in some way, even if by distributed reactances. However, in the actual applications of random-noise analysis, the noise level is always transferred to the remainder of the circuit through a certain finite bandwidth, and this bandwidth plays a significant part in determining the level of the resulting noise, as will be seen shortly.

First, consider a narrow segment of bandwidth Δf. The mean-square noise voltage in this frequency segment will be denoted by $\overline{\Delta v^2}$, and it can be shown from certain results of statistical thermodynamics that the differential open-circuit mean-square voltage across the resistor on a one-sided spectral basis is

$$\overline{\Delta v^2} = 4RkT\Delta f \qquad (9\text{-}1)$$

In this equation, the following parameters are defined:

R = resistance, in ohms

k = Boltzmann's constant = 1.38×10^{-23} joules/kelvin (J/K)

T = absolute temperature, in K

$\overline{\Delta f}$ = frequency increment, in hertz

$\overline{\Delta v^2}$ = mean-square voltage increment, square volts (V^2)

The reader may recall that the absolute temperature T in kelvin is determined from the SI temperature in celsius by the addition of 273 K.

The noise expression of (9-1) applies even when the circuit contains one or more reactive elements, provided that the frequency-dependent resistance $R(f)$ at a given set of terminals is used in place of R. Thus, consider a complex impedance, which may be represented as a function of frequency at a single set of terminals according to the model shown in Fig. 9-2. The impedance $Z(f)$ "seen" at the two terminals can be represented as

$$Z(f) = R(f) + jX(f) \qquad (9\text{-}2)$$

where $R(f)$ is the resistance (real part) and $X(f)$ is the reactance (imaginary part). The differential mean-square open-circuit voltage on a one-sided basis is now

$$\overline{\Delta v^2} = 4R(f)kT\,\Delta f \qquad (9\text{-}3)$$

By the appropriate limiting process, the total mean-square open-circuit voltage $\overline{v^2}$ can be expressed as

$$\overline{v^2} = \int_0^\infty 4R(f)kT\,df \qquad (9\text{-}4a)$$

$$= 4kT \int_0^\infty R(f)\,df \qquad (9\text{-}4b)$$

Consider now the special case where the open-circuit noise voltage is observed over the passband of a unity gain ideal low-pass filter with a rectangular passband characteristic of bandwidth B (on a one-sided basis).

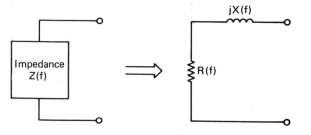

Figure 9-2. Form for complex impedance used in noise analysis.

Assuming that R is constant over this bandwidth, the mean-square open-circuit noise voltage is

$$\overline{v^2} = 4RkTB \qquad (9\text{-}5)$$

The form of (9-5) will be frequently used in many of the developments that appear in the text, even when no filtering elements are actually shown. It will be assumed in such cases that the mean-square open-circuit voltage of concern is the value that *would* be measured over the bandwidth B, so this subtle condition should be understood. This rather standard practice is common in noise analysis, and it simplifies some of the developments and the notation.

It should also be stressed that the preceding equations for voltage refer to the level measured under *open-circuit* conditions; that is, there is no loading by the external circuit used to measure or filter the noise. This point of view will be continued through this section, and the manner in which loading affects the noise will be considered in the next section.

It was stated at the outset that the mean value is $\overline{v} = 0$. This means that the mean-square voltage and the variance associated with the voltage are equal; that is, $\sigma_v^2 = \overline{v^2}$. (The subscript v is necessary here since it is also possible to refer to the variance of the noise current or to the variance of the noise power.) Likewise, the total rms voltage is equal to the ac rms voltage or standard deviation; that is,

$$v_{\text{rms}} = \sigma_v = \sqrt{\overline{v^2}}$$

From a knowledge of the mean and variance, the PDF $f_v(v)$ corresponding to the noise voltage could be constructed and various probability computations could be made. This process will be illustrated in Ex. 9-1.

Observe from (9-5) that the mean-square noise voltage (or variance) increases linearly with the resistance, the temperature, and the bandwidth. On the other hand, the rms voltage increases as the square root of the resistance, the temperature, and the bandwidth.

While the preceding equations hold for any arbitrary temperature T, many noise calculations and measurements utilize what has become a standard reference temperature $T = T_0 = 290$ K. The absolute temperature T_0 corresponds to the "energy conservative" conventional temperature of 17°C or 62.6°F. Development of this "cool" standard predated the current energy dilemma! Apparently, a major reason for this seemingly odd choice is that the product $kT = kT_0 = 4 \times 10^{-21}$ W/Hz, a rather convenient value. (While the units for k are J/K, the units of kT are W/Hz. Don't worry about it if you don't understand why!) In all subsequent developments, the symbol T_0 will be used to represent the standard temperature of 290 K.

A few representative calculations will be made at this point to assist the reader in developing a feel for the order of magnitude of typical noise voltages. Assume a resistance $R = 10$ kΩ, the standard temperature $T = T_0 =$

290 K, and assume a unit gain ideal rectangular bandwidth $B = 10$ kHz over which the noise is to be measured. From (9-5), the mean-square voltage (which is also the variance σ_v^2 of the voltage) is

$$\overline{v^2} = 4 \times 10^4 \times 1.38 \times 10^{-23} \times 290 \times 10^4 \qquad (9\text{-}6)$$
$$= 1.6 \times 10^{-12} \text{ V}^2$$

The rms voltage (or standard deviation σ_v of the voltage) is the square root of (9-6).

$$v_{\text{rms}} = \sqrt{\overline{v^2}} = 1.265 \ \mu\text{V} \qquad (9\text{-}7)$$

This voltage is obviously quite small and would be insignificant for applications in which the signal level was on the order of volts or even millivolts. However, the signal level at the input of a communications receiver is often of the order of microvolts, so this noise voltage could be quite significant in such cases.

If the resistor were changed from 10 kΩ to 1 MΩ, the noise voltage would increase by a factor of $\sqrt{100} = 10$, resulting in an rms value of 12.65 μV. If, in turn, the bandwidth were increased from 10 kHz to 1 MHz, the noise voltage would again increase by a factor of $\sqrt{100} = 10$, and the final rms noise voltage would be 126.5 μV. These values illustrate the potential noise difficulties that can arise in circuits having relatively large bandwidths and/or resistance values.

Before the reader decides to visit a laboratory and attempt to measure the noise voltage produced by a resistor, it should be pointed out that such a measurement is not a simple task at all. First, typical voltage values are obviously quite low and well below the range of most standard voltmeters. Second, if one attempts to amplify the noise to enhance the measurement process, additional noise and stray pickup are introduced by the amplifier itself, which makes it difficult to separate the noise produced by the resistor alone. Third, the bandwidth parameter B is based on an assumed ideal rectangular filter, and some means must be used to correct for the nonideal characteristics of an actual filter. Finally, a true rms instrument with a wide bandwidth is required for an accurate measurement, since the noise is a random wideband process. Thus, unless the reader has access to some rather sophisticated noise-measuring equipment and procedures, it is best at this point to accept the results as given. Special procedures and noise-measuring equipment have been developed for this purpose, and some of the problems at the end of the chapter will be concerned with this matter.

Example 9-1

Consider the first noise voltage calculated in the section, whose mean-square and rms values were given in Eqs. (9-6) and (9-7), respectively. Assume that this noise voltage is applied to the input of a hypothetical ideal noise-free amplifier with a voltage gain of 10^6 and whose bandwidth is exactly the same as in the calculation (i.e., 10 kHz). Determine (a) the rms value of the output open-circuit noise voltage, (b) the mean-square value of the output noise voltage, and (c) the probability that a single random sample of the output noise voltage will exceed a certain threshold level of 2 V.

Solution

(a) Let $v_0(t)$ represent the instantaneous noise voltage at the output of the ideal amplifier. The rms output voltage will be denoted as v_{orms}, and the output mean-square voltage will be denoted as $\overline{v_0^2}$. The rms output noise voltage is simply the rms input noise voltage times the voltage gain. Thus,

$$v_{\text{orms}} = 10^6 \times 1.265 \ \mu V = 1.265 \ V \tag{9-8}$$

(b) The mean-square output voltage is the square of the output rms voltage.

$$\overline{v_{\text{orms}}^2} = (1.265)^2 = 1.6 \ V^2 \tag{9-9}$$

Alternatively, the input mean-square voltage can be multiplied by the voltage gain squared; that is,

$$\overline{v_0^2} = 1.6 \times 10^{-12} \times (10^6)^2 = 1.6 \ V^2 \tag{9-10}$$

It can be inferred from these results that the rms voltage level at the input is multiplied by the voltage gain while the mean-square voltage (or power in some cases) is multiplied by the voltage gain squared (or power gain).

(c) The probability that a random sample of $v_0(t)$ exceeds 2 V is determined from the procedure of the last chapter based on a gaussian variable with a mean $\overline{v_0} = 0$ and a standard deviation $\sigma_{v0} = v_{\text{orms}} = 1.265$ V. A development follows:

$$P(V_0 > 2 \ V) = Q\left(\frac{2-0}{1.265}\right) = Q(1.581) = 0.056 \tag{9-11}$$

within the accuracy of the $Q(y)$ curve.

Example 9-2

Consider the parallel combination of the resistor and capacitor shown in Fig. 9-3. Determine the total mean-square noise voltage $\overline{v^2}$ appearing across the two terminals.

Solution

Due to the presence of the capacitor, the impedance will vary with frequency, so it is necessary to determine the specific variation of the resistance with frequency. The impedance $Z(f)$ is readily determined by a

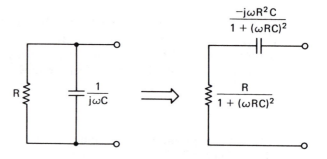

Figure 9-3. Circuit of Example 9-2.

parallel combination of the respective impedances. We have

$$Z(f) = \frac{R\left(\dfrac{1}{j\omega C}\right)}{R + \dfrac{1}{j\omega C}} = \frac{R}{1 + j\omega RC} \tag{9-12}$$

To determine the resistance $R(f)$, it is necessary to expand (9-12) into real and imaginary parts. This is achieved by multiplying numerator and denominator by the conjugate of the denominator, that is, $1 - j\omega RC$. This results in

$$Z(f) = \frac{R}{1 + (\omega RC)^2} - \frac{j\omega R^2 C}{1 + (\omega RC)^2} \tag{9-13}$$

in which $R(f)$ is recognized as the real part and is

$$RCf) = \frac{R}{1 + (\omega RC)^2} = \frac{R}{1 + (2\pi RCf)^2} \tag{9-14}$$

The mean-square voltage $\overline{v^2}$ is then determined from an application of (9-4b).

$$v^2 = 4RkT \int_0^\infty \frac{df}{1 + (2\pi RCf)^2} \tag{9-15}$$

This particular integral is found in most standard tables, and the result is the arctangent function. Without showing all the details, we have

$$\overline{v^2} = \frac{4RkT}{2\pi RC} \tan^{-1} 2\pi RCf \Big|_0^\infty = \frac{2kT}{\pi C}\frac{\pi}{2} = \frac{kT}{C} \tag{9-16}$$

This result may seem puzzling since the total mean-square voltage turns out to be completely independent of the resistance. The reason is that, while the mean-square voltage per unit bandwidth is directly proportional to R, the effective bandwidth over which the noise appears at the terminals is inversely proportional to R, and the two effects cancel.

9-3   COMBINATIONS OF RESISTORS

In the last section, both the noise voltage associated with a single resistance and the noise arising from the resistive part of a complex impedance were analyzed. In this section, the analysis will be extended to the case of any number of resistances connected in an arbitrary arrangement. There are some worthwhile concepts that arise in the modeling and analysis of such configurations.

In all cases considered in the last section, the *open-circuit* noise voltage across the resistor was the quantity of interest. Suppose, however, that there is a loading effect produced by another resistor. This leads us to some more general models for representing the random noise based on standard network theorems.

Both Thevenin's and Norton's theorems may be applied to the resistor, although it is easier to apply Thevenin's theorem first due to the fact that the noise has been stated so far in terms of voltage. In this case the mean-square open-circuit voltage is computed over a hypothetical rectangular bandwidth B as given by (9-5). With this source deenergized, the equivalent impedance is readily determined to be simply the resistance R. Thus, the Thevenin equivalent circuit of the resistor plus thermal noise is as shown in Fig. 9-4a. Observe that the usual plus and minus signs are omitted from the source in this case since they can be misleading. Instead, the $\overline{v^2}$ label identifies the source as a voltage source.

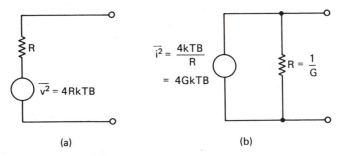

Figure 9-4.   (a) Thevenin and (b) Norton equivalent circuits for thermal noise sources associated with resistor.

Norton's theorem may now be applied directly to the Thevenin model by measuring the short-circuit current. Actually, it is often more convenient in noise analysis to work directly in terms of mean-square values; so let $\overline{i^2}$ represent the mean-square short-circuit current. Since we are working with squared quantities, we have

$$\overline{i^2} = \frac{\overline{v^2}}{R^2} = \frac{4RkTB}{R^2} = \frac{4kTB}{R} = 4GkTB \qquad (9\text{-}17)$$

where $G = 1/R$ is the conductance associated with the resistance. The corresponding Norton equivalent circuit is shown in Fig. 9-4b. Observe that the usual arrow is omitted from the source in this case since it can be misleading. Instead, the $\overline{i^2}$ label identifies that source as a current source.

It should be pointed out that the instantaneous short-circuit noise current $i(t)$ is gaussian distributed and can be analyzed from the same statistical point of view as the instantaneous open circuit voltage $v(t)$. Since $i(t) = v(t)/R$, the form of the current is identical with the form of the voltage and thus obeys all the general properties discussed in the last section for the noise voltage. The only difference, of course, is that the variables are different, and the units for the statistical parameters would appear in terms of amperes or amperes squared. Thus, when the short-circuit current is measured over a rectangular bandwidth B, the variance σ_i^2 of the current is equal to the mean-square value $\overline{i^2}$ (since $\overline{i} = 0$) and is

$$\sigma_i^2 = \overline{i^2} = 4\,GkTB \qquad (9\text{-}18)$$

The two equivalent circuits developed in this section can be used to simplify noise computations in circuits containing several resistors. One important rule should be remembered when combining the effects of sources contained in more than one resistor. The rule is that *the net mean-square (or power) effect produced by more than one independent noise source is obtained by adding individual mean-square (or power) effects.* This concept is based on the fact that the mean-square voltages (or powers) produced by separate resistors are all statistically independent, and the corresponding voltages or currents add incoherently. Thus, one does not simply add, for example, an rms noise voltage of 5 V to another rms voltage of 5 V to obtain an incorrect result of 10 V. Instead one adds 25 V² to 25 V² to obtain 50 V², corresponding to a net rms value of 7.07 V. This is the same process used in circuit theory for determining the net rms value of several sine waves at different frequencies. Because of the manner in which the variables add, it is often easier to work directly with squared voltage or current quantities in analyzing noise in resistive circuits.

To illustrate some of the preceding concepts, consider the simple series connection of two resistors, both at the same temperature T, as shown in Fig. 9-5a. The two resistors can be represented by their Thevenin models as shown in Fig. 9-5b. In view of the simple series connection, both the respective resistances and the mean-square voltages can be added. Thus, the net resistance R is given by

$$R = R_1 + R_2 \qquad (9\text{-}19)$$

and the net mean-square voltage $\overline{v^2}$ is expressed in terms of the separate mean-square voltages as

$$\overline{v^2} = \overline{v_1^2} + \overline{v_2^2} \qquad (9\text{-}20a)$$

$$= 4R_1kTB + 4R_2kTB \qquad (9\text{-}20b)$$

$$= 4RkTB \qquad (9\text{-}20c)$$

A resulting equivalent circuit is shown in Fig. 9-5c.

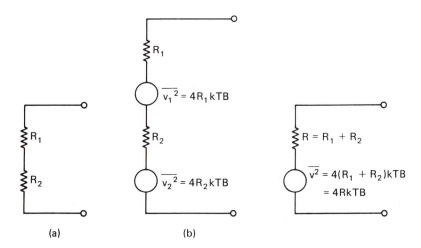

Figure 9-5.   Series combination of two resistors and equivalent models to represent the effect of thermal noise.

Several comments will now be made about this development. First, note in (9-20) that the combined effect of the sources was obtained by adding *mean-square* values, as previously discussed in the text. Polarity is not important for this purpose since the mean-square values are all positive anyway. Next, observe the result of (9-20c). This equation suggests that the net mean-square voltage is the value that would be associated with a single resistance $R = R_1 + R_2$. While this circuit represents a single simple example, it turns out the concept is more general and can be stated as follows: *The net thermal noise effect of any arbitrary combination of simple resistances all at a constant temperature T is the same as that of a single resistance whose value is the equivalent resistance of the combination at the reference terminals of interest.*

This concept reduces considerably the amount of effort involved with analyzing thermal noise in circuits containing several resistors. For example, consider the circuit shown in Fig. 9-6a, and assume that all resistors are at the same temperature T. Assume that a noise model is desired at the two terminals A-A'. The mean-square open-circuit voltage is quickly written as

$$\overline{v^2} = 4R_{eq}kTB \qquad (9\text{-}21)$$

where R_{eq} is the equivalent resistance looking to the left at A-A' and is expressed as

$$R_{eq} = R_4 \parallel (R_3 + R_1 \parallel R_2) \qquad (9\text{-}22)$$

The $\parallel$ symbols in (9-22) represent a parallel combination of the quantities on the two sides of the symbols. The resulting Thevenin equivalent for the combination is shown in Fig. 9-6b.

It was stated that this rather simplified reduction scheme was applicable when the temperature was the same for all the resistors. If the temperatures

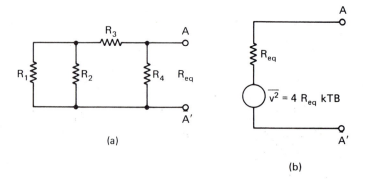

(a)

(b)

Figure 9-6. Combination of several resistors at the same temperature and the Thevenin equivalent circuit for the noise.

are different, the simplified approach is no longer valid. Instead, individual Thevenin and/or Norton models should be used for each resistor, and the effects can be combined by standard circuit analysis methods.

Example 9-3

This problem is designed to demonstrate the equivalence of several different approaches that could be used in the analysis of resistor combinations. Consider the simple parallel combination of two resistors R_1 and R_2 in

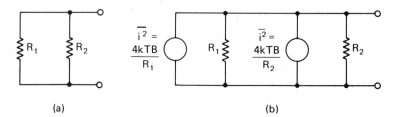

(a)

(b)

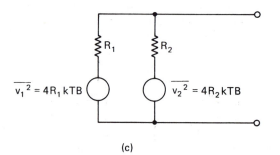

(c)

Figure 9-7. Parallel combination of two resistors and some equivalent circuits used in Example 9-3.

Fig. 9-7a, both of which are assumed to be at the same temperature T. Determine the open-circuit mean-square voltage $\overline{v^2}$ across the two terminals three ways: (a) the "easy" way by obtaining a net equivalent resistance, (b) by using the Norton models, and (c) by using the Thevenin models.

Solution

(a) Since the temperature is the same for both resistors, the combined effect is most easily determined by first obtaining the equivalent resistance R_{eq} of the parallel combination. This value is

$$R_{eq} = \frac{R_1 R_2}{R_1 + R_2} \tag{9-23}$$

The mean-square voltage is then

$$\overline{v^2} = 4R_{eq}kTB \tag{9-24}$$

(b) Norton models are often easier to work with for parallel combinations than Thevenin models, so the corresponding Norton forms for the resistors are shown in Fig. 9-7b. The two sources are readily combined into one equivalent current source whose mean-square current $\overline{i^2}$ is given by

$$\overline{i^2} = \overline{i_1^2} + \overline{i_2^2} \tag{9-25a}$$

$$= \frac{4kTB}{R_1} + \frac{4kTB}{R_2} \tag{9-25b}$$

$$= 4kTB\left(\frac{R_1 + R_2}{R_1 R_2}\right) = \frac{4kTB}{R_{eq}} \tag{9-25c}$$

To determine the mean-square open-circuit voltage, the Norton form is converted back to a Thevenin form. The mean-square voltage $\overline{v^2}$ is readily determined as

$$\overline{v^2} = R_{eq}^2 \overline{i^2} = R_{eq}^2\left(\frac{4kTB}{R_{eq}}\right) = 4R_{eq}kTB \tag{9-26}$$

which agrees with the result of (9-24).

(c) This last approach is the most clumsy of the three but should serve some educational value. Consider the circuit form shown in Fig. 9-7c in which the Thevenin models are employed. Note the points in the circuit across which the voltage is to be measured.

Because of some potential sign problems, the superposition principle is probably the easiest way for obtaining the combined effects of several sources in this type of arrangement. Superposition of mean-square values is valid due to the fact that the different sources are statistically independent. With this approach, the mean-square voltage across the two terminals will be represented as

$$\overline{v^2} = \overline{v_a^2} + \overline{v_b^2} \tag{9-27}$$

where $\overline{v_a^2}$ is the contribution from $\overline{v_1^2}$ with $\overline{v_2^2}$ deenergized and $\overline{v_b^2}$ is the contribution from $\overline{v_2^2}$ with $\overline{v_1^2}$ deenergized. The various calculations follow.

$$\overline{v_a^2} = \frac{R_2^2 \overline{v_1^2}}{(R_1 + R_2)^2} = \frac{R_2^2}{(R_1 + R_2)^2}(4R_1 kTB) \tag{9-28}$$

$$\overline{v_b^2} = \frac{R_1^2 \overline{v_2^2}}{(R_1 + R_2)^2} = \frac{R_1^2}{(R_1 + R_2)^2}(4R_2 kTB) \tag{9-29}$$

$$\overline{v^2} = \frac{(R_2^2 R_1 + R_1^2 R_2)}{(R_1 + R_2)^2}(4kTB)$$

$$= \frac{R_1 R_2 (R_1 + R_2)}{(R_1 + R_2)^2}(4kTB) = \frac{R_1 R_2}{R_1 + R_2}(4kTB) = 4R_{eq}kTB \tag{9-30}$$

which is the same as obtained by the two easier approaches. Observe the manner in which the squared quantities were manipulated throughout the development. Observe also that the superposition effects were *added* directly, since mean-square values are always positive. Thus, polarity is unimportant when combining mean-square and power values produced by different independent sources.

Example 9-4

Consider the circuit of Fig. 9-8a containing two resistors at different temperatures in series. Determine the open-circuit mean-square voltage and a Thevenin equivalent circuit for the combination.

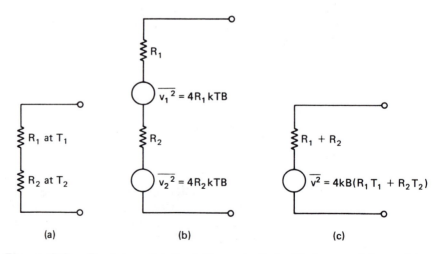

(a)   (b)   (c)

Figure 9-8. Resistor circuit of Example 9-4 with two resistors at two different temperatures.

Solution

The simplest approach of determining the noise voltage directly due to the series resistance combination does not work in this case due to the fact that the two resistors have different temperatures. Instead, Thevenin models for the two separate resistors are first defined as indicated by Fig. 9-8b. The net mean-square voltage is

$$\overline{v^2} = 4R_1 kT_1 B + 4R_2 kT_2 B \qquad (9\text{-}31)$$
$$= 4kB(R_1 T_1 + R_2 T_2)$$

This result is slightly more involved than simply combining the two resistors, a result of the two distinct temperatures.

The net Thevenin equivalent circuit is $\overline{v^2}$ of (9-31) in series with the net resistance $R_1 + R_2$, as shown in Fig. 9-8c. One must be careful about the meaning of this equivalent circuit. While it produces correct results as far as any external circuit is concerned, like all Thevenin and Norton equivalent forms, the internal structure may be different. In this case, the equivalent resistance $R_1 + R_2$ must be considered to be at a fictitious temperature $(R_1 T_1 + R_2 T_2)/(R_1 + R_2)$ in order to satisfy the simplest form of the noise expression of a single resistance as given by (9-5). (See Problem 9-22.)

9-4 POWER SPECTRUM OF NOISE

So far in the chapter the emphasis on thermal noise models has been directed toward voltage and current representations with the bandwidth B over which the quantities were measured understood. These are the easiest forms to understand at the beginning, since they are most related to standard circuit quantities that can theoretically be measured.

For system level analysis of noise effects, the concept of the power spectral density is often more appropriate. In this section, several useful variations of this concept will be developed. In a collective sense, all the different forms in this section will be considered as *power spectral density* functions even though voltage-squared and current-squared functions are included in the considerations.

First, consider again the expression for the mean-square value of the open-circuit voltage across a resistance.

$$\overline{v^2} = 4RkTB \qquad (9\text{-}32)$$

Instead of specifying a hypothetical bandwidth B, a one-sided open-circuit *mean-square voltage spectral density function* can be defined as

$$S_v^{(1)}(f) = 4\,RkT \qquad \text{V}^2/\text{Hz}, \qquad (9\text{-}33)$$

where the superscript (1) implies a one-sided spectral representation. If

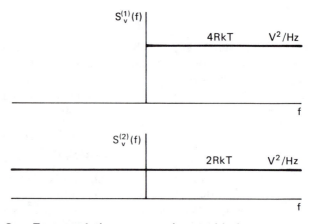

Figure 9-9. Forms of the one- and two-sided mean-square voltage spectral density functions for unfiltered white noise.

desired, the mean-square voltage spectral density can also be represented on a two-sided basis as

$$S_v^{(2)}(f) = 2RkT \qquad \text{V}^2/\text{Hz} \tag{9-34}$$

where the superscript (2) implies a two-sided spectral representation.

The forms of the mean-square voltage representations of (9-33) and (9-34) are compared in Fig. 9-9 for unfiltered thermal white noise. In the one-sided case, the power is assumed to exist only over positive frequencies but has twice the level of the second representation, which is assumed to exist over both positive and negative frequencies.

Assume now that the resistance is connected across the input to a linear system having a voltage transfer function $H(f)$. It is assumed for this presentation that no additional noise is contributed by the system. It is also assumed that either (1) the input to the system is an open circuit so that no loading effect occurs on the noise source, or (2) the transfer function includes the effect of R at the input.

The output one-sided mean-square voltage spectral density $S_{v0}^{(1)}(f)$ is given by

$$S_{v0}^{(1)}(f) = |H(f)|^2 S_v^{(1)}(f) \tag{9-35}$$

$$= |H(f)|^2 4RkT$$

The total output open-circuit mean-square voltage $\overline{v_0^2}$ is

$$\overline{v_0^2} = \int_0^\infty S_{v0}^{(1)}(f)\, df = \int_0^\infty |H(f)|^2 S_v^{(1)}(f)\, df$$

$$= 4RkT \int_0^\infty |H(f)|^2\, df \tag{9-36}$$

A similar result could be expressed on a two-sided basis by using the form of (9-34) as the spectral density, and the integration in that case would have to be performed over both negative and positive frequencies. The reader is invited to verify the equivalence of the two forms.

In a similar fashion, a one-sided short-circuit *mean-square current density function* $S_i^{(1)}(f)$ can be defined as

$$S_i^{(1)}(f) = 4GkT \qquad \text{A}^2/\text{Hz} \qquad (9\text{-}37)$$

The total output short circuit mean-square current $\overline{i_0^2}$ is then

$$\overline{i_0^2} = 4GkT \int_0^\infty |H(f)|^2 \, df \qquad (9\text{-}38)$$

In this case, the transfer function $H(f)$ must be considered as a *current transfer function* relating the output short-circuit current to the input current. It is also assumed that either (1) the input to the system is a short circuit so that all the noise current flows into the system, or (2) the transfer function includes the effect of R at the input. A two-sided form could also be developed if desired.

The integration forms such as (9-36) and (9-38) are most often carried out in practical problems through the use of an equivalent noise bandwidth concept. The details of this approach will be developed in the next section, as this section is devoted primarily to establishing a theoretical basis and to discussing the different spectral forms that can be used.

We next turn to the case of the actual *power* spectral density for the noise. Voltage-squared spectral density is best represented with an *open-circuit* form, and current-squared spectral density is best represented with a *short-circuit* form. Neither of these forms is ideal for power, since no power is actually transferred when a source looks into either an ideal open circuit or an ideal short circuit.

It turns out that the most convenient and widely employed form for power spectral density utilizes the *available-power* concept under the assumption of matched conditions. This concept can be a little confusing at the outset, so some initial clarification and motivation are necessary.

Consider the nonideal source model of Fig. 9-10 containing a voltage source having an rms value V_{rms} in series with a resistance R_s. The source and resistor are assumed to be "in a box" so that neither value can be changed.

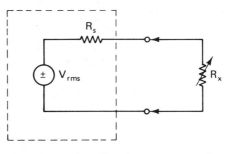

Figure 9-10. Circuit used to develop concept of available power.

Assume now that the external variable load resistance R_x is connected to the terminals and that the power delivered to R_x is measured (or calculated) as a function of R_x. From the maximum power transfer theorem, it is well known that maximum load power will be obtained when $R_x = R_S$. Under this matched condition, the output load power is the largest that it can ever be for the given form of the source. The value of this maximum possible load power will be defined as the *available power* P_a. By simple circuit analysis performed under matched conditions, the available power is seen to be

$$P_a = \frac{V_{rms}^2}{4R_s} \qquad (9\text{-}39)$$

It should be stressed that this is the maximum power that can be obtained *external* to the source V_{rms} in series with R_s. It does not include the internal dissipation in R_s itself, which, incidentally, is equivalent to the external load power under matched conditions.

To summarize, the available power corresponding to a source and an internal resistance is the maximum power that can be obtained in an external load. This available power is obtained only when the load resistance is equal to the internal resistance. (If the internal impedance is complex, the matched condition corresponds to the load impedance being equal to the conjugate of the internal impedance, but these results can be adapted to that case.)

This concept of representing power in terms of available power on a matched basis is important for the following reasons:

1. Many communications systems are designed to exhibit matched impedances at all junctions between stages to minimize reflections and to maximize power transfer. In such cases, available power at each stage truly represents the power transferred on to the next stage.

2. Even in systems having impedance mismatches, the concept of available power can still be used (with some care) for computations involving signal power to noise power ratios. In such cases, niether the signal power nor the noise power computed on an available-power basis may represent the true power levels, but any difference factors affect signal and noise the same way and thus cancel out in the process.

3. As will be seen shortly, the expression for available noise power is quite simple in form and does not specifically require that resistance levels in the system be known for many calculations.

Consider now a resistance R at an absolute temperature T, as shown in Fig. 9-11a. Voltage and current models have already been discussed at length, but the immediate objective is to obtain a form for available power. This can be readily achieved from the circuit form of Fig. 9-11b in which the resistor is connected to an external load R_x and in which the Thevenin internal model for R has been used. Maximum power is transferred to R_x when $R_x = R$.

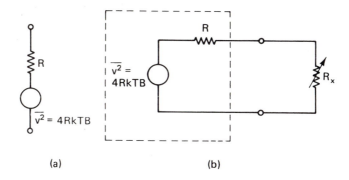

Figure 9-11. Noise model used to develop expression for available noise power.

From the result of (9-39), the available power P_a over a bandwidth B can be expressed as

$$P_a = \frac{\overline{v^2}}{4R} = \frac{4RkTB}{4R} = kTB \qquad (9\text{-}40)$$

This result is rather interesting in that the available power turns out to be completely independent of the resistance! Thus, while the open-circuit noise voltage and the short-circuit noise current are both heavily dependent on the value of the resistance, the actual noise power transferred on a matched basis is completely independent of the actual resistance value. This result is actually simpler in form than those for voltage or current, and it is not even necessary to know the resistance levels in many calculations using the available power concept.

We will next consider the associated forms for power spectral density. Removing the reference of the bandwidth factor B in (9-40), a *one-sided available power spectral density* $S_p^{(1)}(f)$ for the noise can be defined as

$$S_p^{(1)}(f) = \eta = kT \qquad \text{W/Hz} \qquad (9\text{-}41)$$

where the symbol η has been introduced as a constant representing the one-sided power density. The corresponding two-sided density $S_p^{(2)}(f)$ is

$$S_p^{(2)}(f) = \frac{\eta}{2} = \frac{kT}{2} \qquad \text{W/Hz} \qquad (9\text{-}42)$$

Note the use of η and $\eta/2$ in the two respective forms. These terms will appear quite frequently throughout the remainder of the book. The forms of the one-sided and two-sided power spectral density functions for white noise are shown in Fig. 9-12.

The power spectral density $\eta = kT$ (or $\eta/2 = kT/2$) for a resistance is constant over a very wide frequency range from near dc to well above the range for normal RF communications. Eventually, the power spectrum begins

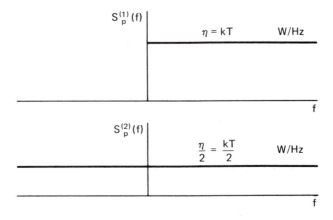

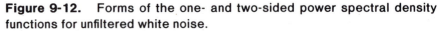

Figure 9-12. Forms of the one- and two-sided power spectral density functions for unfiltered white noise.

to decrease for, otherwise, it would imply an infinite amount of thermal energy present.

Consider now a *matched* transfer function $H(f)$, which is assumed to contain no internal sources of noises. A resistance R is connected to the input port. The fact that the transfer function is matched means that the input impedance is also R. Furthermore, it also means that a load resistance R_L connected to the output is matched to the internal output, and the available output power is delivered to this load. The output one-sided available power density $S_{p0}^{(1)}(f)$ is given by

$$S_{p0}^{(1)}(f) = |H(f)|^2 S_p^{(1)}(f) \tag{9-43}$$
$$= |H(f)|^2 \eta = |H(f)|^2 kT$$

The total output available noise power N_0 is then obtained by the integral

$$N_0 = \int_0^\infty S_{p0}^{(1)}(f)\,df = \eta \int_0^\infty |H(f)|^2\,df = kT \int_0^\infty |H(f)|^2\,df \tag{9-44}$$

Alternatively, the same power expressed in terms of the two-sided density is

$$N_0 = \int_{-\infty}^\infty S_{p0}^{(2)}(f)\,df = \frac{\eta}{2} \int_{-\infty}^\infty |H(f)|^2\,df = \frac{kT}{2} \int_{-\infty}^\infty |H(f)|^2\,df \tag{9-45}$$

Detailed computations of this type will be discussed in the next section using the equivalent noise bandwidth concept.

Having introduced the one- and two-sided forms for the various functions in this section, we will now simplify the notation for further developments. The superscripts (1) and (2) will be eliminated, and the adjectives "one-sided" and "two-sided" will be used when required. Since true power (often the available power) will be the most common spectral function of interest, the subscript p will be dropped, and the adjective "available" will be used as required. However, the subscripts v and i will be retained to define voltage-squared and current-squared density functions. Thus, $S(f)$, $S_v(f)$, and $S_i(f)$ will be used to represent power spectral density, voltage-squared spectral density, and current-squared density functions, respectively.

9-5 EQUIVALENT NOISE BANDWIDTH

In the past section, several forms of power-related spectral density functions were introduced. The first was a mean-square voltage density function defined on an open-circuit basis, the second was a mean-square current density function defined on a short-circuit basis, and the third was an available power spectral density function defined on a matched basis in which load and source resistances were assumed to be equal at all junctions. In each of the three cases, the pertinent total output quantity involved an integration of a transfer function magnitude squared over an infinite frequency range. The three integration formulas of primary interest to this development are (9-36), (9-38), and (9-44), which are the three one-sided forms corresponding to mean-square voltage, mean-square current, and available power, respectively.

The result of either one of these integrations is the total amount of noise that passes through the given system over all frequencies. One could, of course, attempt to carry out the appropriate integration each time the noise is to be evaluated. However, such integrations become rather messy for all but the simplest types of transfer functions. We shall develop an approach here that permits the use of a standard result for a given filter type, meaning that the required integration need be carried out only once for each type.

The approach makes use of a parameter called the *equivalent noise bandwidth* B_n, which will be defined on a one-sided basis. A given transfer function, whether it represents a linear amplifier, a filter, or a combination of several units, can be represented by a maximum voltage squared, current squared, or power gain H_0^2 and an equivalent noise bandwidth B_n. The equivalent noise bandwidth of a transmission system is the bandwidth of a hypothetical rectangular characteristic that produces the same output noise power as the given transmission system when the input noise has a uniform spectral density.

Refer now to Fig. 9-13 for assistance in developing and understanding this concept. While either one of the three forms discussed earlier could be used in this development, the available power spectral density form will be selected. From (9-44), the actual output available power N_0 is

$$N_0 = \eta \int_0^\infty |H(f)|^2 \, df \qquad (9\text{-}46)$$

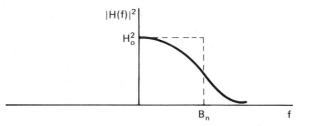

Figure 9-13. Concept of the equivalent noise bandwidth.

The hypothetical rectangular filter has a bandwidth B_n and a maximum power gain level H_0^2. It would produce an output available power N_0' given by

$$N_0' = \eta H_0^2 B_n \qquad (9\text{-}47)$$

From the forms of (9-46) and (9-47), we set $N_0' = N_0$ and obtain

$$B_n = \frac{1}{H_0^2} \int_0^\infty |H(f)|^2 \, df \qquad (9\text{-}48)$$

Once B_n for a particular filter has been determined from (9-48), it may be used for all subsequent noise computations with that particular filter, thus eliminating the need to perform the integration process each time a routine noise calculation is made. The noise bandwidth is, in general, a different value than the usual bandwidth parameters, such as 3-dB bandwidth or 1-dB bandwidth. For filters having very sharp cutoff rates, the noise bandwidth may be very close to the usual bandwidth definition, and the two are often assumed to be the same. However, for relatively slow rolloff rates, it is best to try to determine the equivalent noise bandwidth of the system before attempting any accurate noise calculations.

A point about the level factor H_0^2 should be made. We have assumed that H_0^2 represents the maximum level of the power transfer function. Occasionally, however, the noise bandwidth is defined in terms of a particular level that may not always represent the maximum level. For example, a low-pass filter characteristic could have H_0^2 defined in terms of the dc gain, which is usually (but not always) the maximum gain. For some filter types (e.g., even-order Chebyshev filters), the maximum response occurs at some frequency other than dc so that a different value of B_n would be obtained if H_0^2 were defined as the dc gain. In most applications, H_0^2 is defined as the maximum level, and that assumption will be made throughout this book. The preceding discussion is made only to caution the reader to be alert to the possibility of occasionally finding a different form in some references.

As the number of poles in the filter increases, the ratio of B_n to most conventional bandwidth definitions becomes closer to unity for filter types having relatively flat passband characteristics. Unfortunately, the evaluation of B_n becomes increasingly more difficult in such cases. Fortunately, a number of useful cases have been evaluated and tabulated using contour integration and/or numerical integration methods, and these results are available in the literature. An abbreviated set of noise bandwidth values for low-pass Butterworth filters up to 10 poles is provided in Table 9-1. In each case, the maximum value of H_0^2 occurs at dc and is $H_0^2 = 1$ for a unity gain filter. In each case f_1 is the 3-dB bandwidth, which is also frequently called the *cutoff frequency*.

We are now in a position to summarize the algebraic forms of the equations for determining the output available noise power N_0, the output mean-square voltage $\overline{v^2}$, or the output mean-square current $\overline{i^2}$ in terms of the equivalent noise bandwidth when the input is white noise. The reader should keep in mind that, while the same symbol is used for the transfer function in all three equations, $H(f)$ has a different meaning in each. The equations are

$$N_0 = kTH_0^2 B_n = \eta H_0^2 B_n \tag{9-49}$$

$$\overline{v^2} = 4RkTH_0^2 B_n \tag{9-50}$$

$$\overline{i^2} = 4GkTH_0^2 B_n \tag{9-51}$$

where $R = 1/G$ is the resistance at the input, and the equivalent noise bandwidth B_n is defined on a one-sided basis according to (9-48).

In cases where a two-sided spectral representation is desired, the input spectral density is multiplied by $\frac{1}{2}$ and a two-sided noise bandwidth $2B_n$ is used. The result clearly is the same since it amounts to nothing more than $(\frac{1}{2}) \times (2) = 1$. Yet it is a common source of confusion in communications system analysis, and everyone working in the field (including this author) has likely made errors because of it.

To simplify the notation, the subscript n will be omitted in some of the later developments when it is clear that the bandwidth is the equivalent noise value.

Table 9-1.   Equivalent Noise Bandwidths of Butterworth Filters

Poles	1	2	3	4	5	6	7	8	9	10
B_n/f_1	1.571	1.111	1.047	1.026	1.017	1.012	1.008	1.006	1.005	1.004

Example 9-5

Determine the equivalent noise bandwidth of the *RC* one-pole low-pass filter shown in Fig. 9-14. The input is v_1 and the output is v_2.

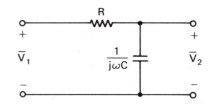

Figure 9-14. *RC circuit of Example 9-5 whose equivalent noise band-width is to be determined.*

Solution

The transfer function $H(f)$ relating the output open-circuit voltage v_2 to the input voltage v_1 must be determined. Working initially with ω and replacing it later by $2\pi f$, we have

$$H(f) = \frac{\overline{V}_2}{\overline{V}_1} = \frac{1/j\omega C}{R + (1/j\omega C)} = \frac{1}{1 + j\omega RC} \qquad (9\text{-}52)$$

$$= \frac{1}{1 + j2\pi RCf} = \frac{1}{1 + j(f/f_c)}$$

where f_c is the 3-dB cutoff frequency given by

$$f_c = \frac{1}{2\pi RC} \qquad (9\text{-}53)$$

The magnitude-squared function $|H(f)|^2$ of the transfer function is

$$|H(f)|^2 = \frac{1}{1 + (f/f_c)^2} \qquad (9\text{-}54)$$

The maximum value of the magnitude-squared function occurs at dc and is simply $H_0^2 = 1$. The equivalent noise bandwidth is then computed from the definition of (9-48). We have

$$B_n = \frac{1}{H_0^2} \int_0^\infty |H(f)|^2 \, df = \int_0^\infty \frac{df}{1 + (f/f_c)^2} \qquad (9\text{-}55)$$

$$= f_c \tan^{-1} \frac{f}{f_c} \Big]_0^\infty = \frac{\pi}{2} f_c = 1.57 f_c$$

Thus, the equivalent noise bandwidth of a one-pole low-pass filter is about 57% larger than the 3-dB bandwidth.

Some alternative forms for B_n of the one-pole filter can be determined by substituting (9-53) in (9-55). They are

$$B_n = \frac{\pi}{2} \left(\frac{1}{2\pi RC} \right) = \frac{1}{4RC} = \frac{1}{4\tau} = \frac{\alpha}{4} \qquad (9\text{-}56)$$

where $\tau = RC$ is the time constant of the RC network, and α is the magnitude of the pole location in the s-plane; that is, $s = -\alpha$ is the pole of the s-plane transfer function.

Example 9-6

Consider the two-amplifier system shown in Fig. 9-15. Assume that the input is a white thermal noise source $n_i(t)$ having a one-sided available power spectral density $S(f) = \eta_i = 1 \text{ pW/Hz} = 1 \times 10^{-12} \text{ W/Hz}$. Assume that all impedances at various junctions in the circuit are matched over the output bandwidth of interest so that maximum power is transferred at each stage. Assume finally that the noise contributions from the amplifiers, as well as all circuit resistances (including the load), are sufficiently small in comparison with the noise produced by the input source that they can be neglected. Calculate (a) the available noise power N_1 at the output of the first stage and (b) the noise power N_2 delivered to the load by the second stage. Note that the decibel power gains of the two stages as well as their equivalent noise bandwidths are shown on the figure.

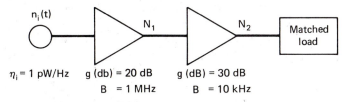

Figure 9-15. Two-amplifier cascaded system of Example 9-6.

Solution

The two gains in decibels must first be converted to the corresponding absolute power gains. The manner in which the gains are given here is typical of manufacturers' specifications, which often provide decibel ratings. The two decibel gain values convert to absolute power gains of $g_1 = 100$ and $g_2 = 1000$, respectively.

(a) The available noise power N_{al} at the output of the first stage is obtained from the following formulation:

$$N_1 = \eta_i g_1 B_1 \tag{9-57}$$
$$= 1 \times 10^{-12} \times 100 \times 1 \times 10^6 = 100 \ \mu\text{W}$$

There is a temptation for us to say that this total amount of power is transferred on the second stage, and this may be the case. However, some caution must be exercised for the following reason: It has been stated that impedances are matched only over the final bandwidth of interest, that is, 10 kHz for the output amplifier. Outside this bandwidth, the input impedance of the last stage may not be matched to the output impedance of the first stage, and so the total available power N_1 may not be accepted by the second stage. Thus, all we can comfortably say is that the available output power of the first stage is 100 μW.

(b) Irrespective of the behavior of the second stage outside its passband, its available output power, which is equal to the power delivered to the load, is obtained from the following formulation:

$$N_2 = \eta_i g_1 g_2 B_2 \tag{9-58}$$
$$= 1 \times 10^{-12} \times 100 \times 1000 \times 1 \times 10^4$$
$$= 1000 \ \mu\text{W} = 1 \ \text{mW}$$

Several interesting observations can be made from this result. The first is that the output noise power of the second stage is only ten times the available power at the output of the first stage, even though the power gain of the second stage is 1000. The reason is that the bandwidth of the second stage is much narrower than that of the first stage and tends to offset the gain factor. The second observation is that in a chain of amplifiers it is the *most narrow* noise bandwidth that establishes the effective bandwidth of the output noise power. For example, if B_1 were smaller than B_2, it would be B_1 that would be used in (9-58) instead of B_2. A little common sense and reasoning is better in this regard than attempting to memorize specific sets of formulas.

When we begin analyzing complete systems later in the book, the type of situation just encountered will frequently arise in a communications receiver. One particular portion of most receivers (specifically the IF amplifier) usually establishes the minimum bandwidth of the receiver, and it is this bandwidth that is used in determining the output noise power.

An alternative, but equivalent, point of view is to first determine the available noise power spectral density at each stage and then multiply by the corresponding bandwidth. Thus, let η_1 and η_2 represent the respective one-sided noise densities over the *flat portions* of their frequency range.

$$\eta_1 = \eta_i g_1 \tag{9-59}$$
$$= 1 \times 10^{-12} \times 100 = 100 \ \text{pW/Hz}$$
$$\eta_2 = \eta_i g_1 g_2 \tag{9-60}$$
$$= 1 \times 10^{-12} \times 100 \times 1000 = 0.1 \ \mu\text{W/Hz}$$

By multiplying the preceding values by their respective noise bandwidths, the preceding values of N_1 and N_2 are again obtained. This approach works equally well, but the reader should bear in mind that the values η_1 and η_2 do not represent the power spectral density values for truly white sources since the noise spectra at the outputs of both stages are now filtered. Instead, they represent the maximum power spectral densities for the hypothetical rectangular filter characteristic used in the equivalent noise bandwidth formulation.

9-6   MODELS FOR INTERNALLY GENERATED NOISE

The emphasis in this section will be directed toward the development and discussion of the means by which the overall effects of noise generated within an amplifier or other linear device may be represented. The actual noise originating in a complex combination of amplifier stages (such as a receiver) arises from a variety of different mechanisms, including thermal noise, shot noise, and special types of noise peculiar to particular active devices. While component designers are likely interested in each of these individual noise sources and how they can be minimized, the communications systems designer is seldom able to separate such effects. Instead, he or she must work with the combined effects of all the noise originating in the unit and deal with the noise on a systems level.

The appropriate approach at the systems level, then, is a model that includes all the noise effects as a composite combination. The noise model is of primary importance at the receiver. The signal levels are so low at the receiver input that noise effects originating within the receiver can have overwhelming effects on the performance if proper conditions are not met.

A number of different ways have been developed for representing the noise produced within a receiver or amplifier. The two most widely employed methods are (1) the *effective noise temperature* method and (2) the *noise figure* method. While either method alone would suffice, in actual practice manufacturers often mix specifications based on the two different concepts, so anyone working with noise analysis of receiver components must be able to work with both forms. This author has a preference for the noise temperature approach, and, as a result, this approach will be considered first.

Effective Noise Temperature Method

The concept of the effective noise temperature will be illustrated by the amplifier block diagram shown in Fig. 9-16. In much of the systems-level work dealing with overall noise effects, available output power, available input power, and available power gain are used almost exclusively. The available power gain g_a is defined as

$$g_a = \frac{\text{available output power}}{\text{available input power}} \qquad (9\text{-}61)$$

Recall that if impedances are matched at both input and output the available input power is the actual input power, the available output power is the actual output power, and the available power gain is the actual power gain.

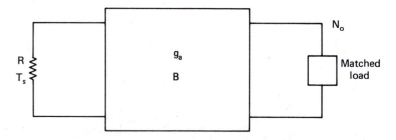

Figure 9-16. Block diagram used in defining noise temperature.

Assume that the input of the amplifier is terminated in a matched resistance R. This resistance represents an external "source" resistance, and its temperature is assumed to be T_s. The available one-sided source power spectral density is simply kT_s W/Hz. If the amplifier contributed no noise at all, the output noise power N_0' would simply be

$$N_0' = g_a k T_s B \tag{9-62}$$

where B is the equivalent noise bandwidth of the amplifier. However, the actual output noise power N_0 is always greater than N_0' due to the additional noise introduced by the amplifier. Thus, let

$$N_0 = N_0' + N_{\text{add}} \tag{9-63}$$

where N_{add} is the additional noise arising in the amplifier. This additional noise is assumed to originate from a separate fictitious input matched source resistor having an *effective noise temperature* T_e such that

$$N_{\text{add}} = g_a k T_e B \tag{9-64}$$

meaning that T_e can be defined as

$$T_e = \frac{N_{\text{add}}}{g_a k B} \tag{9-65}$$

Combining (9-62) through (9-65), the total output noise can be expressed as

$$N_0 = g_a k (T_s + T_e) B \tag{9-66}$$

The result of (9-66) represents the practical form in which most interpretations of noise temperature are utilized. If the effective noise temperature of an amplifier is known, it is *added to the source input temperature* at the amplifier *input* to determine the total effective input temperature. This combined net *input* temperature* will be denoted at T_i, and it is

$$T_i = T_s + T_e \tag{9-67}$$

The net input temperature T_i is then treated the same way in determining the total output noise as the preceding input source temperature was treated when the amplifer was considered noise free.

---

*This temperature is denoted as the *operating* temperature in many references.

One point should be stressed before continuing. The effective noise temperature T_e is *not* a physical temperature. Indeed, some noisy amplifiers exhibit effective noise temperatures of several thousand kelvin, and such physical temperatures would be absurd. Instead, the effective noise temperature can be thought of as the hypothetical physical temperature that a fictitious resistor would have to assume at the input in order to produce the same output noise power as the given amplifier.

As a matter of fact, even the source input temperature T_s need not necessarily be a physical temperature. For sources that contribute noise in excess of the thermal noise corresponding to the physical temperature, the value T_s can be adjusted to represent the combined effects. For example, a properly designed antenna at the input to a receiver acts as a constant resistance over the passband of interest, and it possesses a certain physical temperature. However, depending on its orientation and beamwidth, it can provide additional noise to the receiver input, and so its effective source temperature T_s may be quite different from the physical temperature.

From the preceding few paragraphs, the reader may suddenly decide that "a noise temperature is not really a temperature at all." This statement is both "right and wrong." In its simplest form for representing the true thermal noise produced by a simple resistance, the noise temperature is actually a physical temperature. However, in actual usage, it has taken on a more general meaning, and it is widely used to represent noise effects for situations in which the physical temperature is quite different. Thus, one should be aware of this common discrepancy between physical and noise temperature when making noise calculations.

Noise Figure Method

Next, we consider the concept of the *noise figure F* (also called the *noise factor*). There are a number of variations in the literature on the definition of noise figure depending on whether it is defined on a frequency dependent (or "spot") basis or on an overall average basis. In accordance with most manufacturers' specifications, we will consider only the average basis representation. Refer to Fig. 9-17, which is very similar to Fig. 9-16 except that a

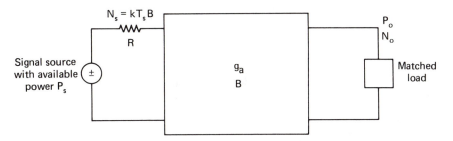

Figure 9-17. Block diagram used in defining noise figure.

few additional quantities are labeled on this figure. Along with the noise source input, a signal source input is also assumed. The available power corresponding to the signal source input is denoted as P_s, and the available signal output power is denoted as P_0. The output noise power N_0 is the same as with the noise temperature development, and $N_s = kT_sB$ is the input available noise source power defined over the equivalent noise bandwidth. Let P_s/N_s represent the signal source power to noise power ratio, and let P_0/N_0 represent the output signal power to noise power ratio.

The noise figure F is defined as

$$F = \frac{P_s/N_s}{P_0/N_0}\bigg]_{T_s = T_0} \tag{9-68}$$

Stated in words, the noise figure is the source signal-to-noise ratio divided by the output signal-to-noise ratio determined *at the standard reference temperature of T_0* = 290 K. Some of the difficulties that many persons experience interpreting and using the noise figure definition are related to this fixed temperature assumption. One has to exercise some care in using the noise figure at operating temperatures other than 290 K. Conversely, the noise temperature definition may be more easily adapted to different operating temperatures.

Since all the quantities in (9-68) are defined over the same bandwidth B, the output noise will never be less than the input noise and will normally be greater by the amount contributed by the amplifier. Consequently, the output signal-to-noise ratio will be less than or equal to the input signal-to-noise ratio, and F is bounded by $1 < F < \infty$. An ideal noise-free amplifier would result in $F = 1$. The larger the value of F, the more the signal-to-noise ratio is degraded by the amplifier. Thus, the amplifier with the lowest noise figure would be the most desirable if all other factors are equal.

In many specifications, the noise figure in decibels is given. Let F_{dB} represent this value, and it is defined by

$$F_{dB} = 10 \log_{10}F \tag{9-69}$$

In this form, the ideal noise-free amplifier would have $F_{dB} = 0$ dB.

We will now develop a relationship between the noise temperature and the noise figure such that if one is known the other can be readily determined. Start with (9-68) and rewrite it as

$$F = \frac{P_s}{P_0}\frac{N_0}{N_s}\bigg]_{T_s = T_0} = \frac{N_0}{g_a k T_0 B} \tag{9-70}$$

where the ratio $P_0/P_s = g_a$ and the value $T_s = T_0$ have been substituted. Next, substitute the expression of (9-66) for N_0 in (9-70) with $T_s = T_0$. This results in

$$F = \frac{g_a k T_0 B + g_a k T_e B}{g_a k T_0 B} \tag{9-71}$$

After cancellation of all the surplus factors, we obtain

$$F = \frac{T_0 + T_e}{T_0} = 1 + \frac{T_e}{T_0} \tag{9-72}$$

The inverse relationship is readily determined as

$$T_e = (F - 1) T_0 \tag{9-73}$$

Thus, (9-72) permits the determination of noise figure from effective noise temperature, and (9-73) permits the determination of effective noise temperature from noise figure. These relationships are very useful in noise analysis and should be carefully noted by the reader.

Example 9-7

A certain high-gain amplifier has a noise figure of 9.03 dB, an available gain of 50 dB, and a one-sided equivalent noise bandwidth of 10 kHz. (a) Determine the effective noise temperature. Determine the available output power when the input is terminated in a matched source resistance whose noise temperature is (b) $T_s = T_0$, (c) $T_s = 10T_0$, and (d) $T_s = 100T_0$.

Solution

(a) First, the noise figure and gain, which are both given in decibels, must be converted to absolute ratios. We obtain $F = 8$ and $g_a = 10^5$ as absolute ratios. The noise computations will first be made on an effective noise temperature basis. Thus T_e for the amplifier is calculated from F using (9-73). We obtain

$$T_e = (F - 1)T_0 = 7T_0 = 2030 \text{ K} \tag{9-74}$$

(b) The output available noise power N_0 for any arbitrary source temperature T_s can now be calculated from (9-66). For $T_s = T_0 = 290$ K, we obtain

$$N_0 = g_a k (T_s + T_e)B \tag{9-75}$$
$$= 10^5 \times 1.38 \times 10^{-23}(290 + 2030) \times 10^4$$
$$= 32 \ \mu\text{W}$$

(c) When $T_s = 10T_0 = 2900$ K, we have

$$N_0 = 10^5 \times 1.38 \times 10^{-23}(2900 + 2030) \times 10^4 \tag{9-76}$$
$$= 68 \ \mu\text{W}$$

Observe from a comparison of (9-76) and (9-75) that, while the source temperature increased by a factor of 10, the output noise power only slightly more than doubled. The reason was that most of the noise in part (a) was produced by the amplifier itself so that this internal noise dominated until the source noise temperature increased by a substantial amount.

(d) When $T_s = 100\ T_0 = 29,000$ K, we have

$$N_0 = 10^5 \times 1.38 \times 10^{-23}(29,000 + 2030) \times 10^4 \qquad (9\text{-}77)$$
$$= 428.2\ \mu\text{W}$$

In this range, the noise produced by the input source is significantly greater than the noise produced by the amplifier itself.

Consider now the possible use of the noise figure parameter F in the analysis. In part (b), first compute the source noise power over the amplifier bandwidth; i.e., $N_s = kT_sB = 1.38 \times 10^{-23} \times 290 \times 10^4 = 40 \times 10^{-18}$ W. The output noise power can then be computed as $N_0 = Fg_aN_s = 8 \times 10^5 \times 40 \times 10^{-18} = 32\mu$W, which is the same result as obtained using the effective noise temperature approach. However, try the same approach in parts (c) and (d) and see what happens. The results are not the same at all! What are we trying to prove?

The point being pursued is the following: The equation $N_0 = Fg_aN_s$ is a correct result only when the temperature of the source power N_s is the same as used in measuring or defining the noise figure F, which is usually $T_s = T_0 = 290$ K. However, this equation is often used (incorrectly) in making noise calculations. Provided that the actual source temperature is near T_0, no serious errors are introduced in the process, and this is probably why this incorrect practice continues to be done. Strictly, however, calculations involving the use of the noise figure parameter are a bit tricky and subject to error if performed at any other temperature besides T_0 unless the equation is modified. This is the major reason why this author prefers to convert noise figures to equivalent noise temperatures first and then carry out the detailed computations using that approach.

9-7 NOISE OF CASCADED SYSTEMS

In the previous section, the concepts of effective noise temperature and noise figure were established as a basis for representing the noise within a given amplifier. The consideration, however, was restricted to a single amplifier. (The complete amplifier itself could have consisted of many different stages of smaller amplifiers, but only the overall noise effects on a single-stage basis were considered.) In this section, the concept will be generalized to include any number of individual amplifiers connected in cascade, each of which has an individual noise temperature or figure. The problem is to determine what the combined effect of all the different noise contributions will be on the overall noise behavior. This analysis will permit a receiver designer, for example, to assess the relative contribution of each component on the overall noise temperature or noise figure of the receiver. It will also assist in identifying those portions that are most critical and will provide guidance in selecting specifications for individual components.

The concept will be developed for the case of an amplifier system with three components. This is sufficiently large to permit the concept to be generalized, and yet it is small enough to permit a straightforward solution. Consider then the system shown in Fig. 9-18 containing three amplifiers and an input noise source at temperature T_s. It is assumed that the three effective temperatures of the individual amplifier stages are T_{e1}, T_{e2}, and T_{e3}, respectively, and the corresponding available power gains are g_{a1}, g_{a2}, and g_{a3}.

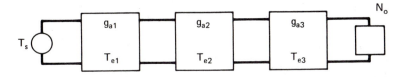

Figure 9-18. Three amplifiers in cascade whose total noise effects are to be determined.

The actual noise output N_0 can be represented as the sum of several separate noise effects as follows: (1) the input source noise amplified by the total gain $g_{a1}g_{a2}g_{a3}$; (2) the effective noise produced by the first stage referred back to its input amplified by the same gain as in (1); (3) the effective noise produced by the second stage referred back to its input multiplied by the gain of the last two stages (i.e., $g_{a2}\,g_{a3}$); and (4) the effective noise of the last stage referred back to its input multiplied by the gain of the last stage g_{a3}. In the same order as just listed, N_0 can be mathematically expressed as

$$N_0 = g_{a1}\,g_{a2}g_{a3}\,k\,T_s B + g_{a1}\,g_{a2}g_{a3}\,k\,T_{e1}B \qquad (9\text{-}78)$$
$$+ g_{a2}g_{a3}\,k\,T_{e2}B + g_{a3}\,k\,T_{e3}B$$

where B is the effective noise bandwidth of the system.

This result is fine, but suppose now we wish to determine a single effective noise temperature T_e for the entire system. In this manner, the entire block of several stages can be represented as a "super-block" with gain $g_a = g_{a1}\,g_{a2}\,g_{a3}$ and temperature T_e. Referring back to (9-66) describing the form of a single stage, a representation of this type must satisfy the following requirement:

$$N_0 = kg_a T_i B = kg_{a1}g_{a2}g_{a3}(T_e + T_s)B \qquad (9\text{-}79)$$

Equating (9-79) to (9-78) and canceling common factors, we obtain

$$T_e = T_{e1} + \frac{T_{e2}}{g_1} + \frac{T_{e3}}{g_1 g_2} \qquad (9\text{-}80)$$

This result is a single effective noise temperature *referred to the input* of the composite amplifier such that, when used with the basic form of (9-66), it predicts the total output noise.

The form of (9-80) can be readily generalized to the case of N stages. For the successive terms in the equation, each is divided by the product of all available power gains from the input up to, but not including, the gain of the particular stage corresponding to that temperature term.

A very important fact concerning the relative effect of different stage temperatures can be deduced from an inspection of (9-80). Note that T_e appears alone, but all subsequent temperature terms are divided by gain factors. Assuming for the moment that all these factors are larger than unity (which is not always the case), we readily deduce that the effective weighting of the various terms becomes increasingly smaller as we move to the right. Basically, this means that *in a cascade of amplifier stages the input stage is usually the most important one in establishing the noise temperature or figure for the system.* Indeed, in the special case where each of the individual gains is large, the total system effective noise temperature may be only slightly larger than that of the first stage. This is not always the case, however. As we will see later, some types of stages (e.g., transmission lines and mixers) exhibit losses that result in multiplicative effects on the effective output noise. Nevertheless, as a general guide, the following design rule is usually a good approach: *The first (or input) stage of a receiver should be selected to have the combination of as low as noise temperature (or figure) and as high a gain as is feasible.*

We will now obtain an expression for the noise figure of the combination in terms of the individual noise figure. The easiest way to accomplish this is to start with (9-80) and substitute for each T_e value the corresponding value of F. Thus, let F_1, F_2, and F_3, represent the three noise figures corresponding to T_{e1}, T_{e2}, and T_{e3}, respectively. Let F represent the composite noise figure corresponding to T_e. Using (9-73) as the basis in each case, we have

$$(F - 1)T_0 = (F_1 - 1)T_0 + \frac{(F_2 - 1)T_0}{g_1} + \frac{(F_3 - 1)T_0}{g_1 g_2} \qquad (9\text{-}81)$$

After cancellation of the common factors and some rearrangement, we obtain

$$F = F_1 + \frac{F_2 - 1}{g_1} + \frac{F_3 - 1}{g_1 g_2} \qquad (9\text{-}82)$$

A form similar to (9-80) for the overall noise temperature is obtained. However, it is noted that all F terms on the right except the F_1 term have the -1 form appearing in their numerators. There is an interesting reason for this, and it is based on the fact that the noise figure definition includes both the source noise and the amplifier noise in its definition. Since the source is applied only to the first stage, all noise figures except F_1 have to be reduced to compensate.

The next topic is to consider the effect of an attenuation element on the overall noise. For the purpose of this discussion, the term *matched attenuation network* will include any network having both an input and an output that satisifies the following requirements: (1) Impedances are matched both at the input and the output. (2) The network is a "passive" network in the conventional sense; that is, the only source of energy in the network is that produced by thermal noise effects. (3) Some of the power absorbed by the network at the input is dissipated within the network so that the power delivered to the load is less than the power accepted at the input.

A block diagram of such a system is shown in Fig. 9-19. The most common element that produces this type of effect in a communications receiver is a *transmission line* (or waveguide at microwave frequencies). A passive transmission line is required to couple between the antenna and the receiver, and the latter two units are often physically separated by some distance. Because of the losses in the transmission line, it acts as a "matched attenuator" if the line is sufficiently long.

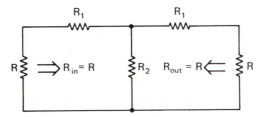

Figure 9-19. Illustration of matched attenuator system.

The most convenient way to represent the losses in a matched attenuator is through the use of an *insertion power loss* factor L. Let P_{ai} represent the input available power, and let P_{ao} represent the output available power. The relationship between these powers is

$$P_{ao} = \frac{P_{ai}}{L} \tag{9-83}$$

The factor L for a true attenuator network must satisfy $1 \leq L \leq \infty$, with the lower bound corresponding to no attenuation at all, and the upper bound corresponding to complete absorption of the signal in the network. The factor L is often given as a decibel value L_{dB}, where

$$L_{dB} = 10 \log_{10} L \tag{9-84}$$

The factor $1/L$ for a matched attenuator is treated in much the same way as available gain g_a for an amplifier. This association will help in some of the results that occur later.

Now consider the situation depicted in Fig. 9-20 with the following conditions imposed: (1) A resistance with an effective source temperature T_s is connected to the input. (2) The complete attenuator network has a constant physical temperature T_p. The source could represent an antenna or other source whose effective temperature is a result of a number of different

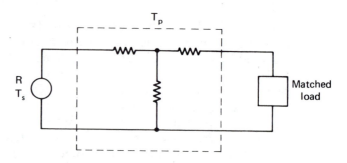

Figure 9-20. Matched attenuator at a physical temperature T_p excited by noise input source with effective temperature T_s.

contributions. However, the attenuation network is assumed to have no pickup noise so that the only way it produces noise is through its internally generated noise corresponding to T_p.

The total noise power N_0 at the output arises from two effects: (1) input noise from the source at temperature T_s attenuated by the network and (2) additional noise produced by the network at temperature T_p. A rigorous development of the manner in which these noise effects combine is a bit tricky and involves some thermodynamic reasoning. We will sidestep this approach and use a simpler argument, which, if not as convincing, is at least intuitively straightforward.

As it turns out, the output noise over a noise bandwidth B can be expressed in the form

$$N_0 = \alpha_1 k T_s B + \alpha_2 k T_p B \qquad (9\text{-}85)$$

where α_1 and α_2 are relative weighting factors for the two available noise powers. From the more rigorous thermodynamic viewpoint, it can be shown that $\alpha_1 + \alpha_2 = 1$. Furthermore, the first factor is $\alpha_1 = 1/L$ in view of the basic definition of the loss factor. The second factor must then be $1 - (1/L)$. Substituting these terms in (9-85), we obtain

$$N_0 = \frac{k T_s B}{L} + \left(1 - \frac{1}{L}\right) k T_p B \qquad (9\text{-}86)$$

However, assume now that an effective temperature T_e is defined such that

$$N_0 = \frac{1}{L} k (T_s + T_e) B \qquad (9\text{-}87)$$

which is the standard form for evaluating the output noise power in terms of the effective temperature with the "gain" set as $1/L$. Equating (9-87) and (9-86), we obtain

$$T_e = (L - 1) T_p \qquad (9\text{-}88)$$

The result of (9-88) indicates that the matched attenuator can be handled in the same way as an amplifier by defining an effective noise temperature at the input. Note that T_e increases linearly with changes in L. This means that the greater the attenuation, the greater the effective noise temperature resulting from the attenuation.

The noise figure F corresponding to (9-88) is obtained from the application of (9-72). The result is

$$F = 1 + (L - 1)\frac{T_p}{T_0} \qquad (9\text{-}89)$$

An interesting special case occurs when $T_p = T_0$. In this case (9-89) reduces simply to

$$F = L \qquad (9\text{-}90)$$

For the special case then when the attenuator is at the standard temperature of 290 K, the noise figure is equal to the insertion power loss factor. Because of the simplicity of this relationship and the fact that in some systems it is not possible to specify precisely the temperature of a transmission line, for example, there is a temptation to assume the simplified form of (9-90) whenever it can be reasonably justified.

One more form of the noise temperature will be considered. The result of (9-88) provides an effective noise temperature at the input. In some cases, it is desirable to refer to the output of the attenuator. For example, if the source is an antenna and the attenuator represents a transmission line, it may be more desirable to refer all noise calculations to the input of the receiver, which is the output of the attenuator. In this case, the effective noise temperature of the "pure receiver" without the attenuator is used. However, the effective input source temperature is no longer T_s of the antenna but instead is the effective noise temperature measured at the output of the attenuator. Letting T_s' represent this quantity, it is obtained by combining T_e of the attenuator as given by (9-88) with T_s and multiplying by $1/L$ for the attenuator to transfer the effect to the output. The result is

$$T_s' = \frac{T_s}{L} + \left(1 - \frac{1}{L}\right)T_p \qquad (9\text{-}91)$$

At the receiver input, let T_{er} and g_{ar} represent the effective noise temperature and available gain of the *receiver alone*, respectively. The output noise N_0 is then

$$N_0 = g_{ar}k(T_s' + T_{er})B \qquad (9\text{-}92)$$

Thus, in this form the noise produced by the losses in the matched attenuator has been taken care of by increasing the effective temperature as indicated by T_s'.

Up to this point, the term T_s has been used to represent any noise source since it could represent a resistance, a noise generator, or any other arbitrary source of noise. In much of the later work, the noise of interest will be the total noise at the output of an antenna connected to a receiver. In that specific case, the term T_a will be used to represent the net noise temperature as "seen" by the antenna. Thus, T_a will replace T_s in the various noise expressions.

Example 9-8

Consider the two forms of a receiving system shown in Fig. 9-21. Besides the antenna, the subsystem consists of the transmission line and two amplifier modules. Various specifications on the modules are shown on the figure. (a) Determine the effective noise temperature T_e at the input to the transmission line and the output noise power for the system as connected in Fig. 9-21a. (b) Assume amplifier 1 has sufficient weather protection that it can be mounted

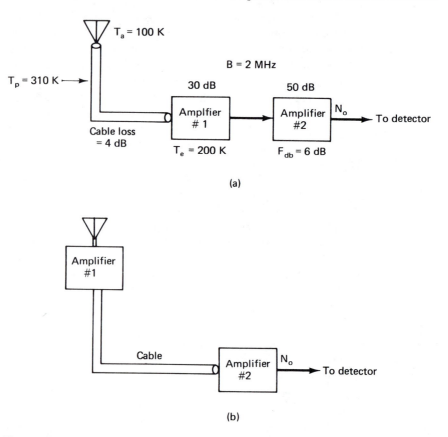

(a)

(b)

Figure 9-21. Receiving system of Example 9-8 connected in two possible ways.

right at the antenna and ahead of the transmission line, as noted in Fig. 9-21b. Determine again the effective noise temperature at the system input and the output noise power. (c) How does the signal power at the output change between parts (a) and (b)? (d) Which configuration is preferable from a noise point of view? It is assumed that impedances are matched at all junctions.

Solution

(a) Although all the different forms of the various parameters are not needed, Table 9-2 has been prepared for instructional purposes. The reader can verify the table as obtained from the given data. Note that the noise figure for the cable (4.177 dB) varies slightly from the insertion loss (4 dB) for the cable. If the cable had been at a cooler temperature of 290 K, the two values would have been exactly the same, as noted earlier in the text.

Table 9-2. Data for Example 9-8

Component	g_a (dB)	g_a	F_{dB}	F	T_e (K)
Transmission Line	− 4 dB	0.398 ($L = 2.512$)	4.177 dB	2.616	468.7
Amplifier 1	30 dB	10^3	2.28 dB	1.690	200
Amplifier 2	50 dB	10^5	6 dB	3.981	864.5

The effective input noise temperature for this configuration is determined by the development that follows.

$$T_e = 468.7 + \frac{200}{0.398} + \frac{864.5}{(0.398)(10^3)} \tag{9-93}$$

$$= 468.7 + 502.51 + 2.17 = 973.4 \text{ K}$$

Note that division by the "gain" of 0.398 in the second and third terms corresponds to multiplication by the insertion loss factor of $1/0.398 = 2.512$. The output noise N_0 is given by

$$N_0 = gk(T_e + T_a)B \tag{9-94}$$

$$= (0.398)(10^3 \times 10^5) \times 1.38 \times 10^{-23} \times (973.4 + 100) \times 2 \times 10^6$$

$$= 1.179 \ \mu\text{W}$$

(b) The system is now restructured as in Fig. 9-21b. The effective noise temperature in this case is determined as follows:

$$T_e = 200 + \frac{468.7}{10^3} + \frac{864.5}{(0.398)(10^3)} \tag{9-95}$$

$$= 200 + 0.47 + 2.17 = 202.64 \text{ K}$$

The effective noise temperature is less than one-fourth the value obtained in part (a). Computation of the output noise now yields

$$N_0 = 0.398 \,(10^3 \times 10^5) \times 1.38 \times 10^{-23} \times (202.64 + 100) \times 2 \times 10^6 \quad (9\text{-}96)$$
$$= 0.332 \,\mu\text{W}$$

The output noise is less than one-third the value obtained in part (a). (Why doesn't the output noise reduce in exactly the same proportion as the noise temperature?)

(c) The total system gain is exactly the same as before since the only change has involved reversing the order of two of the stages. Consequently, the output signal would be the same in both cases.

(d) Obviously, the second configuration is preferable *if* it is feasible to mount the amplifier at the antenna output. This may or may not be feasible in a given situation, so the answer depends partly on other considerations. However, this problem clearly illustrates how the effective output noise of a system can be reduced significantly by providing a reasonable amount of gain at the input before any attenuation is encountered.

PROBLEMS

9-1 (a) Determine the rms noise voltage produced by a 47-kΩ resistor in a 60-kHz bandwidth at the standard temperature $T = T_0 = 290$ K.

(b) Determine the rms noise voltage if the resistance is changed to 470 kΩ.

(c) Determine the rms noise voltage if the resistance is changed to 4.7 MΩ.

(d) With $R = 4.7$ MΩ, determine the rms noise voltage if the bandwidth is changed to 600 kHz.

(e) With $R = 4.7$ MΩ, determine the rms noise voltage if the bandwidth is changed to 6 MHz.

(f) With the values of R and B of part (e), determine the rms noise voltage if the temperature is increased to 310 K.

9-2 Assume that the 4.7-MΩ resistor of Prob. 9-1e is connected to an ideal noise-free amplifier having a voltage gain of 10^4 and an equivalent noise bandwidth of 6 MHz. Determine (a) the rms value of the output noise voltage, (b) the mean-square value of the output noise voltage, and (c) the probability that a single random sample of the output noise voltage will exceed 10 V.

9-3 Consider the parallel *RLC* circuit shown in Fig. P9-3.

(a) Determine an expression for the impedance $Z(f)$ as a function of frequency.

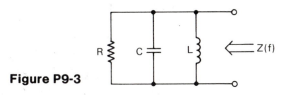

Figure P9-3

(b) Determine an expression for the net resistance $R(f)$.

(c) Set up an integral expression for determining the mean-square noise voltage $\overline{v^2}$ appearing across the network. You need not attempt to evaluate.

9-4 Determine the net rms noise voltage appearing across the terminals for each of the resistor combinations in Fig. P9-4. The bandwidth over which the noise is measured is 10 kHz, and all resistors are assumed to be at the temperature $T = T_0 = 290$ K.

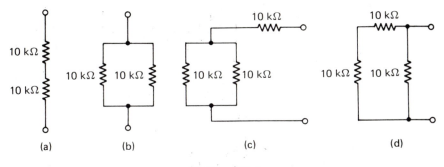

Figure P9-4

9-5 Consider the parallel combination of two resistors at different temperatures as shown in Fig. P9-5. Derive an expression for the mean-square voltage across the terminals over a bandwidth B.

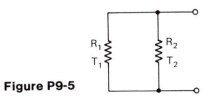

Figure P9-5

9-6 For a resistance $R = 10$ kΩ with $T = T_0 = 290$ K, determine an expression and sketch the form for (a) the one-sided open-circuit mean-square spectral density function $S_v^{(1)}(f)$ and (b) the two-sided open-circuit mean-square voltage spectral density function $S_v^{(2)}(f)$.

9-7 For a resistance $R = 10$ kΩ with $T = T_0 = 290$ K, determine an expression and sketch the form for (a) the one-sided short-circuit mean-square current spectral density function $S_i^{(1)}(f)$ and (b) the two-sided short-circuit mean-square current spectral density function $S_i^{(2)}(f)$.

9-8 Calculate the *available* noise power from a 10-kΩ resistor at a temperature $T = T_0 = 290$ K and a bandwidth 10 kHz. How would the result change if the resistance were changed to 100 kΩ?

9-9 For any resistance with $T = T_0 = 290$ K, determine an expression and sketch the form for (a) the one-sided available power spectral density $S_p^{(1)}(f)$ and (b) the two-sided available power spectral density $S_p^{(2)}(f)$.

9-10 Assume that a resistor at temperature $T = T_0 = 290$ K is connnected to the input of an ideal noise-free very wide bandwidth amplifier with an available power gain of 120 dB. Impedances are matched at the input and output ports.

    (a)   Determine an expression for the output one-sided available power spectral density $S_0(f)$ over the passband.

    (b)   If the passband characteristic is an ideal rectangular shape with $B = 10$ MHz, determine the output noise power N_0.

    (c)   If the load resistance is 50 Ω, determine the output rms noise voltage.

9-11 Using the results of Ex. 9-5, determine the noise bandwidth of the *RC* circuit shown in Fig. P9-11 for each of the following combinations of R and C: (a) $R = 100$ Ω, $C = 100$ pF; (b) $R = 100$ kΩ, $C = 0.01$ μF; (c) $R = 1$ MΩ, $C = 1$ μF.

9-12 A 3-pole Butterworth filter is excited by a noise voltage having a one-sided open-circuit mean-square voltage spectral density function of $S_v(f) = 10^{-8}$ V^2/Hz over a wide frequency range. The 3-dB bandwidth of the filter is 10 kHz, and the dc gain is unity. Using Table 9-1, determine the rms value of the output voltage of the filter.

9-13 Derive an expression for the noise bandwidth of the *RL* filter shown in Fig. P9-13.

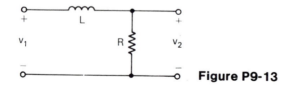

Figure P9-13

9-14 Consider the three-amplifier system of Fig. P9-14. Assume that the input is a white thermal noise source $n_i(t)$ having a one-sided power spectral density $S_i(f) = \eta_i = 1 \times 10^{-20}$ W/Hz. Assume that all impedances at various junctions in the circuit are matched over the output bandwidth of interest,

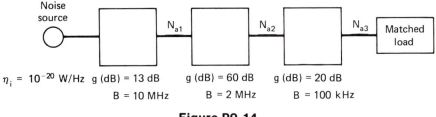

Figure P9-14

and assume that the noise contributions from the amplifiers are small in comparison with the noise produced by the input source. Calculate (a) the available noise power N_{a1} at the output of the first stage, (b) the available noise power N_{a2} at the output of the second stage, and (c) the noise power N_{a3} delivered to the load by the third stage.

9-15 Calculate the noise figure F and its decibel value F_{dB} for each of the following effective noise temperature values: (a) $T_e = 0$ K, (b) $T_e = 145$ K, (c) $T_e = 435$ K, (d) $T_e = 580$ K, (e) $T_e = 2500$ K, and (f) $T_e = 10T_0$.

9-16 Calculate the effective noise temperature T_e for each of the following noise figures: (a) $F = 1$, (b) $F = 2$, (c) $F = 3$, (d) $F_{dB} = 12$ dB, and (e) $F_{dB} = 20$ dB.

9-17 In Prob. 9-10, assume that the amplifier has an effective noise temperature $T_e = 2T_0 = 580$ K. Repeat the calculations with respect to the net output noise.

9-18 A certain amplifier has a noise figure $F_{dB} = 12$ dB, an available gain of 40 dB, and a one-sided equivalent noise bandwidth of 100 kHz. Determine the available output noise power when the input is terminated in a matched source whose noise temperature is (a) $T_s = T_0$, (b) $T_s = 10T_0$, and (c) $T_s = 100T_0$;

9-19 In Ex. 9-7, determine the source temperature T_s such that the output noise power is exactly twice the value when $T_s = T_0$.

9-20 In Prob. 9-14, assume that the three amplifiers have the following values for effective noise temperatures: $T_{e1} = 300$ K, $T_{e2} = 400$ K, and $T_{e3} = 500$ K. Repeat the calculations with respect to the net noise at each point.

9-21 (a) For the three-amplifier system of Prob. 9-14 and with the noise temperatures of Prob. 9-20, determine the effective noise temperature T_e referred to the input.

     (b) Using the value obtained in part (a), determine the net available output noise N_{a3} in one step and compare with the result of Prob. 9-20.

9-22 Assume N resistors with values $R_1, R_2, \ldots, R_N$ connected in series. Assume that the temperatures are $T_1, T_2, \ldots, T_N$. Show that for noise analysis purposes the combination may be represented as a single resistance

R_{eq} whose value is the sum of the N resistors at a fictitious temperature given by

$$T = \frac{1}{R_{eq}} \sum_1^N R_n T_n$$

9-23 (a) Determine the effective noise temperature of the three-amplifier system shown in Fig. P9-23 referred to the input.

(b) Using the result of part (a), determine the overall noise figure of the system.

(c) Determine the overall noise figure by first determining the individual noise figures and combining their effects.

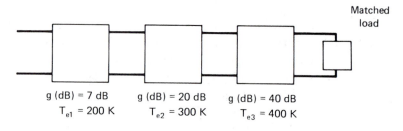

Matched load

g (dB) = 7 dB   g (dB) = 20 dB   g (dB) = 40 dB
T_{e1} = 200 K T_{e2} = 300 K T_{e3} = 400 K

Figure P9-23

9-24 (a) Determine the overall noise figure of the three-amplifier system shown in Fig. P9-24.

(b) Using the result of part (a), determine the effective noise temperature referred to the input.

(c) Determine the effective noise temperature by first determining the individual noise temperatures and combining their effects.

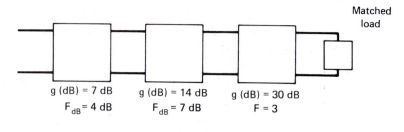

Matched load

g (dB) = 7 dB   g (dB) = 14 dB   g (dB) = 30 dB
F_{dB} = 4 dB F_{dB} = 7 dB   F = 3

Figure P9-24

9-25 (a) Determine the effective noise temperature T_e of the system of Fig. P9-25 referred to the input of the lossy transmission line.

(b) Determine the effective noise temperature T_e' at the input to the receiver (i.e., at the output of the transmission line).

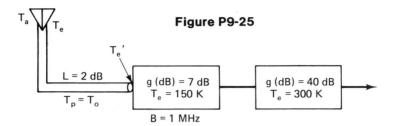

Figure P9-25

(c) Assume that the antenna effective source temperature is $T_a = 200$ K. Compute the system output noise power N_o using both the results of parts (a) and (b). When using the results of part (b), T_e' will be added to a value T_a' referred to the same point. The equivalence of the two output values should demonstrate that both approaches yield identical results if interpreted properly.

9-26 Assume that a given receiver has an effective noise temperature T_{e1}. Assume that a transmission line with a power loss factor L and a physical temperature T_p is added to the input. Show that the modified noise temperature referred to the system input is $T_{e1}' = T_{e1} + (L - 1)(T_p + T_{e1})$.

9-27 Assume that the effective noise temperature of a certain receiver referred to the input is 150 K, and assume that the physical temperature of the transmission line connecting the receiver to the antenna is $T_p = T_o = 290$ K. Show that for each increase of 0.1 dB in the attenuation of the transmission line, the effective noise temperature of the system increases by about 10 K.

9-28 The receiving system shown in Fig. P9-28 consists of an antenna with noise temperature T_a, a cable or wave guide with power loss L_c and physical temperature T_p, an RF amplifier with gain G_{RF} and noise temperature T_{RF}, a mixer with power loss L_M and noise temperature T_M, and an IF amplifier with gain G_{IF} and noise temperature T_{IF}.

(a) Show that the total input noise temperature of the system referred to the cable input is

$$T_i = T_a + (L_c - 1)T_p + L_c T_{RF} + \frac{L_c T_M}{G_{RF}} + \frac{L_c L_M}{G_{RF}} T_{IF}$$

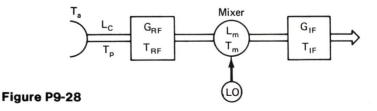

Figure P9-28

(b) Compute T_i and the IF output noise power N_0 for the following component values: T_a = 120 K; T_p = T_o = 290 K; $L_C(\text{dB})$ = 1 dB; $G_{RF}(\text{dB})$ = 30 dB; T_{RF} = 200 K; $L_M(\text{dB})$ = 9 dB, T_M = 1700 K; $G_{IF}(\text{dB})$ = 40 dB; T_{IF} = 300 K; and B = 10 MHZ

9-29 Consider the system of Prob. 9-28 with reference to the component values in part (b). Assume that to reduce costs an inexperienced designer replaces the 30-dB gain RF amplifier with a lower-cost 10-dB gain unit having the same noise temperature (200 K). Since the system now has 20 dB less gain than before, he finds that the gain of the IF amplifier can be increased from 40 to 60 dB to compensate with virtually no additional cost. The noise temperature of the IF amplifier turns out to still be 300 K. The designer is happy with his cost-effective change until tests on the new receiver design are performed.

(a) By computing the output noise and comparing with the result of Prob. 9-28, demonstrate the difficulty produced by the new design.

(b) What does this conclusion indicate about the RF amplifier requirement when a noisy lossy mixer is used?

9-30 For the system of Ex. 9-8, assume that the available signal power at the antenna is 500 fW, and assume that the effective noise temperature of the antenna is 100 K. The effective noise bandwidth of the system is 2 MHz.

(a) Calculate the antenna source signal power to noise power ratio P_s/N_a referred to the system bandwidth.

(b) Calculate the output signal power to noise power ratio P_o/N_o for the configuration of Ex. 9-8a.

(c) Calculate the output signal power to noise power ratio P_o/N_o for the configuration of Ex. 9-8b.

9-31 One technique that is used for determining the effective noise temperature of a receiver is the Y factor method. To employ this method, a matched resistor whose physical temperature can be accurately controlled is connected to the receiver input terminals. With the temperature of the resistor set to T_1, the output noise power N_1 is measured. The temperature is then changed to T_2, and the output noise power N_2 is measured. Define $Y = N_2/N_1$. Show that the effective noise temperature T_e of the receiver is given by

$$T_e = \frac{T_2 - YT_1}{Y - 1}$$

Noise Effects in Modulation
10   Systems

10-0 INTRODUCTION

Various modulation techniques have been considered throughout the book. While a few qualitative remarks concerning their performance characteristics have been made, no detailed quantitative comparison has been attempted. Such a comparison depends on an understanding of the various modulation methods, as well as the means by which the noise effects can be represented. Now that the basic background in noise theory has been established, consideration of modulation systems in the presence of disturbing noise can be achieved. This process leads to a comparison between different modulation systems in terms of their relative immunity to noise, as well as other system operating characteristics.

Much of the work in this chapter will consist of derivations pertaining to signal and noise effects occurring during frequency translation and detection processes. This material is important if one is to analyze the behavior of communications systems performance at a fundamental level. On the other hand, it is possible with some care to utilize the results of these derivations in systems planning and design without necessarily following the details of their developments. A general systems overview will be presented in Chapter 11, and the use of various derivations developed in this chapter will be made. Readers who are interested only in applying the results of modulation performance analysis to practical system implementation and design may omit much of the detail of this chapter, particularly the portions from Sec. 10-3 on. However, it is recommended as a minimum that Secs. 10-1 and 10-2 be read in reasonable detail and that some overview be made of the remainder of the chapter.

10-1 BAND-PASS NOISE (SIMPLE FORMS)

Various noise sources were considered in Chapter 9, and means for describing some of the important properties of noise were introduced. Although there were many different types of contributors, we saw that the

435

most basic and inherent sources of noise can be represented in terms of thermal-noise models. Consequently, the developments in this chapter will emphasize the thermal-noise point of view. It will be assumed that the effective noise temperature of the receiver is known and that the equivalent noise level from the antenna can be estimated. The approximate total input noise level will thus be assumed to be known within a reasonable bound. If additional sources of noise (e.g., atmospheric) are expected, additional allowance might have to be made in the form of a worst-case upper bound.

As was noted in Chapter 5, in the discussion of superheterodyne receivers, the IF bandwidth B_{IF} of a modern receiver should normally be chosen in the ideal case to be just sufficient to pass all components of the desired signal. Since the noise power admitted to the detector increases linearly with bandwidth, the minimum value of bandwidth is a limiting factor in determining the overall noise level reaching the detector. Consequently, computations concerning the noise level of interest in detector performance are usually made on the basis of the IF bandwidth.

Strictly, noise computations should be made on the basis of the equivalent noise bandwidth B_n of a system, as was shown in Chapter 9. In practice, however, a good communications receiver will employ an IF amplifier with a relatively flat passband characteristic and very sharp skirts, so the assumption that $B_n \simeq B_{IF}$ is reasonable and will be made in this chapter and through the remainder of the book. The transmission bandwidth of modulated signals has been designated as B_T earlier in the book, and it is also assumed that the IF bandwidth is just adequate to accommodate the signal. To simplify the notation in further work, we will drop subscripts and define B as $B = B_T = B_{IF} = B_n$ under the constraints previously discussed. Thus, when the symbol B appears without subscripts, it will be assumed to represent the transmission bandwidth, the IF bandwidth, and the noise bandwidth, all of which are equal in the ideal case.

The most simplified forms of a band-pass noise power spectrum $S(f)$ are shown in Fig. 10-1. The function of Fig. 10-1a is the one-sided form whose power spectral density is η W/Hz and the function of Fig. 10-1b is the two-sided form whose power spectral density is $\eta/2$ W/Hz. Whenever modulation operations are to be performed, the two-sided form is most useful, as will be demonstrated in later sections.

Band-pass noise of this type arises from the filtering operations occurring in the receiver. The RF amplifier band-pass characteristic first filters the wideband "white" noise at the receiver input to produce band-limited noise. However, the RF amplifier output noise is not nearly as rectangular in form as the function of Fig. 10-1 owing to the lower selectivity of the RF stage. The nearly rectangular IF band-pass characteristic produces a noise spectrum at its output having the shape of the passband power transfer function, so the noise spectrum of Fig. 10-1 could represent a close approximation to the form encountered at the output of the IF amplifier.

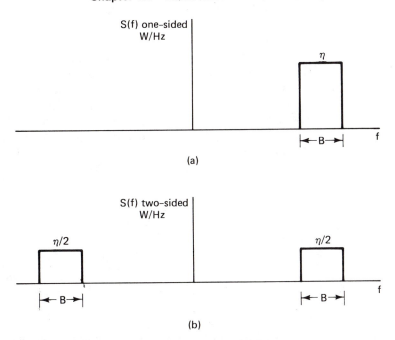

Figure 10-1. Simplified forms of a band-pass noise power spectrum: (a) one-sided form and (b) two-sided form.

Let N represent the available noise power represented by the spectrum of Fig. 10-1. This quantity can be calculated as the area under the curve from either Fig. 10-1a or b as follows:

$$N = \eta B \qquad \text{from Fig. 10-1a} \qquad (10\text{-}1a)$$

or

$$N = 2 \times \left(\frac{\eta}{2}\right) B = \eta B \qquad \text{from Fig. 10-1b} \qquad (10\text{-}1b)$$

Obviously, both results are identical, and it is primarily a matter of "bookkeeping."

The amplitude distribution of the noise in a large number of cases will be gaussian with a mean value zero. The variance is

$$\sigma^2 = N = \eta B \qquad (10\text{-}2)$$

In problems where the preceding various quantities are computed at different points in a system, subscripts may be used as required.

In Chapter 9, the symbol T_s was used to represent the source temperature for noise computations. In subsequent work, the noise "source" will be assumed to be the available noise at the antenna; so the term T_a will be used to represent this noise.

Example 10-1

The system of Figure 10-2 represents the front end of a certain communications receiver. The effective noise temperature T_e of the complete receiver referred to the input is $T_e = 600$ K, and the antenna input temperature is $T_a = 400$ K. The RF amplifier has a gain of 26 dB, the mixer has a *loss* of 6 dB, and the IF amplifier has a gain of 80 dB. The IF amplifier has a nearly rectangular filter response centered at 5 MHz with a bandwidth of 100 kHz. Compute the available IF noise output power, and sketch the form of the spectrum of the band-pass noise.

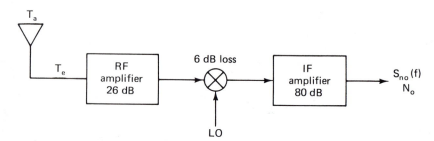

Figure 10-2. Front end of communications receiver considered in Example 10-1.

Solution

The total input one-sided noise power spectral density η_i is

$$\eta_i = k(T_a + T_e) = 1.38 \times 10^{-23} (600 + 400) \qquad (10\text{-}3)$$
$$= 1.38 \times 10^{-20} \text{ W/Hz}$$

The net gain between the input to the RF amplifier and the output of the IF mixer is $26 - 6 + 80 = 100$ dB, which corresponds to an absolute ratio of 10^{10}. The one-sided power spectral density $S_{no}(f)$ at the output of the IF amplifier in the IF bandwidth is thus

$$S_{no}(f) = 1.38 \times 10^{-20} \times 10^{10} = 1.38 \times 10^{-10} \text{ W/Hz} = \eta_o \quad (10\text{-}4)$$

This result is illustrated in both one-sided and two-sided forms in Fig. 10-3a and b.

The available noise power N_o at the output of the IF amplifier is determined by multiplying the power spectral density by the IF bandwidth. The output noise power N_o is thus

$$N_o = 1.38 \times 10^{-10} \times 10^5 = 13.8 \ \mu\text{W} \qquad (10\text{-}5)$$

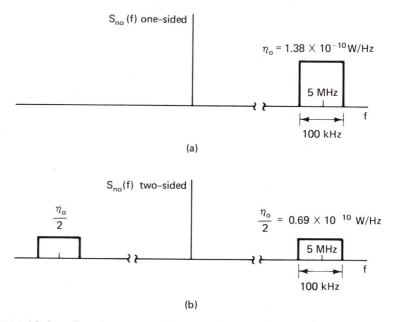

Figure 10-3. Band-pass noise spectra at output of IF amplifier in Example 10-1.

10-2   NOISE MODULATION PRODUCTS

In this section, the effect of mixing a noise signal with a sinusoidal carrier in a balanced modulator will be considered. The result of this analysis will be significant in predicting the effects of frequency translation and detection processes on the noise appearing with a signal in a receiver.

Consider the block diagram shown in Fig. 10-4. A noise function $n_i(t)$ appears as one input to an ideal multiplier (modulator), and the other input is a unit magnitude sinusoid. The output function $n_o(t)$ can be readily expressed as

$$n_o(t) = n_i(t) \cos \omega_0 t \qquad (10\text{-}6)$$

We will momentarily restrict the noise functions to be energy signals. Let $\overline{N}_i(f)$ represent the Fourier transform of $n_i(t)$, and let $\overline{N}_o(f)$ represent the Fourier transform of $n_o(t)$. Application of the Fourier modulation theorem of Chapter 2 to (10-6) yields

$$\overline{N}_o(f) = \tfrac{1}{2}\,\overline{N}_i(f - f_0) + \tfrac{1}{2}\,\overline{N}_i(f + f_0) \qquad (10\text{-}7)$$

The form of this result should be very familiar to us from the work with DSB in Chapter 4. It indicates that the output spectrum is obtained by shifting the input spectrum both to the right and to the left by f_0 and multiplying by $\tfrac{1}{2}$. It

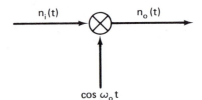

Figure 10-4. Mixing of noise function with sinusoid in a balanced modulator.

differs from the work of Chapter 4 only in that we are considering noise here rather than signal.

There are, however, some subtle differences that suggest a slightly different approach when dealing with noise. First, noise is a power signal rather than an energy signal and can only be rigorously handled using a limiting form, such as was discussed in Sec. 2-1. In addition, noise is a random signal and its instantaneous behavior can only be described in statistical terms. For these reasons, the spectrum of noise is best represented through a power spectrum rather than a linear spectrum.

Let $S_{ni}(f)$ represent the power spectral density at the input to the balanced modulator just discussed, and let $S_{no}(f)$ represent the corresponding output power spectral density. Using a certain concept known as the autocorrelation function, it can be rigorously justified that

$$S_{no}(f) = \tfrac{1}{4} S_{ni}(f - f_0) + \tfrac{1}{4} S_{ni}(f + f_0) \qquad (10\text{-}8)$$

This result is rather easy to visualize and can be illustrated by the forms shown in Fig. 10-5. The original power spectrum shown is shifted to the right by f_0 and to the left by f_0. Each component is multiplied by $\tfrac{1}{4}$ for the case where the mixing sinusoid has unit magnitude. (If the sinusoid has a magnitude A, each component would be multiplied by $A^2/4$.)

While we have not considered the derivation of the power spectrum of the modulated signal, one can very easily remember the relationship between the components of (10-7) and (10-8). The key is to note that the linear terms in (10-7) are replaced by power terms in (10-8), and the $\tfrac{1}{2}$ factor in (10-7) is squared in (10-8) and thus becomes $\tfrac{1}{4}$.

In the example shown in Fig. 10-5, the components of the shifted spectra do not overlap. However, when there is overlap, the resulting power densities add. For deterministic signals, voltages or currents add, meaning that power levels increase as the square of voltages or currents. However, in this case the noise components are incoherent with respect to each other so that power levels add.

The available input noise power (or variance) $N_i = \sigma_i^2$ is the area under the input power density curve, and the available output noise power $N_0 = \sigma_0^2$ is the area under the output power density curve. For the case of a unit sinusoid and

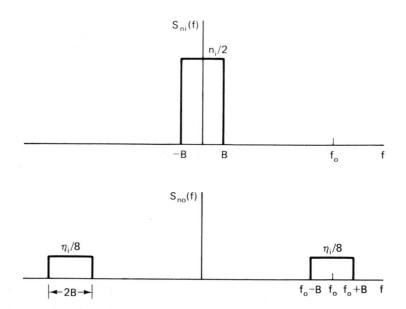

Figure 10-5.   Input and output noise power spectra of balanced modulator.

no filtering of the output, the illustration of Fig. 10-5 readily leads to the conclusion

$$\sigma_o^2 = \frac{\sigma_i^2}{2} \tag{10-9}$$

Thus, the noise variance has been reduced by one-half.

It will be emphasized that the two-sided form of the power spectrum should be used when modulation operations are to be applied. Some of the new frequency components resulting from the modulation operations can be considered as arising from the negative frequencies, as can be observed in Fig. 10-5. Thus, while the one-sided form is perfectly fine for simple noise computations, the two-sided form is most appropriate when modulation operations are involved. Both the input and output spectra in Fig. 10-5 are two-sided forms. Observe also that the two-sided power density of the output noise over the applicable frequency range could be expressed as $\eta_o/2 = \eta_i/8$ if desired.

Example 10-2

Consider the product demodulator shown in Fig. 10-6. Assume that the input $n_i(t)$ is a band-pass noise function whose input power spectrum $S_{ni}(f)$ is shown in Fig. 10-7a. (a) Plot the form of the spectrum $S_{no}'(f)$ at the output

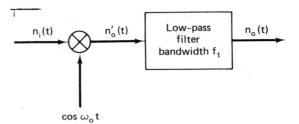

$\cos \omega_o t$

Figure 10-6. Product demodulator of Example 10-2.

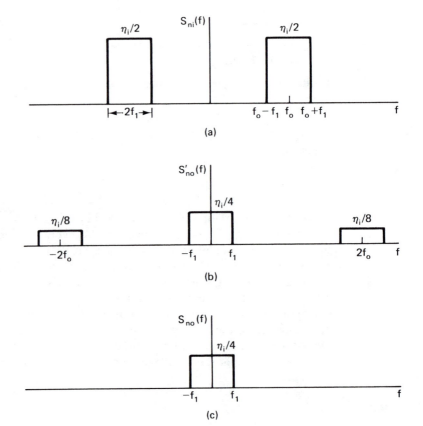

Figure 10-7. Noise spectra at different points in the system of Figure 10-6.

of the mixer. (b) Plot the form of the spectrum $S_{no}(f)$ at the output of the low-pass filter. (c) Determine the available noise power (variance) at each of the three points under consideration.

Solution

(a) Since the sinusoid is assumed to have unit magnitude, the output $n'_o(t)$ of the ideal multiplier can be expressed as

$$n'_o(t) = n_i(t) \cos \omega_0 t \qquad (10\text{-}10)$$

In accordance with (10-8), the power spectrum $S'_{no}(f)$ at the output of the multiplier is

$$S'_{no}(f) = \tfrac{1}{4} S_{ni}(f - f_0) + \tfrac{1}{4} S_{ni}(f + f_0) \qquad (10\text{-}11)$$

Referring to Fig. 10-7a and b, the input power spectrum $S_{ni}(f)$ is multiplied by $\tfrac{1}{4}$ and shifted both to the left and to the right by f_0. The positive frequency portion of $S_{ni}(f)$ centered at f_0 is translated to a center frequency $2f_0$ on the right shift, and the negative frequency portion of $S_{ni}(f)$ centered at $-f_0$ is translated to a center frequency $-2f_0$ on the left shift. On the other hand, the right shift of the negative frequency component of $S_{ni}(f)$ and the left shift of the positive frequency component of $S_{ni}(f)$ both appear centered at dc. In view of the additive nature of power for noise components, the resulting power density near dc is twice as large as the components at $\pm 2f_c$, but half the size of the original spectrum.

(b) The low-pass filter removes the components centered at $\pm 2f_0$, and the resulting power spectrum $S_{no}(f)$ contains only low-frequency noise, as shown in Fig. 10-7c. This result indicates one of several possible ways in which unwanted band-pass noise in a receiver can be translated down to low frequencies through the detection process.

(c) Let $N_i = \sigma_i^2$ represent the input available noise power or variance. From the plot in Fig. 10-7a, this value is determined as

$$N_i = \sigma_i^2 = 2 \times \left(\frac{\eta_i}{2} \times 2f_1 \right) = 2\eta_i f_1 \qquad (10\text{-}12)$$

The available power or variance of $n'_o(t)$ will be designated as $N_{o1} = \sigma_{o1}^2$. From the plot in Fig. 10-7b, this value is

$$N_{o1} = \sigma_{o1}^2 = \frac{\eta_i}{4} \times 2f_1 + 2 \times \left(\frac{\eta_i}{8} \times 2f_1 \right) = \eta_i f_1 \qquad (10\text{-}13)$$

Finally from Fig. 10-7c, the net available output noise power or variance $N_o = \sigma_o^2$ is

$$\sigma_o^2 = \frac{\eta_i}{4} \times 2f_1 = \frac{\eta_i f_1}{2} \qquad (10\text{-}14)$$

Comparing (10-12), (10-13), and (10-14), we note that the demodulation process results in halving of the noise power, and the output filtering removes half of the remaining noise power.

10-3 QUADRATURE REPRESENTATION OF BAND-PASS SIGNALS

The simplified approach considered so far in this chapter is adequate for representing band-pass noise in many applications. However, in certain cases, it fails to provide adequate insight to analyze the complete effect. A form that is useful in certain applications is the *quadrature representation*. This form may seem rather perplexing at first introduction, but it turns out to be quite useful. To provide the reader with as much insight as possible, the process will be developed in some detail.

The quadrature representation may be applied to any arbitrary band-pass signal or noise function, but since the focus of this development is on noise, a band-pass noise $n(t)$ will be assumed. We will also restrict the band-pass function to have a constant power spectral density over the passband. This

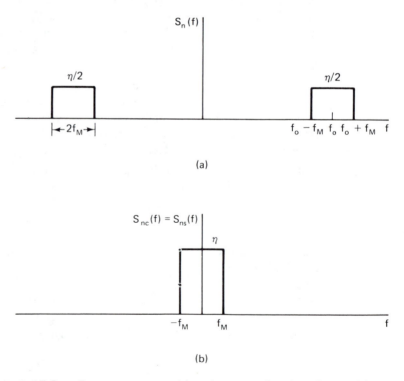

(a)

(b)

Figure 10-8. Power spectra of band-pass noise function and low-pass quadrature functions.

form is adequate to represent all cases of interest in this text, and it is the easiest to deal with mathematically. Thus, assume a noise signal $n(t)$ with a power spectral density $S_n(f)$ of the form shown in Fig. 10-8a. The noise is assumed to be band-limited from $f_0 - f_M$ to $f_0 + f_M$ as indicated, and the value of the noise power density is $\eta/2$ on a two-sided basis. The bandwidth is $2f_M$ and the net noise power N (or variance σ_n^2) is

$$N = \sigma_n^2 = 2 \times \left(\frac{\eta}{2} \times 2f_M \right) = 2\eta f_M \qquad (10\text{-}15)$$

In most cases of practical interest, the band-pass bandwidth $2f_M \ll f_0$.

Although $S_n(f)$ is shown as a continuous spectrum, a discrete spectrum will be assumed in the derivation that follows since it is easier to manage mathematically. The form of the noise function is assumed to be

$$n(t) = \sum_{m=-M}^{M} C_m \cos[(\omega_0 + m\omega_1)t + \theta_m] \qquad (10\text{-}16)$$

where f_1 is the frequency increment between spectral components and $f_M = Mf_1$. The cosine terms in the series will now be expanded by a basic trigonometric identity as follows:

$$n(t) = \sum_{-M}^{M} [C_m \cos \omega_0 t \cos (m\omega_1 t + \theta_m) \qquad (10\text{-}17a)$$
$$- C_m \sin \omega_0 t \sin (m\omega_1 t + \theta_m)]$$

$$= \left[\sum_{-M}^{M} C_m \cos (m\omega_1 t + \theta_m) \right] \cos \omega_0 t \qquad (10\text{-}17b)$$
$$- \left[\sum_{-M}^{M} C_m \sin (m\omega_1 t + \theta_m) \right] \sin \omega_0 t$$

Let

$$n_c(t) = \sum_{-M}^{M} C_m \cos (m\omega_1 t + \theta_m) \qquad (10\text{-}18)$$

and

$$n_s(t) = \sum_{-M}^{M} C_m \sin (m\omega_1 t + \theta_m) \qquad (10\text{-}19)$$

Substituting the definitions of (10-18) and (10-19) in (10-17b), the band-pass noise $n(t)$ can be expressed in the form

$$n(t) = n_c(t) \cos \omega_0 t - n_s(t) \sin \omega_0 t \qquad (10\text{-}20)$$

We will now investigate the nature of the components $n_c(t)$ and $n_s(t)$. From (10-18) and (10-19), it is observed that the highest frequency contained in both signals is $Mf_1 = f_M$. The reader may be initially puzzled by the apparent "two-sided" nature of (10-18) and (10-19), which is usually a characteristic of exponential series. However, the components corresponding to negative m actually fold over and add algebraically to the corresponding components for positive m. For example, the pair of terms corresponding to $m = \pm k$ for both cosine and sine terms may be expressed as

$$C_k \cos (k\omega_1 t + \theta_k) + C_{-k} \cos (-k\omega_1 t + \theta_{-k}) \tag{10-21}$$
$$= C_k \cos (k\omega_1 t + \theta_k) + C_{-k} \cos (k\omega_1 t - \theta_{-k})$$

and

$$C_k \sin (k\omega_1 t + \theta_k) + C_{-k} \sin (-k\omega_1 t + \theta_{-k}) \tag{10-22}$$
$$= C_k \sin (k\omega_1 t + \theta_k) - C_{-k} \sin (k\omega_1 t - \theta_{-k})$$

This combined sum of the two cosine terms and the two sine terms represents the net spectral contribution at the positive frequency kf_1. Thus, the spectral coefficients corresponding to (10-18) and (10-19) are actually one-sided forms after the foldover has been considered. (Note, however, that two-sided forms are shown in Fig. 10-8. This is a situation where the concept is easier to derive on a one-sided basis, but the final result is easier to use on a two-sided basis.)

From an inspection of (10-18) and (10-19) with the preceding discussion in mind, it can be concluded that $n_c(t)$ and $n_s(t)$ are low-pass functions having spectra over the frequency range $0 \le f \le f_M$ in terms of positive frequencies (or the frequency range $-f_M \le f \le f_M$ in terms of both negative and positive frequencies). It can also be seen from the same equations that all spectral components of $n_c(t)$ have exactly 90° phase shifts with respect to the corresponding components of $n_s(t)$.

The two functions $n_c(t)$ and $n_s(t)$ are referred to as the *low-pass quadrature* components of the band-pass noise $n(t)$. The band-pass function can be expressed in terms of the low-pass quadrature components by the form of (10-20). Each of the terms in (10-20) is a DSB function centered at f_0. Thus, a band-pass function can be considered to be generated by two quadrature low-pass functions modulating two carrier terms located in the center of the passband. The carrier terms also differ in phase by 90°.

At this point, the skeptical reader might wonder what possible value could result from this exercise. It turns out that in many band-pass systems it is easier to study various transmission and modulation effects on the low-pass components than on the actual band-pass functions. As we will see, some of the properties of the band-pass function are preserved in the low-pass components. In particular, some forms of noise at the output of detectors may be expressed directly in the quadrature forms.

The next point of interest is that of the power contained in the noise and how it relates to the power contained in the low-pass components. An expression for the noise power N was developed in (10-15) from the point of view of the spectrum. Consider now the general band-pass representation as given by (10-20). The power can also be derived from this result by squaring

$n(t)$ and determining the mean value of the result. The development is somewhat lengthy and will not be carried out here, but the result is rather simple and is

$$N = \sigma_n^2 = \frac{1}{2} N_c^2 + \frac{1}{2} N_s^2 = \frac{1}{2} \sigma_{nc}^2 + \frac{1}{2} \sigma_{ns}^2 \qquad (10\text{-}23)$$

where $N_c = \sigma_{nc}^2$ is the noise power or variance corresponding to $n_c(t)$, and $N_s = \sigma_{ns}^2$ is the same result corresponding to $n_s(t)$. In view of the symmetry of $n_c(t)$ and $n_s(t)$, $N_c = N_s$, and (10-23) can be expressed as

$$N = N_c = N_s \qquad (10\text{-}24)$$

This result indicates that the power contained in each of the low-pass components is the same as the power contained in the band-pass function. If the band-pass noise has a gaussian amplitude distribution (which is usually the case), the quadrature components also have gaussian distributions. Thus, the primary statistical properties of the band-pass noise are preserved in the low-pass components.

Since the power contained in each of the low-pass components must be the same as that of the band-pass function, the power spectral density of each of the low-pass components must be twice the value of the band-pass function in the applicable frequency regions. This property can be deduced from the fact that the low-pass bandwidths are one-half the bandwidth of the band-pass function, and since the areas under the two spectral curves must be equal, the amplitudes of the low-pass functions must be twice that of the band-pass function.

The form of the low-pass power spectral density function $S_{nc}(f) = S_{ns}(f)$ is shown in Fig. 10-8b. In accordance with the preceding discussion, the power is

$$N_c = N_s = 2f_M \times \eta = 2\eta f_M \qquad (10\text{-}25)$$

which is the same as N calculated in (10-15) for the band-pass function. The reader should carefully note the forms of the band-pass and low-pass power spectral density functions in Fig. 10-8.

The most important properties of the low-pass quadrature components representing a band-pass signal will now be summarized:

1. Each of the functions $n_c(t)$ and $n_s(t)$ has a low-pass bandwidth equal to one-half the bandwidth of the band-pass function $n(t)$.
2. When the power spectral density of the band-pass function is constant over the passband, the power spectral density of each of the low-pass quadrature components is constant over the passband and numerically equal to twice the band-pass value.
3. The power contained in each of the low-pass components is equal to the power contained in the band-pass functions.

4. The phase difference between the two low-pass quadrature functions at corresponding frequencies is exactly 90°.
5. If the band-pass signal is a noise function having a gaussian distribution, the two low-pass functions also have gaussian distributions.

10-4 AMPLITUDE MODULATION DETECTION PERFORMANCE

In this section, the detection of three forms of AM in the presence of noise will be analyzed. The three forms considered are SSB, DSB, and conventional AM. Synchronous detection will be assumed for SSB and DSB, and envelope detection will be assumed for AM.

The emphasis in this section will be on the basic derivations of the various signal and noise levels. Detailed explanations and practical interpretations of the results will be deferred to Chapter 11, in which the overall performance of different types of systems will be considered.

To simplify the mathematics involved, a single-tone modulating signal will be assumed. However, the noise will be assumed to be spread over a sufficient RF bandwidth B to allow a baseband modulating bandwidth W.

The model chosen to analyze SSB and DSB is shown in Fig. 10-9. This form represents a simplified representation of a synchronous detector. The imput is an RF (or IF) modulated signal plus band-pass noise, and the output is the detected signal plus baseband noise down-converted through the detection process. While this model is obviously far short of a complete receiver, the major effects resulting from the detection process in a complete receiver can be inferred from this simplified model.

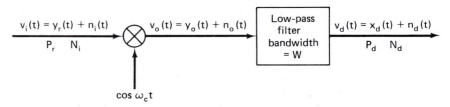

Figure 10-9. Block diagram of synchronous detector used in determining performance of DSB and SSB in the presence of noise.

The various quantities on the figure and in the analytical development are defined as follows:

$y_r(t)$ = received input modulated signal based on single-tone modulating signal

$n_i(t)$ = effective instantaneous noise input with assumed uniform power spectrum over passband

$v_i(t)$ = composite input function = $y_r(t) + n_i(t)$

P_r = received input signal power

η_i = one-sided noise power density over passband

N_i = total input noise power

R_i = input signal power to noise power ratio = P_r/N_i

$v_o(t)$ = composite output of synchronous detector before filtering = $y_o(t) + n_o(t)$

$y_o(t)$ = signal output of synchronous detector before filtering

$n_o(t)$ = instantaneous noise output of synchronous detector before filtering

$v_d(t)$ = composite detected output after filtering = $x_d(t) + n_d(t)$

$x_d(t)$ = detected output signal after filtering

$n_d(t)$ = detected output noise after filtering

P_d = detected output signal power

N_d = detected output noise power

R_d = detected output signal power to noise power ratio = P_d/N_d

R_d/R_i = receiver processing gain

The results for SSB and DSB will now be analyzed separately.

Single Sideband

The received input tone modulated signal is assumed to be

$$y_r(t) = A \cos (\omega_c + \omega_m)t \qquad (10\text{-}26)$$

which represents the USB modulation of a carrier of radian frequency ω_c with a sinusoidal tone of radian frequency ω_m. The composite input function consisting of signal plus noise can be expressed as

$$v_i(t) = A \cos (\omega_c + \omega_m)t + n_i(t) \qquad (10\text{-}27)$$

The received input normalized signal power based on a 1-Ω reference is given by

$$P_r = \frac{A^2}{2} \qquad (10\text{-}28)$$

The total input noise power N_i is

$$N_i = \eta_i B = \eta_i W = kT_i W \qquad (10\text{-}29)$$

where the minimum IF bandwidth $B = W$ for SSB has been assumed. The input signal-to-noise ratio R_i is thus

$$R_i = \frac{P_r}{N_i} = \frac{P_r}{\eta_i W} = \frac{A^2}{2\eta_i W} \qquad (10\text{-}30)$$

The composite output $v_o(t)$ of the synchronous detector before filtering can be expressed as

$$v_o(t) = v_i(t) \cos \omega_c t \tag{10-31}$$
$$= y_r(t) \cos \omega_c t + n_i(t) \cos \omega_c t$$
$$= y_o(t) + n_o(t)$$

First, the signal component $y_o(t)$ will be inspected. This quantity is

$$y_o(t) = y_r(t) \cos \omega_c t = A \cos (\omega_c + \omega_m)t \cos \omega_c t \tag{10-32a}$$

$$= \frac{A}{2} \cos \omega_m t + \frac{A}{2} \cos (2\omega_c + \omega_m)t \tag{10-32b}$$

The first term in (10-32b) represents the desired signal, while the second term is a high-frequency component resulting from the multiplication operation. After low-pass filtering, the detected low-pass signal output $x_d(t)$ is simply

$$x_d(t) = \frac{A}{2} \cos \omega_m t \tag{10-33}$$

The detected output signal power based on a 1-Ω reference is then

$$P_d = \frac{1}{2} \left(\frac{A}{2} \right)^2 = \frac{A^2}{8} \tag{10-34}$$

The noise appearing at the multiplier output can be analyzed using the technique of Sec. 10-2. Referring to Fig. 10-10, the power spectrum $S_{no}(f)$ is determined from $S_{ni}(f)$ by multiplying by ¼ and shifting both to the left and to the right as shown in Fig. 10-10b. The power spectrum $S_{nd}(f)$ at the output of the filter is shown in Fig. 10-10c. The resulting detected output noise power is determined to be

$$N_d = 2W \times \frac{\eta_i}{8} = \frac{\eta_i W}{4} \tag{10-35}$$

The detected output signal-to-noise ratio R_d is

$$R_d = \frac{P_d}{N_d} = \frac{A^2/8}{\eta_i W/4} = \frac{A^2}{2\eta_i W} = R_i \tag{10-36}$$

The result of (10-36) is quite significant in that it indicates that the *signal-to-noise ratio at the output of an ideal SSB detector is the same as the signal-to-noise ratio at the input*. The levels of both signal and noise are different at the output than at the input, but the ratios are the same. We can now see why the assumption of simplified gain constants in the analysis is reasonable, since the signal and noise would both be amplified by the same constant, but their levels relative to each other would not be changed.

The preceding result not only has significance for SSB detection, but it may also be applied with some care in the analysis of frequency translation processes such as encountered in superheterodyne receivers. If all significant

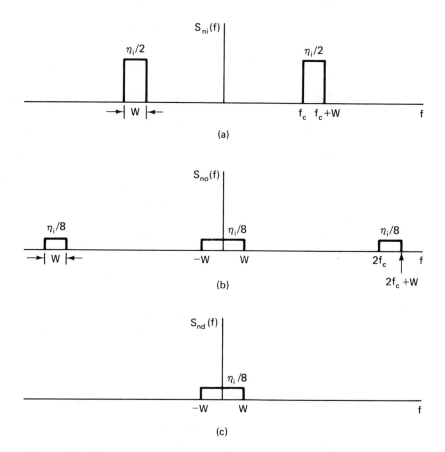

Figure 10-10.   Noise spectra at different points for SSB detection.

noise sources precede a mixer, the signal-to-noise ratio in the passband of the up-converted or down-converted component at the output of the mixer is equal to the corresponding ratio at the input. However, if the translation process occurs at a very low level stage, this concept must be used with care due to the manner in which the effective noise temperature or noise figure is defined, particularly when the mixer has a loss.

Double Sideband

The received input tone modulated signal is assumed to be

$$y_r(t) = \sqrt{2}\, A \cos \omega_m t \cos \omega_c t \qquad (10\text{-}37)$$

where the $\sqrt{2}$ factor has been chosen to make the signal power the same for this case as for SSB. This point is verified by expanding $y_r(t)$ as follows:

$$y_r(t) = \frac{\sqrt{2}}{2} A \cos (\omega_c - \omega_m)t + \frac{\sqrt{2}}{2} A \cos (\omega_c + \omega_m)t \qquad (10\text{-}38)$$

The received input signal power is the sum of the powers associated with the two components and is

$$P_r = \frac{1}{2}\left(\frac{\sqrt{2}A}{2}\right)^2 + \frac{1}{2}\left(\frac{\sqrt{2}A}{2}\right)^2 = \frac{A^2}{2} \qquad (10\text{-}39)$$

The required bandwidth of the DSB signal is $2W$; so the input noise power is

$$N_i = 2\eta_i W \qquad (10\text{-}40)$$

The input signal-to-noise ratio R_i is then

$$R_i = \frac{P_r}{N_i} = \frac{P_r}{2\eta_i W} = \frac{A^2}{4\eta_i W} \qquad (10\text{-}41)$$

Comparing (10-41) with (10-30), we note that R_i for DSB is one-half the value for SSB for a given signal power and a given noise density, which is a direct result of the necessity to provide twice the bandwidth.

Combining the instantaneous signal with the instantaneous noise, the composite output of the synchronous detector before filtering is

$$v_o(t) = v_i(t) \cos \omega_c t \qquad (10\text{-}42)$$
$$= y_r(t) \cos \omega_c t + n_i(t) \cos \omega_c t$$
$$= y_o(t) + n_o(t)$$

The signal component $y_o(t)$ is

$$y_o(t) = \sqrt{2} A \cos \omega_m t \cos^2 \omega_c t \qquad (10\text{-}43a)$$

$$= \frac{\sqrt{2}A}{2} \cos \omega_m t + \frac{\sqrt{2}A}{2} \cos \omega_m t \cos 2\omega_c t \qquad (10\text{-}43b)$$

The first term in (10-43b) represents the desired signal, while the second term is a high-frequency component.

After low-pass filtering, the detected low-pass signal output $x_d(t)$ is

$$x_d(t) = \frac{\sqrt{2}A}{2} \cos \omega_m t \qquad (10\text{-}44)$$

The detected output signal power P_d is then

$$P_d = \frac{1}{2}\left(\frac{\sqrt{2}A}{2}\right)^2 = \frac{A^2}{4} \qquad (10\text{-}45)$$

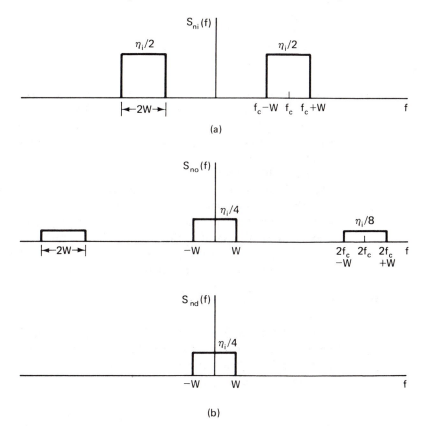

Figure 10-11. Noise spectra at different points for DSB detection.

The noise appearing at the multiplier output is analyzed in Fig. 10-11. The power spectrum $S_{no}(f)$ is determined from $S_{ni}(f)$ in Fig. 10-11a by multiplying by $\frac{1}{4}$ and shifting both to the left and to the right as shown in Fig. 10-11b. In contrast to the case of SSB, however, two components overlap, and their power spectra add. The power spectrum $S_{nd}(f)$ at the output of the filter is shown in Fig. 10-11c. The resulting detected output noise power is determined to be

$$N_d = 2W \times \frac{\eta_i}{4} = \frac{\eta_i W}{2} \tag{10-46}$$

The detected output signal-to-noise ratio is

$$R_d = \frac{A^2/4}{\eta_i W/2} = \frac{A^2}{2\eta_i W} = 2R_i \tag{10-47}$$

The ratio of R_d to R_i is

$$\frac{R_d}{R_i} = \frac{A^2/2\eta_i W}{A^2/4\eta_i W} = 2 \qquad (10\text{-}48)$$

Comparing (10-47) with (10-36), it follows that *the detected signal-to-noise ratio for DSB is the same as for SSB on the basis of equal signal powers and equal noise power densities at the input*. Thus, while the input signal-to-noise ratio for DSB is one-half the value for SSB, the signal-to-noise ratio for DSB is doubled by the detection process.

The doubling process of the signal-to-noise ratio for DSB can be explained qualitatively as follows: The two signal sidebands have similar phase shifts, and the two components add in a coherent manner during the detection process. However, the noise components have random phase shifts with respect to each other and add in an incoherent fashion. The result is that the signal is enhanced with respect to the noise during the detection process, thus compensating for the lower signal-to-noise ratio at the input.

Conventional AM

The detection process for AM will be developed using the simplified envelope detector model shown in Fig. 10-12. All components will be assumed to be ideal, and the same notation used in analyzing DSB and SSB will be assumed as much as possible.

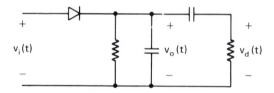

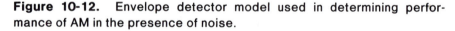

Figure 10-12. Envelope detector model used in determining performance of AM in the presence of noise.

A particular assumption not considered thus far will be employed in analyzing AM. The envelope detection process is one of a number of processes to be encountered that exhibits a *threshold effect*. This process will be explained in greater detail later, but the concept means that the input signal-to-noise ratio must exceed some minimum value before linear extraction of the intelligence without distortion can be assumed. Such an assumption was not necessary for either SSB or DSB, since synchronous detection is theoretically a linear process for any range of the signal-to-noise ratio.

The received input tone modulated AM signal is assumed to be

$$y_r(t) = A(1 + m \cos \omega_m t) \cos \omega_c t \qquad (10\text{-}49)$$

where m is the modulation factor as defined in Chapter 4. The composite input signal can be expressed as

$$v_i(t) = y_r(t) + n_i(t) \tag{10-50a}$$

$$= A(1 + m \cos \omega_m t) \cos \omega_c t + n_c(t) \cos \omega_c t - n_s(t) \sin \omega_c t \tag{10-50b}$$

where the quadrature representation for the noise with center frequency ω_c has been assumed. As we will shortly see, the quadrature form is the most appropriate one for dealing with the AM envelope detector.

The input signal power is determined by first momentarily expanding $y_r(t)$ into carrier plus sidebands as

$$y_r(t) = A \cos \omega_c t + \frac{mA}{2} \cos (\omega_c - \omega_m)t + \frac{mA}{2} \cos (\omega_c + \omega_m)t \tag{10-51}$$

The received input signal power based on a 1-Ω reference is then determined as the sum of the separate powers for the three components in (10-51).

$$P_r = \frac{A^2}{2} + \frac{1}{2}\left(\frac{mA}{2}\right)^2 + \frac{1}{2}\left(\frac{mA}{2}\right)^2 \tag{10-52}$$

$$= \frac{A^2}{2} + \frac{m^2 A^2}{4} = \frac{A^2}{2}\left(1 + \frac{m^2}{2}\right)$$

The input noise power is the same as for DSB and is

$$N_i = 2\eta_i W \tag{10-53}$$

The input signal-to-noise ratio is

$$R_i = \frac{P_r}{N_i} = \frac{(A^2/2)(1 + m^2/2)}{2\eta_i W} = \frac{A^2(2 + m^2)}{8\eta_i W} \tag{10-54}$$

The modulated phasor model for the input $v_i(t)$ will now be employed. Let

$$v_i(t) = \text{Re}\,[\overline{V}_i(t)\epsilon^{j\omega_c t}] \tag{10-55}$$

where

$$\overline{V}_i(t) = A(t)\underline{/\phi(t)} \tag{10-56}$$

in accordance with the procedure of Sec. 5-5. The rectangular form of $\overline{V}_i(t)$ can be determined directly from (10-50b). The result can be expressed in either of the forms

$$\overline{V}_i(t) = A(1 + m \cos \omega_m t) + n_c(t) + jn_s(t) \tag{10-57}$$

or

$$\overline{V}_i(t) = A + \frac{mA}{2}\underline{/-\omega_m t} + \frac{mA}{2}\underline{/\omega_m t} + n_c(t) + jn_s(t) \tag{10-58}$$

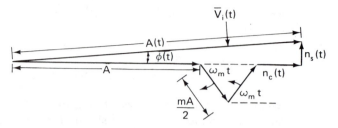

Figure 10-13. Modulated phasor representation of AM signal plus noise.

The form of (10-58) is shown in Fig. 10-13. The individual phasors shown in the figure represent the five terms in (10-58).

Assume now that the ratio of signal power to noise power is at least moderately large. With this condition, the contribution of the quadrature noise component $n_s(t)$ to the net magnitude should be negligible. The amplitude $A(t)$ can then be very closely approximated by the real part of $\overline{V}_i(t)$, and this result is also the composite output $v_o(t)$ of the detector before filtering. Thus

$$v_o(t) = A(t) \simeq A(1 + m \cos \omega_m t) + n_c(t) \qquad (10\text{-}59)$$

Following removal of the dc component A, the detected output signal is

$$x_d(t) = mA \cos \omega_m t \qquad (10\text{-}60)$$

The normalized power corresponding to this signal is

$$P_d = \frac{m^2 A^2}{2} \qquad (10\text{-}61)$$

The detected output noise power is the power contained in $n_c(t)$, and since the bandwidth is $2W$, this noise power is

$$N_d = 2\eta_i W \qquad (10\text{-}62)$$

The detected output signal-to-noise ratio is

$$R_d = \frac{P_d}{N_d} = \frac{m^2 A^2 / 2}{2\eta_i W} = \frac{m^2 A^2}{4\eta_i W} \qquad (10\text{-}63)$$

The ratio of the detected output signal-to-noise ratio R_d to the input signal-to-noise ratio R_i is determined from dividing (10-63) by (10-54). This results in

$$\frac{R_d}{R_i} = \frac{m^2 A^2 / 4\eta_i W}{A^2(2 + m^2)/8\eta_i W} = \frac{2m^2}{2 + m^2} \qquad (10\text{-}64)$$

This result is heavily dependent on the modulation factor m, but the maximum value is ⅔, which occurs for 100% modulation ($m = 1$). This means that the detected output signal-to-noise ratio for AM is always less than the input signal-to-noise ratio. This property is a result of the fact that power must be "wasted" in the carrier to produce the desired results with AM.

A more detailed discussion along with various practical interpretations of the preceding derivations will be made in Chapter 11. The intent of this section was to provide a development of the various results rather than to employ the results in practical problems.

10-5 ANGLE-MODULATION DETECTION PERFORMANCE

We will now analyze the noise performances of the two major forms of angle modulation, that is, phase modulation (PM) and frequency modulation (FM). The same notation established at the beginning of Sec. 10-4 for AM will be employed with a few minor modifications as required. Each of the two types of angle modulation will be analyzed separately.

Phase Modulation

Assume that the form of the received signal is

$$y_r(t) = A \cos [\omega_c t + \phi_i(t)] \qquad (10\text{-}65)$$
$$= A \cos (\omega_c t + \Delta\phi \cos \omega_m t)$$

as established in Chapter 5 for a PM signal modulated with a single tone of the form $\cos \omega_m t$.

The composite input signal can be expressed as

$$v_i(t) = y_r(t) + n_i(t) \qquad (10\text{-}66)$$
$$= A \cos (\omega_c t + \Delta\phi \cos \omega_m t) + n_i(t)$$

The received input signal power is

$$P_r = A^2/2 \qquad (10\text{-}67)$$

The input noise power is

$$N_i = \eta_i B$$

where B is the required transmission bandwidth for PM. Recall from Chapter 5 that B can be approximated from Carson's rule for PM as

$$B = 2(\Delta\phi + 1)W \qquad (10\text{-}68)$$

The input signal-to-noise ratio is then

$$R_i = \frac{P_r}{N_i} = \frac{P_r}{\eta_i B} = \frac{A^2}{2\eta_i B} = \frac{A^2}{4\eta_i(\Delta\phi + 1)W} \qquad (10\text{-}69)$$

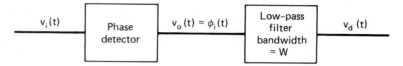

Figure 10-14. Simplified phase detector model used in determining performance of PM in the presence of noise.

The simplified phase detector system assumed for the analysis is shown in Fig. 10-14. The block on the left represents a circuit that produces an output voltage proportional to the instantaneous signal phase. For simplicity, the gain constant for this block is normalized to unity, so its output $v_o(t)$ is defined as

$$v_o(t) = \phi_i(t) \tag{10-70}$$

where $\phi_i(t)$ is the composite signal phase of the input function.

Both PM and FM are nonlinear processes, and as a result the assumption of superposition that we have previously used in analyzing the separate signal and noise effects for AM is not strictly valid with angle modulation. Indeed, a complete mathematical analysis of the precise noise effects is most difficult. Many research studies devoted to determining the properties of angle modulation in the presence of noise have been made, and only a limited view of these results is feasible in this book.

Fortunately, some reasonable approximations have been found to yield satisfactory results for applied engineering requirements. First, it is assumed that the input signal-to-noise ratio exceeds a certain minimum level (roughly 10 dB or more) so that operation above threshold is assumed. (The concept of FM threshold will be discussed in detail in Chapter 11.) When the signal-to-noise ratio exceeds this minimum requirement, it has been verified that the following approximately linear approach is valid: (1) Momentarily consider the signal alone without the noise and determine the detected output. (2) Consider the combination of the noise and the unmodulated carrier, and determine the detected output. The ratio of the result of (1) to that of (2) is the approximate detected signal-to-noise ratio.

Note that it is necessary to consider the presence of the carrier when analyzing the noise. This assumption is necessary with both forms of angle modulation, since the presence of the carrier results in a "quieting" effect compared to the noise level above. The reader can readily verify this phenomenon by turning an FM receiver across the dial and noting the significant reduction in the noise level when a strong station appears even when there is no modulation. (An FM station was used for illustration here due to its familiarity even though we are discussing PM. This phenomenon occurs in both cases.)

Following the steps of the proposed analysis process, we first determine the detected signal output. From the form of the detector assumed, the output is

$$y_o(t) = \Delta\phi \cos \omega_m t = x_d(t) \qquad (10\text{-}71)$$

where it has been recognized that the signal $y_o(t)$ at this point has the form of the desired detected signal $x_d(t)$. (The purpose of the filter is to limit the bandwidth of the noise, as we will see shortly.) The detected output signal power is

$$P_d = \frac{(\Delta\phi)^2}{2} \qquad (10\text{-}72)$$

To determine the detected noise output, we will assume the presence of the carrier $A \cos \omega_c t$ and a band-pass representation of the noise at the input. Calling this function $v_n(t)$, we have

$$v_n(t) = A \cos \omega_c t + n_c(t) \cos \omega_c t - n_s(t) \sin \omega_c t \qquad (10\text{-}73)$$

To visualize the process involved, we will convert to a modulated phasor form. Let

$$v_n(t) = \text{Re}\,[\overline{V}_n(t)\epsilon^{j\omega_c t}] \qquad (10\text{-}74)$$

where

$$\overline{V}_n(t) = A_n(t)\,\underline{/\phi_n(t)} \qquad (10\text{-}75)$$

The complex phasor $\overline{V}_n(t)$ can be expressed in rectangular form as

$$\overline{V}_n(t) = A + n_c(t) + jn_s(t) \qquad (10\text{-}76)$$

A graphical representation of this function is shown in Fig. 10-15.

The angle $\phi_n(t)$ can be expressed from the trigonometry of the triangle as

$$\phi_n(t) = \tan^{-1} \frac{n_s(t)}{A + n_c(t)} \qquad (10\text{-}77)$$

We will not bother computing the net amplitude $A_n(t)$ since the phase detector should not respond to amplitude variations in the ideal case.

The assumption has been made that the signal-to-noise ratio is reasonably high. In this case $A \gg |n_c(t)|$ in (10-77), and, in addition, the approximation

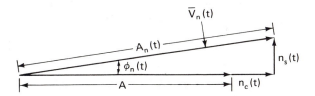

Figure 10-15. Modulated phasor representation of unmodulated PM carrier plus noise.

$\tan \phi_n \simeq \phi_n$ will be made since $\phi_n \ll \pi/2$. Consequently, $\phi_n(t)$ reduces to

$$\phi_n(t) \simeq \frac{n_s(t)}{A} \tag{10-78}$$

Thus, the detected output noise $n_o(t)$ before filtering is given by

$$n_o(t) \simeq \frac{n_s(t)}{A} \tag{10-79}$$

Since the bandwidth of $n_s(t)$ is $B/2$ on a positive frequency basis, the total noise power at the detector output could be calculated on the basis of this RF (or IF) bandwidth. However, at this point, only a baseband bandwidth W need be provided since the signal only occupies this much bandwidth. A low-pass filter can be used to reject the portion of the detected noise spectrum outside the baseband bandwidth if necessary.

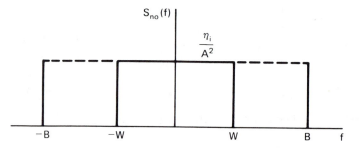

Figure 10-16. Form of noise power spectrum at output of PM detector.

From the work of Sec. 10-3, it is recalled that if the two-sided power spectral density of the band-pass noise is $\eta_i/2$, the two-sided noise density of $n_s(t)$ is η_i. In view of the $1/A$ factor in (10-79), the form of $S_{no}(f)$ is shown in Fig. 10-16. The detected output noise power N_d in the baseband bandwidth W is then given by

$$N_d = 2W \times \frac{\eta_i}{A^2} = \frac{2\eta_i W}{A^2} \tag{10-80}$$

An interesting form of (10-80) is obtained by substituting the definition of P_r from (10-67) in (10-80). This results in

$$N_d = \frac{\eta_i W}{P_r} \tag{10-81}$$

This result indicates that the output noise power is inversely proportional to the input carrier power, a concept that certainly illustrates both the nonlinear nature of angle modulation as well as the quieting effect.

The detected output signal-to-noise ratio is given by

$$R_d = \frac{P_d}{N_d} = \frac{(\Delta\phi)^2/2}{\eta_i W/P_r} = \frac{(\Delta\phi)^2 P_r}{2\eta_i W} \qquad (10\text{-}82)$$

The ratio of the output detected signal-to-noise ratio to the input signal-to-noise ratio is determined as

$$\frac{R_d}{R_i} = \frac{(\Delta\phi)^2 P_r/2\eta_i W}{P_r/\eta_i B} = \frac{(\Delta\phi)^2}{2}\left(\frac{B}{W}\right) \qquad (10\text{-}83)$$

Frequency Modulation

Recall from Chapter 5 that the general form of a received FM signal can be expressed as

$$y_r(t) = A \cos\left(\omega_c t + \Delta\omega \int_0^t x(t)\,dt\right) \qquad (10\text{-}84)$$

where $\Delta\omega\, x(t)$ represents the intelligence, and $x(t)$ is a normalized modulating signal. In accordance with preceding developments, a single-tone modulating signal of the form $\cos \omega_m t$ will be assumed, and the received FM signal is then expressed as

$$y_r(t) = A \cos(\omega_c t + \beta \sin \omega_m t) \qquad (10\text{-}85)$$

where $\beta = \Delta\omega/\omega_m = \Delta f/f_m$ is the modulation index. The composite input signal can be expressed as

$$v_i(t) = y_r(t) + n_i(t) \qquad (10\text{-}86)$$
$$= A \cos(\omega_c t + \beta \sin \omega_m t) + n_i(t)$$

The received input signal power is

$$P_r = \frac{A^2}{2} \qquad (10\text{-}87)$$

The input noise power is

$$N_i = \eta_i B \qquad (10\text{-}88)$$

where B is the required transmission bandwidth for FM. Recall from Chapter 5 that B can be approximated from Carson's rule as

$$B = 2(D + 1)W \qquad (10\text{-}89)$$

where $D = \Delta f/W$ is the deviation ratio. The input signal-to-noise ratio is then

$$R_i = \frac{P_r}{N_i} = \frac{P_r}{\eta_i B} = \frac{A^2}{2\eta_i B} = \frac{A^2}{4\eta_i(D + 1)W} \qquad (10\text{-}90)$$

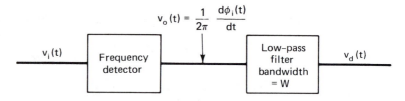

Figure 10-17. Simplified frequency detector model used in determining performance of FM in the presence of noise.

The simplified frequency detection system assumed for this analysis is shown in Fig. 10-17. The block on the left represents a circuit that produces an output voltage proportional to the instantaneous signal frequency. For simplicity, the gain constant for this block is normalized to a value that produces unity output for an input signal frequency deviation of 1 Hz. This relationship can be expressed as

$$v_o(t) = \frac{1}{2\pi} \frac{d\phi_i(t)}{dt} \tag{10-91}$$

The approach taken with PM of considering the signal alone and then considering the noise plus an unmodulated carrier is equally valid for FM. As in the case of PM, the input signal-to-noise ratio must be assumed to be greater than a minimum threshold level (say 10 dB).

Considering the signal only first, the output of the frequency detector is

$$y_o(t) = \frac{\Delta\omega}{2\pi} \cos \omega_m t = \Delta f \cos \omega_m t = x_d(t) \tag{10-92}$$

The detected output signal power is

$$P_d = \frac{(\Delta f)^2}{2} \tag{10-93}$$

To determine the detected noise output, we will assume the presence of the carrier $A \cos \omega_c t$ and a band-pass representation of the noise at the input. The form at this point is identical to that for PM, so the results of Eqs. (10-74) through (10-78) and the phasor diagram of Fig. 10-15 apply directly in this case.

The first point at which the procedure for FM differs is after Eq. (10-78) is established. The detected output noise $n_o(t)$ must be determined by differentiating (10-78) in accordance with (10-91). The result is

$$n_o(t) = \frac{1}{2\pi} \frac{d\phi_i(t)}{dt} = \frac{1}{2\pi A} \frac{dn_s(t)}{dt} \tag{10-94}$$

Since $n_s(t)$ is a random noise function, it is not possible to write a deterministic expression to describe it. Instead, the power spectrum will be determined.

In general, the differentiation theorem for Fourier transforms is

$$\mathscr{F}\left[\frac{dx(t)}{dt}\right] = j\omega \overline{X}(f) \qquad (10\text{-}95)$$

where $\overline{X}(f)$ is the Fourier transform of $x(t)$. Since power is proportional to the magnitude of voltage or current squared, the power spectral density $S_{no}(f)$ corresponding to $n_o(t)$ can be expressed as

$$S_{no}(f) = \frac{\omega^2}{(2\pi A)^2} S_{ns}(f) = \frac{f^2}{A^2} S_{ns}(f) \qquad (10\text{-}96)$$

Substituting $S_{ns}(f) = \eta_i$ and the definition of P_r from (10-67), the detected output power spectral density can be expressed as

$$S_{no}(f) = \frac{\eta_i f^2}{2P_r} = \text{for } -\frac{B}{2} < f < \frac{B}{2} \qquad (10\text{-}97)$$

This function is illustrated in Fig. 10-18 and represents an important property of an FM detector. *The noise at the output of an ideal FM detector has a parabolic power spectrum.*

The detected output noise N_d is the total noise power contained in the frequency range from $-W$ to W (on a two-sided basis) in the parabolic power spectrum. This power is the area under the curve of $S_{no}(f)$ over that range. The reader is invited to show (Prob. 10-4) that the result is

$$N_d = \frac{\eta_i W^3}{3P_r} \qquad (10\text{-}98)$$

Comparing (10-98) with (10-81), it is observed that PM detected noise power increases linearly with the baseband bandwidth, while FM detected noise increases as the third power of baseband bandwidth. However, the total FM

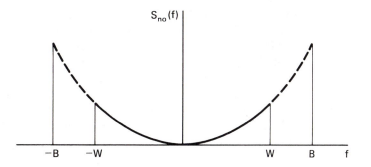

Figure 10-18. Form of the parabolic noise power spectrum at the output of an FM detector.

noise power is inversely proportional to the carrier power, a property that was previously shown to be the same for PM.

The detected output signal-to-noise ratio R_d is

$$R_d = \frac{P_d}{N_d} = \frac{(\Delta f)^2/2}{\eta_i W^3/3P_r} = \frac{3(\Delta f)^2 P_r}{2\eta_i W^3} \qquad (10\text{-}99)$$

The ratio of the detected signal-to-noise ratio to the input signal-to-noise ratio is

$$\frac{R_d}{R_i} = \frac{3(\Delta f)^2 P_r/2\eta_i W^3}{P_r/\eta_i B} = \frac{3}{2}\left(\frac{B}{W}\right)\left(\frac{\Delta f}{W}\right)^2 = \frac{3}{2} D^2\left(\frac{B}{W}\right) \qquad (10\text{-}100)$$

where the definition of the deviation ratio $D = \Delta f/W$ has been used. Various practical interpretations of the results for PM and FM will be discussed in Chapter 11.

10-6 PREEMPHASIS

The result of Eq. (10-97), which was shown graphically in Fig. 10-18, indicates that the power spectrum of noise at the output of an FM detector has a parabolic characteristic and thus increases as the square of the frequency. While the overall or average signal-to-noise ratio may be readily increased with FM, the effect of the increasing noise level at higher modulation frequencies, coupled with the fact that the higher-frequency components of many modulating signals have less energy that the low-frequency components, results in a reduction of the detected signal-to-noise ratio in that range.

It is possible to partially compensate for the decreased signal-to-noise ratio at higher modulation frequencies by a process called *preemphasis*. Preemphasis, in general, is a process in which the signal level in one or more frequency ranges is artificially boosted or increased relative to other frequencies, the purpose being to "emphasize" those frequencies that would be most susceptible to noise. Various preemphasis forms are used in phonograph and tape recording processes and in many other areas. The particular form of preemphasis used with FM involves boosting the higher frequency content of the modulating signal above a certain frequency range so as to render that frequency range less susceptible to noise.

The form of the preemphasis transfer function $H_p(f)$ used in most FM systems is given by

$$H_p(f) = K\left(1 + j\frac{f}{f_1}\right) \qquad (10\text{-}101)$$

where K is a constant and f_1 is the frequency at which the response is boosted 3 dB. The shape of the magnitude response $|H_p(f)|$ in decibels on a semilog

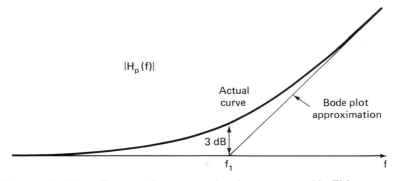

Figure 10-19. Form of the preemphasis curve used in FM systems.

scale is shown in Fig. 10-19, along with the Bode-plot approximation. A circuit that will produce this function over a specific frequency range will be explored in Prob. 10-5.

The magnitude response $|H_p(f)|$ is nearly flat over the lower portion of the frequency range. However, as the frequency approaches f_1, the response begins to rise. Well above f_1, the response approaches a 6 dB/octave slope in which the amplitude response increases linearly with frequency. Thus, the signal level can be boosted above the increasing detected noise level in this critical frequency range.

In boosting the signal level at high frequencies, we have also introduced distortion in the original signal spectrum. For example, an audio signal will display a noticeable amount of high-frequency content and will not sound natural. Fortunately, it is relatively easy to compensate by the addition of a *deemphasis* network following the detector at the receiver.

The deemphasis network is designed with a transfer function

$$H_d(f) = \frac{1}{1 + j(f/f_1)} \qquad (10\text{-}102)$$

The form of the deemphasis magnitude response $|H_d(f)|$ is shown in Fig. 10-20. A simple *RC* low-pass filter can readily achieve this response, and this concept will be explored in Prob. 10-6. Note that the combined transmission effect of the preemphasis transfer function $H_p(f)$ and the demphasis transfer function $H_d(f)$ is.

$$H_p(f)H_d(f) = K\left(1 + j\frac{f}{f_1}\right) \times \frac{1}{1 + j(f/f_1)} = K \qquad (10\text{-}103)$$

which is a constant independent of frequency. Thus, the deemphasis network compensates for the unnatural boost of the high frequencies by attentuating the high frequencies in a controlled manner. Note, however, that the deemphasis does not "undo" the enhancement that was performed to boost the signal level above the noise level, since that process was achieved back at the

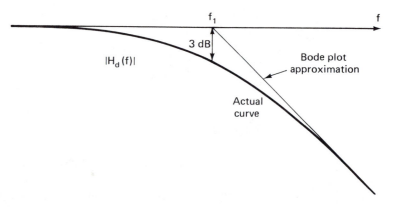

Figure 10-20. Form of the deemphasis curve used in FM systems.

transmitter, while the deemphasis curve is applied *after* the detection process.

It should be understood that the 3-dB frequencies for the preemphasis and deemphasis curves must be exactly the same for the compensation process to work, and this was stipulated by the choice of f_1 for both functions. Should the 3-dB frequencies be different, the detected signal will be distorted, so it is necessary to design the system as a total unit.

We now turn to the difficult task of determining quantitatively the improvement resulting from the preemphasis process. Although it is well known that preemphasis produces an improvement, the exact value of the improvement depends on some factors that are not so easy to define or control. There are some variations in the literature concerning the value of the improvement because of the differences in the assumptions made.

The approach taken here is based on the observation that many baseband modulating signals have spectral forms that tend to display a significant rolloff as the frequency increases. Specifically, the form of the power or energy spectrum $S_x(f)$ of a typical modulating signal will be assumed to be

$$S_x(f) = \frac{S_o}{1 + (f/f_2)^2}, \qquad \text{for } -W \le f \le W \qquad (10\text{-}104)$$

where S_o is the dc value of the power spectrum and f_2 is some reference frequency representing the point at which the power spectrum is down 3 dB. Next assume that enough information is known about the spectrum so that the preemphasis and deemphasis frequency f_1 can be selected to be the same as the frequency f_2; that is, $f_1 = f_2$.

When the emphasis curve is applied to the intelligence signal, the total power or energy (area under the curve) of the net modulating signal will be increased if K in (10-101) is unity. It is shown in more advanced treatments of FM that the bandwidth of a wideband random FM signal is directly proportional to the mean-square value of the modulating signal. If the transmitter is adjusted for a certain frequency deviation based on the unmodified modulating signal, the deviation would be increased with preemphasis unless K is lowered.

Based on the preceding discussion, the strategy assumed will be that when preemphasis is applied the value of K will be set at a value such that the mean-square value of the preemphasized signal is the same as that of the unprocessed signal. Let P_{r1} represent a reference normalized power (1-Ω reference) without preemphasis, and let P_{r2} represent the same signal with preemphasis. The value of P_{r1} is obtained by determining the area under the curve of (10-104). We have

$$P_{r1} = \int_{-\infty}^{\infty} S_x(f)\, df = \int_{-W}^{W} \frac{S_o\, df}{1 + (f/f_2)^2} = 2f_2 S_o \tan^{-1} \frac{W}{f_2} \quad (10\text{-}105)$$

The preemphasized signal power spectrum is obtained by multiplying (10-104) by the power transfer function corresponding to (10-101). The new power P_{r2} is then obtained by determining the area under the curve of the resulting power spectrum. Thus,

$$P_{r2} = \int_{-W}^{W} S_x(f) K^2 \left[1 + \left(\frac{f}{f_1} \right)^2 \right] df \quad (10\text{-}106a)$$

$$= \int_{-W}^{W} \frac{K^2 S_o [1 + (f/f_1)^2]\, df}{[1 + (f/f_2)^2]} \quad (10\text{-}106b)$$

Assume now that $f_1 = f_2$, in which case (10-106b) simplifies to

$$P_{r2} = \int_{-W}^{W} K^2 S_o\, df = 2WK^2 S_o \quad (10\text{-}107)$$

We require that $P_{r1} = P_{r2}$. Substituting $f_2 = f_1$ in (10-105) and equating that result to (10-107), we obtain

$$K^2 = \frac{f_1}{W} \tan^{-1} \frac{W}{f_1} \quad (10\text{-}108)$$

It turns out that $K < 1$ in the range of interest, which in effect means that the low-frequency level of the signal must be reduced when the high frequency is preemphasized in order to maintain the same total power.

Let P_{d1} represent the detected power of the signal without preemphasis. On a normalized reference basis without considering any gain constants, this function can be equated to the input power P_{r1}. Thus,

$$P_{d1} = P_{r1} = 2f_1 S_o \tan^{-1} \frac{W}{f_1} \quad (10\text{-}109)$$

where $f_2 = f_1$ has been substituted. Let N_{d1} represent the detected noise without preemphasis. This value was derived in the preceding section, and the result of (10-98) with $P_r = P_{r1}$ is repeated here:

$$N_{d1} = \frac{\eta_i W^3}{3 P_{r1}} \quad (10\text{-}110)$$

Let P_{d2} represent the detected signal power resulting from the combination of preemphasis at the transmitter followed by deemphasis at the receiver. The preemphasis applied to the signal at the transmitter is offset by the deemphasis factor at the receiver, and the net detected signal power is modified by the constant K^2. Thus,

$$P_{d2} = K^2 P_{d1} \qquad (10\text{-}111)$$

Let N_{d2} represent the detected noise following deemphasis. The deemphasized noise power spectrum is obtained by multiplying the detected power spectral density function of (10-97) by the deemphasis power transfer function obtained from (10-102). The resulting noise power may be expressed as

$$N_{d2} = \int_{-W}^{W} \frac{(\eta_i f^2/2P_{r1})\,df}{1 + (f/f_1)^2} \qquad (10\text{-}112)$$

The exact details of this integration will be left as an exercise for the interested reader (Prob. 10-7). The result is

$$N_{d2} = \frac{\eta_i f_1^3}{P_{r1}}\left[\frac{W}{f_1} - \tan^{-1}\frac{W}{F_1}\right] \qquad (10\text{-}113)$$

The improvement ratio IR will be defined as

$$\text{IR} = \frac{P_{d2}/N_{d2}}{P_{d1}/N_{d1}} = \frac{P_{d2}}{P_{d1}}\frac{N_{d1}}{N_{d2}} \qquad (10\text{-}114)$$

Substituting (10-109), (10-110), (10-111), and (10-113) in (10-114) and simplifying, there results

$$\text{IR} = \frac{(W/f_1)\tan^{-1}(W/f_1)}{3[1 - (f_1/W)\tan^{-1}(W/f_1)]} \qquad (10\text{-}115)$$

In most cases of practical interest, f_1 is substantially smaller than W. With this assumption, the expression of (10-115) is not significantly different from the much simpler result

$$\text{IR} \simeq \frac{\pi}{6}\frac{W}{f_1} \qquad (10\text{-}116)$$

The process leading to this simpler result is left as a guided exercise for the reader (Prob. 10-8). While (10-116) is not necessarily an extremely close approximation to (10-115), the various assumptions employed in the derivation are sufficiently inexact to justify this much simpler form as a reasonable estimate.

The improvement ratio in FM resulting from preemphasis is an additional enhancement in the signal-to-noise ratio over that achieved by the receiver processing gain. For example, if the receiver processing gain is 20 and the improvement ratio is 5, the overall gain in the signal-to-noise ratio with FM plus preemphasis would be $20 \times 5 = 100$. The same data expressed in decibels are a processing gain of 13 dB, an improvement ratio of 7 dB, and an overall improvement in signal-to-noise ratio of $13 + 7 = 20$ dB. The use of preemphasis in some typical system calculations will be illustrated in Chapter 11.

Before leaving the subject of preemphasis, an interesting comparison between FM with preemphasis and PM will be made. It will be recalled that the major difference between PM and FM is that in the former the phase deviation is proportional to the modulating signal, while in the latter the frequency deviation is proportional to the modulating signal. For a constant-amplitude single-tone input, the instantaneous frequency for PM increases linearly with the modulating frequency. When a modulating signal is preemphasized, the same effect occurs in the frequency range where preemphasis is applied. Thus, FM with preemphasis is, in a sense, a combination of FM and PM. The combination is capable of producing a better result in many systems than either PM or FM alone.

10-7   DETECTION OF BASEBAND PULSES IN NOISES

The general principles of PCM, in which a signal can be represented by a series of binary numbers or words, was introduced in Chapter 7. It was noted that a signal could theoretically be reconstructed to within a quantization error range of uncertainty at the receiver provided that a minimum sampling rate is employed and provided that the receiver is able to distinguish between a transmitted 0 and a transmitted 1. Thus, when the noise level is negligible, the receiver circuitry need only "look at" the signal during specific intervals of time and reconstruct a clear replica of the sample value based on the outcome of the observations.

Unfortunately, many digital communications systems of practical interest are required to operate under conditions in which the noise level may be high. The random spikes of noise may then cause errors in detection. It is possible for a positive spike of noise to make a transmitted 0 appear as a 1 and vice versa. The same phenomenon occurs in many types of radar systems and may result in errors in the detection of critical pulse returns.

Errors arising from the failure to correctly determine proper levels in pulse or digital signals due to the presence of noise are called *decision errors*. The space program stimulated a significant body of research aimed at reducing decision errors in digital data systems. Various mathematical techniques have been employed for predicting the error rate in different types of digital systems, and procedures have been developed for lowering the error rate subject to given power levels.

Several complete books and hundreds of journal articles dealing with various aspects of the digital decision problem in noise have been written. Much of this material is highly specialized and somewhat obscure to the hardware engineer. The intent will be to acquaint the reader with the general nature of the problem and to provide a base foundation of the concept. This will be achieved through an analysis of several simplified cases. The concept will then be extended to several representative cases of practical importance. While no pretense will be made that anyone will become an expert from studying this section and the next few, the cases considered will lead to a reasonable appreciation of the dimensions of the problem, and the results provided will be of value in analyzing complete systems in Chapter 11.

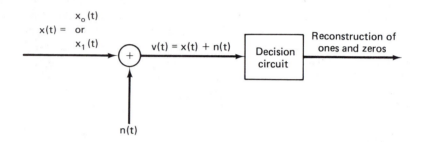

Figure 10-21. Baseband signal plus additive noise model used to study detection of pulses with noise.

A quantitative analysis of several simplified cases will now be made. Consider the block diagram shown in Fig. 10-21. A baseband digital signal $x(t)$ and additive noise $n(t)$ are assumed. The composite channel output function $v(t)$ is then

$$v(t) = x(t) + n(t) \qquad (10\text{-}117)$$

The signal $x(t)$ is assumed to have two possible forms, $x_0(t)$ and $x_1(t)$, representing the two possible binary states. Where it is necessary to delineate the possible forms of the composite received signal, we can write

$$v_o(t) = x_o(t) + n(t) \qquad (10\text{-}118)$$
$$v_1(t) = x_1(t) + n(t) \qquad (10\text{-}119)$$

The exact forms for $x_o(t)$ and $x_1(t)$ will be discussed shortly. The channel noise is assumed to be gaussian with a mean $\bar{n} = 0$ and an rms value σ_n (or noise power σ_n^2).

We will consider in this section the effect of a decision based on the level of the signal at the point of observation. All pulses will be assumed to be reproduced perfectly; that is, the effects of band-limiting will not be considered. The two cases of unipolar and bipolar digital pulses will be considered separately.

Unipolar Pulses

With unipolar pulses, a 1 is represented as a positive pulse of amplitude A and a 0 is represented as no pulse at all. Thus, $x_o(t)$ and $x_1(t)$ may be expressed as

$$x_o(t) = 0 \qquad (10\text{-}120)$$

and

$$x_1(t) = A \qquad (10\text{-}121)$$

The problem at the receiver is to determine if a 0 or a 1 is received. If noise were not present, such a decision would obviously be quite trivial. However, the presence of additive noise can result in two types of errors.

Referring to Fig. 10-22a, a positive spike can trigger the receiver threshold circuit even when no pulse is transmitted, or, as in Fig. 10-22b, a negative noise spike can suppress a positive pulse, thus causing the net signal to be lower than the threshold level. (In radar terminology, the first type of error is called a *false alarm,* and the second is called a *miss* for somewhat obvious reasons.)

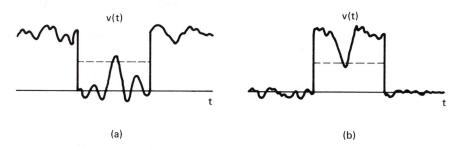

(a)   (b)

Figure 10-22. (a) Positive spike of noise exceeding threshold when no pulse is actually present (false alarm). (b) Negative spike of noise suppressing pulse below threshold (miss).

The following terms are defined:

P_o = probability that a 0 is transmitted
P_1 = probability that a 1 is transmitted
P_{eo} = probability that a decision error occurs when a 0 is transmitted (i.e., a 0 is read as 1)
P_{e1} = probability that a decision error occurs when a 1 is transmitted (i.e., a 1 is read as a 0)
P_e = total (or average) probability of error

An error can arise in the following manner: A 0 is transmitted *and* the 0 is read as a 1 *or* a 1 is transmitted *and* the 1 is read as a 0. The statements on

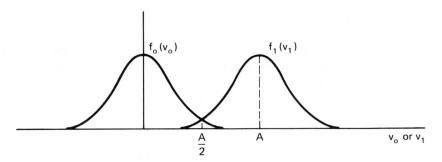

Figure 10-23. Probability density functions for signal plus noise for unipolar transmission.

either side of the "or" are mutually exclusive, so the preceding process can be formulated using the notation defined earlier, as

$$P_e = P_o P_{eo} + P_1 P_{e1} \qquad (10\text{-}122)$$

Since the noise is gaussian and additive, the two possible probability density functions at the receiver are shown in Fig. 10-23. The function labeled $f_o(v_o)$ represents the PDF corresponding to no transmitted pulse (noise only), and $f_1(v_1)$ represents the PDF corresponding to a transmitted pulse of amplitude A (plus channel noise). Mathematically, these functions can be expressed as

$$f_o(v_o) = \frac{1}{\sqrt{2\pi}\,\sigma_n}\,\epsilon^{-v_o^2/2\sigma_n^2} \qquad (10\text{-}123)$$

$$f_1(v_1) = \frac{1}{\sqrt{2\pi}\,\sigma_n}\,\epsilon^{-(v_1-A)^2/2\sigma_n^2} \qquad (10\text{-}124)$$

In the most general case, ones and zeros are transmitted with unequal probabilities, and an optimum threshold level is sought such that the average error rate is minimum. However, the most common case of practical interest in digital communications is the situation in which zeros and ones are transmitted with equal probabilities.

From this point forward, we will assume $P_o = P_1 = 1/2$. It can be shown in this case that the optimum value for the threshold decision level is $A/2$. This means that the receiver will be designed to sense the level of a received signal and assign the value 1 if the received signal is greater than $A/2$ and the value 0 if the received signal is less than $A/2$.

The error probabilities can be expressed as

$$P_{eo} = P\left[V_o > \frac{A}{2}\right] = \int_{A/2}^{\infty} f_o(v_o)\,dv_o \qquad (10\text{-}125)$$

$$P_{e1} = P\left[V_1 < \frac{A}{2}\right] = \int_{-\infty}^{A/2} f_1(v_1)\,dv_1 \qquad (10\text{-}126)$$

We will leave as an exercise for the reader (Prob. 10-9) to show that the preceding two probabilities are equal and can be expressed as

$$P_{eo} = P_{el} = Q\left(\frac{A}{2\sigma_n}\right) \tag{10-127}$$

where the $Q(y)$ function was defined in Chapter 8.

Combining the result of (10-127) with the preceding assumption that $P_0 = P_1 = 1/2$, the probability of error P_e from (10-122) can be expressed as

$$P_e = Q\left(\frac{A}{2\sigma_n}\right) \tag{10-128}$$

To relate this result to the signal-to-noise ratio, let

$$R = \frac{\text{average signal power}}{\text{mean-square noise power}} \tag{10-129a}$$

$$= \frac{A^2/2}{\sigma_n^2} = \frac{A^2}{2\sigma_n^2} \tag{10-129b}$$

where a 1-Ω reference is assumed. Note that since the level is A for about half the time, which corresponds to a power of A^2, and since the level is 0 for about half the time, which corresponds to a power of 0, the average power is $A^2/2$.

Substitution of the parameter R in (10-128) results in the following expression for the probability of error:

$$P_e = Q\left(\sqrt{\frac{R}{2}}\right) \tag{10-130}$$

This function is shown in Fig. 10-24 with abscissa I corresponding to the value of $R_{dB} = 10 \log_{10} R$. (The other two abscissas will be explained later.) This curve indicates how the probability of error depends on the signal-to-noise ratio.

Many systems are designed to operate with P_e in the neighborhood of 10^{-5}. This means that one error would occur for about every 10^5 decisions (or pulse intervals). The signal-to-noise ratio required for $P_e = 10^{-5}$ with ideal unipolar pulses and a single-level decision process is $R_{dB} \simeq 15.6$ dB.

Bipolar Pulses

With bipolar pulses, a 1 is represented as a positive pulse of amplitude A, and a 0 is represented as a negative pulse of amplitude $-A$. Thus, $x_o(t)$ and $x_1(t)$ may be expressed as

$$x_o(t) = -A \tag{10-131}$$

and

$$x_1(t) = +A \tag{10-132}$$

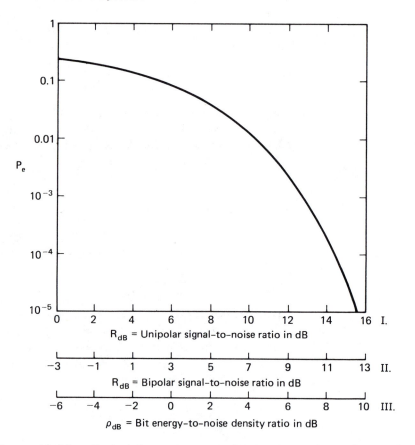

Figure 10-24. Probability of error versus signal-to-noise ratios for three cases developed in Sections 10-7 and 10-8.

The two possible received signals have the forms

$$v_o(t) = -A + n(t) \qquad (10\text{-}133)$$

$$v_1(t) = \quad A + n(t) \qquad (10\text{-}134)$$

The corresponding PDFs are shown in Fig. 10-25.

The development from this point on follows very closely the development for the unipolar case. The major difference is that the optimum threshold level in this case is at a level of zero, assuming an equal frequency of zeros and ones. Thus, the receiver will assign a value of 1 to the signal at the decision time if the received signal is positive and a value of 0 if the received signal is negative. We will leave as an exercise for the reader (Prob. 10-10) to show that the individual probabilities of error in this case are

$$P_e = P_{eo} = P_{e1} = Q\left(\frac{A}{\sigma_n}\right) \qquad (10\text{-}135)$$

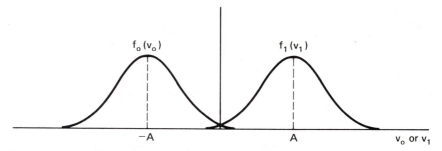

Figure 10-25. Probability density functions for signal plus noise in bipolar transmission.

In this case, the average signal power is A^2. Why? Thus, the signal-to-noise ratio R for the bipolar case is

$$R = \frac{A^2}{\sigma_n^2} \qquad (10\text{-}136)$$

The probability of error can then be expressed as

$$P_e = Q(\sqrt{R}) \qquad (10\text{-}137)$$

This function is of the same form as for unipolar pulses but differs only in that the argument $\sqrt{R/2}$ for the unipolar case is replaced by $\sqrt{R}$ for the bipolar case. The effect is that the same probability of error for bipolar pulses is achieved for a signal-to-noise ratio 3 dB less than for unipolar pulses. The curve of Fig. 10-24 thus applies to this case, but the abscissa is now line II. The signal-to-noise ratio for $P_e = 10^{-5}$ is $R_{dB} \approx 12.6$ dB.

10-8 PULSE DETECTION WITH A MATCHED FILTER

In the previous section, we investigated the problem of detecting ideal pulse signals in the presence of additive gaussian noise using simple amplitude threshold detection. In this section we will consider the same problem, but with one important addition. The received signal will be first processed through a special filter circuit. It will be seen that this process will result in a significant improvement in the detectability of the signal.

A diagram indicating a model of the receiving filter is shown in Fig. 10-26. This particular circuit is called an *integrate and dump filter,* which will be denoted as an ID filter in the remainder of this section. The circuit consists of an integrator, which integrates the resulting signal for a period of T seconds, where T is the pulse duration. At the end of the interval, a threshold circuit is used to formulate a decision on the basis of the integrated signal rather than the unprocessed signal, as was the case in the previous section. The capacitor in the integrator is then quickly discharged, and the process is repeated.

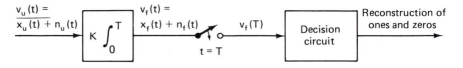

Figure 10-26. Integrate and dump filter used as a matched filter for baseband bipolar pulses.

Much of the notational form introduced in the last section will be continued here. In this case $v_u(t)$ represents the composite unfiltered input to the filter, and $v_f(t)$ represents the filtered output. These functions will be represented as

$$v_u(t) = x_u(t) + n_u(t) \tag{10-138}$$

$$v_f(t) = x_f(t) + n_f(t) \tag{10-139}$$

where $x_u(t)$ and $n_u(t)$ are the unfiltered input signals and noise, and $x_f(t)$ and $n_f(t)$ are the filtered output signal and noise, respectively.

The ID filter is linear, and thus superposition may be applied to analyze the effects on signal and noise separately. A decision will be made at time $t = T$, so the values of the output signal and noise evaluated at that time are most pertinent.

Only the case of bipolar pulse transmission will be considered. Assume that the value $-A$ is transmitted, and assume that the gain constant for the integrator is K. The filtered output signal at time $t = T$ is

$$x_f(T) = K \int_0^T x_u(t) \, dt = K \int_0^T -A \, dt = -KAT \tag{10-140}$$

Similarly, when the value $+A$ is transmitted, the filtered output at time T is

$$x_f(T) = K \int_0^T A \, dt = KAT \tag{10-141}$$

The output noise is not so easily obtained since the input noise is a gaussian random process. However, it can be evaluated by first assuming that the input noise spectrum $S_{nu}(f)$ is flat and utilizing the frequency-domain approach. Let $S_{nu}(f) = \eta_i/2$ represent the input noise power spectral density on a two-sided basis.

It is necessary to determine the frequency response of the integrate and dump filter in order to determine the noise level at the output. This is most easily achieved by determining the impulse response in the time domain and taking its Fourier transform. Consider the unit area impulse function shown symbolically in Fig. 10-27a. If this function were applied at $t = 0$ to the ID filter, the output would instantly assume the value K as a result of the integration. The output would remain at the value until $t = T$, at which time it would be "dumped." Thus, the impulse response is a pulse of the form shown in Fig. 10-27b.

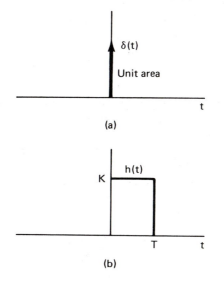

(a)

(b)

Figure 10-27. Unit area impulse function applied to ID filter and the corresponding output.

The Fourier transform of this function is the transfer function, and it is readily determined from the results of Chapter 2 as

$$H(f) = KT\frac{\sin \pi fT}{\pi fT}\epsilon^{-j\pi fT} \tag{10-142}$$

The power transfer function $|H(f)|^2$ is

$$|H(f)|^2 = K^2T^2\left(\frac{\sin \pi fT}{\pi fT}\right)^2 \tag{10-143}$$

The filtered output power spectral density $S_{nf}(f)$ is obtained by weighting the unfiltered power spectral density with the power transfer function; that is,

$$S_{nf}(f) = |H(f)|^2 S_{nu}(f) \tag{10-144a}$$

$$= \frac{\eta_i}{2}K^2T^2\left(\frac{\sin \pi fT}{\pi fT}\right)^2 \tag{10-144b}$$

The noise power or mean-square value of the output noise at the end of the integrate period will be denoted as $N_f(T)$, and it is determined as

$$N_f(T) = \int_{-\infty}^{\infty} S_{nf}(f)\, df = \frac{\eta_i K^2 T^2}{2}\int_{-\infty}^{\infty}\left(\frac{\sin \pi fT}{\pi fT}\right)^2 df \tag{10-145}$$

An integral found in most standard integral tables is

$$\int_0^\infty \left(\frac{\sin x}{x}\right)^2 dx = \frac{\pi}{2} \tag{10-146}$$

We will leave as an exercise for the reader (Prob. 10-11) to show that the application of (10-146) to (10-145) yields

$$N_f(T) = \frac{K^2 \eta_i T}{2} \qquad (10\text{-}147)$$

The threshold decision process can now be visualized as shown in Fig. 10-28. In one sense, this is the same concept as encountered in the previous section for bipolar pulses, but there is an important difference. Rather than make a decision on the basis of the "raw" signal in the presence of noise, the composite signal has been integrated for T seconds before a decision is made. As we will see shortly, this results in a more reliable basis for a decision.

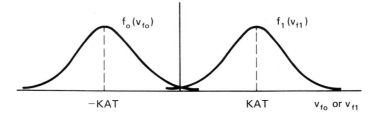

Figure 10-28.   Probability density function at output of ID matched filter.

The details on determining the probability of error P_e follow the same mathematical development as utilized in the last section for the bipolar case, except the signal level of concern is the filtered composite signal at time T, that is, $v_f(T)$, and the filtered noise power is now $N_f(T)$. The probability density functions corresponding to the two possible transmitted signals will be denoted as $f_o(v_{fo})$ and $f_1(v_{f1})$. These fuctions are illustrated in Fig. 10-28.

Assuming equal probabilities of zeros and ones and a threshold level of zero, the symmetry employed in the past section leads to the following expression for the probability of error:

$$P_e = P[V_{fo} > 0] = Q\left[\frac{KAT}{\sqrt{N_f(T)}}\right] \qquad (10\text{-}148)$$

The expression appearing in the argument of the Q function can be manipulated as follows:

$$\frac{KAT}{\sqrt{N_f(T)}} = \frac{KAT}{K\sqrt{\eta_i T/2}} = \sqrt{\frac{A^2 T^2}{\eta_i T/2}} = \sqrt{\frac{A^2 T}{\eta_i/2}} \qquad (10\text{-}149)$$

A convenient and useful definition in this analysis is the *bit energy* E_b. Using a normalized 1-Ω reference, E_b is

$$E_b = A^2 T \qquad (10\text{-}150)$$

We now define a quantity ρ as

$$\rho = \frac{\text{bit energy}}{\text{one-sided noise power density}} \qquad (10\text{-}151a)$$

$$= \frac{E_b}{\eta_i} = \frac{A^2 T}{\eta_i} \qquad (10\text{-}151b)$$

Recognizing this form in (10-149), the probability of error can be expressed as

$$P_e = Q(\sqrt{2\rho}) \qquad (10\text{-}152)$$

The value of P_e has the same form as for the two cases considered in the last section. However, the argument now is $\sqrt{2\rho}$, which compares with $\sqrt{R}$ and $\sqrt{R/2}$ for the two cases of a decision based on an unfiltered amplitude observation. In one sense, we are comparing "apples with oranges," since the parameter R referred to a power ratio, while the parameter ρ refers to an energy ratio. However, this is a result of the difference in which the two signals are processed. For the present case, the same probability of error occurs for a ρ value 3 dB less than the corresponding R value for the bipolar case and 6 dB less than the corresponding R value for the unipolar case when the signal is unfiltered. The exact form of P_e is inferred from Fig. 10-24 by the use of scale III. The signal-to-noise ratio corresponding to $P_e = 10^{-5}$ is $\rho_{dB} \approx 9.6$ dB.

The integrate and dump filter is a special case of a *matched filter*. A matched filter is a special processing circuit that maximizes the output signal-to-noise ratio for a given signal plus noise input. Equivalently, the probability of error is minimized for given input signal power and noise levels.

The concept of the matched filter requires that the filter response and the two possible binary waveforms all be selected in accordance with an optimization process. When the two possible binary waveforms are positive and negative pulses, respectively, the optimum matched filter is the integrate and dump filter previously discussed. This is a most common case and the one that will serve as the basis for our work. It should be emphasized, however, that the probability of error P_e shown in Fig. 10-24 and utilizing scale III applies to a variety of other possible cases in which the matched filter concept is utilized.

When RF pulses are considered, the matched filter is often implemented as a *correlator*. Consider, for example, digital transmission with phase-shift keying (PSK) signals as developed in Chapter 7. A matched filter can be realized through the scheme of Fig. 10-29.

The three cases of pulse detection considered in this section and the last should illustrate to the reader the general nature of the problem and the concept by which the error rate is a function of the signal-to-noise ratio in a communications channel. In particular, the case of pulse detection at the

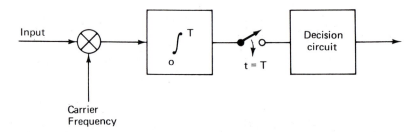

Figure 10-29. Matched filter correlator used in the detection of RF PSK digital data.

output of a matched filter is a rather general result and appears in a number of different types of data systems that are optimum in a theoretical sense. There are certain other forms that arise in specific data systems, and some of these will be presented in Chapter 11.

10-9 NOISE IN PCM SYSTEMS

The performance of PCM systems in the presence of noise will be analyzed in this section. There are two primary types of noise that arise in the generation and detection of PCM signals: (1) quantization noise and (2) decision noise. *Quantization noise* is a natural phenomenon that arises due to the encoding of the analog signal into a finite number of amplitudes. This concept was discussed at some length in Chapter 7 and will be reviewed in this section as it relates to the overall performance analysis.

Decision error noise is the modification that results from incorrect decisions concerning the received bits due to the presence of background noise. This problem is a function of the actual noise in the channel and is related to the probability of error of the particular encoding method. The foundation for this concept was established earlier in Secs. 10-7 and 10-8.

Quantization noise is a function only of the number of bits used in the encoding process and is independent of the actual channel noise. However, decision noise increases as the channel noise increases, and a certain minimum power must be used to maintain this effect to a tolerable level.

We will first derive an expression for the signal-to-noise ratio based only on quantization noise effects, since this is the condition for which system operation is desired. Thus, assume initially that the power level is sufficiently high that decision errors are negligible. To give the results most general validity, we will assume a bipolar analog signal $x(t)$ and assume it is normalized to the range $-1 < x < 1$ in accordance with the work of Chapter 7. Let ΔX represent the step size or increment between successive levels. For n bits, the number of levels $m = 2^n$ and

$$\Delta X = \frac{2}{2^n} = 2^{-n+1} \tag{10-153}$$

Assuming rounding in the A$^{\circ}$/D converter, the peak quantization error is $\pm\Delta X/2 = \pm 2^{-n}$. From a statistical point of view, the mean-square error is more meaningful than the peak error. Irrespective of the type of noise present (most often gaussian), it is usually reasonable to assume that the amplitude distribution is approximately constant over the range of one step change. Consequently, a uniform distribution for the quantization noise is normally assumed.

Let n_q represent the quantization noise. The PDF $f_{nq}(n_q)$ is shown in Fig. 10-30. Note that the range of n_q is assumed to be from $-\Delta X/2$ to $\Delta X/2$. The mean-square quantization noise $\overline{n_q^2}$ is determined as follows:

$$\overline{n_q^2} = \int_{-\Delta X/2}^{\Delta X/2} \frac{1}{\Delta X} n_q^2 \, dn_q = \frac{(\Delta X)^2}{12} \tag{10-154}$$

This corresponds to an rms error of $\Delta X/(2\sqrt{3}) = 0.577 \times$ peak noise.

In accordance with earlier assumptions employed in system analysis, assume that the signal $x(t)$ is a unit magnitude sinusoid

$$x(t) = \cos \omega_m t \tag{10-155}$$

The signal power P_r on a 1-Ω basis is then

$$P_r = \frac{1}{2} \tag{10-156}$$

Let R_q represent the signal power to quantization noise power. This quantity is the ratio of (10-156) to (10-154):

$$R_q = \frac{1/2}{(\Delta X)^2/12} = \frac{6}{(\Delta X)^2} \tag{10-157}$$

Substitution of (10-153) in (10-157) yields

$$R_q = 1.5(2)^{2n} = 1.5m^2 \tag{10-158}$$

This result is more readily interpreted when converted to decibels. Let $R_q(\text{dB})$ represent the corresponding value in decibels, which is

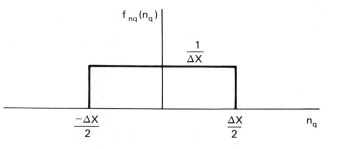

Figure 10-30. Probability density function of quantization noise in PCM.

$$R_q(\text{dB}) = 10 \log_{10} R_q = 10 \log [1.5(2)^{2n}] \qquad (10\text{-}159)$$
$$= 6.02n + 1.76 \text{ dB}$$

The practical implications of this result will be discussed at some length in Chapter 11. For the moment, it will be emphasized that R_q is a constant for a given number of bits and is not a function of the power, provided, of course, that the power is sufficiently large that decision errors are negligible.

The next step in this development is the introduction of decision noise effects. This phenomenon is somewhat more tedious to deal with because it is necessary to relate the decision error for a bit to the level of the resulting decoded signal. To illustrate this process, consider the 4-bit word 1111, and assume for the sake of this discussion that each step in the 16 possible levels corresponds to 1 V for some reference analog signal. If the LSB is incorrectly read at the receiver, an error of 1 V results. If an incorrect decision is made on the next bit, the error is 2 V, and if an error is made on the third bit the error is 4 V. Finally, a decision error on the MSB results in an error of 8 V, which is 8 times the error of the LSB! Thus, the magnitude of the error depends very heavily on the location of the bit in the given data word.

In view of the properties just discussed, much research on error detection and correction has been performed to develop methods whereby a limited number of bit errors can be detected and/or corrected. Such schemes involve redundancy by adding extra check bits to detect and/or correct possible errors, so the data rate is necessarily reduced for a given bandwidth by the necessity to transmit the extra bits. Another well-known scheme is the use of Gray codes in which successive levels always differ by one bit value so that, when one bit is incorrectly read, the signal will never be more than one level removed from the desired one. With natural binary encoding, an error in one bit can drastically change the value of the error as was noted with the example.

The concentration here will be on the natural binary representation, since this represents a worst-case bound. Let ϵ_d represent a decision error random variable indicating the difference between the actual reconstructed analog signal and the theoretical quantized analog signal without decision errors. The manner in which ϵ_d depends on the bit location is summarized as follows:

For LSB, $\epsilon_d = \Delta X$

For next bit, $\epsilon_d = 2\Delta X$

For next bit, $\epsilon_d = 4\Delta X$

.
.
.

For MSB, $\epsilon_d = 2^{n-1}\Delta X$

Without any form of error detection or correction, it is reasonable to assume that an error can occur with equal likelihood for any of the n bits.

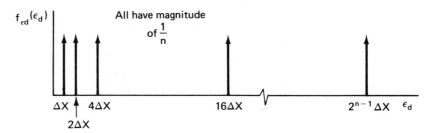

Figure 10-31. Probability density function of word decision errors in PCM.

Consequently, a PDF can be constructed to enhance the statistical analysis of the error. The quantity ϵ_d is a discrete variable since it has only n possible values, but no further integer subscripts will be used in order to keep the notation as simple as possible. The form of the PDF $f_{\epsilon d}(\epsilon_d)$ is shown in Fig. 10-31. Due to the marked increase in the steps as ϵ_d increases, only a few of the lines can be shown.

The mean-square value $\overline{\epsilon_d^2}$ of the decision noise is determined from the summation

$$\overline{\epsilon_d^2} = \sum_1^n \epsilon_d^2 f_{\epsilon d}(\epsilon_d) = \frac{1}{n} \sum_1^n \epsilon_d^2 \tag{10-160a}$$

$$= \frac{1}{n} [(\Delta X)^2 + 4(\Delta X)^2 + 16(\Delta X)^2 + \cdots + 2^{2(n-1)}(\Delta X)^2] \tag{10-160b}$$

The quantity in the brackets of (10-160b) is a geometric series and can be expressed in closed form. The result is

$$\overline{\epsilon_d^2} = \frac{(2^{2n} - 1)(\Delta X)^2}{3n}, \qquad \text{for } n \geq 1 \tag{10-161}$$

The preceding development provides the mean-square decision error based on a binary word. However, this result must in turn be related to the probability of error P_e based on a single bit. For a bit error of P_e, there is an average of 1-bit error per $1/P_e$ bits. However, since each word has n bits, this means that there will be about one error per $1/(nP_e)$ words. The net weighted mean-square decision noise $\overline{n_d^2}$ is then determined by multiplying the mean-square error per word $\overline{\epsilon_d^2}$ from (10-161) by the fraction of the total words that contain an error (i.e., nP_e). Thus

$$\overline{n_d^2} = nP_e \overline{\epsilon_d^2} = \frac{(2^{2n} - 1)(\Delta X)^2}{3} P_e \tag{10-162}$$

Another way to look at this result is that the factor nP_e weights an error in a word on the same averaging basis as quantization noise.

We can now combine the effects of quantization noise and decision noise since they are being weighted on the same basis. Assuming statistical independence between these two sources, the net noise power or variance N_d is given by the sum of (10-154) and (10-162).

$$N_d = \frac{(\Delta X)^2}{12} + \frac{(2^{2n} - 1)(\Delta X)^2 P_e}{3} \qquad (10\text{-}163)$$

For a reasonable number of levels $2^{2n} - 1 \simeq 2^{2n} = m^2$. In addition, $\Delta X = 2^{-n+1}$, and N_d is approximately

$$N_d \simeq \frac{1}{3m^2} (1 + 4m^2 P_e) \qquad (10\text{-}164)$$

based on a range of x from -1 to $+1$. The net detected output signal-to-ratio R_d is then

$$R_d = \frac{P_d}{N_d} = \frac{1.5m^2}{1 + 4m^2 P_e} \qquad (10\text{-}165)$$

This result obviously depends on P_e so that specific types of schemes would have to be assumed before quantitative results can be determined. Observe that, as $P_e \rightarrow 0$, R_d in (10-165) approaches the value of R_q of (10-158), representing the maximum possible signal-to-noise ratio for PCM. Some useful interpretations of these results and their practical applications will be discussed in Chapter 11.

PROBLEMS

10-1 Consider a system of the form of Ex. 10-1 as illustrated in Fig. 10-2, but with the following changes: $T_e = 300$ K, $T_a = 100$ K, RF amplifier gain $= 34$ dB, IF amplifier gain $= 86$ dB, mixer loss $= 10$ dB. The IF amplifier is centered at 30 MHz and has a bandwidth of 1 MHz.

    (a) Compute the available IF noise output power, and sketch the form of the band-pass noise.

    (b) If the total available noise output power is dissipated in a 50-Ω matched load, compute the rms output noise voltage.

10-2 Consider the band-pass power spectral density function of Ex. 10-1 at the output of the IF amplifier, that is, Eq. (10-4). Assume that the output of the IF amplifier is applied as one input to a mixer and assume that a unit amplitude sinusoid with frequency f_0 is applied as the second input. Sketch and label the forms of the power spectra at the output of the mixer for each of the following values of f_0: (a) 4.95 MHz, (b) 5 MHz, (c) 5.05 MHz. [*Hint:* The result of Eq. (10-4) should be first converted to a two-sided spectral form.]

10-3 Consider the band-pass power spectral density function of Ex. 10-1 at the output of the IF amplifier, that is, Eq. (10-4). Draw the form of the power spectrum $S_{nc}(f) = S_{ns}(f)$ of the low-pass quadrature components on a two-sided basis. Label the magnitude and the frequency limits. [*Hint:* The result of Eq. (10-4) should be first converted to a two-sided spectral form.]

10-4 Consider the power spectrum of noise at the output of an FM detector as given by Eq. (10-97). Show that the total noise power contained in this function over the applicable frequency range ($-W$ to W) is given by Eq. (10-98).

10-5 A relatively simple circuit that can be used to achieve the form of the FM preemphasis curve over a specified frequency range is shown in Fig. P10-5. Assume that $R_1 \gg R_2$. Let $R_{eq} = R_1 \| R_2$, $f_1 = 1/(2\pi R_1 C)$, and $f_2 = 1/(2\pi R_{eq} C)$. Show that for $f \ll f_2$,

$$H_p(f) \approx \frac{R_2}{R_1 + R_2}\left(1 + j\frac{f}{f_1}\right)$$

[*Note:* Any arbitrary value of K in Eq. (10-101) can be achieved by combining this circuit with additional amplification or attenuation.]

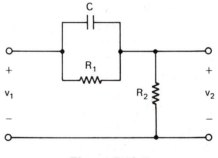

Figure P10-5

10-6 The simple RC low-pass circuit shown in Fig. P10-6 can be used for an FM deemphasis circuit. Show that the transfer function is

$$H_d(f) = \frac{1}{1 + j\dfrac{f}{f_1}}$$

where $f_1 = 1/2\pi RC$.

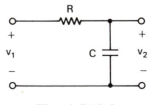

Figure P10-6

10-7 Starting with Eq. (10-112), carry out the integration and verify the result of Eq. (10-113). *Hint:* It is best to first expand the variable form in the integral by the relationship

$$\frac{x^2}{x^2 + 1} = 1 - \frac{1}{1 + x^2}$$

10-8 The simplification leading from Eq. (10-115) to Eq. (10-116) utilizes one of the two arctangent approximations, which are

$$\tan^{-1} x \simeq x \qquad \text{for} \quad x \ll 1$$

$$\tan^{-1} x \simeq \frac{\pi}{2} - \frac{1}{x}, \qquad \text{for} \qquad x \gg 1$$

Specifically, the second form is used. Based on the assumption that $W \gg f_1$, expand (10-115) with this approximation. Neglecting certain additional terms, show that (10-116) results.

10-9 Show that the two integrals of Eqs. (10-125) and (10-126) for unipolar pulses both yield identical results, which can be expressed in the form of Eq. (10-127).

10-10 Formulate two integrals for the probabilities of error $P_{eo} = P_{e1}$ for the bipolar pulse case. These correspond to Eqs. (10-125) and (10-126) for the unipolar case. Show that both yield identical results, which can be expressed in the form of Eq. (10-135).

10-11 Show that the application of Eq. (10-146) to Eq. (10-145) yields the result of Eq. (10-147).

10-12 Threshold for PCM systems is sometimes defined as the point at which the detected signal-to-noise ratio is 1 dB below the theoretical limit in which the only noise effect is quantization noise. Using Eq. (10-165), show that the probability of error at this threshold point is given by $P_e = 0.06473/m^2$.

Performance of Complete Systems
11

11-0 INTRODUCTION

The primary objective of this chapter is to present an overview of the performance of complete communications systems in the presence of noise. This will be achieved by utilizing the results of modulation performance analysis coupled with link analysis. The results derived in the last chapter will be interpreted from a practical point of view and extensively used. The concepts of link analysis dealing with antenna parameters will be presented. Both one- and two-way (radar) link results will be discussed. Finally, a relative comparison of a number of different modulation systems will be made.

11-1 GENERAL DISCUSSION

The major objective of Chapter 10 was the development of various signal, noise, and signal-to-noise parameters for different types of modulation systems, both analog and digital. As stated at the beginning of Chapter 10, much of those details could be omitted without loss of continuity for readers wishing to concentrate only on using the results in solving practical problems, rather than on understanding their theoretical basis. Consequently, a portion of this chapter will be devoted to a summary and interpretation of the results obtained from the various derivations of Chapter 10. This material is not intended as a rehash of Chapter 10, but rather it complements the analytical efforts of that chapter by providing practical implications of the results. Following this discussion, these results will then be utilized in dealing with the performance of some complete systems.

It will be emphasized that all the modulation performance analysis developments were based on a number of arbitrary assumptions, some of which may not always be valid. In all cases, the modulation was assumed to be a single-tone modulating signal for simplicity. The results may or may not be reasonable for all operating conditions. It is believed that the results given here are meaningful in making comparisons and in approximating the absolute operation for systems, but these results must be tempered with good judgment and reasonable additional allowances for worst-case performance. In particular, the use of a single-tone modulation tends to create a false impression of the "potency" of a signal as compared, for example, with signals having "noiselike" character and a wide dynamic range.

The reader who searches through other communications books will find some that compare more different methods than are presented here. The choice was made in this text to limit system comparisons to a relatively few cases for several reasons. First, these comparisons are difficult to make if too many different types of systems are mixed together, and the results are often difficult to interpret. Second, the systems used here represent two distinctly different, but very important, classes of systems. The traditional analog systems, such as the various AM processes and FM, are widely used and will no doubt continue to be used for many years to come. In particular, FM is one of the best performers in the presence of noise and deserves much attention. In contrast, reasonable attention is given to the performance of PCM and digital modulation in the presence of noise. It is this writer's opinion that this general area represents the trend of the future in the evolution of new communications technology and, likewise, deserves to be considered as much as possible.

11-2 ANALOG SYSTEM COMPARISONS

The block diagram shown in Figure 11-1 will be used to illustrate the various signal and noise parameters used in analyzing receiver performance for analog modulation systems. The quantity P_r represents the received average signal power, and the quantity N_a represents the average noise power

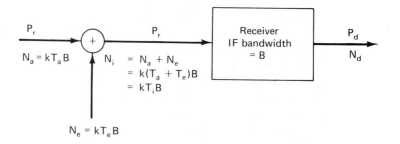

Figure 11-1. Block diagram illustrating parameters used in analog receiving system comparisons.

at the antenna output. (The antenna output noise becomes the effective noise "source.") The effective noise power of the receiver referred back to the input is N_e, and this quantity is added to N_a to produce the total input noise power N_i; that is,

$$N_i = N_a + N_e = kT_aB + kT_eB \qquad (11\text{-}1)$$

The IF bandwidth B will be assumed to be adjusted to a value just sufficient to pass a signal whose baseband bandwidth is W, and the selectivity will be assumed to be sufficiently high that the equivalent noise bandwidth and the conventional bandwidth values are the same. If the impedances are assumed to be matched at the input, the total input noise power in the IF bandwidth B can be expressed as

$$N_i = kT_aB + kT_eB = k(T_a + T_e)B = kT_iB = \eta_iB \qquad (11\text{-}2)$$

where

$$T_i = T_a + T_e \qquad (11\text{-}3)$$

and

$$\eta_i = kT_i = k(T_a + T_e) \qquad (11\text{-}4)$$

Recall from Chapter 9 that T_a is the antenna noise temperature, T_e is the effective receiver temperature referred to the input, and T_i is the total input temperature to the receiver. The one-sided total input noise power density is η_i.

The net signal-to-noise power ratio at the receiver input will be designated as R_i, and it is

$$R_i = \frac{P_r}{N_i} = \frac{P_r}{\eta_iB} = \frac{P_r}{kT_iB} = \frac{P_r}{k(T_a + T_e)B} \qquad (11\text{-}5)$$

The actual noise power entering the receiver will usually be greater than N_i since the RF stage bandwidth is normally greater than the minimum IF bandwidth B. However, after suitable conversion and filtering, the bandwidth is reduced to B, so the result of (11-5) is the final signal-to-noise ratio appearing at the detector input.

The detector portion of the receiver extracts the intelligence from the carrier and produces a useful baseband signal. However, noise will also appear at the detector output. The resulting detected signal power will be denoted as P_d, and the detected noise power will be denoted as N_d, as shown in Fig. 11-1. The net signal-to-noise power ratio at the detector output will be designated as R_d, and it is

$$R_d = \frac{P_d}{N_d} \qquad (11\text{-}6)$$

Depending on the type of modulation, the type of detector, and other factors, the signal-to-noise ratio at the detector output may be less than, equal to, or greater than the input signal to noise ratio. It is convenient to define a quantity called the *receiver processing gain G_r* as

$$G_r = \frac{R_d}{R_i} \qquad (11\text{-}7)$$

If $G_r > 1$, the detection process actually enhances the signal-to-noise ratio; that is, the detected signal level relative to the noise level is greater than the ratio of the RF signal power to the input noise level. If $G_r = 1$, the relative strength of the signal in comparison to the noise remains the same, and if $G_r < 1$, the "gain" is actually a loss.

One other ratio of interest is a quantity that will be defined as the *baseband reference gain G_b*, and it is

$$G_b = \frac{R_d}{R_b} \qquad (11\text{-}8)$$

where R_b is the ratio of the received power to a noise power defined as follows: Consider the noise power density at the receiver input (i.e., η_i), and compute the noise power based on the modulating signal bandwidth W. Thus,

$$R_b = \frac{P_r}{\eta_i W} = \frac{P_r}{k T_i W} \qquad (11\text{-}9)$$

The quantity R_b represents the signal-to-noise ratio that could be achieved with simple baseband direct transmission (e.g., over a telephone line) with the same signal power and with the same noise density as at the receiver. If $G_b > 1$, the system is superior to baseband as far as signal-to-noise ratio is concerned.

The receiver processing gain is useful in analyzing the actual performance of a given receiver as part of an operating communications systems. On the other hand, the baseband reference gain is more useful in making general comparisons between different types of modulation systems since all parameters are referred to the same standard.

Most of the various quantities just discussed were derived for different forms of analog systems in Chapter 10, and the results are summarized in Table 11-1. The definitions of the symbols are summarized at the bottom of the table. It should be stressed again that since these results were derived on the basis of single-tone modulation under ideal conditions, the results tend to be on the optimistic side as compared with realistic signals. Additional allowances should be made for realistic signal conditions, as well as other imperfections in the receiver. Note that, while a single-tone modulating signal is used, a sufficient bandwidth is allowed for a baseband bandwidth W in order to provide a valid noise level.

Table 11-1. Signal-to-Noise Parameters for Analog Modulation Methods

Modulation Method	R_i	R_d	$G_r = \dfrac{R_d}{R_i}$	$G_b = \dfrac{R_d}{R_b}$
SSB	$\dfrac{P_r}{\eta_i W}$	$\dfrac{P_r}{\eta_i W}$	1	1
DSB	$\dfrac{P_r}{2\eta_i W}$	$\dfrac{P_r}{\eta_i W}$	2	1
AM	$\dfrac{P_r}{2\eta_i W}$	$\dfrac{m^2 P_r}{(2 + m^2)\eta_i W}$	$\dfrac{2m^2}{2 + m^2}$	$\dfrac{m^2}{1 + m^2}$
AM $(m = 1)$	$\dfrac{P_r}{2\eta_i W}$	$\dfrac{P_r}{3\eta_i W}$	$\dfrac{2}{3}$	$\dfrac{1}{3}$
PM	$\dfrac{P_r}{\eta_i B}$	$\dfrac{(\Delta\phi)^2 P_r}{2\eta_i W}$	$\dfrac{(\Delta\phi)^2}{2}\left(\dfrac{B}{W}\right)$	$\dfrac{(\Delta\phi)^2}{2}$
FM	$\dfrac{P_r}{\eta_i B}$	$\dfrac{3(\Delta f)^2 P_r}{2\eta W^3}$	$\dfrac{3}{2}D^2\left(\dfrac{B}{W}\right)$	$\dfrac{3}{2}D^2$
FM (preemphasis)	$\dfrac{P_r}{\eta_i B}$	R_d, G_r, and G_b are each multiplied by $\dfrac{\pi}{6}\left(\dfrac{W}{f_1}\right)$		

A few observations from the table will now be made. The discussion will focus on interpretations of the receiver processing gain and the baseband reference gain for different types of modulation.

R_i = detector signal power to noise power ratio; this ratio may be formed at receiver input if all noise effects are referred to input and B is IF bandwidth

R_d = detected output signal power to noise power ratio

R_b = baseband reference signal power to noise power ratio = $P_r/\eta_i W$

G_r = receiver processing gain

G_b = baseband reference gain

P_r = received signal power

η_i = one-sided net input noise power spectral density

W = baseband bandwidth of modulating signal

B = transmission bandwidth for modulating signal (assumed equal to IF bandwidth)

$\Delta\phi$ = maximum phase deviation for PM

Δf = maximum frequency deviation for FM

D = deviation ratio for FM = $\Delta f/W$

f_1 = 3-dB frequency for preemphasis and deemphasis curves

The results for SSB are the simplest of all the forms given. The receiver processing gain and the baseband reference gain both have values of unity.

This is a result of the fact that the signal-to-noise ratio at the detector output is the same as at the input, and this value in turn is the same as for baseband transmission.

DSB has a receiver processing gain $G_r = 2$. However, the signal-to-noise ratio at the input has half the value for SSB with all other factors equal, which is a result of the fact that the noise power is distributed over two sidebands. These two factors cancel, and the baseband reference gain $G_b = 1$, so the overall performance of DSB is the same as for SSB and baseband. This case illustrates how the values of G_r and G_b may be quite different and must be interpreted properly.

The results for AM are heavily dependent on the modulation factor m. Observe that both G_r and G_b are less than unity for all values of m, and both values may become very small for small values of m. This means that the detected signal-to-noise ratio for AM is less than either the input signal-to-noise ratio or the reference baseband signal-to-noise ratio. This property results from the fact that considerable power is wasted in the carrier, which cannot be converted into useful detected power.

Both forms of angle modulation are capable of realizing a receiver processing gain and a baseband reference gain greater than unity. Observe that for PM, the two gain ratios are heavily dependent on the peak phase deviation $\Delta\phi$, and for FM, the factors are equally dependent on the deviation ratio D. The receiver processing gain in both cases contains the factor B/W, but this factor actually offsets the lower signal-to-noise ratio at the input resulting from a wider noise bandwidth. From the point of view of baseband reference gain, the signal-to-noise ratio at the output increases with $(\Delta\phi)^2$ for PM and with D^2 for FM. Curves of the receiver processing gain in decibels for both PM and FM are shown in Fig. 11-2.

The use of preemphasis with FM results in an additional improvement in the detected signal-to-noise ratio as can be seen from the table. The quantity f_1 represents the frequency at which the preemphasis curve at the transmitter is shifted upward by 3 dB from the very low frequency gain value. (See Sec. 10-6.) Above f_1, the transfer curve of the transmitter preemphasis network approaches a 6 dB/octave boost. It is assumed that the receiver employs a deemphasis network having the inverse characteristic of the preemphasis network. There are also some subtle assumptions concerning the form of the spectrum of the modulating signal, and these conditions were discussed in Sec. 10-6, in which the preemphasis improvement factor was derived. Because of these assumptions and some further approximations made in developing the simplified result of Table 11-1, the preemphasis improvement factor should be considered at best as a reasonable approximation rather than as an exact result.

The use of various results in the table will be illustrated with a number of examples at the end of the next section. However, before effective use can be made of these results, it is necessary to consider the threshold effect, which

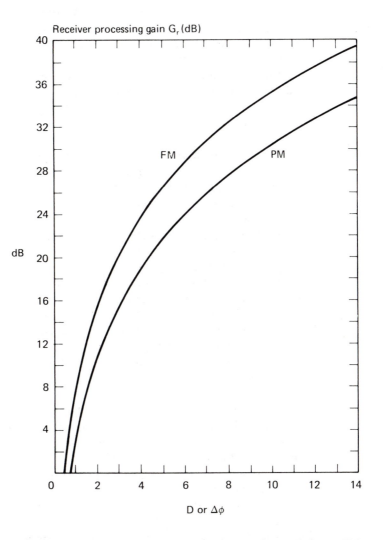

Receiver processing gain G_r (dB)

dB

D or $\Delta\phi$

Figure 11-2. Receiver processing gains for PM and FM.

occurs with some systems. This phenomenon will be considered in the next section, and the examples following that section will be used to illustrate the concepts of both Secs. 11-2 and 11-3.

11-3   THRESHOLD EFFECTS IN ANALOG SYSTEMS

Before considering the actual performances of different types of communications systems, it is necessary to understand a phenomenon called the *threshold effect*. A number of different types of communications processes exhibit this phenomenon, and it is a function of the type of detector in some cases. Qualitatively, the threshold effect is a phenomenon that arises if the signal-to-noise ratio at the detector input decreases below some critical level, and the result is that the detected signal becomes severely distorted or mutilated by the noise. The detected signal-to-noise ratio decreases very quickly with further decreases in signal level, and the result will likely be unintelligible.

All angle-modulation systems exhibit the threshold effect. Consider the two curves shown in Fig. 11-3. The abscissa represents the input signal-to-noise ratio, and the ordinate represents the detected signal-to-noise ratio. Thus, the levels of any curves shown are dependent on the receiver processing gain functions.

The straight line has a slope of unity, meaning that the detected signal-to-noise ratio is the same as the input signal-to-noise ratio. This corresponds to SSB with synchronous product detection and there is no threshold effect.

The more complex curve represents the form of receiver processing gain for either PM or FM at some particular value of $\Delta\phi$ or D. Let R_{it} represent the threshold value. For $R_i > R_{it}$, this curve is essentially a straight line (when the scales are in decibels), and it is above the curve for SSB. This means, of course, that the detected signal-to-noise ratio is greater than the input signal-to-noise ratio in accordance with the appropriate results of Table 11-1. The difference between the two curves in this region is the processing gain converted to decibels.

Observe, however, the phenomenon that arises if R_i decreases below R_{it}. The detected signal-to-noise ratio drops very quickly and is soon below the unity signal processing gain line. In addition, as previously noted, the signal is mutilated and will likely not be usable.

Various mathematical developments to predict the exact behavior of the threshold phenomenon in angle modulation have been made, but the results are too involved to attempt to cover in this text. Instead, a simplified practical guideline will be given. It has been found experimentally and verified theoretically that a threshold level for the input signal-to-noise ratio of approximately 10 dB is a reasonable rule of thumb. Since 10 dB corresponds

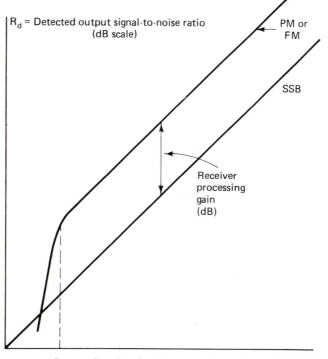

Figure 11-3. Threshold effect in the receiver processing gain for a PM or FM detector and comparison with SSB.

to an absolute ratio of 10, we will subsequently assume for angle modulation an approximate threshold level of

$$R_{it} = 10 \qquad \text{for angle modulation} \qquad (11\text{-}10)$$

This assumption means that the receiver processing gain results of Table 11-1 are considered to be valid for PM and FM only if $R_i \geq 10$. Since operation below threshold is almost always undesirable, it is necessary to specify sufficient signal-to-noise ratio at the detector input to ensure operation above threshold.

Because of the processing gain associated with angle modulation, an unusual phenomenon occurs as the input signal-to-noise ratio drops below the critical threshold level. Suppose, for example, the processing gain under a given set of circumstances is 30 dB, and the input signal-to-noise ratio is slightly more than 10 dB. The detected signal-to-noise ratio is above 40 dB, so the net signal will exhibit very little noise. However, if the signal strength starts to decrease slightly, a sudden drop in the detected signal-to-noise ratio will result, and the signal quickly becomes obliterated. The significant phenomenon here is the sudden change from a highly intelligible signal to one that is likely unintelligible with only a few decibels of input signal change. The reader may have observed this phenomenon with an FM radio in an automobile while traveling. A signal may be very strong and appear to be virtually noise free. As the distance from the transmitter increases (or as a result of any other phenomenon that decreases the signal strength), the noise level suddenly rises above the signal level and renders intelligible reception impossible.

Neither of the AM processes exhibit a threshold effect when *product* detection is used. Since product detection is normally used with SSB and DSB, the threshold effect need not be considered for such systems. However, envelope detection is used most often for AM, and this form of detection does exhibit the threshold effect. However, its *relative* effect is not nearly as important in AM as in angle modulation for reasons that will be discussed in the next paragraph.

For a detected signal to be virtually noise free and to exhibit negligible degradation as a result of noise, the detected signal-to-noise ratio should be 30 to 40 dB or more. The threshold level for an AM envelope detector is in the neighborhood of about 10 dB. Since the processing gain for AM is less than unity, the input signal-to-noise ratio would have to be something greater than 30 dB at a minimum in order to produce a good detected signal-to-noise ratio. This condition automatically guarantees operation above threshold. If the input signal-to-noise ratio were gradually decreased down to the threshold level, the output signal-to-noise ratio would already be degraded significantly by the additive noise before threshold is observed. Thus, the threshold effect in an AM envelope detector is much more gradual than for angle modulation. To reemphasize a point, the normal range for quality detected signal-to-noise ratio in an AM linear envelope detector demands an input signal-to-noise ratio that would exceed the threshold level.

Some digital modulation systems make use of AM-type signals under low signal-to-noise ratio conditions. In such cases, an exact replica of the detected signal is not necessary, but only the recognition of "ones" and "zeros" is important. In such cases, the use of synchronous product detection of AM could be justified, since the threshold effect could then be avoided.

Example 11-1

Assume that the received signal power at the input to a certain receiver is $P_r = 200$ fW. (1 fW = 1 femtowatt = 10^{-15} W.) The antenna noise temperature is $T_a = 300$ K, and the effective receiver temperature referred to the input is $T_e = 425$ K. The IF bandwidth is $B = 50$ kHz.

Determine the detected output signal-to-noise ratio R_d (as an absolute value and in decibels) for each of the following modulation methods and conditions: (a) SSB, (b) DSB, (c) AM with 100% modulation, (d) AM with 50% modulation, (e) PM with $\Delta\phi = 5$, (f) FM with $D = 5$ (no preemphasis), and (g) FM with $D = 5$, preemphasis, and $W/f_1 = 8$. (Since the bandwidth B is assumed to be constant, it is assumed that W varies with different methods.)

Solution

At the outset, the input signal-to-noise ratio must be calculated. The effective input noise power contained in the 50-kHz bandwidth is

$$N_i = kT_iB = k(T_a + T_e)B \qquad (11\text{-}11)$$
$$= 1.38 \times 10^{-23} \times (300 + 425) \times 50 \times 10^3 = 5 \times 10^{-16} \text{ W}$$
$$= 0.5 \text{ fW}$$

The input signal-to-noise ratio is

$$R_i = \frac{P_r}{N_i} = \frac{200 \text{ fW}}{0.5 \text{ fW}} = 400 \qquad \text{(or 26.02 dB)} \qquad (11\text{-}12)$$

The detected output signal-to-noise ratio R_d for each of the various modulation conditions is determined by calculating G_r using Table 11-1 and multiplying this value by $R_i = 400$. (Equivalently, the decibel value for each of the gains could be added to 26.02 dB.) The various calculations follow.

(a) SSB:

$$G_r = 1$$
$$R_d = R_i = 400$$
$$R_d(\text{dB}) = 26.02 \text{ dB}$$

(b) DSB:

$$G_r = 2$$
$$R_d = 2R_i = 2 \times 400 = 800$$
$$R_d(\text{dB}) = 29.03 \text{ dB}$$

(c) AM (100% modulation): The value of R_i is well above the threshold level of approximately 10 dB for envelope detection of AM.

$$G_r = \frac{2}{3}$$

$$R_d = \frac{2}{3} R_i = \frac{2}{3} \times 400 = 266.67$$

$$R_d(\text{dB}) = 24.26 \text{ dB}$$

(d) AM (50% modulation)

$$G_r = \frac{2(0.5)^2}{2 + (0.5)^2} = 0.2222$$

$$R_d = 0.2222 \times 400 = 88.88$$

$$R_d(\text{dB}) = 19.49 \text{ dB}$$

(e) PM: Before considering either of the angle-modulation methods, it must be emphasized that the results in the table are correct only if the threshold level is exceeded. Throughout the text, the value $R_i = 10$ (also 10 dB) is assumed as the approximate value of threshold, and since $R_i = 400$, operation above the threshold level can obviously be assumed.

Note from Table 11-1 that the ratio B/W must be known. Carson's rule applied to PM is

$$B = 2(\Delta\phi + 1)W \qquad (11\text{-}13)$$

Solving for B/W and substituting the value of $\Delta\phi$, we obtain

$$\frac{B}{W} = 2(\Delta\phi + 1) = 2(5 + 1) = 12$$

The values of G_r and R_d are then determined to be

$$G_r = \frac{(5)^2}{2}(12) = 150$$

$$R_d = 150 \times 400 = 60,000$$

$$R_d(\text{dB}) = 47.78 \text{ dB}$$

(f) FM: As in the case of PM, the ratio B/W must be known. Carson's rule applied to FM is

$$B = 2(D + 1)W \qquad (11\text{-}14)$$

Since $D = 5$, the value of B/W is

$$B/W = 2(6) = 12$$

The remaining calculations are

$$G_r = \frac{3}{2}(5)^2(12) = 450$$

$$R_d = 450 \times 400 = 180,000$$

$$R_d(\text{dB}) = 52.55 \text{ dB}$$

(g) FM with preemphasis: The values for G_r and R_d as compared with part (f) are each multiplied by the factor

$$\frac{\pi}{6}\frac{W}{f_1} = \frac{\pi}{6} \times 8 = 4.189$$

Thus,

$$G_r = 450 \times 4.189 = 1885$$
$$R_d = 1885 \times 400 = 754{,}000$$
$$R_d(\text{dB}) = 58.77 \text{ dB}$$

It will be stressed again that this problem assumed a constant transmission bandwidth for all the different methods, so the corresponding baseband bandwidth that could be used varies with the different methods. It will be left as an exercise for the reader (Prob. 11-1) to determine the baseband bandwidth values for this particular system. Somewhat different results will be obtained when the baseband modulating signal bandwidth and the received power are held constant, and the transmission bandwidth is varied as required. Various problems at the end of the chapter illustrate the trade-offs involved.

A final point concerns the use of decibel ratios in the analysis. Any of the receiver processing gain values could be readily stated as decibel values, and this is commonly done in practice. The detected output signal-to-noise ratio in decibels is then determined by adding the processing gain in decibels to the input signal-to-noise ratio in decibels. For example, in part (f), the receiver processing gain of 450 corresponds to 26.53 dB and $R_i(\text{dB}) = 26.02$ dB. The detected output signal-to-noise ratio in decibels is $R_d(\text{dB}) = 26.02 + 26.53 = 52.55$ dB, as determined earlier. The addition of preemphasis in part (g) results in a decibel improvement of $10 \log_{10} 4.189 = 6.22$ dB so that $R_d(\text{dB})$ in part (g) is $52.55 + 6.22 = 58.77$ dB, as determined earlier. These manipulations illustrate the ease and convenience of decibel calculations.

The numerical results of this problem should illustrate to the reader the relative order of the detected signal-to-noise ratio values for different modulation systems. The poorest performer is, of course, AM. FM performs better than PM, and further improvement results with FM when preemphasis is employed.

Example 11-2

The design specifications in a certain communications system require a detected signal-to-noise ratio $R_d(\text{dB}) = 50$ dB. The baseband modulating signal bandwidth is $W = 10$ kHz. The antenna temperature is $T_a = 250$ K, and the effective receiver temperature referred to the input is $T_e = 475$ K. Determine the required signal received power P_r for each of the following modulation methods and conditions: (a) SSB, (b) DSB, (c) AM with 100% modulation, (d) PM with $\Delta\phi = 5$, (e) FM with $D = 5$ (no preemphasis), (f) FM with $D = 5$, preemphasis, and $W/f_1 = 10$, and (g) FM with $D = 10$, preemphasis, and $W/f_1 = 10$. It is assumed that the receiver IF bandwidth is adjusted to match the bandwidth of the given modulated signal in each case.

Solution

While there are several differences between this example and Ex. 11-1, one particularly significant difference is the fact that the modulating signal bandwidth is fixed in the present example. Thus, the bandwidth B over which the noise is computed will vary with different methods. It is convenient at the outset to compute the total input noise power density. On a one-sided basis, we have

$$\eta_i = kT_i = k(T_a + T_e)$$
$$= 1.38 \times 10^{-23} \times (250 + 475) = 1 \times 10^{-20} \text{ W/Hz} \qquad (11\text{-}15)$$

The desired value of detected signal-to-noise ratio is 50 dB, which corresponds to $R_d = 10^5$. From the definition of receiver processing gain, the required input signal-to-noise ratio is determined as

$$R_i = \frac{R_d}{G_r} = \frac{10^5}{G_r} \qquad (11\text{-}16)$$

Once the value of R_i is determined, it can then be expressed as

$$R_i = \frac{P_r}{N_i} = \frac{P_r}{\eta_i B} = \frac{P_r}{10^{-20} B} \qquad (11\text{-}17)$$

which leads to the computation of P_r as

$$P_r = 10^{-20} B R_i \qquad (11\text{-}18)$$

These results could be combined in one equation, but it is preferable not to do so. The reason is that the value of R_i would then be hidden in the equation, and this quantity is of interest. For systems exhibiting a threshold effect, it is necessary that R_i exceed a certain minimum value for the other results to be valid. This will be illustrated in part (g). The preceding results will be used freely in the computations that follow.

(a) SSB:

$$G_r = 1$$
$$R_i = \frac{10^5}{1} = 10^5$$
$$B = W = 10 \text{ kHz}$$
$$P_r = 10^{-20} \times 10^4 \times 10^5 = 10^{-11} \text{ W} = 10 \text{ pW}$$

(b) DSB:

$$G_r = 2$$
$$R_i = \frac{10^5}{2} = 5 \times 10^4$$

$$B = 2W = 20 \text{ kHz}$$
$$P_r = 10^{-20} \times 20 \times 10^3 \times 5 \times 10^4 = 10^{-11} \text{ W} = 10 \text{ pW}$$

(Observe that SSB and DSB both require the same amount of power.)

(c) AM (100% modulation)

$$G_r = \frac{2}{3}$$

$$R_i = \frac{10^5}{2/3} = 1.5 \times 10^5$$

$$B = 2W = 20 \text{ kHz}$$
$$P_r = 10^{-20} \times 20 \times 10^3 \times 1.5 \times 10^5 = 3 \times 10^{-11} \text{ W} = 30 \text{ pW}$$

(d) PM ($\Delta\phi = 5$): The bandwidth is first determined by Carson's rule.

$$B = 2(\Delta\phi + 1)W = 12W = 12 \times 10 \text{ kHz} = 120 \text{ kHz}$$

$$G_r = \frac{(5)^2}{2}(12) = 150$$

$$R_i = \frac{10^5}{150} = 666.67$$

(This value of R_i is considerably greater than the threshold value.)

$$P_r = 10^{-20} \times 120 \times 10^3 \times 666.67 = 8 \times 10^{-13} \text{ W} = 0.8 \text{ pW}$$

(e) FM ($D = 5$): The bandwidth is first determined by Carson's rule.

$$B = 2(D + 1)W = 12W = 12 \times 10 \text{ kHz} = 120 \text{ kHz}$$

$$G_r = \frac{3}{2}(5)^2(12) = 450$$

$$R_i = \frac{10^5}{450} = 222.22$$

(This value of R_i is considerably greater than the threshold value.)

$$P_r = 10^{-20} \times 120 \times 10^3 \times 222.22 = 2.667 \times 10^{-13} \text{ W} = 266.7 \text{ fW}$$

(f) FM ($D = 5$ and preemphasis): The value of G_r in this case is the value in part (e) multiplied by the preemphasis constant. We have

$$G_r = 450 \times \frac{\pi}{6} \times 10 = 2356.19$$

$$R_i = \frac{10^5}{2356.19} = 42.44$$

(This value still exceeds the threshold.)

$$P_r = 10^{-20} \times 120 \times 10^3 \times 42.44 = 5.093 \times 10^{-14} \text{ W} = 50.93 \text{ fW}$$

(g) FM ($D = 10$ and preemphasis)

$$B = 2(10 + 1)W = 22W = 22 \times (10 \times 10^3) = 2.2 \times 10^5 \text{ Hz}$$

$$G_r = \frac{3}{2}(10)^2 \times 22 \times \frac{\pi}{6} \times 10 = 17,278.8$$

$$R_i = \frac{10^5}{17,278.8} = 5.787$$

We must now pause to analyze the implication of this result. The value of R_i obtained is below the threshold level of 10 assumed for FM. If we operated with this value of R_i, the output signal-to-noise ratio would be substantially less than the desired value of 10^5. Furthermore, operation would be very sensitive in this region; that is, a slight fade in the signal strength could cause a significant difference in the quality of the result. To prevent this problem, the signal would have to be increased to a level greater than 10. As an arbitrary assumption for this problem, suppose we set $R_i = 20$ (allowing a 3-dB margin). The value of P_r is then

$$P_r = 10^{-20} \times 2.2 \times 10^5 \times 20 = 4.4 \times 10^{-14} \text{ W} = 44 \text{ fW}$$

Note that by increasing the signal strength to ensure operation above threshold the output detected signal-to-noise ratio increases to

$$R_d = 17,278.8 \times 20 = 3.456 \times 10^5 \quad \text{or} \quad R_d(\text{dB}) = 55.39 \text{ dB}$$

The reader should observe the wide range of power levels required to produce a specified output signal-to-noise ratio. At one extreme, AM requires a received power of 30 pW, while at the other extreme, wideband FM with preemphasis requires a power level of only 44 fW. The ratio between these two power levels is nearly 682 as an absolute ratio or about 28.34 dB. This illustrates the general superiority of wideband FM as compared to AM.

11-4 DIGITAL SYSTEM COMPARISONS

In this section, the performance of digital PCM systems in the presence of noise will be discussed. Meaningful results for predicting the detected signal-to-noise ratio will be summarized. Some of these results were derived in Chapter 11, but the emphasis here will be on their interpretation and application. It will be seen that PCM systems exhibit a threshold effect, which produces a significant degradation in quality at low input signal-to-noise ratios.

A derivation of the detected signal-to-noise ratio in a PCM system based on single-tone modulation was performed in Sec. 10-9. It was noted there that the primary noise effects in PCM were (1) quantization noise and (2) decision noise. Quantization noise is a function of the number of bits used in the encoding process, and decision noise is a function of decoding errors made at the receiver on the data bits. Decision noise can generally be reduced by increasing the input signal-to-noise ratio, while quantization noise is fixed for a given number of bits and can only be reduced by increasing the number of bits.

The detected signal-to-noise ratio R_d for a PCM system operating under ideal conditions with single-tone modulation was given in Eq. (10-165) and will be repeated here.

$$R_d = \frac{P_d}{N_d} = \frac{1.5m^2}{1 + 4m^2P_e} \qquad (11\text{-}19)$$

where $m = 2^n$ is the number of possible levels, n is the number of bits in each word, and P_e is the probability of error in detection. This result depends heavily on m and P_e, but the general shape will always be of the form shown in Fig. 11-4.

At low values of input signal-to-noise ratio, the value of P_e is quite large, and the detected signal-to-noise ratio is a critical function of the signal and noise levels. This region is sometimes denoted as the *thermal noise limited* region, referring to the fact that many decision errors result from thermal

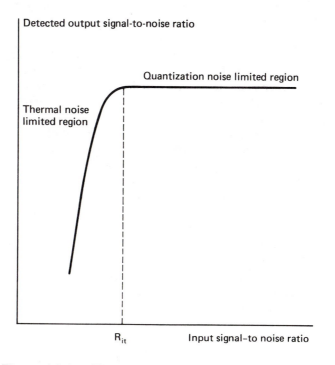

Figure 11-4. Threshold effect in a digital PCM system.

noise in this range. This region corresponds to operation below threshold and would be quite undesirable for system operation.

As the input signal-to-noise ratio increases, P_e eventually reduces to a negligible value, and the detected signal-to-noise ratio approaches a constant limiting value based on the number of bits. This region is sometimes referred to as the *quantization noise limited* region, and operation in this region is almost always desired.

The threshold value is a function of the probability of error P_e and the number of levels m used in the digital encoding process. For techniques possessing relatively high values for the probability of error, the input signal-to-noise ratio must be correspondingly higher to overcome the effects of decision noise. Likewise, as the number of levels increases, the relative effect of decoding errors is greater, so the input signal-to-noise ratio must be increased accordingly. For these reasons, it is more difficult to precisely state a general value for the threshold signal-to-noise ratio. One common criterion for the threshold value is the input signal-to-noise ratio that results in a 1-dB degradation in the detected signal-to-noise ratio from the quantization noise-limited region. The value $R_{it} = 15$ will be assumed as a representative value for certain computations in this chapter, but it should not be interpreted as a general result. Another common criterion is to specify operation at a specific probability of error and then assume that the relative effects of decision noise will be negligible in that region. ($P_e = 10^{-5}$ is a representative value for a moderate word length.)

If source encoding for the purpose of detecting and/or correcting errors is used, the threshold level may be reduced to a relatively low level. Source encoding methods employ redundancy in the form of extra bits to assist in the detection or correction of errors. At one extreme, a parity bit represents a simple form of encoding. More sophisticated methods employ several bits, which represent various combinations of the data pattern. Some of these methods require digital processing at the receiver to detect and/or correct any errors and to sort the desired signal out of the complex encoded signal.

Once the input signal-to-noise ratio exceeds the threshold level and the curve flattens out, the detected signal-to-noise ratio increases no further as the input signal-to-noise ratio continues to increase. This is in sharp contrast to analog systems, where the detected signal-to-noise ratio continues to increase as the input signal-to-noise ratio increases. The detected signal-to-noise ratio in this region is a function only of the number of bits per word and can be increased only by adding more bits.

The detected signal-to-noise ratio R_d in the quantization noise limited region reduces to

$$R_d = 1.5m^2 = (1.5)2^{2n} \tag{11-20}$$

The corresponding value in decibels is

$$R_d(\text{dB}) = 1.76 + 6.02n \tag{11-21}$$

This last result is interesting in that it provides a simple and easily remembered rule of thumb. In the region above threshold, PCM provides a detected signal-to-noise ratio of about *6 dB per bit*.

Since the signal-to-noise ratio in PCM reaches a limiting value above threshold, it is desirable to operate as close to threshold as possible in situations where the transmitted power is at a premium (e.g., space applications). The threshold level is very dependent on the probability of error P_e, so system designers utilize probability of error results extensively in determining appropriate power levels. Certain special cases of probability of error functions were derived in Secs. 10-7 and 10-8. Some of the results of those sections as applied to realistic systems, along with other results that are not derived in this text, are shown in Fig. 11-5.

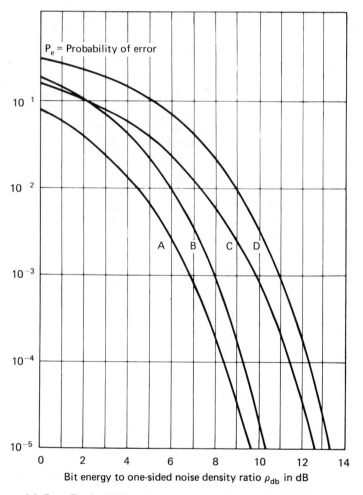

Figure 11-5.   Probability of error curves for several PCM systems.

The abscissa for each of these curves is based on a parameter ρ as defined in Sec. 10-8 and repeated here for convenience.

$$\rho = \frac{E_b}{\eta_i} = \frac{\text{bit energy (J)}}{\text{one-sided noise power density (W/Hz)}} \quad (11\text{-}22)$$

where the bit energy is related to the received power P_r and the bit width T_b by

$$E_b = P_r T_b \quad (11\text{-}23)$$

As in the case of analog systems, the power level at the receiver input may be used provided η_i reflects the *total* effective input noise density at the same point. The actual abscissa in Fig. 11-5 is expressed in terms of $\rho_{dB} = 10 \log_{10} \rho$. The ordinate is the probability of error P_e.

Curve A corresponds to the matched filter phase-shift keying (PSK) detection (also realizable as coherent PSK detection). This curve represents a case of matched filter detection as derived in Sec. 10-8 and is an optimum situation. Curve B represents differential phase-shift keying (DPSK), whose general operation was discussed in Sec. 7-7. Observe that DPSK has a higher probability of error than optimum PSK, which means that a higher power level would be required for PSK to produce the same detected signal-to-noise ratio as compared with matched filter PSK.

Curve C represents the probability of error for matched filter amplitude-shift keying (ASK). The general operation of ASK was discussed in Sec. 7-6, and it was noted there that ASK could be detected by either coherent detection or by envelope detection. The coherent detection process corresponds to matched filter detection and is superior from a signal-to-noise point of view to envelope detection. However, because one of the levels in ASK corresponds to "off" rather than a negative value, there is a 3-dB degradation as compared with matched filter PSK in which the carrier is on all the time. The reader may readily verify this point by noting that curves A and C are shifted apart horizontally by 3 dB.

Curve D represents the results of envelope detection of frequency-shift keying (FSK). Envelope detection is one of several methods for detecting FSK. As a general rule, noncoherent detection processes tend to be inferior to coherent techniques.

These four cases provide a survey of representative probability of error results for PCM systems. There are many other possibilities, but these are some of the most common cases.

Example 11-3

Assume that the signal power at the input to a certain receiver in a PCM data system is $P_r = 200$ fW. The antenna noise temperature is $T_a = 300$ K, and the effective receiver temperature referred to the input is $T_e = 425$ K. (These are the same values assumed in the analog system of Ex. 11-1.) The bandwidth is adjustable to meet the requirements of specific data formats and bit rates.

Based on single-tone modulation, determine the detected output signal-to-noise ratio R_d (as an absolute value and in decibels) for PCM transmission with 64 levels (6 bits) for each of the following conditions: (a) envelope detected FSK and a data rate of 2 Mbits/s, (b) matched filter ASK and a data rate of 2 Mbits/s, (c) DPSK and a data rate of 2 Mbits/s, (d) PSK with matched filter and a data rate of 2 Mbits/s, and (e) PSK with matched filter and a data rate of 8 Mbits/s. In all cases, the data bits are assumed to represent a continuous stream; that is, there are no empty spaces between adjacent bits.

Solution

For parts (a) to (d), the data rate is fixed at 2 Mbits/s, so the bit duration T_b for continuous data transmission is $T_b = 1/(2 \times 10^6) = 0.5$ μs. The bit energy E_b at the input to the receiver is

$$E_b = 200 \times 10^{-15} \times 0.5 \times 10^{-6} = 1 \times 10^{-19} \text{ J (joules)} \qquad (11\text{-}24)$$

The one-sided input noise power spectral density η_i at the receiver input is

$$\eta_i = kT_i = k(T_a + T_e) = 1.38 \times 10^{-23} \times (300 + 425)$$
$$= 1 \times 10^{-20} \text{ W/Hz} \qquad (11\text{-}25)$$

The parameter ρ for parts (a) to (d) is then determined as

$$\rho = \frac{1 \times 10^{-19} \text{ J}}{1 \times 10^{-20} \text{ W/Hz}} = 10 \qquad (11\text{-}26)$$

The procedure in each part that follows will be to first determine P_e for the given modulation method using Fig. 11-5. The detected signal-to-noise ratio may then be determined from Eq. (11-19). The value of ρ will remain the same in the first four parts, but a new value must be computed in part (e) due to the different data rate there. It turns out that the value $\rho = 10$ also corresponds to $\rho_{dB} = 10$ dB, but for any arbitrary value of ρ, the appropriate ρ_{dB} must be first calculated in order to use Fig. 11-5.

(a) From curve D of Fig. 11-5, $P_e \simeq 3.2 \times 10^{-3}$. Substitution of this value and $m = 64$ in (11-19) yields

$$R_d = \frac{1.5 \times (64)^2}{1 + 4 \times (64)^2 \times 3.2 \times 10^{-3}} = 114.99 \text{ (or 20.61 dB)}$$

(b) From curve C of Fig. 11-5, $P_e \simeq 8 \times 10^{-4}$. The value of R_d is

$$R_d = \frac{1.5 \times (64)^2}{1 + 4 \times (64)^2 \times 8 \times 10^{-4}} = 435.52 \text{ (or 26.39 dB)}$$

(c) From curve B of Fig. 11-5, $P_e \simeq 2.2 \times 10^{-5}$. The value of R_d is

$$R_d = \frac{1.5 \times (64)^2}{1 + 4 \times (64)^2 \times 2.2 \times 10^{-5}} = 4516.16 \text{ (or 36.55 dB)}$$

(d) Curve A in Fig. 11-5 is the appropriate one to use for coherent PSK, but the value of P_e for $\rho_{dB} = 10$ dB is not on the curves provided. However, as we will shortly see, P_e is sufficiently small that the system is essentially operating in the quantization limited region, so the exact value does not affect the result significantly. If the curve is extrapolated below the level provided, a rough estimate is $P_e \simeq 0.25 \times 10^{-5}$. Substitution of this value yields

$$R_d = \frac{1.5 \times (64)^2}{1 + 4 \times (64)^2 \times 0.25 \times 10^{-5}} = 5902.24 \text{ (or 37.71 dB)}$$

If the signal power were increased, the value of R_d would approach the limiting value $1.5 \times (64)^2 = 6144$ (or 37.88 dB), which represents the ultimate quantization noise limited case. The value of 37.71 dB calculated in part (d) using a rough estimate of P_e is so close to the limiting value that operation in the quantization limited region can be assumed in this part.

(e) The effect of an increase in the data rate is that the bit energy is reduced. Specifically, if the data rate is increased from 2 Mbits/s to 8 Mbits/s, the bit duration reduces from 0.5 to 0.125 μs, and the bit energy E_b reduces in the same proportion from 1×10^{-19} J to 0.25×10^{-19} J. The new value of ρ is $10/4 = 2.5$ or $\rho_{dB} = 3.98$ dB. From curve A of Fig. 11-5, the probability of error is $P_e \simeq 1.26 \times 10^{-2}$. The value of R_d is then

$$R_d = \frac{1.5 \times (64)^2}{1 + 4 \times (64)^2 \times 1.26 \times 10^{-2}} = 29.62 \text{ (or 14.72 dB)}$$

The system is operating well below threshold now, so the signal-to-noise ratio is very sensitive to changes in the power or noise level. Comparing the results of part (e) with part (d), it is noted how the data rate has a significant effect on the overall performance.

Observe that in this example, there is a difference of more than 17 dB between the best case and the worst case at the same data rate for different modulation methods and a difference of about 23 dB between the best case and the worst case for the same modulation method with a 4-to-1 range in data rate. Yet the number of levels remained constant in all cases, which means that the limiting detected signal-to-noise ratio would be the same in all cases if the input signal-to-noise ratio were sufficiently high. This example was formulated to delineate some of the striking features of PCM systems when probability of error and threshold effects must be considered. With the exception of part (d), none of the cases represents desirable conditions, since operation below threshold is evident.

Example 11-4

The design specifications in a certain 8-bit PCM system call for $P_e = 10^{-5}$. The antenna temperature is $T_a = 250$ K, and the effective receiver temperature referred to the input is $T_e = 475$ K. Determine the required received power P_r for each of the following modulation methods and conditions: (a) envelope detected FSK and a data rate of 1 Mbits/s, (b) matched filter ASK and a data rate of 1 Mbits/s, (c) DPSK and a data rate of 1 Mbits/s, (d) PSK with matched filter and a data rate of 1 Mbits/s, and (e) PSK with matched filter and a data rate of 2 Mbits. In all cases, the data bits are transmitted in a continuous stream.

Solution

In each case, Fig. 11-5 will be used to determine the value of ρ_{dB} at which $P_e = 10^{-5}$. The corresponding value ρ will then be calculated as

$$\rho = 10^{\rho_{dB}/10} \tag{11-27}$$

in accordance with standard decibel computations. The input noise density is

$$\eta_i = 1.38 \times 10^{-23} \times (250 + 475) = 1 \times 10^{-20} \text{ W/Hz} \tag{11-28}$$

Since $\rho = E_b/\eta_i$, the bit energy E_b is determined as

$$E_b = \eta_i \rho = 1 \times 10^{-20} \rho \tag{11-29}$$

Next, the signal power P_r is determined from E_b as

$$P_r = \frac{E_b}{T_b} \tag{11-30}$$

This result is equivalent to multiplying E_b by the data rate in bits per second. The bit duration in parts (a) to (d) is $T_b = 1/(1 \times 10^6) = 1$ μs. However, in part (e), the bit duration is $T_b = 1/(2 \times 10^6) = 0.5$ μs. The various calculations are listed in the steps that follow:

(a) From curve D of Fig. 11-5, $\rho_{dB} \simeq 13.4$ dB for $P_e = 10^{-5}$. The value of ρ is determined from (11-27).

$$\rho = 10^{13.4/10} = 21.88$$

The bit energy is determined from (11-29).

$$E_b = 1 \times 10^{-20} \times 21.88 = 2.188 \times 10^{-19} \text{ J}$$

The necessary received power level is then determined from (11-30).

$$P_r = \frac{2.188 \times 10^{-19}}{10^{-6}} = 218.8 \text{ fW}$$

(b) From curve C of Fig. 11-5, $\rho_{dB} \simeq 12.6$ dB for $P_e = 10^{-5}$. The necessary calculations are

$$\rho = 10^{1.26} = 18.20$$
$$E_b = 1 \times 10^{-20} \times 18.20 = 1.82 \times 10^{-19} \text{ J}$$
$$P_r = \frac{1.82 \times 10^{-19}}{10^{-6}} = 182 \text{ fW}$$

(c) From curve B of Fig. 11-5, $\rho_{dB} \simeq 10.3$ dB for $P_e = 10^{-5}$. The necessary calculations are

$$\rho = 10^{1.03} = 10.72$$
$$E_b = 1 \times 10^{-20} \times 10.72 = 1.072 \times 10^{-19} \text{ J}$$
$$P_r = \frac{1.072 \times 10^{-19}}{10^{-6}} = 107.2 \text{ fW}$$

(d) From curve A of Fig. 11-5, $\rho_{dB} \simeq 9.6$ dB for $P_e = 10^{-5}$. The necessary calculations are

$$\rho = 10^{0.96} = 9.12$$
$$E_b = 1 \times 10^{-20} \times 9.12 = 9.12 \times 10^{-20} \text{ J}$$
$$P_r = \frac{9.12 \times 10^{-20}}{10^{-6}} = 91.2 \text{ fW}$$

(e) This part differs from part (d) only in that $T_b = 0.5 \times 10^{-6}$ s is substituted in the denominator in the computation for P_r. This yields

$$P_r = \frac{9.12 \times 10^{-20}}{0.5 \times 10^{-6}} = 182.4 \text{ fW}$$

which is the same result as for matched filter ASK at the slower data rate.

This problem should illustrate to the reader how the necessary received power level varies as a function of the modulation method for a given probability of error and a given data rate. Specifically, operation with matched filter PSK requires the minimum power level for all the methods considered in this book. Note that as the data rate increases, the necessary received power level increases in direct proportion for a given error rate.

Incidentally, the detected signal-to-noise ratio can be determined for all cases from (11-19), with the recognition that $m = 2^8 = 256$. The result is

$$R_d = \frac{1.5 \times (256)^2}{1 + 4 \times (256)^2 \times 10^{-5}} = 27,145 \text{ (or 44.34 dB)}$$

The quantization limited case signal-to-noise ratio based on 256 levels is readily determined as 98,304 (or 49.93 dB). Thus, the performance is 5.59 dB below the maximum possible signal-to-noise ratio that could ultimately be achieved with this system.

11-5 ANTENNA PRIMER

Any communications system must employ a transmission "channel" in which the signal from the information source is transferred to the destination. The simplest channel is the wire link in which the signal is transmitted over a pair of wires or cable. The most obvious example of this nature is the standard telephone circuit within a local area. (Some long-distance contacts employ other methods.) Another special example of a channel is water in which audio-frequency sonar signals are transmitted.

The channel concept that provides the most overall capability for information transmission is the use of electromagnetic waves in the atmosphere. All circuits with time-varying voltages and currents tend to produce electromagnetic waves, which may or may not be significant in size. However, a system in which electromagnetic radiation is desired employs a device that enhances the conversion of the circuit energy to that of a wave. This device is the *antenna*. The antenna at the transmitter is an interface between the circuit and the outside electromagnetic field. Similarly, at the receiver, an antenna serves to enhance the conversion of the wave back to a voltage or current that can be processed by the receiver circuitry.

In free space, the wavelength λ of a propagating signal is determined from the equation

$$\lambda = \frac{c}{f} \tag{11-31}$$

where c is the velocity of wave propagation in free space, usually taken as $c = 3 \times 10^8$ m/s. For c in meters per second and f in hertz, λ will be expressed in meters. The velocity of wave propagation in air is very slightly less than that of free space, but for the vast majority of engineering calculations, the free space velocity can be assumed for air. On the other hand, when a wave is propagating in a dielectric medium other than air, the velocity of propagation can be considerably smaller and is $c/\sqrt{\epsilon_r}$, where ϵ_r is a parameter called the *dielectric constant* for the medium.

Antennas are rather complicated and specialized devices. The theory of operation at the most fundamental level involves an extensive amount of electromagnetic theory to predict the exact distribution of the field pattern and the associated radiation characteristics. Much of the detailed work in analyzing antennas at a fundamental mathematical level and in developing new antennas has been performed by specialists who have devoted much of their life to establishing the necessary mathematical proficiency required.

Fortunately, the communications systems technologist or engineer can utilize existing antenna designs without understanding all the basic theory involved. It is possible to deal with antennas from the "black box" approach in which specifications relating only to the external behavior are considered. This is the approach that will be taken here.

Some of the most important properties of antennas from a system-level point of view will be considered in this section. Those properties pertinent to the specification, analysis, and operation of an overall communications system will be emphasized. Some of the rather specialized terminology peculiar to antenna theory will be simplified slightly in order to make it more meaningful for our purposes.

No attempt will be made to present a collection of the antenna types available. Such an effort would not only be futile, but it would be misleading in a brief presentation such as this section. The intent here is merely to acquaint the reader with the major properties and terminology so that antennas can be considered in the overall systems approach. A number of fine books devoted solely to antennas have been written, and the reader desiring more detail should consult such references. A few representative antenna types are illustrated in Fig. 11-6.

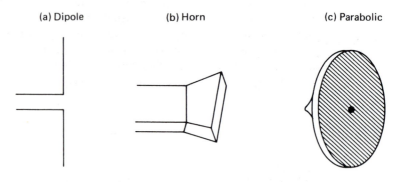

Figure 11-6.   Representative antennas.

Power Density

The relative strength of a signal radiating from an antenna can be specified or measured in terms of its power density at different points in space. The power density concept is illustrated in Fig. 11-7. Assume that an imaginary surface in the shape of a sphere is formed such as to completely enclose the antenna, which is located at the center. At any point on the imaginary sphere, the power density is the power per unit area radiating through an infinitesimal unit of area. Note that the direction of propagation is perpendicular to surface area at a given point. We will use the symbol $\mathcal{P}$ to denote power density, and it is measured in watts per square meter (W/m^2).

The actual power passing through a closed surface can be determined by integrating the power density over the surface area. By conservation of energy, this value of power must equal the actual power radiated by the antenna if the medium is lossless.

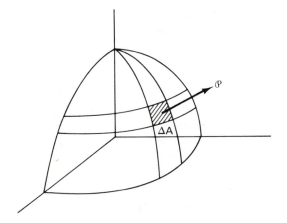

Figure 11-7.   Power density of signal radiating from antenna.

Radiation Pattern

The radiation pattern is a plot of the relative strength or intensity of the antenna radiation as a function of the orientation in a given plane. The quantity used in specifying the relative strength may be the power density previously discussed or basic electromagnetic quantities such as the electric field intensity (in volts per meter) or the magnetic field intensity (in amperes per meter).

Two examples of radiation patterns are shown in Fig. 11-8. The intensity of the radiation of the antenna in Fig. 11-8a is constant at a given distance from the antenna and is independent of the angular orientation with respect to the antenna. This type of pattern is said to be *omnidirectional* in the given plane. On the other hand, the pattern of the antenna in Fig. 11-8b is more directive toward the top of the page. This particular antenna is characterized by a *major lobe* and four *minor lobes* in the radiation pattern. Thus, there will be some radiation to the left and to the right, but the intensity in those directions

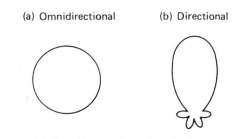

(a) Omnidirectional   (b) Directional

Figure 11-8.   Examples of radiation patterns.

is smaller than for the major lobe. There is almost no radiation in the direction opposite to the peak of the major lobe.

Isotropic Point Radiator

In the analysis of complex antenna systems, reference is frequently made to a fictitious ideal isotropic point radiator. This hypothetical antenna would radiate power equally well in all directions in a volume sense. Thus, the ideal isotropic point radiator would be omnidirectional in all planes. No real antenna can achieve such a pattern in all planes, but the idealized concept is a useful one. The isotropic point radiator model is used in the relative comparison of different types of antennas.

The power density at a radius r from an isotropic point radiator may be determined by dividing the power transmitted P_t by the area of a sphere of radius r enclosing the antenna. Thus

$$\mathcal{P} = \frac{P_t}{4\pi r^2} \qquad (11\text{-}32)$$

Gain

The gain g of an antenna is a parameter that compares the concentration of power in the given antenna to that of some reference antenna. It is defined as

$$g = \frac{\text{maximum power density of given antenna}}{\text{maximum power density of reference antenna}} \qquad (11\text{-}33)$$

where the power densities are measured at the same distance from the antenna. The reference antenna most often selected is the hypothetical isotropic point radiator. In this case, the term "maximum" in the denominator of (11-33) is redundant. However, some antenna gain measurements utilize a "real-life" antenna such as a dipole for the reference, in which case the maximum power density of the reference antenna should be used in the comparison. It should be noted that it is necessary to know what type of reference antenna is used in the comparison when any antenna gain value is specified.

In practice, most antenna gain values are specified in decibels. The decibel gain is defined as

$$g_{dB} = 10 \log_{10} g \qquad (11\text{-}34)$$

The term "gain" may be confusing to the beginner since an antenna does not "create" energy or power, and it is a passive device. Yet the concept of gain might erroneously imply that the output power is greater than the input power. What actually happens is that the power that would be radiated equally well in all directions if the antenna were an isotropic radiator is focused or beamed in a much more limited range of directions, and the resulting power density in this limited range can be much greater than for the completely omnidirectional antenna. In simple terms, power is taken away from some directions and added to the power in other directions, and the result is an effective gain in the direction of maximum radiation.

The gain is a complex function of the type of structure, the number of elements, the spacing between elements, the phasing of elements, and other factors. Waves are phased so as to cancel in certain directions and to reinforce in other directions. The actual gains of realistic antennas vary considerably. The simple half-wave dipole antenna has a theoretical gain of only about 1.64 (2.15 dB), while at the other extreme, some parabolic reflector systems have gains exceeding 10^6 (60 dB).

As a general rule, it is much easier to achieve higher gains in a practical sense at higher frequencies. As the gain increases, the size of the structure relative to the wavelength increases. This results in a practical difficulty in constructing antennas with high gains at low frequencies. This is a major reason why increased attention has been given to operation in the microwave communications region for many aerospace and satellite systems where low transmitter power and high antenna gain in a minimum-sized package are desirable.

Directivity

In a qualitative sense, directivity is the characteristic of an antenna referring to its capability of beaming the energy in a narrow range of directions. The precise definition of directivity in antenna theory is such that it is very closely related to gain and would be equal if the antenna were 100% efficient. In general, however, the gain is less than the directivity because practical antennas are not 100% efficient. All the developments given here will utilize the gain parameter, so the directivity concept will not be pursued further.

Capture Area or Aperture

The capture area or aperture A_e of an antenna is an effective "area" parameter for the antenna. It may or may not correspond to the physical area. It is more convenient to define capture area when the antenna is receiving power, and it is

$$A_e = \frac{\text{received power (W)}}{\text{maximum power density at antenna (W/m}^2\text{)}} \qquad (11\text{-}35)$$

It is readily observed that A_e has the units of area (m²). In some cases (e.g., horn antennas), the capture area may correspond very closely to the physical area. However, some antennas have essentially no physical area (e.g., dipole antennas), but all antennas have an effective capture area. In general, we will say that A_e is an "area parameter" such that, when multiplied by the power density in watts per square meter, it produces the received power in watts.

The capture area and the gain of an antenna are closely related. When the antenna gain is expressed relative to the isotropic radiator, a development in antenna theory has established the following relationship:

$$A_e = \frac{\lambda^2}{4\pi} g \qquad (11\text{-}36)$$

where the wavelength λ and the capture area are based on the same unit system (e.g., if λ is in centimeters, A_e will be in square centimeters).

Beamwidth

The beamwidth is the angular separation between the two half-power points (3 dB down from maximum) of the radiation pattern in a given plane. The concept is illustrated in Fig. 11-9. In general, there is a direct trade-off between the antenna gain and the beamwidth; that is, antennas having greater gains tend to have narrower beamwidths, and vice versa.

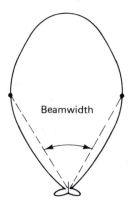

Figure 11-9. Antenna pattern beamwidth.

Antenna Impedance

A given antenna at a given frequency will exhibit a certain input impedance at the feeder point. This impedance may actually be measured with an RF bridge. The impedance will often be highly frequency dependent and will, in general, have both a real part and an imaginary part. In the proper operating range, the imaginary part should ideally be zero, and the real part should ideally be equal to the characteristic resistance of the transmission line

if reflections are to be avoided. It is possible to match the antenna impedance to that of the feeder line with reactive networks when there is a difference between the two levels.

Radiation Resistance

The radiation resistance R_{rad} is the real part of the complex input impedance to the antenna and is the value that accounts for the actual power accepted from the input signal. Thus, the input power P_{in} to an antenna is

$$P_{in} = I^2 R_{rad} \qquad (11\text{-}37)$$

where I is the rms value of the input current to the antenna. If the antenna were 100% efficient, P_{in} would represent the actual power radiated. In actual practice, some power is dissipated as heat in the antenna.

Example 11-5

Some common antennas of the types illustrated back in Fig. 11-6 will be considered in this example. The following absolute gains relative to an isotropic radiator will be assumed:

Antenna	Gain
Half-wave dipole	1.64
Horn	20
Parabola	10^4

(The value for the half-wave dipole is a theoretical fixed value, while the values for the horn and parabola are representative only. Both of the latter values vary directly with the radiating area of the antenna structure.) Determine the gain in decibels and the aperture for each antenna if the frequency is 3 GHz.

Solution

The gain in decibels of each antenna is determined from (11–34). The aperture for each antenna can be determined with (11-36). However, the wavelength λ must first be determined. Since $f = 1$ GHz, we have

$$\lambda = \frac{3 \times 10^8}{3 \times 10^9} = 0.1 \text{ m} = 10 \text{ cm}$$

The computed values are tabulated as follows:

Antenna Type	g	g_{dB} (dB)	A_e at 1 GHz (Cm²)
Half-wave dipole	1.64	2.15	13.05
Horn	20	13	159.15
Parabola	10^4	40	79,577

These values are representative examples for a low-gain antenna (diople), a medium-gain antenna (horn), and a high-gain antenna (parabola). In the case of the horn and the parabola, the values of A_e are close to (but not necessarily equal to) the actual physical areas of the antennas.

11-6 FRIIS TRANSMISSION FORMULA

In this section, a development of the Friis transmission formula will be presented. This formula is most important in establishing a relationship between transmitted power, received power, antenna gain, and distance between transmitter and receiver. Its use permits a trade-off between various system link parameters in establishing proper operating levels for a complete communications system.

The Friis transmission formula applies to a system in which there is a transmitting antenna, a receiving antenna, and a direct electromagnetic path without obstacles between the two antennas, as illustrated in Fig. 11-10. It does not apply to the various types of indirect propagation, such as sky-wave bounce and tropospheric scatter. Such indirect propagation techniques have depended on empirical data to determine proper operating conditions, and such results are not always reliable. Consideration of the various types of indirect wave propagation is outside the scope of this book, but the reader should be aware that there are a number of different types that have been widely employed.

The trend of the future is toward the use of direct wave propagation in as many systems as possible, particularly at higher frequencies. Prior to the advent of communications satellites, direct ray propagation was limited to a very short range owing to the curvature of the earth and the corresponding difficulty of locating antennas at sufficient heights to utilize direct rays.

Figure 11-10. Direct ray model used in developing the Friis transmission formula.

Following the development of satellite repeater systems, new dimensions for long-range communications became feasible. Direct ray propagation is now practical between the earth and a satellite, and the signal can then be retransmitted to points around the globe. As we will see with realistic examples later, the amount of power required is relatively modest when direct ray propagation can be used. Thus, the future should find increasing use of direct ray propagation in conjunction with satellites for many forms of long-range communications.

The development of the several steps leading to the Friis formula will contribute to its understanding. Assume an isotropic point radiator generating a power P_t watts. Consider an imaginary spherical surface of radius d surrounding the radiator, and assume that the transmitter is at the center of the sphere. Let $\mathcal{P}$ represent the power density on the surface at any point. Since the power radiates equally well in all directions, it divides uniformly across the imaginary surface and is

$$\mathcal{P} = \frac{P_t}{4\pi d^2} \tag{11-38}$$

An actual antenna having gain g_t with respect to an isotropic antenna would produce a power density $\mathcal{P}'$ in the direction of maximum radiation given by

$$\mathcal{P}' = g_t \mathcal{P} = \frac{g_t P_t}{4\pi d^2} \tag{11-39}$$

This will be the power density at the location of the receiving antenna in Fig. 11-10. The receiving antenna will "capture" a power P_r given by

$$P_r = A_e \mathcal{P}' = \frac{A_e g_t P_t}{4\pi d^2} \tag{11-40}$$

The receiving antenna capture area can be expressed as

$$A_e = \frac{\lambda^2}{4\pi} g_r \tag{11-41}$$

where g_r is the gain of the receiving antenna. Substitution of (11-41) in (11-40) yields

$$P_r = \frac{\lambda^2 g_t g_r P_t}{(4\pi)^2 d^2} \tag{11-42}$$

The result of (11-42) is the *Friis transmission formula* for one-way direct ray propagation from transmitter to receiver. It is also referred to as the *one-way link equation.* Several deductions can be made from this formula. The power at the receiver is directly proportional to the product of the transmitted power, the transmitter antenna power gain, and the receiver antenna power gain. Within limits, these three parameters may be adjusted without affecting performance provided that the product remains the same. For example, a transmitter with a power output of 2 W having an antenna with a gain of 100 would produce the same received power (in the direction of maximum radiation) as a transmitter with a power output of 50 W and a gain of 4. As a second example, if a receiving antenna is replaced by one having half the previous gain, the original received power could be restored by either doubling the transmitter output power or by replacing the transmitting antenna with one having twice the gain.

The denominator of (11-42) indicates that the received power varies inversely with the square of the distance d. Thus, each time the distance from the transmitted antenna is doubled, the received power is reduced by one-fourth.

The decibel variation as a result of the $1/d^2$ law is interesting. It can be readily shown (Prob. 11-18) that, if the distance between the two antennas is doubled, the received power level is reduced by about 6 dB. If the distance between the two antennas is increased by a factor of 10, the received power level is reduced by exactly 20 dB.

The variation of received power with respect to wavelength λ is more subtle than the other quantities and warrants a special discussion before erroneous impressions are formed. If a direct inspection of (11-42) is made without additional insight, one might deduce that the received power is directly proportional to the square of the wavelength, which would suggest that the received power would increase markedly at lower frequencies. This deduction would be correct *if* all the other parameters in the equation could remain constant as the wavelength increased. The fallacy in the argument is that g_t and g_r relate to the gains of given antennas at a specific operating frequency. As the wavelength increases, it is necessary that antennas specifically designed for the new frequencies be employed to provide optimum conditions for transmission. A general property of antennas is that higher gain demands more complex and formidable structures. As explained in Sec. 11-5, one of the major reasons for operating at frequencies in the microwave region is that high gain antennas are easier to implement at such short wavelengths. The implementation of high-gain antennas at very low frequencies is very difficult. In summary, then, the variation of P_r with λ^2 as suggested directly by (11-42) is valid only if g_t and g_r could be held constant by replacing existing antennas with new ones having the same values. Because of the complexity and many facets of this trade-off, we will avoid making any general conclusions regarding the variation of received power with respect to wavelength (or frequency).

A slightly different form of the Friis formula is obtained by rewriting (11-42) in the form

$$P_r = \frac{g_t g_r P_t}{(4\pi)^2 \, (d/\lambda)^2} \qquad (11\text{-}43)$$

In this form, the quantity d/λ appears as a dimensionless parameter indicating the "number of wavelengths." The author finds this latter form easier to remember and interpret than the earlier form, but this is purely a personal preference.

In its basic form, the reduction of the received power level as predicted by the Friis transmission formula is known as the *free space path loss*. This loss can be closely predicted by use of this formula for direct ray or "line of sight" propagation under ideal conditions. In general, however, other factors have to be considered in a full system design for worst-case conditions. Many of these additional factors are too complex and specialized to be considered in a general treatment such as this, but certain ones will be mentioned.

Losses resulting from atmospheric absorption are important in certain frequency ranges. Such losses occur from both water vapor and oxygen in the atmosphere, each of which can absorb a portion of the electromagnetic energy. These losses are shown as a function of frequency in Fig. 11-11. At frequencies well below 20 GHz, water and oxygen losses are very low. Losses due to water vapor absorption are most significant in the vicinities of 23 and 180 GHz. Similarly, losses due to oxygen absorption are most significant in the vicinities of 60 and 120 GHz. It is interesting, however, that there are certain regions in this general frequency range in which these losses are greatly reduced. Such frequencies are called *windows,* and two window locations are 33 and 110 GHz. Operation above 20 GHz is still quite experimental at the time of this writing, but such future operation might be forced to utilize these window regions in order to minimize losses.

Other factors that might have to be considered are antenna pointing errors, bending of waves due to atmospheric refraction, attenuation due to rainfall, and multipath fading. This latter phenomenon results from the successive reinforcement and cancellation of the signal when two or more rays having different path lengths (along with different phase shifts) combine at the receiving antenna.

Example 11-6

Consider the communications system of Ex. 11-2 in which the detected signal-to-noise ratio was specified as 50 dB, and the required received power levels for different modulation processes were computed. Assume that the proposed transmitter will be located at a distance of approximately 36 km away from the receiver. The following specifications on system components have been established: transmitter antenna gain = 6 dB, receiver antenna gain = 9 dB, operating frequency = 1 GHz, receiver noise temperature = 475 K, and antenna temperature = 250 K. (The last two specifications were given in Ex. 11-2). Determine the transmitter output power that would be required for each of the modulation methods and conditions considered in Ex. 11-2.

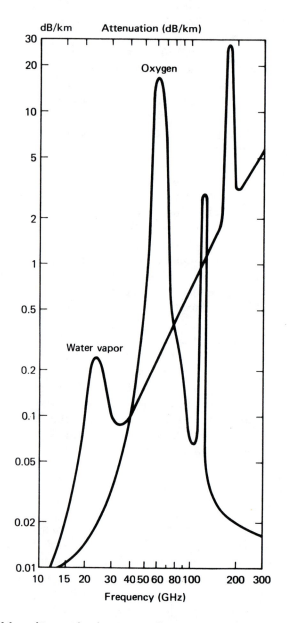

Figure 11-11.   Atmospheric attenuation due to water vapor and oxygen.

Solution

As a first step, the specifications as given will be converted to the form required in the link equation. We have

$$g_t = 3.981$$

$$g_r = 7.943$$

$$\lambda = \frac{3 \times 10^8}{1 \times 10^9} = 0.3 \text{ m}$$

From the Friis transmission formula as given in (11-43), the transmitter power P_t required to produce a given received power P_r is determined to be

$$P_t = (4\pi)^2 \left(\frac{d}{\lambda}\right)^2 \frac{P_r}{g_t g_r} \tag{11-44}$$

All parameters but P_r on the right are constant, so the tabulated values listed earlier may be substituted in (11-44). With a slight realistic rounding of the result, we have

$$P_t = 7.19 \times 10^{10} P_r \tag{11-45}$$

The values of P_r obtained in Ex. 11-2 may now be substituted in (11-45) to determine the required transmitter power values. The results (with realistic rounding) are summarized in the same order as in Ex. 11-2.

(a) SSB: $P_t = 7.19 \times 10^{10} \times 10 \times 10^{-12} = 0.719 \text{ W} = 719 \text{ mW}$

(b) DSB: $P_t = 719 \text{ mW}$

(c) AM (100% modulation): $P_t = 7.19 \times 10^{10} \times 30 \times 10^{-12} = 2.157 \text{ W}$

(d) PM ($\Delta\phi = 5$): $P_t = 7.19 \times 10^{10} \times 0.8 \times 10^{-12} = 57.52 \text{ mW}$

(e) FM ($D = 5$): $P_t = 7.19 \times 10^{10} \times 266.7 \times 10^{-15} = 19.18 \text{ mW}$

(f) FM ($D = 5$ and preemphasis): $P_t = 7.19 \times 10^{10} \times 50.93 \times 10^{-15} = 3.66 \text{ mW}$

(g) FM ($D = 10$ and preemphasis): $P_t = 7.19 \times 10^{10} \times 44 \times 10^{-15} = 3.16 \text{ mW}$

The required transmitter power varies all the way from 3.16 mW to 2.157 W.

Example 11-7

A satellite communications system is to be used to transmit a video signal from the satellite to earth using FM. The following specifications are given:

Center frequency = 3 GHz
Bandwidth of modulating signal = 4 MHz

Satellite antenna gain = 20 dB
Earth station antenna gain = 50 dB
Distance from satellite to earth station = 36,000 km
Receiving antenna noise temperature = 20 K
Receiver noise temperature = 15 K (a cooled maser amplifier is employed)
Required detected signal-to-noise ratio = 40 dB

Assume that the receiver input signal-to-noise ratio is arbitrarily set at 3 dB above the threshold level of 10 dB as a design margin and that receiver processing gain is used to provide the additional signal-to-noise enhancement. Determine the required values of (a) deviation ratio and (b) transmitter power.

Solution

(a) From the specifications $R_i(\text{dB}) = 10 + 3 = 13$ dB. The receiver processing gain must be $G_r(\text{dB}) = 40 - 13 = 27$ dB, which corresponds to $G_r = 501.19$. Based on a baseband bandwidth $W = 4$ MHz, the receiver processing gain from Table 11-1 must satisfy the equation

$$G_r = 501.19 = \frac{3}{2} D^2 \left(\frac{B}{W}\right) = 3D^2 (D + 1) \tag{11-46}$$

The result of (11-46) is a cubic equation in D and could be somewhat difficult to solve. However, the curve of Fig. 11-2 can be readily used, and it is found that a receiver processing gain of 27 dB occurs for $D \simeq 5.2$. This is the required value of deviation ratio.

(b) The receiver IF bandwidth B should be just sufficient for the FM signal with $D = 5.2$ and $W = 4$ MHz. From Carson's rule, we have

$$B = 2(D + 1)W = 2(5.2 + 1) \times 4 \text{ MHz} = 49.6 \text{ MHZ}$$

The net input noise power N_i is based on this bandwidth and is

$$N_i = kT_iB = k(T_a + T_r)B = 1.38 \times 10^{-23} (20 + 15) \times 49.6 \times 10^6$$

$$= 2.396 \times 10^{-14} \text{ W}$$

We require that $R_i(\text{dB}) = 13$ dB or $R_i = 20$. This means that

$$\frac{P_r}{N_i} = \frac{P_r}{2.396 \times 10^{-14}} = 20$$

or

$$P_r = 4.791 \times 10^{-13} \text{ W}$$

The wavelength is $\lambda = 3 \times 10^8/3 \times 10^9 = 0.1$ m. The link equation applied to this problem gives

$$P_r = 4.791 \times 10^{-13} = \frac{P_t g_t g_r}{(4\pi)^2(d/\lambda)^2} = \frac{P_t \times 100 \times 10^5}{(4\pi)^2\,(36 \times 10^6/0.1)^2}$$

Solving for P_t, we obtain

$$P_t = 0.981 \text{ W} = 981 \text{ mW}$$

11-7 RADAR TRANSMISSION FORMULA

In this section, a development and discussion of the radar link transmission formula will be presented. This form of the link equation is a "two-way" version in that power is transmitted along a direct path in one direction, but the ultimate power level of interest is the amount that is returned along the same path in the opposite direction. The simplest form of radar system is one in which the power is transmitted in short bursts, and the time delay is measured for the return energy to be detected back at the source location. From a knowledge of the time delay, T_d, the distance d to the object may be computed as

$$d = \frac{cT_d}{2} \tag{11-47}$$

where c is the speed of propagation of the waves. The factor of 2 in the denominator of (11-47) results from the fact that the actual round-trip distance of the signal is $2d$.

More sophisticated radar systems make use of various signal-processing techniques, such as sawtooth modulated FM signals and tone modulated FM signals. By suitable processing at the receiver, estimates of both the position and velocity of the object under observation (commonly called the "target") may be made.

In designing a radar system, it is important to be able to estimate the approximate signal strength returned to the source so that a suitable power level may be used. As we will see shortly, the returned signal level may be very small in a practical radar system, and there is more uncertainty surrounding the actual signal level than with systems employing one-way transmission.

The model for developing the radar equation is shown in Fig. 11-12. Located at essentially the same position as the transmitter is the radar receiver. The signal is transmitted to the right, and a portion of the energy that encounters the target is backscattered to the left toward the radar receiver. The reader may be tempted to use the term "reflection" to describe

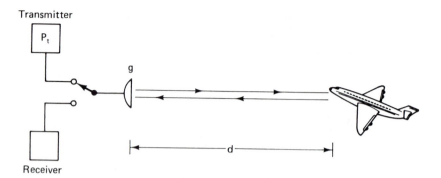

Figure 11-12. Model used in developing the radar range equation.

this process, and this term is commonly used in casual conversation. Strictly, however, a reflected wave has a precise meaning in physical optics and electromagnetic theory in terms of a coherent relationship between the incident and reflected waves. The term "backscattered wave" represents energy that is a combination of both coherent and random noncoherent components, which is the case with most radar signals. Consequently, the term "backscatter" will be used as the basis for discussing the radar return phenomena.

The signal returning to the source location is processed by the radar receiver. Some means must be provided to avoid overloading the low-level receiver by leakage energy from the transmitter. In the simplest classical radar system, the transmitter is turned off during the "listen" interval, but other means of isolation are available.

In this development, we will assume that the same antenna (by means of appropriate switching or isolation) is used for both transmitting and receiving. The first part of the development parallels that of the preceding section. The power density $\mathcal{P}$ of the transmitted signal in the vicinity of the target is of the same form as (11-39) and can be expressed as

$$\mathcal{P} = \frac{gP_t}{4\pi d^2} \tag{11-48}$$

where g is used as the common antenna gain.

We now introduce a term σ called the *backscatter cross section*, which is defined as

$$\sigma = \frac{\text{power backscattered in direction of source (W)}}{\text{incident power density (W/m}^2)} \tag{11-49}$$

Observe that σ has the dimension of square meters, so it is an "area" parameter. The backscattered power P_{bs} is then determined as

$$P_{bs} = \sigma \mathcal{P} = \frac{\sigma P_t g}{4\pi d^2} \qquad (11\text{-}50)$$

Back at the receiver, the power density $\mathcal{P}_1$ resulting from the backscattered energy is

$$\mathcal{P}_1 = \frac{P_{bs}}{4\pi d^2} = \frac{\sigma P_t g}{(4\pi)^2 d^4} \qquad (11\text{-}51)$$

The power P_r captured by the antenna is

$$P_r = \mathcal{P}_1 A_e = \frac{\sigma P_t g A_e}{(4\pi)^2 d^4} \qquad (11\text{-}52)$$

We may then express A_e as

$$A_e = \frac{\lambda^2 g}{4\pi} \qquad (11\text{-}53)$$

Substituting (11-53) in (11-52), we have

$$P_r = \frac{\sigma P_t g^2 \lambda^2}{(4\pi)^3 d^4} \qquad (11\text{-}54)$$

The result of (11-54) is the radar transmission formula, and it provides an insight into the trade-offs possible with radar systems. Observe that the power is proportional to the square of the power gain of the common antenna. On a decibel basis, this is equivalent to a doubling effect. For example, suppose a common antenna having a 20-dB gain is replaced with one having a 24-dB gain. The actual increase in the received power level is not 4 dB, but 8 dB.

Observe that the received power in a radar system varies as $1/d^4$, which is in contrast to the one-way link process where a $1/d^2$ variation was observed. It can be readily shown (Prob. 11-19) that for a radar system the decibel level of the received power is reduced by about 12 dB if the distance between source and target is doubled (assuming all other factors are equal). The same discussion concerning the variation with wavelength made in the last section applies here.

The most nebulous quantity in the radar equation is the radar cross section σ. While this quantity has the dimensions of and is treated as an area, it is actually a complex function of the surface roughness, surface composition, angle of incidence, and many other factors. Much research has been conducted through the years to establish reasonable estimates of the radar cross section for different types of surfaces, and several books devoted to this area have been written. Anyone using the radar equation for system design would have to utilize such information to assist in estimating power return levels. Because of the potential uncertainty in σ, the results predicted from the radar equation would possess a corresponding degree of uncertainty.

11-8 MODULATION SYSTEM COMPARISONS

All considerations up to this point in the chapter have been directed toward the operating performance analysis of different modulation systems as a function of the power levels, signal-to-noise ratios, and other operational parameters of the system. In determining the actual response of receivers, use was made of the "receiver processing gain" values from Table 11-1 and the various probability-of-error curves for PCM.

In this last section of the book, a comparison will be made at a somewhat philosophical level in order to determine a meaningful relative comparison between different types of systems. To that end, the baseband reference gain values of Table 11-1 will be used since the reference is the same for all comparisons. Finally, the concept of the ideal communications system will be introduced, and it will be compared with some actual realizable systems.

In utilizing the baseband reference gain values in Table 11-1, it should be noted again that this quantity is defined as

$$G_b = \frac{R_d}{R_b} \tag{11-55}$$

where R_d is the detected signal-to-noise ratio and R_b is defined as

$$R_b = \frac{P_r}{\eta_i W} \tag{11-56}$$

Thus, irrespective of the actual bandwidth required, the reference baseband signal-to-noise ratio R_b is defined only over the baseband bandwidth. For all systems other than SSB, R_b is larger than the receiver input signal-to-noise ratio R_i, and the value of G_b is less than the corresponding value of G_r. The value of G_b tends to give a better general measure of the performance of the system since the somewhat larger value of G_r may be deceiving. The larger value for G_r is needed to compensate for the smaller value of R_i at the receiver input with wideband systems.

Before considering specific methods, a general relationship between R_i and R_b will be developed. The quantity R_i may be manipulated as follows:

$$R_i = \frac{P_r}{\eta_i B} = \frac{P_r}{\eta_i W}\left(\frac{W}{B}\right) = R_b\left(\frac{W}{B}\right) \tag{11-57}$$

This results in

$$R_b = R_i\left(\frac{B}{W}\right) \tag{11-58}$$

Thus, for any modulation system (11-58) may be used to determine the value of R_b corresponding to any given value of R_i.

With any system involving a threshold effect, the value of R_b at which threshold occurs must be determined. We have previously defined R_{it} as the value of R_i at which threshold occurs, so let R_{bt} represent the corresponding value of R_b at which threshold occurs. For AM linear envelope detection, the value $R_{it} = 10$ is assumed. Use of (11-58) results in $R_{bt} = 20$, since $B = 2W$ for AM.

For angle modulation, $R_{it} = 10$ was assumed. Using Carson's rule separately for PM and FM, the corresponding values of R_{bt} are

$$\text{PM:} \quad R_{bt} = 20(\Delta\phi + 1) \tag{11-59}$$

$$\text{FM:} \quad R_{bt} = 20(D + 1) \tag{11-60}$$

For PCM, we recall that threshold was heavily dependent on the form of encoding. Only the optimum matched filter scheme, which is equivalent to

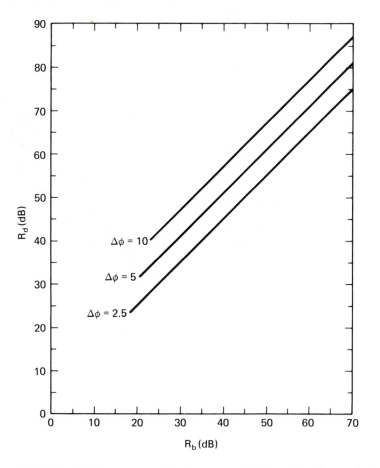

Figure 11-13. Performance of phase modulation for three different maximum phase deviation values.

coherent PSK, will be considered here. In that case $R_{it} = 15$ will be used as a reasonable value. The ratio $B/W = n$ for ideal Nyquist rate sampling will be assumed as a lower bound. Thus, the value for R_{bt} will be assumed as

$$R_{bt} = 15n \qquad (11\text{-}61)$$

We now proceed to the comparison of some results. For analog systems, the result of the R_b column will be used as the function in each case. Comparisons will first be made for a single modulation system with different operating parameters, and different systems will then be compared with each other.

The first system to be considered is PM. Plots of R_d(dB) versus R_b(dB) for $\Delta\phi = 2.5, 5,$ and 10 are shown in Fig. 11-13. The curves begin at the assumed

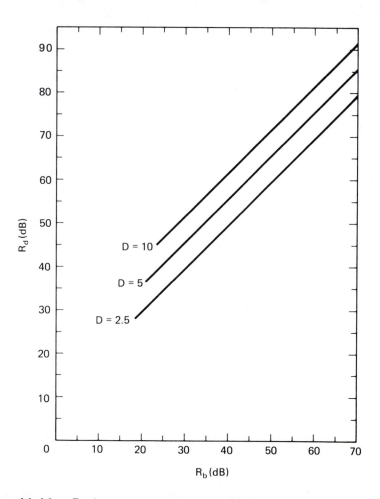

Figure 11-14. Performance of frequency modulation for three different deviation ratios.

threshold point in each case. Observe the improvement in detected signal-to-noise ratio as the value of $\Delta\phi$ increases.

Corresponding curves for FM are shown in Fig. 11-14. The values of deviation ratio are $D = 2.5$, 5, and 10. Observe that the detected signal-to-noise ratio increases as the value of D increases. Since the bandwidth increases as either $\Delta\phi$ or D increases, the trade-off of increased bandwidth for detected signal-to-noise ratio is immediately evident. The improvement for FM with preemphasis is even greater.

We will now compare several different forms of analog modulation on the same scale. In Fig. 11-15, curves of R_d (dB) versus R_b(dB) are given for the following cases: AM ($m = 1$), DSB, SSB, PM with $\Delta\phi = 5$, FM with $D = 5$, and FM with $D = 5$, plus preemphasis with $W/f_1 = 7.07$. (This latter preemphasis ratio corresponds to commercial FM broadcasting.)

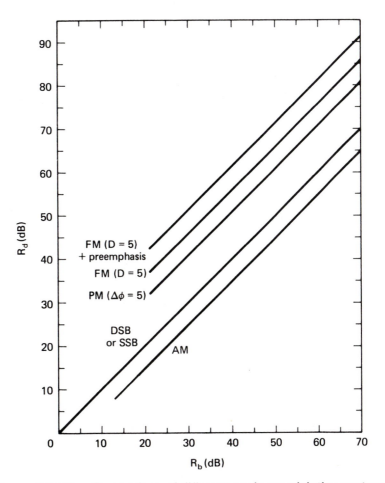

Figure 11-15.   Comparison of different analog modulation systems.

Observe that AM is the poorest performer, and the detected signal-to-noise ratio is less than the baseband reference signal-to-noise ratio. DSB and SSB are both identical when compared in this manner, and both have baseband reference gains of unity. All the angle-modulation-system examples have gains exceeding unity, with FM being superior to PM, and FM with preemphasis being superior to FM with no preemphasis.

For PCM, the results are flat curves, as shown in Fig. 11-16. As noted earlier, the simple rule of thumb of approximately 6 dB/bit quickly identifies the approximate levels of the different curves.

We will now compare some of the "best of two worlds"; that is, FM as the "best" case for analog modulation will be compared directly with PCM, which represents the general form for digital modulation. To make such a comparison valid, the same bandwidth should be assumed for both systems. For this purpose it is convenient to define a parameter

$$\gamma = \frac{B}{W} \tag{11-62}$$

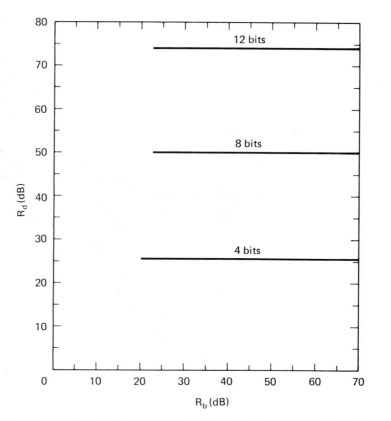

Figure 11-16. Performance of PCM for different word lengths.

The parameter γ represents a *bandwidth expansion factor* and provides a relative measure of the extent to which the bandwidth is broadened to accommodate the given modulation method.

For FM, Carson's rule readily establishes γ as

$$\gamma = 2(D + 1) \qquad (11\text{-}63)$$

For PCM with sampling at the Nyquist rate and with the idealized bandwidth assumptions of Chapters 6 and 7, the value of γ is

$$\gamma = n = \log_2 m \qquad (11\text{-}64)$$

where n is the number of data bits in each word and m is the number of levels.

Curves of FM and PCM for $\gamma = 6$ and 12 are shown in Fig. 11-17. Due to some of the arbitrary assumptions required with preemphasis analysis, only

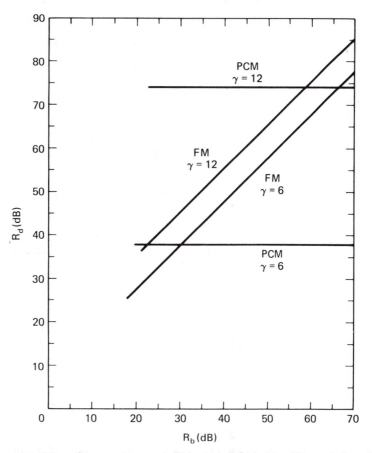

Figure 11-17. Comparison of FM and PCM for different bandwidth expansion ratios.

the case of FM without preemphasis is used. It can be said, however, that preemphasis would result in an additional improvement for FM.

The results of this comparison are interesting and deserve some study. It can be seen that both FM and PCM for a given bandwidth have a region in which each is clearly superior to the other, but neither can be regarded as the "best." At lower signal-to-noise ratios (above threshold, of course), PCM appears to be superior, but as the signal-to-noise ratio increases, FM takes the lead. Both systems are capable of exhibiting outstanding performance, but a number of different factors must be considered in arriving at a suitable decision as to which would be best in a given application.

Because of the superiority of PCM at low signal-to-noise ratios, one can readily determine why much research has been conducted in the space program to lower the threshold level for PCM and thus permit operation at very low signal levels from deep space. In this type of operation, PCM would obviously show superiority.

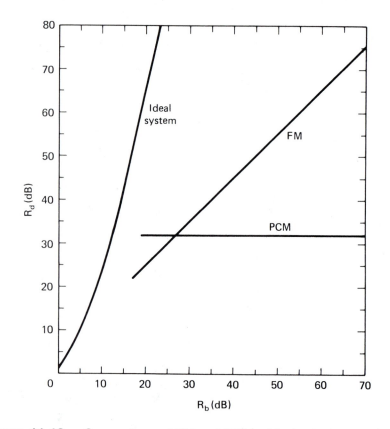

Figure 11-18. Comparison of FM and PCM with the ideal communications system for a baseband expansion ratio $\gamma = 5$.

The last comparison that we will make is based on the concept of the ideal communications system. Based on pioneering theoretical work of Claude Shannon and R. V. Hartley at Bell Telephone Laboratories, the ultimate ideal communications system has been shown to have a detected signal-to-noise ratio given by

$$R_d = \left(\frac{R_b}{\gamma} + 1\right)^{\gamma} - 1 \qquad (11\text{-}65)$$

For $\gamma \gg 1$, this relationship can be approximated as

$$R_d \simeq \left(\frac{R_b}{\gamma}\right)^{\gamma} \qquad (11\text{-}66)$$

It should be noted that no real-life communications system has been found that could equal the performance of this ideal model. The ideal performance is based on theoretical concepts of information theory, and this optimum result stands as a goal to be approached, but one that has not yet been attained in practice.

A comparison of the ideal system, PCM, and FM is shown in Fig. 11-18, all with $\gamma = 5$. The curves for FM and PCM would actually intersect the curve of the ideal system if threshold effects did not exist. However, because of threshold effects, both FM and PCM fall short of the ideal case. Many communications researchers are currently earning their keep by investigating new concepts that will narrow the gap between the ideal and practical cases.

PROBLEMS

11-1 In Ex. 11-1, the transmission bandwidth was assumed to be set at 50 kHz for all the different modulation systems compared. Determine the baseband bandwidth (or equivalently, the highest modulating frequency) W in each case.

11-2 Consider a communications system having the same operating parameters as Ex. 11-1 except that the transmission bandwidth is *not* fixed at 50 kHz. Instead, a baseband bandwidth $W = 10$ kHz is assumed for all cases, and the transmission bandwidth, as well as the receiver IF bandwidth, is assumed to be adjusted in each case to match the particular modulation method. Thus, while the input noise power density η_i will be the same as in Ex. 11-1, the total input noise power, as well as the input signal-to-noise ratio, will vary with the modulation method. Determine the detected output signal-to-noise ratio R_d (as an absolute value and in decibels) for all the modulation methods of Ex. 11-1 based on the new condition.

11-3 Consider a communications system having the same operating parameters as Ex. 11-2 except that the baseband modulating signal bandwidth is not fixed at 10 kHz. Instead, a channel bandwidth (transmission bandwidth) of 50 kHz is assumed to be available, and a study will be made to determine the required received signal power P_r for each of the modulation methods and conditions if the full bandwidth is used in each case. Thus, the baseband bandwidth W will vary with different modulation methods in order to use the available transmission bandwidth. Determine the required signal received power P_r for all the modulation methods of Ex. 11-2 based on the new condition. For part (g), use the 3-dB margin employed in Ex. 11-2.

11-4 Assume that the received signal power at the input to a certain receiver is $P_r(\text{dBf}) = 20$ dBf (i.e., 20 dB above the level of 1 fW). The antenna noise temperature is $T_a = 250$ K, and the noise figure of the receiver is 4 dB. The IF bandwidth is 100 kHz. Determine the output signal-to-noise ratio R_d (as an absolute value and in decibels) for each of the following modulation methods and conditions: (a) SSB, (b) DSB, (c) AM with 100% modulation, (d) PM with $\Delta\phi = 6$, (e) FM with $D = 6$ (no preemphasis), and (f) FM with $D = 6$, preemphasis, and $W/f_1 = 10$. (Since the transmission bandwidth is constant, it is assumed that the baseband bandwidth W varies with different methods.)

11-5 Consider a communications system having the same operating parameters as in Prob. 11-4 except that the transmission bandwidth is *not* fixed at 100 kHz. Instead, a baseband bandwidth $W = 100$ kHz is assumed for all cases, and the transmission bandwidth, as well as the receiver IF bandwidth, are assumed to be adjusted in each case to match the particular modulation method. Determine the detected output signal-to-noise ratio R_d (as an absolute value and in decibels) for the methods of parts (a), (b), and (c) of Prob. 11-4. Show that the input signal-to-noise ratio for parts (d), (e), and (f) is below threshold. Determine the received power required to ensure operation above threshold by a *6-dB* margin. With this power level, determine the corresponding detected signal-to-noise ratios for parts (d), (e), and (f).

11-6 Determine the values of (a) $\Delta\phi$ for PM and (b) D for FM for which the performance is theoretically the same as direct baseband transmission. (The baseband reference gain should be used in establishing the equality of performance.)

11-7 For a given transmission bandwidth and with no preemphasis, show that FM theoretically has a detected signal-to-noise improvement of 4.77 dB over PM.

11-8 Commercial FM broadcasting uses preemphasis corresponding to a time constant $\tau = 75$ μs in the preemphasis curve. [The corresponding frequency is $f_1 = 1/(2\pi\tau)$.] Based on a modulating bandwidth $W = 15$ kHz, determine the theoretical improvement (in decibels) for commercial FM with preemphasis.

11-9 Consider the PCM system of Ex. 11-3, but assume that only 4 bits are to be used for each word. Repeat all the computations of Ex. 11-3 for the modified system. The data rates are to be the same as in Ex. 11-1, so more words can be transmitted in a given amount of time.

11-10 Repeat Prob. 11-9 if 8 bits are to be used for each word. (In this case, fewer words can be transmitted in a given amount of time for the given data rates.)

11-11 Consider the PCM system of Ex. 11-3, but assume that the effective receiver noise temperature is $T_e = 850$ K, and the antenna noise temperature is $T_a = 600$ K. Repeat all the computations of Ex. 11-3 for the modified system.

11-12 Consider the PCM system of Ex. 11-4, but assume that the data rate is increased to 5 Mbits/s. Based on single-tone modulation, determine the required received power P_r for each of the following conditions: (a) envelope detected FSK, (b) matched filter ASK, (c) DPSK, and (d) PSK with matched filter.

11-13 Consider the PCM system of Ex. 11-4, but assume that the probability of error is specified as $P_e = 10^{-4}$. Repeat all the computations of Ex. 11-4 for the modified system.

11-14 Consider a PCM system employing 4-bit words. Compute the detected signal-to-noise ratio (as an absolute value and in decibels) for each of the following probabilities of error: (a) $P_e = 10^{-2}$, (b) $P_e = 10^{-3}$, (c) $P_e = 10^{-4}$, (d) $P_e = 10^{-5}$, (e) $P_e = 10^{-6}$, (f) $P_e = 10^{-7}$, (g) $P_e = 10^{-8}$.

11-15 Repeat Prob. 11-14 if 12-bit words are employed.

11-16 Consider the communications system of Prob. 11-4 in which the received signal level was 20 dBf. Assume that the transmitter will be located at a distance of approximately 100 km from the receiver. The receiving antenna gain is 12 dB, the transmitting antenna gain is 8 dB, and the frequency of operation is 1.5 GHz. Determine the required transmitter output power. Assume direct ray propagation.

11-17 The required signal-to-noise ratio at the input to a certain receiver detector is 13 dB. Determine the required transmitter power P_t for the following system parameters:

Antenna noise temperature = 100 K
Receiver noise figure = 3 dB
IF bandwidth = 1 MHz
Distance to transmitter = 100 km
Transmitter antenna gain = 6 dB
Receiver antenna gain = 12 dB
Frequency of operation = 3 GHz

Assume direct ray propagation.

11-18 For one-way propagation, verify the following two properties by application of the Friis transmission formula:

(a) If the distance between the transmitter and receiver is doubled, the received power level decreases by 6.02 dB (this is usually rounded to 6 dB).

(b) If the distance between the transmitter and receiver is multiplied by 10, the received power level is reduced by 20 dB.

11-19 For two-way radar propagation, verify the following two properties by application of the radar transmission formula:

(a) If the distance between the transmitter and the target is doubled, the received power level decreases by 12.04 dB (this is usually rounded to 12 dB).

(b) If the distance between the transmitter and receiver is multiplied by 10, the received power level is reduced by 40 dB.

11-20 The *effective isotropic radiated power* (denoted as the EIRP) represents the actual transmitted power in the direction of maximum radiation referred to an isotropic radiator. Show that the decibel level of the EIRP can be expressed as

$$\text{EIRP (dBW)} = P_t\,(\text{dBW}) + g_t\,(\text{dB})$$

where $P_t\,(\text{dBW}) = 10\log_{10}P_t$, with P_t in watts, and $g_t\,(\text{dB}) = 10\log_{10}g_t$.

11-21 Utilizing the antenna gains determined in Ex. 11-5 and assuming a transmitter power of 100 W, determine the EIRP values in dBW for each of the following antenna types: (a) hypothetical omnidirectional isotropic radiator, (b) half-wave dipole, (c) horn of the type in Ex. 11-5, and (d) parabola of the type in Ex. 11-5.

11-22 Communications specialists utilize decibel computations extensively in performing link analysis. A number of variations of the link equation expressed in decibel form have been used. One possible form is the following:

$$P_r\,(\text{dBW}) = P_t\,(\text{dBW}) + g_t\,(\text{dB}) + g_r\,(\text{dB}) - \text{path loss (dB)}$$

where $P_r\,(\text{dBW})$ is the effective received power in decibels referred to 1 W, $P_t\,(\text{dBW})$ is the transmitter power in decibels referred to 1 W, $g_t\,(\text{dB})$ is the transmitter antenna gain in decibels, $g_r\,(\text{dB})$ is the receiver antenna gain in decibels, and path loss (dB) is a function of the distance d between transmitter and receiver and the wavelength λ. Show that the link equation may be expressed this way, and show that the path loss may be expressed in either of the following forms:

$$\text{path loss (dB)} = 20 \log_{10} \frac{d \text{ (m)}}{\lambda \text{ (m)}} + 22 \text{ dB}$$

$$= 20 \log_{10} d \text{ (km)} - 20 \log_{10} \lambda \text{ (cm)} + 122 \text{ dB}$$

$$= 20 \log_{10} d \text{ (km)} + 20 \log_{10} f \text{ (GHz)} + 92.44 \text{ dB}$$

11-23 The synchronous orbit for satellites is at a distance of 35,778.3676 km (or 22,236.4 miles) from the earth. At this distance, the centrifugal and gravitational forces balance at a speed equal to the earth's rotational speed, so the satellite appears to be stationary.

    (a)   Compute the path loss (in decibels) at the synchronous orbit for a 15-GHz signal.

    (b)   The synchronous orbit is sometimes rounded off at 36,000 km. Compute the path loss for 15 GHz based on this value and determine the error in decibels by comparing with the result of part (a).

11-24 Prepare a family of curves providing plots of path loss (dB) versus d (km) for each of the following values of f: (a) 10 MHz, (b) 100 MHz, (c) 1 GHz (d) 10 GHz, (e) 100 GHz. For each curve, assume a range of d from 1 km to 100,000 km. Use semilog paper with d on the logarithmic scale and path loss in decibels on the vertical scale. (*Note:* With reasonable care, only a few points need actually be computed, and the general results can be readily deduced.)

11-25 Continuing with the decibel link computations introduced in Prob. 11-22, a common means for specifying the combined effect of the receiving antenna gain and the net input noise temperature is based on a term called the *G/T ratio*, where G refers to the receiver gain g_r and T is the net input noise temperature $T_i = T_a + T_e$. This ratio is commonly expressed in a decibel form as

$$\frac{G}{T} \text{ (dB/K)} = 10 \log_{10} \frac{g_r}{T_i} = g_r(\text{dB}) - 10 \log_{10} T_i$$

Observe that the units assigned to this logarithmic ratio are decibels per kelvin, which might annoy purists, but the practice is widely used. The G/T ratio is used as a relative measure of quality for the antenna and receiver input. The larger the value of the ratio, the greater the achievable input signal-to-noise ratio if all other factors are equal. A slightly different but related parameter is the *C/kT ratio*, where k = Boltzmann's constant = 1.38×10^{-23} J/K. The quantity C is referred to as the "carrier power" and is for our purposes the effective received power P_r. For a number of modulation techniques, P_r is equal to the actual carrier power. The C/kT ratio is commonly expressed in decibel form as

$$\frac{C}{kT} \text{ (dB} \cdot \text{Hz)} = 10 \log_{10} \frac{P_r}{kT_i}$$

Observe that the units assigned are decibels times hertz. The net received power P_r may be expressed in terms of the receiver antenna gain g_r and the power P_r' that would be received with an omnidirectional antenna as

$$P_r = g_r P_r'$$

(a) Show that the preceding quantities are related by

$$\frac{C}{kT}(\text{dB} \cdot \text{Hz}) = \frac{G}{T}(\text{dB/K}) + P_r'(\text{dBW}) + 228.6$$

where $P_r'(\text{dBW}) = 10 \log_{10} P_r'$, with P_r' expressed in watts. The units in the expression for C/kT may seem mixed, but this is a result of the manner in which the terms are grouped.

(b) The net signal power to noise power ratio R_i is denoted in many link calculations as the C/N ratio. Show that for a given transmission bandwidth B (in hertz) the C/N ratio expressed in decibels is

$$\frac{C}{N}(\text{dB}) = \frac{C}{kT}(\text{dB} \cdot \text{W}) - 10 \log_{10} B$$

$$= \frac{G}{T}(\text{dB/K}) + P_r'(\text{dBW}) - 10 \log_{10} B + 228.6$$

11-26 At the time this manuscript was being prepared*, Satellite Television Corporation was proposing a direct-broadcast satellite (DBS) for providing transmission of three special television channels into private homes. The proposed up-link channel from the ground transmitter to the satellite was based on the following link budget:

Earth station EIRP**, 86.6 dBW
Free space loss (17.6 GHz, 48° elevation), 208.9 dB
Assumed rain attenuation, 12 dB
Satellite receiver G/T, 7.7 dB/K
Up-link C/kT [to be determined in part (a)]

The proposed down-link channel from the satellite to the home receiver was based on the following link budget (based on clear weather):

Satellite EIRP [to be determined in part (b)]
Free space loss (12.5 GHz, 30° elevation), 206.1 dB
Atmospheric attenuation, 0.14 dB

---

*This information was provided in *Microwaves*, February 1981, pp. 15–19.
**For definition of EIRP, see Prob. 11-20.

Home receiver G/T (0.75 m), 9.4 dB/K
Receiver pointing loss (0.5° error), 0.6 dB
Polarization mismatch loss (average), 0.04 dB
Down-link C/kT, 88.1 dB · Hz

(a)   The decibel received power P_r' (dBW) that would be captured in an omnidirectional antenna can be determined simply by subtracting the net losses in decibels from the EIRP in dBW. In addition, make use of the results of Prob. 11-25 to determine the up-link C/kT expressed in dB · Hz.

(b)   Employing the results of Prob. 11-20, determine the required satellite EIRP in dBW for clear weather conditions.

(c)   Under certain rain conditions, the atmospheric attenuation is estimated to be 5 dB and the G/T ratio for the home receiver is estimated to drop to 8.1 dB/K. If all other parameters are the same, determine the down-link C/kT ratio in dB · Hz for rainy conditions.

(d)   If the required transmission bandwidth is 16 MHz, determine the C/N ratio in decibels for clear weather.

(e)   Repeat part (d) for rainy weather. If the signal is FM, what result is evident?

11-27   In this problem, a brief study of PCM signal-to-noise ratio as a function of the bandwidth expansion ratio will be made. The result should illuminate the exponential trade-off between bandwidth and signal-to-noise ratio. Consider a PCM system with n bits, and make the following simplifying assumptions: (a) sampling rate $= 2W$, (b) no spacing between pulses, and (c) minimum bandwidth (i.e., $0.5/\tau$) for pulses. Recognize that these assumptions are somewhat unrealistic, but they are useful in establishing a lower bound. Assume operation in the quantization noise limited region. Compute enough data to plot a curve of the detected signal-to-noise ratio in decibels as a function of the bandwidth expansion ratio B/W. While your result will be computed with specific values of n, extrapolate between points to yield a continuous curve.

General References

In a field as diverse and complex as communications, it is virtually impossible to provide a complete list of the numerous books that have been written through the years. Therefore, this list is not intended to be complete, but it represents a few selected references. The author was tempted to organize the list in a simple alphabetical order, which would have represented a "safe" approach to categorization. However, many readers prefer to have some guidance in selecting books to complement this book for particular reasons. Thus, a decision was made to attempt to categorize these references in order to assist readers in researching the field.

It should be clearly understood that these categories are quite arbitrary and do not in any way suggest any sort of "quality" ranking. This breakdown then is totally a matter of personal opinion, but the categories should assist readers in determining possible sources for further information when needed.

1. GENERAL REFERENCES: DESCRIPTIVE CATEGORY

Compared with this book, most books in the descriptive category are less mathematical and generally have more detailed qualitative descriptions of actual circuit details. They are "general" in that they deal with a broad coverage of the communications field. The mathematical level in most cases (but not all) is restricted to algebra and trigonometry. Some of these books are widely used in two-year associate level electronic engineering technology programs.

DeFrance, J. J., *Communications Electronic Circuits,* 2nd Ed. Rinehart Press, San Francisco, Calif., 1972.

Kellejian, R. *Applied Electronic Communication.* Science Research Associates, Inc., Palo Alto, Calif., 1980.

Kennedy, G., *Electronic Communication Systems,* 2nd Ed. McGraw-Hill Book Co., New York, 1977.

Lapatine, S. *Electronics in Communications.* John Wiley & Sons, Inc., New York, 1978.

Mandl, M., *Principles of Electronic Communications.* Prentice-Hall, Inc., Englewood Cliffs, N.J., 1973.

Miller, G. M., *Modern Electronic Communication.* Prentice-Hall, Inc., Englewood Cliffs, N.J., 1978.

Roddy, D., and J. Coolen, *Electronic Communications,* 2nd Ed. Reston Publishing Company, Reston, Va. 1981.

Shrader, R. L., *Electronic Communication,* 4th Ed. McGraw-Hill Book Co., New York, 1980.

Temes, L., *Communication Electronics for Technicians.* McGraw-Hill Book Co., New York, 1974.

2. GENERAL REFERENCES: ANALYTICAL CATEGORY

Compared with this book, most books in the analytical category utilize a higher level of mathematical sophistication. More advanced treatments of certain of the topics in this book appear in some of these references. Certain of these books are widely used for communications courses in senior and graduate level courses in electrical engineering.

Carlson, A. Bruce, *Communications Systems,* 2nd Ed. McGraw-Hill Book Co., New York, 1975.

Gregg, W. D., *Analog and Digital Communication.* John Wiley & Sons, Inc., New York, 1977.

Haykin, Simon, *Communication Systems.* John Wiley & Sons, Inc., New York, 1978.

Lathi, B. P., *Signals, Systems and Communication.* John Wiley & Sons, Inc., New York, 1965.

Lathi, B. P., *Communications Systems.* John Wiley & Sons, Inc., New York, 1968.

Roden, M. S., *Introduction to Communication Theory.* Pergamon Press, Inc., Elmsford, N.Y., 1972.

Sakrison, D. J., *Communication Theory: Transmission of Waveforms and Digital Information.* John Wiley & Sons, Inc., New York, 1968.

Schwartz, M. *Information Transmission, Modulation, and Noise*. McGraw-Hill Book Co., New York, 1980.

Schwartz, M., W. R. Bennett, and S. Stein, *Communication Systems and Techniques*. McGraw-Hill Book Co., New York, 1966.

Simpson, R. S., and R. C. Houts, *Fundamentals of Analog and Digital Communication Systems*. Allyn & Bacon, Boston, 1971.

Stremler, F., *Introduction to Communication Systems*. Addison-Wesley Publishing Co., Inc., Reading, Mass., 1977.

Taub, H., and D. L. Schilling, *Principles of Communications Systems*. McGraw-Hill Book Co., New York, 1971.

Ziemer, R. E., and W. H. Tranter, *Principles of Communications Systems, Modulation, and Noise*. Houghton Mifflin Co., Boston, 1976.

3. COMMUNICATIONS CIRCUIT DESIGN

The following books deal with the detailed design and analysis of various circuits and components used in communications systems.

Clarke, K. K., and D. T. Hess, *Communication Circuits: Analysis and Design*. Addison-Wesley Publishing Co., Inc., Reading, Mass., 1971.

Krauss, H. L., C. W. Bostian, and F. H. Raab, *Solid State Radio Engineering*. John Wiley & Sons, Inc., New York, 1980.

4. COMMUNICATIONS HARDWARE

The books in this category are written at the "hands-on" level and include actual construction and implementation details of certain types of communications components and systems. While the focus of these references is directed toward the hobbyist (particularly the radio amateur operator), there is much practical information that would be of value to anyone working in the communications field. The level of mathematics rarely exceeds elementary algebra, but instead most of the detailed quantitative information is conveyed through tables and curves.

Orr, W. I., *Radio Handbook*. Howard W. Sams & Co., Inc., Indianapolis. (New editions are published periodically.)
Radio Amateurs Handbook. American Radio Relay League, Newington, Conn. (New editions are published periodically.)

5. DATA COMMUNICATIONS

The following application level books are concerned with the rapidly developing area of digital data communications, data transmission standards, and the associated hardware.

Held, G., *Data Communication Components.* Hayden Book Co., Inc., Rochelle Park, N.J., 1979.

McNamara, J. E., *Technical Aspects of Data Communications.* Digital Equipment Company, Maynard, Mass., 1977.

Techo, R., *Data Communications: An Introduction to Concepts and Design.* Plenum Publishing Corp., New York, 1980.

6. NOISE

The following books provide coverage of electrical noise and detailed aspects of system noise computations:

Blachman, N. M., *Noise and Its Effect on Communication.* McGraw-Hill Book Co., New York, 1966.

Mumford, W. W., and E. H. Scheibe, *Noise Performance Factors in Communications Systems.* Horizon House-Microwave, Inc., Dedham, Mass., 1968.

Van der Ziel, A., *Noise: Sources, Characterization, Measurement.* Prentice-Hall, Inc., Englewood Cliffs, N.J., 1970.

7. STATISTICAL ANALYSIS AND PROBABILITY

In view of the large number of applications of statistics, numerous books written from the viewpoints of various disciplines are available. The abbreviated list that follows reflects certain general references on the subject, as well as some books that emphasize applications to electrical signals. Two suggested references for the beginner are the text by Cooper and McGillem and the Schaum outline book by Spiegel. While the latter reference has very few direct applications to electrical signals, it has hundreds of worked out problems covering a wide variety of areas.

Cooper, G. R., and C. D. McGillem, *Probabilistic Methods of Signal and System Analysis.* Holt, Rinehart & Winston, Inc., New York, 1971.

Davenport, W. B., Jr., and W. L. Root, *Introduction to Random Signals and Noise*. McGraw-Hill Book Co., New York, 1958.

Lathi, B. P., *An Introduction to Random Signals and Communication Theory*. International Textbook, Scranton, Pa., 1968.

Papoulis, A., *Probability, Random Variables, and Stochastic Processes*. McGraw-Hill Book Co., New York, 1965.

Parzen, E., *Modern Probability Theory and Its Applications*. John Wiley & Sons, Inc., New York, 1960.

Spiegel, M. R., *Statistics*. Schaum Publishing Company, New York, 1961.

8. COMMUNICATIONS SPECIALTY AREAS

The books in this category include various treatments of specific segments of the communications field. Some are analytical in nature, while others are more applications oriented. The areas are usually evident from the titles.

Abramson, N., *Information Theory and Coding*. McGraw-Hill Book Co., New York, 1963.

Blake, L. V., *Antennas*. John Wiley & Sons, Inc., New York, 1966.

Blanchard, A., *Phase-Locked Loops: Application to Coherent Receiver Design*. John Wiley & Sons, Inc., New York, 1976.

Gardner, F. M., *Phaselock Techniques,* 2nd Ed. John Wiley & Sons, Inc., New York, 1979.

Hansen, G. L., *Introduction to Solid-State Television Systems*. Prentice-Hall, Inc., Englewood Cliffs, N.J., 1969.

Kraus, J. D., *Antennas*. McGraw-Hill Book Co., New York, 1950.

Lin, S., *An Introduction to Error-Correcting Codes*. Prentice-Hall, Inc., Englewood Cliffs, N.J., 1970.

Nichols, M. H., and L. L. Rauch, *Radio Telemetry,* 2nd Ed. John Wiley & Sons, Inc., New York, 1956.

Panter, P. F., *Communications Systems Design: Line-of-Sight and Tropo-scatter Systems*. McGraw-Hill Book Co., New York, 1972.

Pappenfus, E. W., W. B. Bruene, and E. O. Schoenike, *Single Sideband Principles and Circuits*. McGraw-Hill Book Co., New York, 1964.

Peterson, W. W., and E. J. Weldon, Jr., *Error Correcting Codes,* 2nd Ed. M.I.T. Press, Cambridge, Mass., 1972.

Skolnik, M. I. (editor), *Radar Handbook*. McGraw-Hill Book Co., New York, 1970.

Stiltz, H. L. (editor), *Aerospace Telemetry*. Prentice-Hall, Inc., Englewood Cliffs, N.J., 1961.

Stiltz, H. L. (editor), *Aerospace Telemetry, Vol. II*. Prentice-Hall, Inc., Englewood Cliffs, N.J., 1966.

9. BROAD REFERENCE

The single reference book listed in this category was selected because of the volume of tabulated data pertaining to communications components, systems, operating standards, frequencies, and many other areas.

International Telephone and Telegraph, *Reference Data for Radio Engineers*. Howard W. Sams & Co., Inc., Indianapolis. (New editions are published periodically.)

Appendix A

Signal x(t)	Fourier series
Square wave 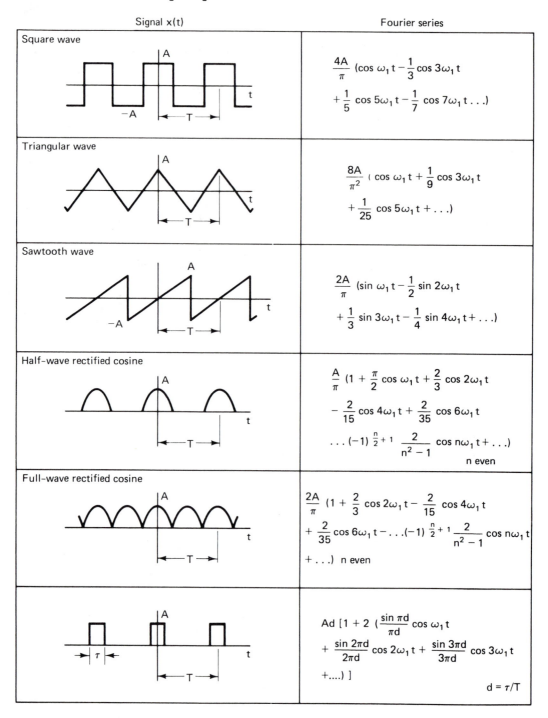	$\dfrac{4A}{\pi}$ ($\cos \omega_1 t - \dfrac{1}{3} \cos 3\omega_1 t$ $+ \dfrac{1}{5} \cos 5\omega_1 t - \dfrac{1}{7} \cos 7\omega_1 t \dots$)
Triangular wave	$\dfrac{8A}{\pi^2}$ ($\cos \omega_1 t + \dfrac{1}{9} \cos 3\omega_1 t$ $+ \dfrac{1}{25} \cos 5\omega_1 t + \dots$)
Sawtooth wave	$\dfrac{2A}{\pi}$ ($\sin \omega_1 t - \dfrac{1}{2} \sin 2\omega_1 t$ $+ \dfrac{1}{3} \sin 3\omega_1 t - \dfrac{1}{4} \sin 4\omega_1 t + \dots$)
Half-wave rectified cosine	$\dfrac{A}{\pi}$ ($1 + \dfrac{\pi}{2} \cos \omega_1 t + \dfrac{2}{3} \cos 2\omega_1 t$ $- \dfrac{2}{15} \cos 4\omega_1 t + \dfrac{2}{35} \cos 6\omega_1 t$ $\dots (-1)^{\frac{n}{2}+1} \dfrac{2}{n^2-1} \cos n\omega_1 t + \dots$) $\quad$ n even
Full-wave rectified cosine	$\dfrac{2A}{\pi}$ ($1 + \dfrac{2}{3} \cos 2\omega_1 t - \dfrac{2}{15} \cos 4\omega_1 t$ $+ \dfrac{2}{35} \cos 6\omega_1 t - \dots (-1)^{\frac{n}{2}+1} \dfrac{2}{n^2-1} \cos n\omega_1 t$ $+ \dots$) n even
	$Ad [1 + 2 (\dfrac{\sin \pi d}{\pi d} \cos \omega_1 t$ $+ \dfrac{\sin 2\pi d}{2\pi d} \cos 2\omega_1 t + \dfrac{\sin 3\pi d}{3\pi d} \cos 3\omega_1 t$ $+ \dots)]$ $\quad\quad d = \tau/T$

Figure A-1. Some common periodic signals and their Fourier series.

Appendix B

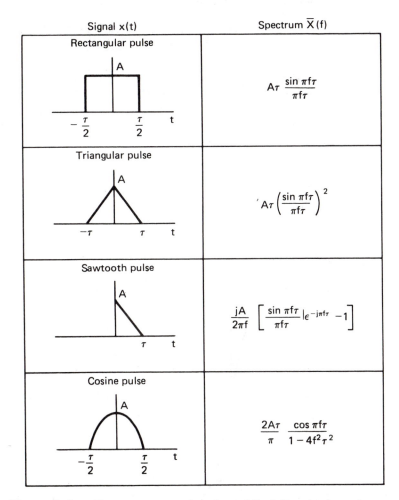

	Signal x(t)	Spectrum $\overline{X}(f)$
Rectangular pulse		$A\tau \dfrac{\sin \pi f\tau}{\pi f\tau}$
Triangular pulse		$A\tau \left(\dfrac{\sin \pi f\tau}{\pi f\tau}\right)^2$
Sawtooth pulse		$\dfrac{jA}{2\pi f} \left[\dfrac{\sin \pi f\tau}{\pi f\tau}\big|_{\epsilon^{-j\pi f\tau}} - 1\right]$
Cosine pulse		$\dfrac{2A\tau}{\pi} \dfrac{\cos \pi f\tau}{1 - 4f^2\tau^2}$

Figure B-1. Some common signals and their Fourier transforms.

Appendix C
Effects of Band-limiting of Common Waveforms with an Ideal Low-Pass Filter

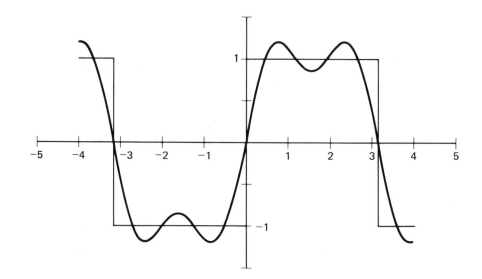

Figure C-1.   Square-wave Fourier representation based on components up through third harmonic (period = 2π).

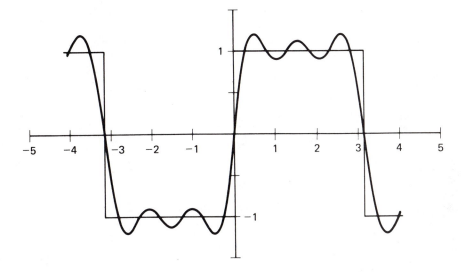

Figure C-2. Square-wave Fourier representation based on components up through fifth harmonic (period = 2π).

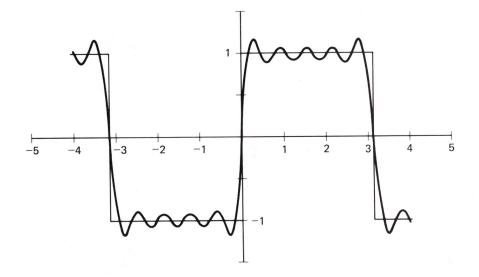

Figure C-3. Square-wave Fourier representation based on components up through ninth harmonic (period = 2π).

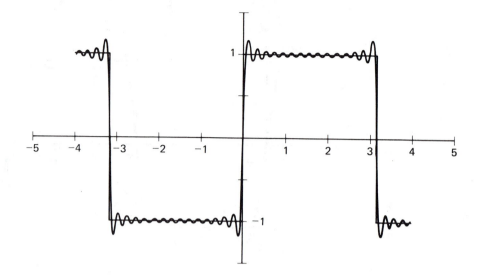

Figure C-4. Square-wave Fourier representation based on components up through twenty-ninth harmonic (period = 2π).

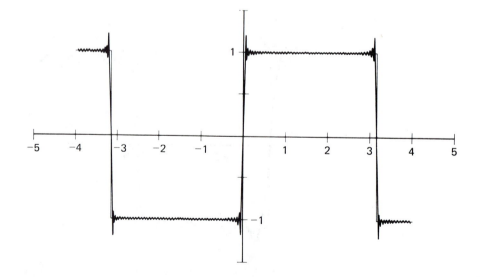

Figure C-5. Square-wave Fourier representation based on components up through ninety-ninth harmonic (period = 2π).

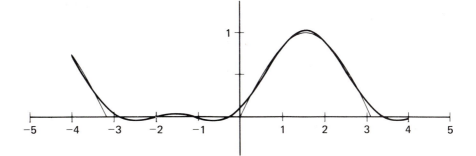

Figure C-6. Sinusoidal pulse Fourier representation based on components up through third harmonic (period = 2π).

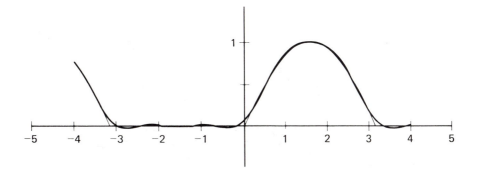

Figure C-7. Sinusoidal pulse Fourier representation based on components up through fifth harmonic (period = 2π).

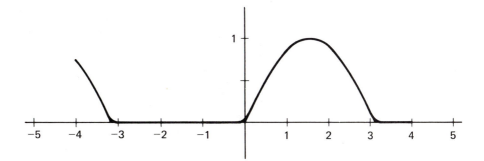

Figure C-8. Sinusoidal pulse Fourier representation based on components up through tenth harmonic (period = 2π).

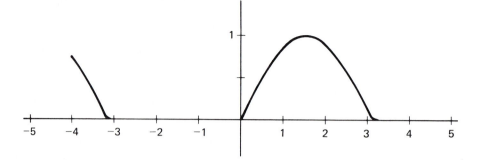

Figure C-9. Sinusoidal pulse Fourier representation based on components up through thirtieth harmonic (period $= 2\pi$).

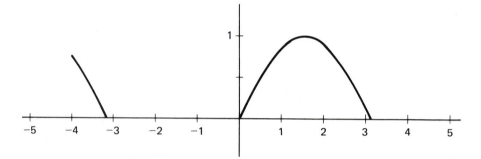

Figure C-10. Sinusoidal pulse Fourier representation based on components up through one-hundredth harmonic (period $= 2\pi$).

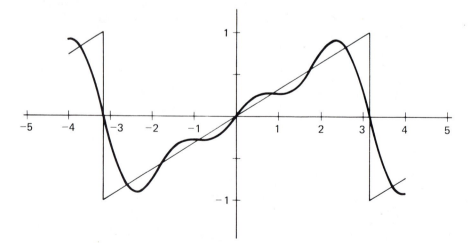

Figure C-11. Sawtooth pulse Fourier representation based on components up through third harmonic (period $= 2\pi$).

554

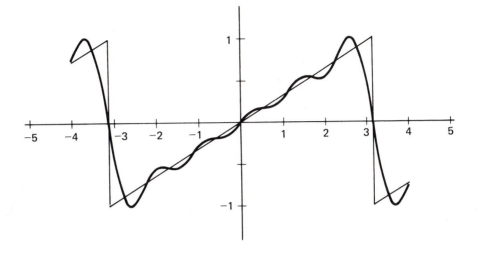

Figure C-12.   Sawtooth pulse Fourier representation based on components up through fifth harmonic (period $= 2\pi$).

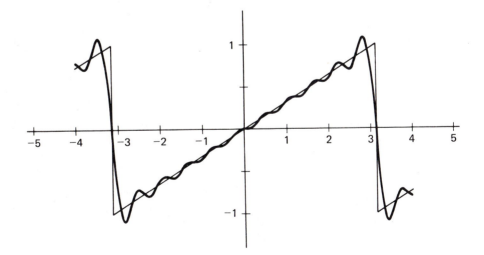

Figure C-13.   Sawtooth pulse Fourier representation based on components up through tenth harmonic (period $= 2\pi$).

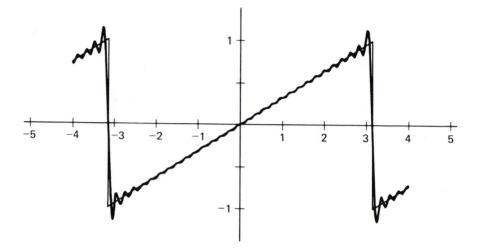

Figure C-14. Sawtooth pulse Fourier representation based on components up through thirtieth harmonic (period $= 2\pi$).

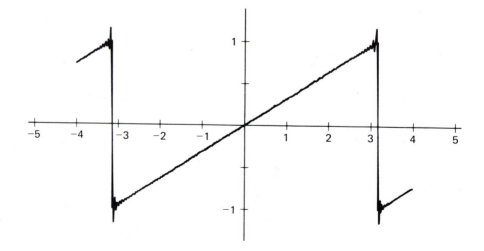

Figure C-15. Sawtooth pulse Fourier representation based on components up through one-hundredth harmonic (period $= 2\pi$).

Appendix D
American National Standard Code for Information Interchange

This material is reproduced with permission from American National Standard Code for Information Interchange ANSI X3.4, copyright 1977 by the American National Standards Institute. Copies of this standard may be purchased from the American National Standards Institute at 1430 Broadway, New York, N.Y. 10018.

1. Scope

This coded character set is to be used for the general interchange of information among information processing systems, communications systems, and associated equipment.

2. Standard Code

b_4	b_3	b_2	b_1	COLUMN / ROW	0	1	2	3	4	5	6	7
0	0	0	0	0	NUL	DLE	SP	0	@	P	`	p
0	0	0	1	1	SOH	DC1	!	1	A	Q	a	q
0	0	1	0	2	STX	DC2	"	2	B	R	b	r
0	0	1	1	3	ETX	DC3	#	3	C	S	c	s
0	1	0	0	4	EOT	DC4	$	4	D	T	d	t
0	1	0	1	5	ENQ	NAK	%	5	E	U	e	u
0	1	1	0	6	ACK	SYN	&	6	F	V	f	v
0	1	1	1	7	BEL	ETB	'	7	G	W	g	w
1	0	0	0	8	BS	CAN	(	8	H	X	h	x
1	0	0	1	9	HT	EM	)	9	I	Y	i	y
1	0	1	0	10	LF	SUB	*	:	J	Z	j	z
1	0	1	1	11	VT	ESC	+	;	K	[	k	{
1	1	0	0	12	FF	FS	,	<	L	\	l	|
1	1	0	1	13	CR	GS	−	=	M	]	m	}
1	1	1	0	14	SO	RS	.	>	N	^	n	~
1	1	1	1	15	SI	US	/	?	O	___	o	DEL

3. Character Representation and Code Identification

The standard 7-bit character representation, with b_7 the high-order bit and b_1 the low-order bit, is shown below:

Example: The bit representation for the character "K," positioned in column 4, row 11, is

$$b_7 \quad b_6 \quad b_5 \quad b_4 \quad b_3 \quad b_2 \quad b_1$$
$$1 \quad 0 \quad 0 \quad 1 \quad 0 \quad 1 \quad 1$$

The code table for the character "K" may also be represented by the notation "column 4, row 11" or alternatively as "4/11." The decimal equivalent of the binary number formed by bits b_7, b_6, and b_5, collectively, forms the column number, and the decimal equivalent of the binary number formed by bits b_4, b_3, b_2, and b_1, collectively, forms the row number.

The standard code may be identified by the use of the notation ASCII.

The notation ASCII (pronounced as-key) should ordinarily be taken to mean the code prescribed by the latest edition of this standard. To explicitly designate a particular (perhaps prior) edition, the last two digits of the year of issue may be appended, as "ASCII 68" or "ASCII 77."

4. Legend

4.1 Control Characters

Col/Row	Mnemonic and Meaning[1]		Col/Row	Mnemonic and Meaning[1]	
0/0	NUL	Null	1/0	DLE	Data Link Escape (CC)
0/1	SOH	Start of Heading (CC)	1/1	DC1	Device Control 1
0/2	STX	Start of Text (CC)	1/2	DC2	Device Control 2
0/3	ETX	End of Text (CC)	1/3	DC3	Device Control 3
0/4	EOT	End of Transmission (CC)	1/4	DC4	Device Control 4
0/5	ENQ	Enquiry (CC)	1/5	NAK	Negative Acknowledge (CC)
0/6	ACK	Acknowledge (CC)	1/6	SYN	Synchronous Idle (CC)
0/7	BEL	Bell	1/7	ETB	End of Transmission Block (CC)
0/8	BS	Backspace (FE)	1/8	CAN	Cancel
0/9	HT	Horizontal Tabulation (FE)	1/9	EM	End of Medium
0/10	LF	Line Feed (FE)	1/10	SUB	Substitute
0/11	VT	Vertical Tabulation (FE)	1/11	ESC	Escape
0/12	FF	Form Feed (FE)	1/12	FS	File Separator (IS)
0/13	CR	Carriage Return (FE)	1/13	GS	Group Separator (IS)
0/14	SO	Shift Out	1/14	RS	Record Separator (IS)
0/15	SI	Shift In	1/15	US	Unit Separator (IS)
			7/15	DEL	Delete

4.2 Graphic Characters

Column/Row	Symbol	Name
2/0	SP	Space (Normally Nonprinting)
2/1	!	Exclamation Point
2/2	"	Quotation Marks (Diaeresis)[2]
2/3	#	Number Sign[3]
2/4	$	Dollar Sign
2/5	%	Percent Sign
2/6	&	Ampersand
2/7	'	Apostrophe (Closing Single Quotation Mark; Acute Accent)[2]
2/8	(	Opening Parenthesis
2/9	)	Closing Parenthesis
2/10	*	Asterisk
2/11	+	Plus
2/12	,	Comma (Cedilla)[2]
2/13	–	Hyphen (Minus)
2/14	.	Period (Decimal Point)
2/15	/	Slant
3/0 to 3/9	0 . . . 9	Digits 0 through 9
3/10	:	Colon
3/11	;	Semicolon
3/12	<	Less Than
3/13	=	Equals
3/14	>	Greater Than
3/15	?	Question Mark
4/0	@	Commercial At[3]
4/1 to 5/10	A . . . Z	Uppercase Latin Letters A through Z
5/11	[	Opening Bracket[3]
5/12	\	Reverse Slant[3]
5/13	]	Closing Bracket[3]
5/14	^	Circumflex[3]
5/15	_	Underline
6/0	`	Opening Single Quotation Mark (Grave Accent)[2,3]
6/1 to 7/10	a . . . z	Lowercase Latin letters a through z
7/11	{	Opening Brace[3]
7/12	|	Vertical Line[3]
7/13	}	Closing Brace[3]
7/14	~	Tilde[2,3]

[2] The use of the symbols in 2/2, 2/7, 2/12, 5/14, 6/0, and 7/14 as diacritical marks is described in A5.2 of Appendix A.

[3] These characters should not be used in international interchange without determining that there is agreement between sender and recipient. (See Appendix B5.)

Index